Date: 6/23/20

581.9759 WUN
Wunderlin, Richard P.,
Flora of Florida.
Dicotyledons, vitaceae

Flora of Florida, Volume III

UNIVERSITY PRESS OF FLORIDA

Florida A&M University, Tallahassee
Florida Atlantic University, Boca Raton
Florida Gulf Coast University, Ft. Myers
Florida International University, Miami
Florida State University, Tallahassee
New College of Florida, Sarasota
University of Central Florida, Orlando
University of Florida, Gainesville
University of North Florida, Jacksonville
University of South Florida, Tampa
University of West Florida, Pensacola

Flora
of ❧
Florida

VOLUME III

DICOTYLEDONS, VITACEAE
THROUGH URTICACEAE

Richard P. Wunderlin and Bruce F. Hansen

University Press of Florida

Gainesville · Tallahassee · Tampa · Boca Raton

Pensacola · Orlando · Miami · Jacksonville · Ft. Myers · Sarasota

This book may be available in an electronic edition.

21 20 19 18 17 16   6 5 4 3 2 1

A record of cataloging-in-publication data is available from the Library of Congress.
ISBN 978-0-8130-6121-4

The University Press of Florida is the scholarly publishing agency for the State
University System of Florida, comprising Florida A&M University, Florida Atlantic
University, Florida Gulf Coast University, Florida International University, Florida
State University, New College of Florida, University of Central Florida, University of
Florida, University of North Florida, University of South Florida, and University of
West Florida.

University Press of Florida
15 Northwest 15th Street
Gainesville, FL 32611-2079
http://www.upf.com

Dedicated to the memory of Duane Isely (1918–2000),
for his contributions to our knowledge of the large
and important Fabaceae, a family featured in this work.

Contents

Acknowledgments

Review and helpful suggestions were kindly provided by Loran Anderson, Kathleen Burt-Utley, Kris DeLaney, Frederick B. Essig, and Alan R. Franck.

The facilities and collections of many herbaria were utilized in preparing this volume. The courtesies extended and the loan of specimens by the curators are gratefully appreciated. These include the Florida Museum of Natural History (FLAS), Florida State University (FSU), Harvard Herbaria (A, GH), Marie Selby Botanical Garden (SEL), New York Botanical Garden (NY), University of Central Florida (FTU), and University of North Carolina—Chapel Hill (NCU). We are especially grateful to Kent Perkins (FLAS), Loran Anderson (FSU), and Austin Mast (FSU) for their continuous support.

The *Flora of Florida* project has been strongly supported by the University of South Florida Institute for Systematic Botany.

Introduction

Volume 1 of the *Flora of Florida* provides background information on the physical setting, vegetation, history of botanical exploration, and systematic treatments of the pteridophytes and gymnosperms. Volumes 2 through 7 will contain the dicotyledons and volumes 8 through 10, the monocotyledons.

This volume contains the taxonomic treatments of 13 families in 4 orders of the dicotyledons (see table of contents).

ORGANIZATION OF THE FLORA

Taxa Included

Florida, with 4,300 taxa, has the third most diverse vascular plant flora of any state in the United States. The *Flora of Florida* is a treatment of all indigenous and naturalized vascular plant taxa currently known to occur in the state. Naturalized is defined as those nonindigenous taxa growing outside of cultivation and naturally reproducing. This includes plants that have escaped from cultivation as well as those that were intentionally or accidentally introduced by human activities in post-Columbian times. Taxa that have not been recently recollected and may no longer exist in the wild in Florida are formally treated both for historical completeness and on the premise that they may be rediscovered in the future.

A taxon is formally treated in this flora if (1) an herbarium specimen has been seen to document its occurrence in Florida, or (2) a specimen is cited from Florida in a monograph or revision whose treatment is considered sound.

Taxa Excluded

Literature reports of taxa attributed to Florida that are considered to be erroneous or highly questionable and therefore to be excluded from this flora are listed following the treatment for the genus, or in the case of genera not otherwise treated, at the end of the family. The reason for exclusion is given in each case. Most commonly, the taxon is excluded because it is based on a misidentified specimen(s), lack of documentation by means of a specimen, or it is based on a misapplied name, that is, a name correctly applied to a plant not found in Florida.

XII / FLORA OF FLORIDA

Systematic Arrangement

Recent studies have demonstrated that the traditional dicotyledons are paraphyletic and that the monophyletic monocotyledons are derived from within the dicotyledons, with several families of aquatic herbs as a probable sister group. We believe that the arrangement as proposed by the Angiosperm Phylogeny Group III (APG III, 2009) has merit and is followed in this work with slight modifications. The linear sequence of families used here essentially follows that proposed by Haston et al. (2009). For convenience, the genera and species within each family are arranged alphabetically.

Descriptions

Descriptions are based on Florida material and are given for each family, genus, species, and infraspecific taxon.

Common Names

Non-Latinized names given for the taxa are derived from published sources as well as from our own experience. No attempt is made to list all names that have been applied to a taxon, or to standardize names with a specific source, or to supply a name for species where one is not in general usage. For plants lacking a common name, the generic name may be used as is the usual practice.

Derivation of Latin Names

The derivation of the generic name and that of each specific and infraspecific epithet is given.

Synonymy

A full literature citation is given for each species, infraspecific taxon, and synonym. Synonyms listed are only those that have been cited for Florida in manuals, monographic treatments, and technical papers. Also included is the basionym and all homotypic synonyms of a name introduced into synonymy. The homotypic synonyms are listed in chronological order in a single paragraph, and the paragraphs of synonyms are put in chronological order according to the basionym of each. If the type of a taxon is a Florida collection and is known, this information is given. We do not attempt to lectotypify the numerous Florida taxa needing lectotypification in the belief that this is best left to monographers.

For families and genera, only the author and date of publication is given. Family and generic synonyms listed are those that have been used in the major publications pertinent to the Florida flora.

Citation of periodical literature conforms to that cited in *Botanico-Periodicum-Huntianum* (Lawrence et al., 1968) and *Botanico-Periodicum-Huntianum/Supplementum* (Bridson and Smith, 1991). Other literature citations conform to that cited in *Taxonomic Literature*, edition 2 (Stafleu and Cowan, 1976, et seq.). Author abbreviations are those listed in *Authors of Plant Names* (Brummitt and Powell, 1992).

Habitat

The terminology used for plant communities generally follows that of Myers (volume 1), but may vary.

Distribution

The global distribution is given for each family and genus where native and naturalized. Relative abundance in Florida (ranked as common, frequent, occasional, or rare) and the distribution is given for each species and infraspecific taxon. The format for distribution of species and infraspecific taxa is: Florida; North America (Continental United States, Canada, and Greenland); tropical America (West Indies, Mexico, Central America, South America); Old World (Europe, Africa, Asia, Australia, Pacific Islands). For taxa occurring in all of these areas, the phrase *nearly cosmopolitan* is used. For taxa of limited distribution in Florida, range statements by county are usually given. For taxa of wide distribution in Florida, the range is given in general terms: *panhandle*—from the Suwannee River west to Escambia County; *peninsula*—east of the Suwannee River and south of the Georgia line southward through the Florida Keys. Because of the vast floristic differences in peninsular Florida, this region is often further subdivided into northern, central, and southern regions and the keys. The northern region is east of the Suwannee River and south of the Georgia line southward through Gilchrist, Alachua, Putnam, and Flagler Counties. The central region extends from Levy, Marion, and Volusia Counties southward through Lee, Hendry, and Palm Beach Counties. The southern peninsula consists of the southernmost four counties (Collier, Broward, Monroe, and Miami-Dade). The Florida Keys consist of the chain of islands from Key Largo to the Marquesas Keys and the Dry Tortugas. Politically, they are part of Monroe County. The panhandle is subdivided into eastern, central, and western regions. The eastern region consists of the counties west of the Suwannee River west through Jefferson County, the central region extends from Leon and Wakulla Counties west through Holmes, Washington, and Bay Counties while the western region consists of the westernmost four counties (Walton, Okaloosa, Santa Rosa, and Escambia).

Endemic or Exotic Status

Endemic taxa are those whose global distribution is confined to the political boundary of Florida. If a taxon is an exotic (non-native, nonindigenous, or alien), the region of nativity is given. Exotic taxa are those that are known to have become part of the flora following the occupation by Europeans in the sixteenth century.

Admittedly, this is an arbitrary starting point. Several species are believed to have been introduced by Paleo-Indians before 1513. Technically, these are considered as native. Another problem in interpretation arises when propagules arrive after 1513 by some means other than human activity (that is, hurricanes, storms, sea-drift, or animals) and the species becomes established. Again technically, these are considered as exotics. It is sometimes difficult to determine whether a widespread species is native or an exotic and our opinion may differ from that of others.

Reproductive Season

The sexual reproductive (flowering) season for each species and infraspecific taxon is given. The reproductive seasons are broadly defined as follows: spring—March through May; summer—June through September; fall—October through November; winter—December through February. Species "flowering out of season" are sometimes encountered and wide ranging species will usually bloom earlier in the southern part of the state than in the northern.

Hybrids

Named hybrids are listed along with the putative parents, nomenclature, usually with comment concerning distribution, and sometimes their distinguishing characteristics.

References

Major monographs, revisions, and other pertinent literature, other than those cited in the nomenclature, are given at the end of the volume.

TAXONOMIC CONCEPTS

Taxonomic interpretations and nomenclature are generally in accord with recent monographs or revisions for the various groups except where it is believed that recent evidence necessitates a change. Citation of a monograph or revision implies consideration of the work during the preparation of the treatment, but not necessarily acceptance. Where a difference of opinion exists among published treatments or the treatment in this work deviates from that of the reference cited, a discussion of alternative opinions is often provided.

Species, subspecies, and varieties are considered as entities with a high degree of population integrity. Color forms and minor morphotypes that occur within a species and that may be formally recognized as *forma* by other authors are accorded no formal recognition in this work.

No nomenclatural innovations are intentionally published in the *Flora*.

LITERATURE CITED

Angiosperm Phylogeny Group III (APG III). 2009. An update of the angiosperm Phylogeny Group classification for the orders and families of flowering plants: APG III. Bot. J. Linn. Soc. 161: 105–21.

Bridson, G. D. R., and E. R. Smith. 1991. Botanico-Periodicum-Huntianum/Supplementum. Pittsburgh: Hunt Botanical Library.

Brummitt, R. K., and C. E. Powell. 1992. Authors of Plant Names. Royal Botanical Gardens, Kew. Basildon: Her Majesty's Stationery Office.

Haston, E., J. E. Richardson, P. F. Stevens, M. W. Chase, and D. J. Harris. 2009. The linear Angiosperm Phylogeny Group (LAPG) III: a linear sequence of the families in APG III. Bot. J. Linn. Soc. 161: 128–31.

Lawrence, G. H. M., A.F.G. Buchheim, G. S. Daniels, and H. Dolezal. 1968. Botanico-Periodicum-Huntianum. Pittsburgh: Hunt Botanical Library.

Stafleu, F. A., and R. S. Cowan. 1976 et seq. Taxonomic Literature. Edition 2. Utrecht: Bohn, Scheltema, and Holkema.

Systematic Treatments

Keys to Major Vascular Plant Groups

1. Plant reproducing by spores ....PTERIDOPHYTES (volume I)
1. Plant reproducing by seeds.
    2. Leaves with a single midvein or with simple or sometimes dichotomously branched veins, these closely parallel and lacking secondary interconnecting cross-veinlets; seeds borne on the surface of specialized bract-scale structures aggregated into woody or fleshy cones or a single seed partly or wholly surrounded by a fleshy aril and drupelike or berrylike; perianth lacking .... GYMNOSPERMS (volume I)
    2. Leaves with parallel veins with secondary interconnecting cross-veinlets or with reticulate veins; seeds borne enclosed within specialized structures (carpels); perianth usually present.
        3. Vascular bundles occurring in a ring or in concentric cylinders; cotyledons 2; flower parts usually in other than whorls of 3 or multiples thereof; leaves usually reticulate-veined...DICOTYLEDONS (volumes II–VII)
        3. Vascular bundles scattered (or rarely single); cotyledon 1; flower parts often in whorls of 3 or multiples thereof; leaves usually parallel-veined (sometimes with midvein only) (some plants diminutive, floating aquatics, the plant body thalloid, not differentiated into stems and leaves, rootless or with 1–few unbranched roots or plants partly or wholly submersed aquatics, the leaves, flowers, and fruits often much reduced)...........................MONOCOTYLEDONS (volumes VIII–X)

Dicotyledons

VITACEAE Juss., nom. cons. 1789. GRAPE FAMILY

Deciduous, woody or herbaceous vines, with axillary stem-tendrils with or without adhesive disks. Leaves alternate, simple or pinnately or palmately compound, petiolate, stipulate. Flowers in axillary or terminal cymes, bracteolate, actinomorphic, bisexual or unisexual (plants polygamodioecious or polygamomonoecious); sepals 4–5, connate; petals 4–5, free or apically coherent and dropping as a unit at anthesis; stamens 4–5, opposite the petals, free, the anthers 2-locular, introrse, longitudinally dehiscent; nectariferous disk intrastaminal; carpels 2, connate, superior, the style 1, the stigma simple, the ovules 2 in each locule. Fruit a 1- to 4-seeded berry.

A family of about 14 genera and about 850 species; nearly cosmopolitan.

Selected reference: Brizicky (1965).

1. Leaves bipinnate .... **Ampelopsis**
1. Leaves simple, 3-foliolate, or palmately compound.
   2. Leaves palmately compound .... **Parthenocissus**
   2. Leaves simple or 3-foliolate.
      3. Petals cohering at their apex and deciduous as a cap .... **Vitis**
      3. Petals free and spreading.
         4. Leaves chartaceous; flowers 5-merous .... **Ampelopsis**
         4. Leaves succulent or coriaceous; flowers 4-merous .... **Cissus**

Ampelopsis Michx. 1803. PEPPERVINE

Woody vines, the tendrils 2-branched, without adhesive disks. Leaves simple or pinnately compound, petiolate, stipulate. Flowers in axillary, compound, thyrsoid cymes, bisexual; sepals 5; petals 5, free; stamens 5; nectariferous disk cupular, the lower part adnate to the base of the ovary. Fruit a 1- to 4-seeded berry.

A genus of about 25 species; North America, West Indies, Mexico, Central America, and Asia. [From the Greek *ampelos*, grapevine, and *opsis*, likeness, in reference to a likeness to *Vitis vinifera*.]

1. Leaves pinnate ....**A. arborea**
1. Leaves simple ....**A. cordata**

Ampelopsis arborea (L.) Koehne [Tree, in reference to its high-climbing habit.]
PEPPERVINE.

Vitis arborea Linnaeus, Sp. Pl. 1: 203. 1753. *Hedera arborea* (Linnaeus) Walter, Fl. Carol. 102. 1788. *Ampelopsis bipinnata* Michaux, Fl. Bor.-Amer. 1: 160. 1803, nom. illegit. *Cissus stans* Persoon, Syn. Pl. 1: 143. 1805. *Cissus bipinnata* Elliott, Sketch Bot. S. Carolina 1: 304. 1817, nom. illegit. *Nekemias bipinnata* Rafinesque, Sylva Tellur. 87. 1838, nom. illegit. *Vitis bipinnata* Torrey & A. Gray, Fl. N. Amer. 1: 243. 1838, nom. illegit. *Cissus arborea* (Linnaeus) Des Moulins, in Durand, Actes Soc. Linn. Bordeaux 24: 156. 1862 ("1861"); non Forsskal, 1775; nec Willdenow ex Roemer & Schultes, 1827; nec Blanco, 1845. *Ampelopsis arborea* (Linnaeus) Koehne, Deut. Dendrol. 400. 1893.

Woody vine; stem with 2-branched tendrils. Leaves unevenly 2- to 3-pinnate, the leaflets short-petiolulate, the blade 1–3(7) cm long, ovate, pinnate-veined, the apex acute, the base rounded, truncate, or cuneate, the margin irregularly and coarsely toothed, the upper surface glabrous, the petiole, petiolules, and major veins on the lower surface sparsely pubescent. Flowers in an axillary, thyrsoid cyme; sepals minute; petals 1–3 mm long, greenish yellow, arched-reflexed. Fruit subglobose, 8–14 mm long and wide, shiny black.

Floodplain forests, cypress swamps, and hammocks. Frequent; nearly throughout. Maryland south to Florida, west to Missouri, Oklahoma, and New Mexico; West Indies and Mexico. Native to North America and Mexico. Spring–fall.

Ampelopsis cordata Michx. [Heart-shaped, in reference to the leaves.] HEARTLEAF PEPPERVINE.

Ampelopsis cordata Michaux, Fl. Bor.-Amer. 1: 159. 1803. *Cissus ampelopsis* Persoon, Syn. Pl. 1: 142. 1805, nom. illegit. *Vitis cordata* (Michaux) Dumont de Courset, Bot. Cult., ed. 2. 4: 619. 1811. *Vitis indivisa* Willdenow, Berlin. Baumz., ed. 2. 538. 1811, nom. illegit. *Cissus indivisa* Des Moulins, in Durand, Actes Soc. Linn. Bordeaux 24: 156. 1862 ("1861"). *Vitis heterophylla* Thunberg var. *cordata* (Michaux) Regel, Trudy Imp. S.-Peterburgsk. Bot. Sada 2: 392. 1873.

Woody vine; stem with 2-branched tendrils. Leaves simple, ovate, to 9 cm long, pinnate- or subpalmate-veined, the apex short-acuminate, the base truncate to subcordate, the upper surface sparsely pubescent, becoming glabrate, the lower surface sparsely pubescent, the margin coarsely and irregularly serrate, occasionally with 1–2 small lateral lobes, the petiole about as long as the blade, sparsely pubescent. Flowers in an axillary, thyrsoid cyme; sepals minute; petals 1–3 mm long, greenish yellow, erect. Fruit subglobose, 7–10 mm long and wide, blue, sometimes iridescent, with low brownish excrescences.

Floodplain forests. Occasional; central panhandle. Connecticut south to Florida, west to Nebraska, Kansas, Oklahoma, and Texas; Mexico. Spring.

Cissus L. 1753. TREEBIND

Woody or herbaceous vines, the tendrils simple, without adhesive disks. Leaves simple or 3-foliolate, petiolate, stipulate. Flowers in axillary, compound, umbelliform cymes, bisexual or unisexual (plants polygamomonoecious); sepals 4; petals 4, free; stamens 4; nectariferous disk cuplike and adnate to the ovary nearly to its summit. Fruit a 1(2)-seeded berry.

A genus of about 350 species; North America, West Indies, Mexico, Central America, Africa, Asia, and Australia. [From the Greek *kissos*, an ancient name for ivy.]

1. Leaves 3-foliolate...**C. trifoliata**
1. Leaves simple... **C. verticillata**

Cissus trifoliata (L.) L. [With three leaflets.] SORRELVINE; MARINEVINE.

Sicyos trifoliata Linnaeus, Sp. Pl. 2: 1013. 1753. *Cissus trifoliata* (Linnaeus) Linnaeus, Syst. Nat., ed. 10. 897. 1759. *Cissus acida* Linnaeus, Sp. Pl., ed. 2. 170. 1762, nom. illegit. *Cissus parvifolia* Salisbury, Prodr. Stirp. Chap. Allerton 66. 1796, nom. illegit. *Kemoxis acida* Rafinesque, Sylva Tellur. 86. 1838, nom. illegit. *Vitis acida* Chapman, Fl. South. U.S. 70. 1860. *Vitis trifoliata* (Linnaeus) Baker, in Martius, Fl. Bras. 14(2): 212. 1871; non Thunberg, 1825.

Vitis incisa Nuttall ex Torrey & A. Gray, Fl. N. Amer. 1: 243. 1838. *Cissus incisa* (Nuttall ex Torrey & A. Gray) Des Moulins, in Durand, Actes Soc. Linn. Bordeaux 24: 156. 1862 ("1861").

Herbaceous or woody vine, to 10 m; stem with simple tendrils, the young stems herbaceous and succulent, the older stems woody, the bark warty. Leaves with the blade broadly ovate or oblong, 3-divided to 3-foliolate, rarely undivided, to ca. 8 cm long and wide, the blade divisions or leaflets ovate to oblong, the apex acute, the base cuneate, the margin coarsely and irregularly toothed, the upper and lower surfaces glabrous, the petiole 1–3.5 cm long. Flowers in an axillary, umbelliform cyme, glabrous, the peduncle exceeding the leaves; sepals minute; petals 1–3 mm long, greenish, creamy yellow, whitish, or purplish, spreading. Fruit ovoid or obovoid, 6–8 mm long, blue-black.

Coastal hammocks and dunes. Frequent; Flagler County, central and southern peninsula, central panhandle, Escambia County. Missouri and Kansas south to Florida and Texas, west to Arizona; West Indies, Mexico, Central America, and South America. Spring–fall.

Cissus verticillata (L.) Nicolson & C. E. Jarvis [In whorls, in reference to the smut-infested inflorescence branches of the type specimen.] SEASONVINE; POSSUM GRAPE.

Viscum vertillatum Linnaeus, Sp. Pl. 2: 1023. 1753. *Phoradendron verticillatum* (Linnaeus) Druce, Bot. Exch. Club Soc. Brit. Isles 3: 422. 1914. *Cissus verticillata* (Linnaeus) Nicolson & C. E. Jarvis, Taxon 33: 727. 1984.

Cissus sicyoides Linnaeus, Syst. Nat., ed. 10. 897. 1759. *Cissus ovata* Lamarck, Tabl. Encycl. 1: 331. 1792, nom. illegit. *Cissus pallida* Salisbury, Prodr. Stirp. Chap. Allerton 66. 1796, nom. illegit. *Irsiola sicyoides* (Linnaeus) Rafinesque, Sylva Tellur. 86. 1838. *Vitis sicyoides* (Linnaeus) Miquel, Ann. Mus. Bot. Lugduno-Batavum 1: 83. 1863. *Vitis sicyoides* (Linnaeus) Miquel var. *ovata* Baker, in Martius, Fl. Bras. 14(2): 203. 1871, nom. inadmiss. *Cissus sicyoides* Linnaeus forma *ovata* Planchon, in A. de Candolle, Monogr. Phan. 5: 526. 1887, nom. inadmiss. *Vitis vitiginea* (Linnaeus) Kuntze var. *sicyoides* (Linnaeus) Kuntze, Revis. Gen. Pl. 1: 139. 1891. *Vitis vitiginea* (Linnaeus) Kuntze forma *ovata* Kuntze, Revis. Gen. Pl. 1: 139. 1891.

Cissus sicyoides Linnaeus forma *floridana* Planchon, in A. de Candolle, Monogr. Phan. 5: 530. 1887. TYPE: FLORIDA: Monroe Co.: Cape Sable, *Curtiss 457* (holotype: P; isotypes: A, NA).

Woody vine, to 20 m; stem with simple tendrils. Leaves with the blade suborbicular-ovate to oblong-ovate, undivided, 5–15 cm long and wide, often asymmetrical, the apex obtuse to acuminate, the base rounded to truncate or cordate, the margin coarsely to finely serrate, the upper and lower surfaces densely pubescent to glabrate, the petiole to ca. 6 cm long. Flowers in an axillary, umbelliform cyme, pubescent; sepals minute; petals 1–3 mm long, green or yellowish green, spreading. Fruit globose-obovoid, 6–9 mm long, black.

Hammocks. Occasional; Brevard and Indian River Counties southward. Florida; West Indies, Mexico, Central America, and South America. All year.

Cissus verticillata is commonly infested with the smut fungus *Mycosyrinx cissi* (Poiret) G. Beck that causes a "witches broom" of the inflorescence. The type of *Viscum verticillatum* is based on such a specimen, and the name was rejected as a "monstrosity" under Article 71 of earlier nomenclatural codes. That article was deleted by the Leningrad Congress in 1975, which led to the combination in *Cissus* by Nicolson and Jarvis (1984).

Parthenocissus Planch., nom. cons. 1887. CREEPER

Woody vines, the tendrils branched, with adhesive disks. Leaves palmately compound, petiolate, stipulate. Flowers in axillary, paniculiform cymes, bisexual; sepals 5; petals 5, free; stamens 5; nectariferous disk obscure, fused with the ovary base. Fruit a 1- to 4-seeded berry.

A genus of about 13 species; North America, Mexico, Central America, and Asia. [From the Greek *parthenos*, virgin, and *kissos*, ivy, an equivalent of *vigne-vièrge*, the French name for the type species.]

Parthenocissus quinquefolia (L.) Planch. [Five-leaved.] VIRGINIA CREEPER; WOODBINE.

Hedera quinquefolia Linnaeus, Sp. Pl. 1: 202. 1753. *Vitis hederacea* Ehrhart, Beitr. Naturk. 6: 85. 1791. *Vitis quinquefolia* (Linnaeus) Lamarck, Tabl. Encycl. 2: 135. 1797; non Naronha, 1790. *Ampelopsis quinquefolia* (Linnaeus) Michaux, Fl. Bor.-Amer. 1: 160. 1803. *Cissus hederacea* (Ehrhart) Persoon, Syn. Pl. 1: 143. 1805, nom. illegit. *Ampelopsis hederacea* (Ehrhart) de Candolle, Prodr. 1: 633. 1824, nom. illegit. *Cissus quinquefolia* (Linnaeus) Desfontaines, Tabl. Ecole Bot., ed. 3. 238. 1829; non Solander ex Sims, 1823. *Quinaria hederacea* (Ehrhart) Rafinesque, Med. Fl. 2: 122. 1830, nom. illegit. *Parthenocissus quinquefolia* (Linnaeus) Planchon, in A. de Candolle, Monogr. Phan. 5: 448. 1887. *Parthenocissus quinquefolia* (L.) Planchon var. *typica* Planchon, in A. de Candolle, Monogr. Phan. 5: 449. 1887, nom. inadmiss. *Psedera quinquefolia* (Linnaeus) Greene, Leafl. Bot. Observ. Crit. 1: 220. 1906.

Vitis hederacea Ehrhart var. *hirsuta* Pursh, Fl. Amer. Sept. 170. 1814. *Ampelopsis hirsuta* (Pursh) Donn ex Schultes, in Roemer & Schultes, Syst. Veg. 5: 321. 1819. *Cissus hirsuta* (Pursh) Steudel, Nomencl. Bot. 1: 199. 1821. *Quinaria hirsuta* (Pursh) Rafinesque, Med. Fl. 2: 122. 1830. *Parthenocissus quinquefolia* (Linnaeus) Planchon var. *hirsuta* (Pursh) Planchon, in A. de Candolle, Monogr. Phan. 5: 449. 1887. *Parthenocissus hirsuta* (Pursh) Graebner, Gartenflora 49: 274. 1900. *Psedera hirsuta* (Pursh) Greene, Leafl. Bot. Observ. Crit. 1: 220. 1906. *Psedera quinquefolia* (Linnaeus) Greene var. *hirsuta* (Pursh) Rehder, Rhodora 10: 26. 1908. *Parthenocissus quinquefolia* (Linnaeus) Planchon forma *hirsuta* (Pursh) Fernald, Rhodora 41: 429. 1939.

Ampelopsis hederacea (Ehrhart) de Candolle var. *murorum* Focke, Abh. Naturwiss. Vereine Bremen 4: 560. 1875. *Parthenocissus quinquefolia* (Linnaeus) Planchon var. *murorum* (Focke) Rehder, Mitt. Deutsch. Dendrol. Ges. 14: 133. 1905.

Vitis inserta Kerner, Pflanzenleben 1: 658, f. 1. 1887. *Parthenocissus inserta* (Kerner) Fritsch, Excursionsfl. Oesterreich., ed. 3. 321. 1922.

High-climbing vine; stem sometimes forming adventitious roots, the tendrils 3- to 8-branched, with adhesive disks at the branch tips. Leaves (3)5-foliolate, the leaflets lanceolate to elliptic or ovate, sometimes oblanceolate to obovate, to 18 cm long, to 5 cm wide, sessile or subsessile, the apex acuminate, the base cuneate, the margin coarsely and irregularly serrate, the upper and lower surfaces glabrous or occasionally short hirsute, the petiole (2)6–8(11) cm long. Flowers in an axillary, paniculiform cyme; sepals minute; petals 1–3 mm long, reddish with greenish margins, hooded, strongly reflexed. Fruit subglobose to obovoid, 6–8 mm long, dark blue to black.

Hammocks and floodplain forests. Frequent; nearly throughout. Quebec south to Florida, west to Saskatchewan, Utah, and Texas; Mexico and Central America; Asia. Native to North America, Mexico, and Central America. Spring.

Vitis L. 1753. GRAPE

Woody vines, the tendrils unbranched or 2- to 3-branched, without adhesive disks. Leaves simple, petiolate, stipulate. Flowers in thyrsoid panicles, unisexual and bisexual (plants polygamodioecious); sepals 5; petals 5, coherent at the apex, separating at the base and falling as a unit at anthesis; stamens 5; nectariferous disk composed of 5 free or coherent glands. Fruit 3- to 4-seeded.

A genus of about 60 species; North America, West Indies, Mexico, Central America, Europe, and Asia. [Classical Latin name of *Vitis vinifera*.]

Muscadinia (Planch.) Small, 1903.

Selected references: Moore (1991); Ward (2006).

1. Tendrils simple; bark tightly adherent; pith continuous at the nodes .... **V. rotundifolia**
1. Tendrils 2- to 3-branched; bark loose, shedding; pith with diaphragms at the nodes.
    2. Lower surface of the mature leaves uniformly densely white- or rusty-tomentose **V. shuttleworthii**
    2. Lower surface of the mature leaves with light to rusty cobwebby trichomes (deciduous in age) or glabrate.
        3. Mature leaves glaucous on the lower surface; stem nodes often glaucous................. **V. aestivalis**
        3. Mature leaves not glaucous on the lower surface; stem nodes never glaucous.
            4. Branchlets of the season angled, arachnoid and/or hirtellous-pubescent to glabrate; stem nodes frequently with a red-pigmented band; mature fruit less than 8 mm in diameter....... .... **V. cinerea**
            4. Branchlets of the season terete, glabrous or arachnoid; stem nodes usually lacking a red-pigmented band; mature fruit usually more than 8 mm in diameter.
                5. Nodal diaphragm more than 2.5 mm wide; leaf apex usually long-acuminate; branchlets of the season suffused with purplish red...**V. palmata**
                5. Nodal diaphragm less than 2.5 mm wide; leaf apex usually acute to short-acuminate; branchlets of the season gray, green, brown, or with a purplish pigmentation only on the side ....**V. vulpina**

Vitis aestivalis Michx. [Of summer.] SUMMER GRAPE.

Vitis aestivalis Michaux, Fl. Bor.-Amer. 2: 230. 1803. *Vitis labrusca* Linnaeus var. *aestivalis* (Michaux) Regel, Trudy Imp. S.-Peterburgsk. Bot. Sada. 2: 396. 1873. *Vitis vinifera* Linnaeus var. *aestivalis* (Michaux) Kuntze, Revis. Gen. Pl. 1: 13. 1891.

Vitis simpsonii Munson, Bull. Div. Pomol. U.S.D.A. 3: 12. 1890; non Munson, 1887. *Vitis smalliana* L. H. Bailey, Gentes Herb. 3: 207. 1934. *Vitis aestivalis* Michaux var. *smalliana* (L. H. Bailey) Comeaux, in

Comeaux & Fantz, Sida 12: 286. 1987. TYPE: FLORIDA: Manatee Co.: collected in Florida by J. H. Simpson, cultivated in Texas, 25 May 1890, *Munson s.n.* (lectotype: PH; isotypes: MO). Lectotypified by Moore (1991: 347).

Vitis rufotomentosa Small, Fl. S.E. U.S. 756, 1334. 1903. TYPE: FLORIDA: Lake Co.: vicinity of Eustis, 16–30 Apr 1894, *Nash 525* (holotype: NY; isotypes: PH, US).

Vitis gigas Fennell, J. Wash. Acad. Sci. 30: 15, f. 1. 1940. TYPE: FLORIDA: Indian River Co.: Sebastian River, 20 Jul 1938, *Fennell 713* (holotype: US).

High-climbing woody vine; bark exfoliating in shreds on the mature stem, the branchlets of the season terete, arachnoid-flocose to glabrate, the pith brown, interrupted by diaphragms at the nodes, the tendrils 2-branched, the nodes frequently glaucous, not banded with red pigmentation. Leaves with the blade cordate to orbicular, 10–30 cm long and wide, 3- to 5-lobed or rarely unlobed, when lobed, the apex and the lobes usually acute, the sinuses rounded to acute, the base cordate, the margin crenate to dentate, the upper surface glabrous or puberulent, the lower surface glaucous, arachnoid-flocose with whitish to rusty trichomes, frequently hirtellous along the veins and in the vein axils, the petiole about as long as the blade, pubescent or glabrate. Flowers in a panicle 5–16 cm long. Fruit globose, 5–12 mm long and wide, black, glaucous, the lenticels absent; seeds pyriform, 3–6 mm long, tan to brown.

Hammocks. Frequent; nearly throughout. Maine south to Florida, west to Ontario, Nebraska, Kansas, Oklahoma, and Texas, also California. Spring.

Vitis cinerea var. **floridana** Munson [Ashy, in reference to the canescent leaves; of Florida.] FLORIDA GRAPE.

Vitis simpsonii Munson, Proc. Annual Meeting Soc. Promot. Agric. Sci. 8: 59. 1887. *Vitis cinerea* (Engelmann) Engelmann ex Millardet var. *floridana* Munson, Bull. Div. Pomol. U.S.D.A. 3: 12. 1890. *Vitis austrina* Small, Fl. S.E. U.S. 755, 1334. 1903, nom. illegit. TYPE: FLORIDA: Manatee Co.: collected by J. H. Simpson in Florida, cultivated in Texas, 1890, *Munson s.n.* (lectotype: MO; isolectotype: MO). Lectotypified by Moore (1991: 351).

Vitis sola L. H. Bailey, Gentes Herb. 3: 203, f. 116. 1934. TYPE: FLORIDA: Duval Co.: swamp near Jacksonville, 20 Sep 1894, *Curtiss 4791* (lectotype: NY; isolectotype: NY). Lectotypified by Moore (1991: 351).

High-climbing woody vine; bark exfoliating in shreds on the mature stem, the branchlets slightly to distinctly angled, slightly to densely arachnoid, usually not evidently hirtellous (if present, then concealed by arachnoid trichomes), the pith brown, interrupted by diaphragms at the nodes, the tendrils 2- to 3-branched, the nodes not glaucous, often banded with red pigmentation. Leaves with the blade cordate, 8–20 cm long and wide, unlobed or ocsasionally 3-lobed, the apex and the lobes acute to acuminate, the base cordate, the margin crenate to dentate, the upper surface pubescent or glabrous, the lower surface densely arachnoid, without hirtellous trichomes on the veins, or if present, then only sparsely so, the petiole about as long as the blade, hirtellous, usually thinly arachnoid. Flowers in a panicle 10–20 cm long. Fruit globose, 4–9 mm long and wide, black, with little or no glaucescence, lenticels absent; seeds obovoid, 2–4 mm long, brown.

Hammocks. Frequent; nearly throughout. Maryland south to Florida, west to Louisiana. Spring–summer.

Vitis palmata Vahl [Palmate, in reference to the leaves.] CATBIRD GRAPE.

Vitis palmata Vahl, Symb. Bot. 3: 42. 1794.

High-climbing woody vine; bark exfoliating in shreds on the mature stem, the branchlets sub-terete, glabrous or thinly arachnoid, the pith brown, interrupted by diaphragms at the nodes, the tendrils 2-branched, the nodes not glaucous, not banded with red pigmentation. Leaves with the blade cordate, 5–20 cm long and wide, usually deeply 3(5)-lobed, the apex and lobes acuminate, the sinuses acute to rounded, the base cordate, the margin dentate-serrate, the upper surface glabrous, the lower surface not glaucous, glabrous or slightly hirtellous along the veins and in the vein axils, the petiole somewhat shorter than the blade, glabrous or puberulent. Flowers in a panicle 6–15 cm long. Fruit globose, 5–8 mm long and wide, bluish black to black, sometimes slightly glaucous, lenticels absent; seeds globose, 4–7 mm long, dark brown.

Floodplain forests. Rare; Suwannee County, central and western panhandle. Illinois and Indiana south to Florida and Texas, also Connecticut and New Jersey. Spring.

Vitis rotundifolia Michx. [Round-leaved.] MUSCADINE.

Vitis rotundifolia Michaux, Fl. Bor.-Amer. 2: 231. 1803. *Muscadinia rotundifolia* (Michaux) Small, Fl. S.E. U.S. 757, 1335. 1903. TYPE: "a Virginia ad Floridam," without data, *Michaux s.n.* (lectotype: P). Lectotypified by Uttal (1984: 62).

Vitis munsoniana J. H. Simpson ex Planchon, in de Candolle, Monogr. Phan. 5: 615. 1887. *Muscadinia munsoniana* (J. H. Simpson ex Planchon) Small, Fl. S.E. U.S. 757, 1335. 1903. *Vitis rotundifolia* Michaux var. *munsoniana* (J. H. Simpson ex Planchon) M. O. Moore, Sida 14: 345. 1991. *Muscadina rotundifolia* (Michaux) var. *munsoniana* (J. H. Simpson ex Planchon) Weakley & Gandhi, J. Bot. Res. Inst. Texas 5: 452. 2011. TYPE: FLORIDA: Manatee Co.: collected along the Manatee River, 1883, 1885, 1887 by J. H. Simpson, cultivated in Texas, 1890, *Munson s.n.* (lectotype: PH). Lectotypified by Moore (1991: 345).

Vitis rotundifolia Michaux var. *pygmaea* McFarlin ex D. B. Ward, Phytologia 88: 219. 2006. *Muscadina rotundifolia* (Michaux) Small var. *pygmaea* (McFarlin ex D. B. Ward) Weakley & Gandhi, J. Bot. Res. Inst. Texas 5: 452. 2011. TYPE: FLORIDA: Highlands Co.: Lake Jackson, Sebring, 15 Aug 1931, *McFarlin 6524* (holotype: US).

Climbing woody vine; bark of young stem tight, that of the older stem exfoliating in plates, the branchlets slightly angled, thinly grayish or rusty arachnoid, the pith brown, continuous through the nodes, the tendrils unbranched, the nodes not glaucous, but often banded with red pigmentation. Leaves with the blade cordate to suborbicular or reniform, 4–9 cm long and wide, very rarely lobed, the apex very short-acuminate, the base cordate to truncate, the margin crenate to dentate, the upper surface glabrous and shiny, the lower surface glabrous or sparsely hirtellous along the veins and in the vein axils, not glaucous, the petiole usually as long as the blade, glabrous or glabrate. Flowers in panicles 3–8 cm long. Fruit globose, 8–25 mm long and wide, black or purplish or occasionally bronze, glaucescent, with tan circular lenticels; seeds oval to ellipsoidal, 5–8 mm long, brown.

Wet to dry hammocks and scrub. Common; nearly throughout. Delaware south to Florida, west to Missouri, Oklahoma, and Texas; West Indies. Spring.

This is an easily recognizable, but highly variable species in Florida. Plants of the peninsula with usually smaller fruit (less than 12 mm long), with usually more than 12 berries per infructescence, and with the leaf blades usually less than 5 cm long have been referred to *Vitis munsoniana* by several authors or to *V. rotundifolia* var. *munsoniana*, following Moore (1991). However, many specimens are intermediate, often making an assignment to variety arbitrary.

Vitis shuttleworthii House [Commemorates Robert James Shuttleworth (1810–1874), British botanist, conchologist, and patron of Ferdinand Rugel (1806–1878), German-born American botanical explorer, pharmacist, and surgeon who collected the type specimen.] CALLOOSE GRAPE.

> *Vitis shuttleworthii* House, Amer. Midl. Naturalist 7: 129. 1921. *Vitis coriacea* Shuttleworth ex Planchon, in A. de Candolle, Monogr. Phan. 5: 345. 1887; non Miquel, 1863. *Vitis candicans* Engelmann var. *coriacea* L. H. Bailey, in A. Gray, Syn. Fl. N. Amer. 1(2): 429. 1897. TYPE: FLORIDA: Manatee Co.: borders of the Manatee River, Jun 1845, *Rugel 111* (holotype: BM; isotype: GH).

Moderately high-climbing woody vine; bark exfoliating in shreds on the mature stem, the branchlets oval to terete, densely tomentose, the pith brown, interrupted by diaphragms at the nodes, the tendrils 2- to 3-branched, the nodes not glaucous, not banded with red pigmentation. Leaves with the blade broadly cordate to nearly reniform, 3–10 cm long and wide, unlobed or 3- to 5-lobed, when lobed, the lobes acute, the sinuses rounded, the base cordate to truncate, the margin with shallow, obtuse teeth or nearly entire, the upper surface floccose or glabrous, the lower surface densely white to rusty tomentose, not glaucous, the petiole about ½ to ¾ the length of the blade. Flowers in a panicle 4–10 cm long. Fruit globose, 9–18 mm long and wide, dark red to purple-black, with little or no glaucescence, lenticels absent; seeds ovoid to globose, 5–6 mm long, dark brown.

Moist hammocks. Frequent; Flagler County, central and southern peninsula. Florida; West Indies. Spring.

Vitis vulpina L. [Fox.] FROST GRAPE.

> *Vitis vulpina* Linnaeus, Sp. Pl. 1: 203. 1753. *Vitis muscadina* Rafinesque, Med. Fl. 2: 132. 1830, nom. illegit.
>
> *Vitis cordifolia* Michaux, Fl. Bor.-Amer. 2: 231. 1803. TYPE: "a Pensylvania ad Floridam," without data, *Michaux s.n.* (lectotype: P). Lectotypified by Uttal (1984: 62).
>
> *Vitis cordifolia* Michaux, var. *sempervirens* Munson, Rev. Vitic. 5: 165. 1896. *Vitis illex* L. H. Bailey, Gentes Herb. 3: 217. 1934. TYPE: FLORIDA: Manatee Co.: collected in Manatee Co., cultivated in Texas, 10 May 1890, *Munson s.n.* (lectotype: BH). Lectotypified by Moore (1991: 353).

High-climbing woody vine; bark exfoliating in shreds on the mature stem, the branchlets slightly angled, sparsely arachnoid or glabrous, the pith brown, interrupted by diaphragms at the nodes, the tendrils 2-branched, the nodes not glaucous, not banded with red pigment. Leaves with the blade cordate, 5–20 cm long and wide, 3-lobed, the apex of the lobes acute to short-acuminate, the base cordate to truncate, the margin irregularly dentate-serrate, the upper surface glabrous or sparsely hirtellous, the lower surface hirtellous along the veins and in the vein axils, sometimes sparsely arachnoid, not glaucous. Flowers in a panicle 10–17 cm long. Fruit globose, 5–10 mm long and wide, black, sometimes glaucous, lenticels absent; seeds ovoid, 3–5 mm long, dark brown.

Wet hammocks. Occasional; northern counties south to Manatee County. Massachusetts and New York south to Florida, west to Ontario, Nebraska, Kansas, Oklahoma, and Texas. Spring.

DOUBTFUL AND EXCLUDED TAXA

Vitis bracteata Rafinesque—Type "Carolina to Florida." Based on Florida material, but Moore (1991) was unable to place the name in synonymy.

Vitis cinerea (Engelmann) Engelmann ex Millardet—Although often reported for Florida, such as by Chapman (1897), Small (1903, 1913a, 1913e, 1933), Correll and Johnston (1970), Godfrey and Wooten (1981), Clewell (1985), Godfrey (1988), Wunderlin (1998), and Wunderlin and Hansen (2003), no specimens of the typical variety have been seen from the state. The name was generally misapplied to material of *V. cinerea* var. *floridana*.

Vitis floridana Rafinesque—Type: Florida. Based on Florida material, but Moore (1991) was unable to place the name in synonymy.

Vitis glareosa Rafinesque—Type: Florida. Based on Florida material, but Moore (1991) was unable to place the name in synonymy.

Vitis labrusca Linnaeus—Reported by Radford et al. (1964, 1968), apparently by mistake. No Florida specimens known; excluded from Florida by Moore (1991).

Vitis latifolia Rafinesque—Type: "Canada to Florida and Louisiana." Based on Florida material, but Moore (1991) was unable to place the name in synonymy.

Vitis tiliifolia Willdenow ex Roemer & Schultes—Reported by Chapman (1860, 1883, both as *V. caribaea* de Candolle) and Small (1903, 1913a, both as *V. caribaea* de Candolle; 1913b, 1913e), the name probably misapplied to *V. cinerea* var. *floridana*.

ZYGOPHYLLACEAE R. Br., nom. cons. 1814. CALTROP FAMILY

Annual or perennial herbs or trees. Leaves opposite, even-pinnately compound, the leaflets inequilateral, petiolate, stipulate. Flowers terminal or pseudaxillary, actinomorphic or slightly zygomorphic, bisexual; sepals 5, free or slightly connate basally; petals 5, free; intrastaminal disk usually present, glandular; stamens 10, in 2 whorls of 5 each, the filaments free or the outer whorl adnate to the petals basally, frequently glandular basally, the anthers 2-loculate, sub-basifixed to versatile, introrse, longitudinal dehiscent; carpels 2, 5, or 10, superior, connate, the ovary 2-, 5-, or 10-loculate, the style terminal, simple. Fruit a septicidal capsule or a schizocarp splitting lengthwise into 5–10 tuberculate or spiny mericarps; seeds 1 per locule.

A family of 22 genera and about 285 species; nearly cosmopolitan.

Selected reference: Porter (1972).

1. Tree or shrub; corolla blue or purple; fruit a 2- to 5-lobed capsule..**Guaiacum**
1. Herb; corolla yellow; fruit a schizocarp separating at maturity into 5 or 10 mericarps.
    2. Fruit with tubercles, separating into 10 mericarps ....**Kallstroemia**
    2. Fruit with spines, separating into 5 mericarps...**Tribulus**

Guaiacum L., nom. cons. 1753. LIGNUMVITAE

Trees. Leaves persistent, with 2–6 pairs of opposite, entire, subsessile leaflets, the base inequilateral. Flowers pseudaxillary, slightly irregular, pedunculate; sepals slightly connate basally; petals clawed and twisting basally; intrastaminal disk annular; stamens free; ovary stipitate, the stigma subulate. Fruit a 2- or 5-carpellate, septicidal capsule; seeds surrounded by a fleshy aril.

A genus of 4–5 species; Florida, West Indies, Mexico, Central America, and South America. [From *guayacán*, "medicine gum," a Carib Indian name for one of the species; the sap used as a treatment for various ailments.]

1. Corolla tomentulose; fruit body obcordate, with 2 winglike angles; leaflets in 2(3) pairs ....
....**G. officinale**
1. Corolla glabrous; fruit obovoid, with 5 winglike angles; leaflets 3–6 pairs .... **G. sanctum**

Guaiacum officinale L. [Of medicinal use.] COMMON LIGNUMVITAE.

Guaiacum officinale Linnaeus, Sp. Pl. 1: 381. 1753. *Guaiacum bijugum* Stokes, Bot. Mat. Med. 2: 486. 1812, nom. illegit.

Tree, to 8 m; stem smooth, light gray, mottled. Leaves with the leaflets in 2(3) pairs, these obovate-elliptic, 1.5–3.5 cm long, ca. 2.5 cm wide, coriaceous, sessile, the apex rounded to obtuse, the base rounded to obtuse, the upper and lower surfaces glabrous, the petiole to 1 cm long, the stipules ca. 1 mm long. Flowers solitary or a few clustered distally; sepals suborbicular, 4–5 mm long and wide, pubescent; petals ca. 1.2 cm long, blue or white, puberulous, the apex obtuse. Fruit 1.5–2 cm long, yellow, 2-carpellate and -winged; seeds ovoid to ellipsoid, ca. 1 cm long, dark brown or black, the fleshy aril red.

Disturbed hammocks. Rare; Miami-Dade County. Florida; West Indies and South America. Escaped from cultivation. Native to the West Indies and South America. Spring–fall.

Guaiacum officinale was last collected in the wild in Florida in 1944 (*Rhoads s.n.*, FLAS) and 1956 (*Craighead s.n.*, FLAS).

Guaiacum sanctum L. [Sanctified or holy.] HOLYWOOD LIGNUMVITAE.

Guaiacum sanctum Linnaeus, Sp. Pl. 1: 382. 1753. *Guaiacum multijugum* Stokes, Bot. Mat. Med. 2: 488. 1812, nom. illegit.

Tree, to 10 m; stem light gray, mottled. Leaves with the leaflets in 3–6 pairs, these oblong to obovate or oblanceolate, 2–3.5 cm long, coriaceous, sessile, the apex obtuse or rounded, usually apiculate, the base cuneate, the upper and lower surfaces glabrous or sparsely sericeous, the petiole to 2 cm long; stipules ca. 3 mm long, pubescent, caducous. Flowers solitary or a few clustered distally; sepals obovate oblong-obovate, 5–7 mm long, pubescent; petals broadly obovate, 7–12 mm long, blue or purple, glabrous, the apex rounded. Fruit broadly obovoid, ca. 1.5 cm long, yellow or orange, 5-carpellate and -winged; seeds ellipsoid, ca. 1 cm long, dark brown or black, the fleshy aril red.

Hammocks. Rare; Monroe County keys where native, Miami-Dade County where escaped from cultivation. Florida; West Indies, Mexico, and Central America. All year.

Guajacum sanctum is listed as endangered in Florida (Florida Administrative Code, Chapter 5B-40).

Kallstroemia Scop. 1777. CALTROP

Annual herbs. Leaves with 3–5(6) pairs of opposite, entire, subsessile leaflets, those on 1 side of the rachis slightly smaller, the base inequilateral. Flowers solitary, pseudaxillary, actinomorphic, pedunculate; sepals free; petals convolute; intrastaminal disk annular, 10-lobed; 5 outer stamens basally adnate to the petals, the 5 inner ones free, subtended by a small bilobed gland; ovary sessile, the stigma capitate. Fruit a 10-carpellate schizocarp dividing into 10 1-seeded mericarps; seeds not arillate.

A genus of 17 species; North America, West Indies, Mexico, Central America, South America, Africa, and Asia. [Commemorates Anders Kallström (1733–1812), a contemporary of Giovanni Antonio Scopoli (1723–1778).]

Selected Reference: Porter (1969).

1. Ovary and fruit glabrous or sometimes strigose..**K. maxima**
1. Ovary and fruit pubescent..**K. pubescens**

Kallstroemia maxima (L.) Hook. & Arn. [The largest.] BIG CALTROP.

> *Tribulus maximus* Linnaeus, Sp. Pl. 1: 386. 1753. *Kallstroemia tribulus* Meisner, Pl. Vasc. Gen. 43. 1837, nom. illegit. *Tribulus decolor* MacFadyen, Fl. Jamaica 186. 1837, nom. illegit. *Kallstroemia maxima* (Linnaeus) Hooker & Arnott, Bot. Beechey Voy. 282. 1838.
> *Tribulus trijugatus* Nuttall, Gen. N. Amer. Pl. 1: 277. 1818. *Tribulus dimidiatus* Rafinesque, Autik. Bot. 176. 1840, nom. illegit.

Annual herb; stem prostrate to decumbent, to 1 m, sericeous and sparsely hirsute, becoming glabrate. Leaves obovate, 1–6 cm long, 1.5–5 cm wide, the leaflets 3–4(6) pairs, 5–29 mm long, 3–14 mm wide, the terminal pair larger, broadly oblong to elliptic, appressed hirsute to glabrate, the margins and veins sericeous; stipules 3–5 mm long, ca. 1 mm wide. Flowers with the peduncle at first shorter than the subtending leaves, equaling them or longer in fruit, 1–5 cm long; sepals ovate, 3–8 mm long, 2–3 mm wide, hirsute; petals obovate, 5–12 mm long, to 10 mm wide, white to pale orange, the base white to yellow-green, green, or rarely red, fading to white or bright orange; stamens as long as the style; ovary ovoid, ca. 1 mm in diameter, usually glabrous but occasionally strigose at the base or rarely to the base of the style, the style 2–3 mm long, cylindric, the base slightly conical, the stigma capitate, obscurely 10-lobed. Fruit ovoid, 5–6 mm long and wide, usually glabrous but occasionally strigose at the base or rarely to the base of the beak, the beak 3–7 mm long, glabrous, the base widely conic, the mericarps 3–4 mm long, ca. 1 mm wide, tuberculate, cross-ridged, and slightly keeled, the sides pitted.

Disturbed sites. Occasional; peninsula, Escambia County. South Carolina, Georgia, Florida, Alabama, and Texas; West Indies, Mexico, Central America, and South America. Native to tropical America. All year.

Kallstroemia pubescens (G. Don) Dandy [Pubescent.] CARIBBEAN CALTROP.

> *Tribulus pubescens* G. Don, Gen. Hist. 1: 769. 1831. *Kallstroemia minor* Hooker f., in Hooker, Niger Fl. 269. 1849, nom. illegit. *Kallstroemia pubescens* (G. Don) Dandy, in Keay, Kew Bull. 10: 138. 1955.

Annual herb; stem prostrate to decumbent, to 1 m, sparsely to densely hirsute and sericeous. Leaves obovate, 1–6 cm long, 1.5–5 cm wide, the leaflets (2)4(4) pairs, elliptic to obovate, 3–26 mm long, 5–17 mm wide, the terminal pair the largest, appressed hirsute to glabrous, the margins and veins usually sericeous; stipules 2–6 mm long, 1–2 mm wide. Flowers with the peduncle shorter or equaling the subtending leaves, 1–3.5 cm long; sepals lanceolate, 4–8 mm long, 2–3 mm wide, hispidulous; petals obovate, 6–11 mm long, 5–8 mm wide, white to pale orange, the base green fading to white or yellow; stamens as long as the style; ovary pyramidal, 3–5 mm long including the style, densely appressed short-pilose, the style stout, conical, the stigma capitate, obscurely 10-lobed. Fruit ovoid, 5–6 mm long and wide, densely appressed short-pilose, the beak 5–8 mm long, ca. as long as the fruit body, short-pilose to glabrous, cylindric, the base conic, the mericarps 3–4 mm long, ca. 1 mm wide, cross-ridged, tuberculate to rugose, the sides pitted.

Disturbed sites. Rare; Franklin County. Florida; West Indies, Mexico, Central America, and South America; Africa and Asia. Native to tropical America. Summer–fall.

Kallstroemia pubescens has not recently been collected in Florida.

Tribulus L. 1753. PUNCTUREVINE

Annual or perennial herbs. Leaves with one of each pair alternately smaller than the other, each with 3–7 pairs of opposite, entire, sessile leaflets, the base inequilateral. Flowers solitary, pseudaxillary, actinomorphic, pedunculate; sepals free; petals deciduous; intrastaminal disk annular, 10-lobed; 5 outer stamens basally adnate to the petals, the inner 5 subtended by nectariferous glands, these free or connate into a ring surrounding the ovary base; ovary sessile, the stigma 5-lobed. Fruit a 5-carpellate schizocarp dividing into 5 1-seeded mericarps; seeds not arillate.

A genus of about 24 species; North America, Africa, and Asia. [From the Greek *tribolos*, a kind of caltrop, an iron instrument with four spines arranged so that one always projects upward, used to impede cavalry.]

1. Petals 8–22 mm long; peduncle 2–3 cm long, usually longer than the subtending leaves; intrastaminal glands connate, forming a 5-lobed ring around the ovary base ....**T. cistoides**
1. Petals 3–5 mm long; peduncle 0.5–1 cm long, usually shorter than the subtending leaves; intrastaminal glands free ....**T. terrestris**

Tribulus cistoides L. [To resemble the genus *Cistus* (Cistaceae), in reference to the flowers.] BURRNUT; JAMAICAN FEVERPLANT.

Tribulus cistoides Linnaeus, Sp. Pl. 1: 387. 1753. *Kallstroemia cistoides* (Linnaeus) Endlicher, Ann. Wiener Mus. Naturgesch. 1: 184. 1836. *Tribulus terrestris* Linnaeus var. *cistoides* (Linnaeus) Oliver, Fl. Trop. Afr. 1: 284. 1868.

Perennial herb; stem prostrate to ascending, to 0.5 m, pilose or strigose. Leaves with the blade to 8 cm long, the leaflets in 5–10 pairs, obliquely oblong to elliptic, 5–15 mm long, ca. 8 mm wide, the apex obtuse to subacute, the upper surface glabrate to sericeous, the lower surface sericeous, the petiole to 1 cm long; stipules subulate, 5–8 mm long. Flower 2–4 cm in diameter, the peduncle 2–3 cm long, usually longer than the subtending leaves; sepals narrowly lanceolate, acute to acuminate, ca. 1 cm long, ca. 3 mm wide; petals obovate, 8–22 mm long, 5–16 mm wide, yellow; intrastaminal glands connate into a 5-lobed ring around the ovary base; ovary hirsute. Fruit to 1.5 cm long, the spines 5–8 mm long.

Disturbed sites. Frequent; central and southern peninsula, Okaloosa County. Georgia, Florida, Louisiana, and Texas; West Indies, Mexico, Central America, and South America; Africa, Asia, Australia, and Pacific Islands. Native to Africa, Asia, and Australia. Spring–fall.

Tribulus cistoides is listed as a Category II invasive species in Florida by the Florida Exotic Pest Plant Council (FLEPPC, 2015).

Tribulus terrestris L. [Of the earth, in reference to its prostrate growth.] PUNCTURE-WEED.

> *Tribulus terrestris* Linnaeus, Sp. Pl. 1: 287. 1753. *Tribulus muricatus* Stokes, Bot. Mat. Med. 2: 496. 1812, nom. illegit.

Annual herb; stem procumbent, to 0.5 m, pilose or strigose. Leaves with the blade 1.5–4.5 cm long, the leaflets in 3–6 pairs, subsessile, ovate to oblong, 4–11 mm long, 1–4 mm wide, the upper and lower surfaces silky-pubescent, at least when young, the petiole to 1 cm long; stipules subulate and often falcate, 1–5 mm long. Flower 0.5–1 cm in diameter, the peduncle usually shorter than the subtending leaves; sepals ovate to ovate-lanceolate, 2–4 mm long, the apex acute, hirsute; petals oblong, 3–5 mm long, 2–3 mm wide, yellow; intrastaminal glands free. Fruit ca. 1.5 cm long, the spines 4–7 cm long.

Disturbed sites. Occasional; northern and central peninsula, western panhandle. Nearly throughout North America and tropical America; Europe, Africa, Asia, Australia, and Pacific Islands. Native to Europe, Africa, Asia, and Australia. Spring–fall.

KRAMERIACEAE Dumort., nom. cons. 1829. RATANY FAMILY

Perennial herbs. Leaves alternate, simple, sessile, estipulate. Flowers axillary, solitary, racemose, or paniculate, bracteate, pedicellate, zygomorphic, bisexual; sepals 5, free, petaloid; petals 5, dimorphic, 3 upper ones petaloid, clawed, the claws connate, the 2 lower ones sessile, glandular; stamens 4, inserted on the base of the 3 connate petaloid claws, the anthers basifixed, opening by apical pores; carpels 2, connate, only 1 fertile, the style obliquely terminal. Fruit indehiscent, 1-seeded.

A family of 1 genus and 18 species; North America, West Indies, Mexico, Central America, and South America.

Selected References: Robertson (1973); Simpson (1989).

Krameria Loefl. 1758. RATANY

Perennial herbs. Leaves simple, entire, sessile, estipulate. Flowers in secund, axillary racemes or panicles terminating the branches; sepals 5, petaloid, slightly unequal; upper 3 petals clawed, the claws connate, 2 lower ones sessile, fleshy and gland-like; stamens in 2 pairs, equal in length, inserted on the base of the connate petal claws; the style obliquely terminal, the stigma terminal. Fruit with retrorsely barbed spines.

A genus of 18 species; North America, West Indies, Mexico, Central America, and South America. [Commemorates Johann Georg Heinrich Kramer (1684–1744), an Austrian Army physician and botanist, or his son William Heinrich Kramer (d. 1765), Austrian physician and naturalist.]

Krameria lanceolata Torr. [Lance-shaped, in reference to the leaves.] SANDSPUR; TRAILING RATANY.

> *Krameria lanceolata* Torrey, Ann. Lyceum Nat. Hist. New York 2: 168. 1827. *Dimenops lanceolata* (Torrey) Rafinesque, Atl. J. 144. 1832. *Krameria secundiflora* de Candolle var. *lanceolata* (Torrey) Chodat, Arch. Sci. Phys. Nat., ser. 3. 24: 498. 1890.
>
> *Krameria spathulata* Small ex Britton, N. Amer. Fl. 23: 197. 1930. TYPE: FLORIDA: Lake Co.: near Eustis, 12 May 1900, *Curtiss 6612* (holotype: NY; isotypes: G, GH, K, P, US).

Sprawling, decumbent, perennial herb; stem to 1 m, densely tomentose to sparsely strigose. Leaves linear to linear-lanceolate or rarely spatulate near the stem base, 5–22 mm long, 1–4 mm wide, the apex acute, mucronate, the margin entire, the upper and lower surfaces strigose. Flowers in a secund, racemose (rarely paniculate) inflorescence terminating most branches, these 10–30 mm long, the buds obliquely lanceolate, slightly expanded ventrally; bracts linear-lanceolate, 10–15 mm long, 0.5–2 mm wide; sepals lanceolate to ovate, deep purple, the margin entire, densely strigose dorsally, the uppermost one 8–15 mm long, 3–5 mm wide, the lowermost one 10–16 mm long, 3–6 mm wide, the lateral sepals narrower than the lowermost; petaloid petals 3, clawed, 5–7 mm long, basally connate for ca. 4 mm, the connate portion cream-colored or pink, the blade reniform, 1.5–2 mm long, 2–3 mm wide, purple or pink with purple edge, the margin crenulate, the glandular petals cuneate, 1.5–3 mm long, pink, apricot, or reddish, with secretory blisters on the upper half or quarter of the dorsal surface, often with a plicate and convolute edge; stamens 3–4 mm long, the lateral pairs connate 1–2.5 mm beyond the point of insertion, pale pink; ovary ovoid, 3–4 mm long, densely strigose, the style 2–4 mm long, glabrous. Fruit subglobose, 5.5–8 mm long and wide (excluding the spines), with a short ridge extending from the tip on each side toward the pedicel, pubescent, with stout, blunt, yellow, retrorsely barbed spines 2–5 mm long.

Sandhills. Occasional; northern and central peninsula, central and western panhandle. Georgia and Florida, also from Arkansas west to Colorado and Arizona; Mexico. Spring.

EXCLUDED TAXON

> *Krameria secundiflora* Mociño & Sessé y Lacasta ex de Candolle—This Mexican species was reported for Florida by Small (1903, 1913a), misapplied to material of *K. lanceolata*.

FABACEAE Lindl., nom. cons. 1836. PEA FAMILY

Trees, shrubs, perennial or annual herbs, or vines. Leaves alternate, pinnately, bipinnately, or palmately compound or simple through reduction or fusion, stipulate or estipulate. Flowers solitary or in terminal and/or axillary panicles, racemes, spikes, heads, or glomerules, bisexual or unisexual (plants monoecious, dioecious, or polygamodioecious), actinomorphic or zygomorphic; sepals 4–5; corolla 4–5; stamens 1–many, free, monadelphous, or diadelphous; ovary 1, the ovules 1–many, the style simple. Fruit a dehiscent or indehiscent legume, loment, or fleshy or berry-like drupe.

A family of about 730 genera and about 19,400 species; nearly cosmopolitan.

Caesalpiniaceae R. Br., nom. cons. 1814; *Cassiaceae* Vest 1818; *Leguminosae* Juss., nom. cons. et nom. alt. 1789; *Mimosaceae* R. Br., nom. cons. 1814.

Selected references: Elias (1974); Isely (1990, 1998); Robertson and Lee (1976).

1. Flowers radially symmetrical .... KEY 1
1. Flowers bilaterally symmetrical or asymmetrical (in Dalea, flowers appearing symmetrical, but with 1 petal arising from the receptacle and 4 petals epistemonous).
  2. Corolla not papilionaceus, i.e., not differentiated into standard, wing, and keel petals (in *Cercis* the corolla pseudopapilionaceous with the uppermost petal in front of the lateral petals rather than behind them)... KEY 2
  2. Corolla papilionaceous, i.e., differentiated into standard, wing, and keel petals (reduced to the standard petal in *Amorpha*, and in *Dalea* the flowers with 1 petal arising from the receptacle and 4 petals epistemonous).
    3. Leaves paripinnate.. KEY 3
    3. Leaves 1-foliolate, 3-foliolate, imparipinnate, or palmate.
      4. Leaves 1-foliolate or palmate .... KEY 4
      4. Leaves 3-foliolate or imparipinnate.
        5. Leaves imparipinnate .... KEY 5
        5. Leaves 3-foliolate .... KEY 6

KEY 1

1. Stamens more than 10.
  2. Stamens free.
    3. Leaves simple, leaflike phyllodes .... **Acacia**
    3. Leaves bipinnate.
      4. Petiole glandular, if eglandular, then the flowers in a spike .... **Vachellia**
      4. Petiole eglandular .... **Acaciella**
  2. Stamens connate into a staminal sheath.
    5. Leaves eglandular; fruit valves separating from the apex and recurving sideways .... **Calliandra**
    5. Leaves with glands on the petiole and/or rachis; fruit indehiscent or if dehiscent, the valves not separating as above.
      6. Petiole with a gland near the base .... **Albizia**
      6. Petiole lacking a gland near the base (sometimes with a gland medial or distal on the petiole).
        7. Pinnae 8–12 pairs; fruit indehiscent, woody, circinately coiled.................... **Enterolobium**
        7. Pinnae 1–4 pairs; fruit various but not as above.
          8. Stipules conspicuous (at least on young growth)................................. **Lysiloma**
          8. Stipules spinose, or if not, then inconspicuous..................................... **Pithecellobium**
1. Stamens 10 or fewer.
  9. Plant with short shoots terminated by thorns... **Dichrostachys**

9. Plant with internodal prickles or unarmed.
    10. Leaves with 1-several sessile glands.
        11. Spreading or prostrate herb ....**Desmanthus**
        11. Tree or shrub...**Leucaena**
    10. Leaves eglandular.
        12. Tree; flowers in slender racemes; fruit with the valves contorted on dehiscence ....
        .... **Adenanthera**
        12. Herb or shrub; flowers in spheroid or short-cylindric heads; fruit with valves remaining straight or breaking into segments on dehiscence.
            13. Flowers yellow or greenish yellow; fruit dehiscent by separation of the valves through 1 or both sutures...**Neptunia**
            13. Flowers pinkish; fruit dehiscent by separation of the valves from the sutures ....**Mimosa**

KEY 2

1. Leaves 1-foliolate, or if 2-foliolate, then a woody vine with tendrils; calyx distinctly united.
    2. Flowers pseudopapilionaceous; stamens 10; fruit tardily dehiscent, dorsally winged ....**Cercis**
    2. Flowers bilaterally symmetrical, but not pseudopapilionaceous; stamens 3 or 5; fruit elastically dehiscent, not dorsally winged.
        3. Woody vine with tendrils ....**Phanera**
        3. Tree or shrub.. **Bauhinia**
1. Leaves several-pinnate, or if 2-foliolate, then other than a woody vine with tendrils; calyx free or nearly so.
    4. Perianth of 2 similar whorls, greenish yellow or greenish white; flowers in catkin-like inflorescences ....**Gleditsia**
    4. Perianth of dissimilar calyx and corolla, corolla brightly colored (red, yellow, or orange-yellow); flowers not in catkinlike inflorescences.
        5. Leaves bipinnate, but 1–3 clustered on spurs and appearing pinnate due to the absence of a common petiole, small leaflets deciduous and leaves then phyllodial; plant with nodal spines.........
        ....**Parkinsonia**
        5. Leaves distinctly pinnate or bipinnate, not phyllodial; plant unarmed or with prickles.
            6. Leaves bipinnate.
                7. Corolla red (upper petal varicolored); petals 4–7 cm long; fruit 3–6 dm long; rachis with pubescent projections between the pinnae.. **Delonix**
                7. Corolla yellow or orange-yellow; petals to 3 cm long; fruit to 1.2 dm long; rachis lacking pubescent projections between the pinnae.
                    8. Fruit samaroid ....**Peltophorum**
                    8. Fruit not samaroid ....**Caesalpinia**
            6. Leaves pinnate.
                9. Petals 3 ....**Tamarindus**
                9. Petals 5.
                    10. Filaments of 3 lower stamens sigmoid-curved and elongate.............................. **Cassia**
                    10. Filaments of all stamens straight.
                        11. Fruit elastically dehiscent; bracteoles present; stipules conspicuously striate, persistent .... **Chamaecrista**
                        11. Fruit indehiscent or if dehiscent, not elastically; bracteoles absent; stipules not conspicuously striate, caducous .... **Senna**

KEY 3

1. Tendrils terminating some or all of the leaves.
   2. Stipules foliaceous, usually larger than the leaflets; style longitudinally folded .... **Pisum**
   2. Stipules, if foliaceous, then not larger than the leaflets; style laterally compressed or terete.
      3. Style laterally compressed, bearded laterally or on the upper side ....**Lathyrus**
      3. Style terete, with an apical tuft or trichomes or bearded on the lower side (rarely glabrous).......
      ....**Vicia**
1. Tendrils lacking.
   4. Fruit a dehiscent loment .... **Aeschynomene**
   4. Fruit other than a loment, or if a loment, then not dehiscent.
      5. Fruit to 2 cm long; keel petals with a pouch in the blade base......................................**Indigofera**
      5. Fruit 3 cm long or longer; keel petals without a pouch in the blade base.
         6. Leaves with 2 pairs of pinna ....**Arachis**
         6. Leaves with 8 or more pairs of pinna.
            7. Vine; seeds red and black ....**Abrus**
            7. Erect herb or shrub; seeds tan to dark brown..**Sesbania**

KEY 4

1. Leaves palmate.
   2. Inflorescence a spike with large bracts partially hiding the flowers; fruit a loment................ **Zornia**
   2. Inflorescence various, but if a spike, then lacking large bracts; fruit not a loment.
      3. Fruit 1-seeded, indehiscent; stamens diadelphous; leaves glandular ....**Orbexilum**
      3. Fruit several-seeded, dehiscent; stamens monadelphous; leaves not glandular ....**Lupinus**
1. Leaves 1-foliolate.
   4. Sprawling shrub ....**Dalbergia**
   4. Herb or vine.
      5. Leaves glandular.
         6. Fruit 1-seeded, indehiscent...**Orbexilum**
         6. Fruit (1)2-seeded, dehiscent..**Rhynchosia**
      5. Leaves not glandular.
         7. Stamens free.. **Baptisia**
         7. Stamens monadelphous or diadelphous.
            8. Vine; leaves sagittate...**Centrosema**
            8. Herb; leaves not sagittate.
               9. Fruit a loment; stamens diadelphous; stipules striate ....**Alysicarpus**
               9. Fruit not a loment; stamens monadelphous; stipules, if present, not striate.
                  10. Fruit laterally compressed...**Lupinus**
                  10. Fruit inflated ....**Crotalaria**

KEY 5

1. Corolla reduced to 1 petal (standard).. **Amorpha**
1. Corolla of 5 petals.
   2. Flowers with 1 petal arising from the receptacle and 4 petals epistemonous............................ **Dalea**
   2. Flowers with all petals arising from the receptacle.
      3. Keel petals with a pouch in the blade base...**Indigofera**
      3. Keel petals without a pouch in the blade base.

4. Tree, shrub, or woody vine.

  5. Woody vine.

    6. Flowers in pendent racemes; corolla standard blade with 2 basal calluses ......**Wisteria**

    6. Flowers in erect or spreading panicles; corolla standard blade lacking basal calluses .. ..**Callerya**

  5. Tree or shrub.

    7. Leaflets alternate on the rachis...**Dalbergia**

    7. Leaflets opposite on the rachis.

      8. Stamens free ....**Sophora**

      8. Stamens monadelphous or diadelphous.

        9. Fruit dry, several-seeded, with 4 papery wings.......................................**Piscidia**

        9. Fruit not as above.

          10. Calyx conspicuously lobed; fruit dehiscent.......................................**Robinia**

          10. Calyx truncate or merely toothed; fruit indehiscent.

            11. Flowers in a raceme ....**Gliricidia**

            11. Flowers in a pseudoraceme or panicle.

              12. Fruit elliptic to half-elliptic, 1-seeded.....................................**Millettia**

              12. Fruit suborbicular to oblong, 1- to several-seeded ......**Lonchocarpus**

4. Erect, spreading, or prostrate herb.

13. Standard 3.5–5 cm long, this greatly exceeding the other petals ....**Clitoria**

13. Standard less than 3.5 cm long, this subequaling or only slightly exceeding the other petals.

  14. Inflorescence a bracteate umbel.

    15. Leaflets 5, the lower pair stipular in position; fruit longitudinally dehiscent **Lotus**

    15. Leaflets 11–19, the lower pair not stipular in position; fruit a dehiscent loment...... ....**Coronilla**

  14. Inflorescence not a bracteate umbel.

    16. Fruit a loment.

      17. Loment subterete; plant erect...**Chapmannia**

      17. Loment flattened; plant prostrate ....**Aeschynomene**

    16. Fruit a dehiscent pod.

      18. Plant erect, ascending or decumbent.

        19. Fruit laterally compressed; inflorescence a pseudoraceme............**Tephrosia**

        19. Fruit 3-angled or dorsiventrally compressed; inflorescence a raceme........... ....**Astragalus**

      18. Plant twining or prostrate.

        20. Leaflets with numerous parallel, straight, lateral nerves...............**Tephrosia**

        20. Leaflets lacking numerous parallel, straight, lateral nerves.

          21. Corolla purple-red to brownish maroon; keel scythe-shaped or spirally incurved; inflorescence dense..**Apios**

          21. Corolla white with red striations; keel only moderately incurved; inflorescence sparsely flowered...**Galactia**

KEY 6

1. Leaflets denticulate (if subentire, the lateral nerves extending to the margin and usually slightly exserted).

  2. Fruit curved or spirally coiled.

    3. Fruit curved...**Trigonella**

    3. Fruit spirally coiled...**Medicago**

  2. Fruit straight.

    4. Petals persistent, concealing the mature fruit...**Trifolium**

    4. Petals deciduous, not concealing the mature fruit...**Melilotus**

1. Leaflets entire or lobed (if denticulate, the lateral nerves not extending to the margin and slightly exserted).

  5. Corolla with 1 petal (standard) arising from the receptacle and 4 petals (keel and wings) epistemonous ....**Dalea**

  5. Corolla with all five petals arising from the receptacle.

    6. Fruit indehiscent, either 1-seeded or a several-seeded loment.

      7. Petiole fused most of its length with the amplexicaul stipules, the leaves thus appearing subsessile or short-petiolate; inflorescence a spike; corolla yellow..........................**Stylosanthes**

      7. Petiole free and the stipules not amplexicaul; inflorescence a raceme or pseudoraceme; corolla other than yellow.

        8. Fruit (1)2- to several-segmented loment with uncinate trichomes.

          9. Loment stipe long exserted from the persistent calyx; leaflets estipulate**Hylodesmum**

          9. Loment stipe included or only slightly exserted from the persistent calyx; leaflets stipellate ....**Desmodium**

        8. Fruit 1-seeded, glabrous or variously pubescent, but never with uncinate trichomes.

          10. Fruit included in the enlarged calyx tube except for the persistent broad beak...........
....**Pediomelum**

          10. Fruit not included in the enlarged calyx tube.

            11. Plant glandular-punctate ....**Orbexilum**

            11. Plant eglandular.

              12. Flowers solitary..**Kummerowia**

              12. Flowers numerous or 2–3 in a cluster..**Lespedeza**

    6. Fruit dehiscent, (1)2- to several-seeded.

      13. Leaves with small yellow resinous dots on the surface..**Rhynchosia**

      13. Leaves lacking small yellow resinous dots on the surface.

        14. Plant with prickles ....**Erythrina**

        14. Plant without prickles.

          15. Keel of the corolla spirally twisted.

            16. Leaves with hooked trichomes; fruit glabrous or nearly so.....................**Phaseolus**

            16. Leaves without hooked trichomes; fruit pubescent.

              17. Wing petals much longer than the other petals; flowers deep red to purpleblack ....**Macroptilium**

              17. Wing petals subequaling the other petals; flowers yellow, pale lavender, or purple ....**Leptospron**

          15. Keel of the corolla not spirally twisted.

            18. Leaflets estipellate.

19. Keel petals with a pouch in the blade base..**Indigofera**
19. Keel petals lacking a pouch in the blade base.
    20. Fruit flat.
        21. Leaflets entire; fruit with the ventral margin 3-nerved; style glabrous ....
        ....**Canavalia**
        21. Leaflets coarsely dentate or palmately lobed; fruit with ventral margin
        not nerved; style barbellate..**Pachyrhizus**
    20. Fruit inflated (sometimes slightly so).
        22. Stamens free, all similar ....**Baptisia**
        22. Stamens monadelphous or diadelphous, dimorphic.
            23. Calyx with the 4 upper lobes connate................................**Lotononis**
            23. Calyx deeply 5-lobed..**Crotalaria**
18. Leaflets stipellate.
    24. Stipules conspicuous and persistent.
        25. Calyx 10–12 mm long ....**Pueraria**
        25. Calyx less than 10 mm long.
            26. Bracteoles obsolete; style glabrous....................................**Amphicarpaea**
            26. Bracteoles calycine, usually persistent to anthesis; style bearded.
                27. Stipules medifixed, extending below the zone of attachment, some-
                times short-auriculate..**Vigna**
                27. Stipules basifixed, not extending below the point of attachment.
                      28. Flowers white to pink, the tip of the keel petals dark purple ....
                      ....**Strophostyles**
                      28. Flowers blue or purple, the tip of the keel petals similar in color .
                      ....**Sigmoidotropis**
    24. Stipules inconspicuous or caducous.
        29. Corolla 2.5–5 cm long, resupinate, the standard exceeding other petals.
            30. Calyx lobes equaling or exceeding the tube ....**Centrosema**
            30. Calyx lobes shorter than the tube.
        29. Corolla not as above.
            31. Plant erect, usually evidently glandular-dotted ....**Cajanus**
            31. Plant prostrate or twining, not glandular-dotted.
                32. Corolla 3–4 cm long..**Mucuna**
                32. Corolla 2 cm long or less.
                      33. Style barbellate; bracteoles calycine, persistent to anthesis; ventral
                      suture of the fruit nodose ....**Lablab**
                      33. Style glabrous; bracteoles caducous or only briefly persistent; ven-
                      tral suture of the fruit not nodose ....**Galactia**

Abrus Adans. 1763.

Vines. Leaves alternate, odd-pinnately compound, petiolate, stipulate. Flowers in axillary or terminal, pedunculate racemes; sepals 5; corolla papilionaceous; stamens 9, monadelphous. Fruit a 2-valved legume, flat, dehiscent.

A genus of 4 species; North America, West Indies, Mexico, Central America, South America,

Africa, Asia, Australia, and Pacific Islands. [From the Greek *habras*, graceful, in reference to the leaflets.]

Selected reference: Breteler (1960).

Abrus precatorius L. [A cursed plant, doubtlessly because of poisonous compounds in the seed.] ROSARY PEA; BLACKEYED SUSAN.

Glycine abrus Linnaeus, Sp. Pl. 2: 753. 1753. *Abrus precatorius* Linnaeus, Syst. Nat., ed. 12. 472. 1767. *Abrus abrus* (Linnaeus) W. Wight, Contr. U.S. Natl. Herb. 9: 171. 1905, nom. inadmiss.

Vine; stem woody below and herbaceous above, the young branches sparsely pubescent. Leaves 5–13 cm long, the rachis sparsely pubescent, the pinnae 8–15 pairs, oblong to obovate, 1–2 cm long, 3–6 cm wide, the apex rounded to mucronate, the base rounded, the margin entire, the upper surface glabrous, the lower surface sparsely pubescent, the petiolule short, the petiole to 1 cm long. Flowers in a raceme 3–8 cm long; calyx subtruncate, 2–4 mm long, the lobes very short; corolla papilionaceous, the petals red to purple or rarely white, 8–12 mm long, the standard ovate, with a short, broad claw, the wings oblong, falcate, the keel larger than the wings, curved; stamens 9, monadelphous; ovary subsessile, the style short, the stigma terminal. Fruit oblong, 2–4 cm long, 1–1.5 cm wide, beaked, pubescent, 3- to 5-seeded; seeds 5–7 mm long, scarlet with a black base.

Disturbed sites. Frequent; central and southern peninsula. Georgia, Alabama, Florida, and Arkansas; West Indies, Mexico, Central America, and South America; Africa, Asia, Australia, and Pacific Islands. Native to Africa, Asia, Australia, and Pacific Islands. Summer.

Abrus precatorius is listed as a Category I invasive species in Florida by the Florida Exotic Pest Plant Council (FLEPPC, 2015).

Acacia Mill., nom. cons. 1754.

Shrubs or trees. Leaves alternate, a simple phyllode, petiolate, stipulate. Flowers numerous in globose heads or cylindrical spikes, actinomorphic, bisexual; sepals 4 or 5, connate; petals 4 or 5, connate; stamens 20–100, free, exserted. Fruit a legume, compressed, dehiscent; seeds arillate.

A genus of about 1,300 species; North America, West Indies, Central America, South America, Africa, Asia, Australia, and Pacific Islands. [From the Greek *ake*, a point, in reference to the stipular spines of some species.]

Acacia was previously treated as a genus of about 1,700 species. Recent cladistic analyses have shown that it is not monophyletic, and subsequently it is currently divided into five genera; three of these (*Acacia*, *Acaciella*, and *Vachellia*) are members of our flora.

1. Phyllodes with parallel venation; flowers in cylindric spikes.. **A. auriculiformis**
1. Phyllodes with pinnate venation; flowers in globose heads.
    2. Basal phyllodal gland prominent, disciform ....**A. saligna**
    2. Basal phyllodal gland not obvious, nor disciform...**A. retinodes**

Acacia auriculiformis A. Cunn. ex Benth. [Ear-shaped, in reference to the fruit.] EARLEAF ACACIA.

Acacia auriculiformis A. Cunningham ex Bentham, London J. Bot. 1: 377. 1842. *Racosperma auriculiforme* (A. Cunningham ex Bentham) Pedley, Bot. J. Linn. Soc. 92: 247. 1986.

Tree, to 12(30) m; branchlets gray to reddish brown, slightly ridged, glabrous, usually somewhat glaucous. Leaves a simple phyllode, this linear to narrowly elliptic, 10–20 cm long, 1–3(5) cm wide, falcate, the venation parallel, usually with 3–5 veins prominent, the apex obtuse, the base cuneate, the margin entire, the upper and lower surfaces glabrous, with a phyllodial gland ca. 3 mm above the pulvinus, the pulvinus 3–5 mm long. Flowers yellow, in a narrow, cylindric spike 5–8 cm long, the spike solitary or 2–3 in the leaf axil, rarely in a pseudoraceme; calyx campanulate, ca. 1 mm long, glabrous; corolla campanulate, ca. 2 mm long, glabrous; stamens 3–4 mm long. Fruit oblong, 3–15 cm long, 1–1.5 cm wide, flattened, not constricted between the seeds, irregularly twisted, woody.

Scrub and dry, disturbed sites. Occasional; Brevard and Charlotte Counties southward. Escaped from cultivation. Florida; West Indies, Mexico, Central America, and South America; Africa, Asia, Australia, and Pacific Islands. Native to Asia and Australia. Summer–fall.

Acacia auriculiformis is listed as a Category I invasive species in Florida by the Florida Exotic Pest Plant Council (FLEPPC, 2015).

Acacia retinodes Schltdl. [*Retz*, network, in reference to the fine network of veins in the phyllodia.] WATER WATTLE.

Acacia retinodes Schlechtendal, Linnaea 20: 664. 1847. *Racosperma retinodes* (Schlechtendal) Pedley, Austrobaileya 6: 484. 2003.

Shrub or tree, to 10 m; branchlets reddish brown, often ridged or flattened, glabrous. Leaves a simple phyllode, this linear-lanceolate to oblanceolate, 3–20 cm long, 5–15 cm wide, slightly falcate, the venation pinnate, the secondary veins obscure, the apex acuminate, apiculate, the base narrowly cuneate, the upper and lower surfaces glabrous, sometimes glaucous, with a conspicuous phyllodial gland 1–10 mm above the pulvinus, the pulvinus 1–3 mm long. Flowers yellow to cream, in a globose head 5–7 mm long, the head solitary or 5–9 in a pseudoraceme in the leaf axil; calyx campanulate, ca. 1 mm long, glabrous; corolla campanulate, 1–2 mm long, glabrous; stamens 3–4 mm long. Fruit linear, 4–15 cm long, 4–7 mm wide, flattened, slightly constricted between the seeds, woody.

Disturbed sites. Rare; Glades County, Monroe County keys. Escaped from cultivation. Florida and California; Australia. Native to Australia. Spring.

Acacia saligna (Labill.) H. L. Wendl. [Willow-like.] COOJONG; GOLDEN WREATH WATTLE.

Mimosa saligna Labillardière, Nov. Holl. Pl. 2: 86, pl. 235. 1807 ("1806"). *Acacia saligna* (Labillardière) H. L. Wendland, Comm. Acac. Aphyll. 26. 1820. *Racosperma salignum* (Labillardière) Pedley, Astrobaileya 2(4): 355. 1987.

Shrub or small tree, to 6 m; branchlets bluish to purplish, pendulous, slightly ridged, glabrous. Leaves a simple phyllode, this linear to elliptic, 7–25 cm long, 0.5–5 cm wide, straight or slightly falcate, the venation pinnate, the secondary veins obscure, the apex acuminate, apiculate, the base cuneate, the margin entire, the upper and lower surfaces glabrous, with a discoid phyllodial gland ca. 3 mm above the pulvinus, the pulvinus 1–3 mm long. Flowers yellow, in a globose head 8–13 mm long, the heads 2–10 in a pseudoraceme or solitary in the distal axils; calyx campanulate, 1–2 mm long, glabrous; corolla campanulate, 3–4 mm long, glabrous; stamens 5–6 mm long. Fruit linear, 8–14 cm long, 5–8 mm wide, constricted between the seeds, woody.

Disturbed sites. Monroe County keys. Escaped from cultivation. Florida and California; Africa and Australia. Native to Africa and Australia. Spring.

Acaciella Britton & Rose 1928.

Shrubs, trees, or perennial herbs. Leaves alternate, even-bipinnately compound, petiolate, stipulate. Flowers in axillary, globose, pedunculate heads, the heads sometimes in terminal pseudoracemes or pseudopanicles, actinomorphic, bisexual; sepals 5, connate; petals 5, connate; stamens numerous. Fruit a legume, tardily dehiscent.

A genus of 15 species; North America, Mexico, Central America, and South America. [The genus *Acacia*, and the Latin *ella*, diminutive, in reference to the small stature of many of the species.]

Selected references: Britton and Rose (1928); Rico Arce and Bachman (2006).

Acaciella angustissima (Mill.) Britton & Rose [The narrowest, in reference to the leaflets.] PRAIRIE ACACIA.

Mimosa angustissima Miller, Gard. Dict. ed. 8. 1768. *Acacia angustissima* (Miller) Kuntze, Revis. Gen. Pl. 3: 47. 1898. *Acaciella angustissima* (Miller) Britton & Rose, in Britton, N. Amer. Fl. 23: 100. 1928. *Acacia angustissima* (Miller) Kuntze subsp. *typica* Wiggins, Contr. Dudley Herb. 3: 229. 1942, nom. inadmiss. *Senegalia angustissima* (Miller) Pedley, Bot. J. Linn. Soc. 92(3): 238. 1986.

Acacia hirta Nuttall, in Torrey & A. Gray, Fl. N. Amer. 1: 404. 1840. *Acacia angustissima* (Miller) Kuntze var. *hirta* (Nuttall) B. L. Robinson, Rhodora 10: 33. 1908. *Acaciella hirta* (Nuttall) Britton & Rose, in Britton, N. Amer. Fl. 23: 102. 1928.

Acacia texensis Nuttall, in Torrey & A. Gray, Fl. N. Amer. 1: 404. 1840. *Acaciella texensis* (Nuttall) Britton & Rose, in Britton, N. Amer. Fl. 23: 100. 1928. *Acaciella filicioides* (Cavanilles) Trelease ex Branner & Coville var. *texensis* (Nuttall) Small, Bull. New York Bot. Gard. 2: 93. 1901. *Acacia angustissima* Miller var. *texensis* (Nuttall) Isely, Sida 3: 372. 1969. *Acaciella angustissima* Miller var. *texensis* (Nuttall) L. Rico, Fl. Guerrero 25: 44. 2005.

Shrub or small tree (sometimes a suffrutescent perennial herb), to 12 m; bark dark brown to gray-brown, smooth, the branchlets glabrous or white-puberulent to -hirsute. Leaves bipinnately compound, the pinnae 9–12(15) pairs, the leaflets 18–30 pairs, the blade asymmetrical, oblong, 3–4 mm long, the apex acute to obtuse, the base rounded to broadly cuneate, the margin entire, glabrous or ciliate, the upper and lower surfaces glabrous or white-puberulent; stipules caducous. Flowers in a globose or elliptic head, distinctly pedicellate, the head solitary or 2–3 in the leaf axil or in a terminal pseudoraceme or pseudopanicle, the peduncle 0.5–2 cm long; calyx campanulate, ca. 1 mm long, greenish white, glabrous; corolla campanulate, 2–4 mm long, greenish white, glabrous; stamens 3–7 mm long, white. Fruit linear, 3–5 cm long, 6–9(12) mm wide, flattened, membranaceous, glabrate or puberulent, stipitate.

Sandhills and disturbed sites. Rare; eastern panhandle, Alachua County south to Citrus County. Florida, Missouri, and Kansas south to Louisiana, Texas, New Mexico, and Arizona; Mexico, Central America, and South America.

Acaciella angustissima is a weedy, widespread, and polymorphic species that has been divided into many segregate taxa by various workers. The one occurring in Florida, formerly known and listed as *Acacia angustissima* var. *hirta* is considered endangered (Florida Administrative Code, Chapter 5B-40).

EXCLUDED TAXA

Acaciella angustissima var. *filicioides* (Cavanilles) L. Rico—Reported for Florida by Chapman (1897, as *Acacia filicina* Willdenow) and Small (1903, 1913, both as *Acacia filicioides* (Cavanilles) Trelease ex Branner & Coville; 1933), the name misapplied to material of var. *angustissima*.

Adenanthera L. 1753. BEADTREE

Trees. Leaves alternate, bipinnately compound, petiolate, stipulate. Flowers in axillary terminal or subterminal racemes or terminal panicles, actinomorphic, bisexual; sepals 5, connate; petals 5, free or slightly connate basally; stamens 10, free, the anthers with a deciduous apical, penicellate gland. Fruit a legume.

A genus of about 10 species; North America, West Indies, Central America, South America, Africa, Asia, and Pacific Islands. [From the Greek *aden*, gland, and *antheros*, anther, in reference to the deciduous pedicellate gland on the anthers.]

Adenanthera pavonina L. [Peacock-like, in reference to its brightly colored seeds.]
RED BEADTREE; RED SANDALWOOD.

Adenanthera pavonina Linnaeus, Sp. Pl. 1: 384. 1753.

Tree, to 15 m; branches glabrous. Leaves bipinnate, 30–40 cm long, the pinnae in 2–5 pairs, 10–20 cm long, the leaflets with the petiolules 1–3 mm long, in 6–8, usually alternate pairs, the blade ovate-elliptic to oblong-elliptic, 2–4 cm long, 1–2 cm wide, the apex obtuse to rounded or sometimes emarginate, the base obliquely obtuse, the upper and lower surfaces glabrous, the petiole to ca. 1.5 cm long. Flowers yellow to orange-red, in an axillary raceme or terminal panicle 10–30 cm long, the pedicel to ca. 3 mm long; calyx campanulate, ca. 1 mm long, minutely lobed; petals free or slightly connate basally, ca. 3 mm long; stamens slightly exserted. Fruit narrowly oblong, laterally compressed, ca. 25 cm long, ca. 1 cm wide, usually curved, the valves thin and twisting; seeds ovoid, compressed, 6–10 mm long, scarlet red, lustrous.

Disturbed sites. Rare; Lee and Miami-Dade Counties, Monroe County keys. Escaped from cultivation. Florida; West Indies, Central America, South America; Africa, Asia, Australia, and Pacific Islands. Native to Asia, Australia, and Pacific Islands. Spring.

Adenanthera pavonina is listed as a Category II invasive species in Florida by the Florida Exotic Pest Plant Council (FLEPPC, 2015).

Aeschynomene L. 1753. JOINTVETCH

Annual or perennial herbs. Leaves alternate, odd-pinnately compound, petiolate, stipulate, the leaflets alternate or subopposite, pulvinate, estipellate. Flowers in axillary racemes, bracteate and bractiolate; sepals 5, basally connate; petals 5, papilionaceous, the 2 wing and 2 keel petals auriculate basally; stamens 10, monadelphous, the sheath splitting along the upper side, the anthers dorsifixed, longitudinally dehiscent. Fruit a loment.

A genus of about 150 species; North America, West Indies, Mexico, Central America, South America, Asia, Africa, Australia, and Pacific Islands. [Greek for "ashamed," in reference to the response of the leaves of some species to fold when touched.]

Secula Small, 1913.

Selected reference: Rudd (1955).

1. Plant a floating aquatic .... **A. fluitans**
1. Plant terrestrial.
   2. Plant prostrate; stipules attached at the base, not peltate; calyx with 5 subequal lobes.
      3. Leaflets 3–7(9), 4–12(18) mm long; plant glandular-hispid ....**A. viscidula**
      3. Leaflets 8–18(30), 3–4(6) mm long; plant canescent...**A. histrix**
   2. Plant erect or ascending; stipules peltate, appendiculate proximal to the point of attachment; calyx bilabiate, the upper lobe 2-fid, the lower lobe 3-fid.
      3. Leaflets with 2–4 longitudinal nerves ....**A. americana**
      3. Leaflets with 1 longitudinal nerve.
         4. Loment distinctly crenate on both margins, the isthmi distinctly narrower than the width of the segments.. **A. pratensis**
         4. Loment entire or only slightly undulate on the upper margin, crenate on the lower margin, the isthmi nearly as wide as the segments.
            5. Plant glabrous to sparsely hispid; flowers 2–10 mm long...................................... **A. indica**
            5. Plant distinctly hispid; flowers 8–15 mm long .... **A. rudis**

Aeschynomene americana L. [Of America.] SHYLEAF.

Aeschynomene americana Linnaeus, Sp. Pl. 2: 713. 1753.

Erect or spreading annual or perennial herb, to 2.5 m; stem densely hirsute with yellowish, somewhat viscid-glandular trichomes. Leaves 2–8 cm long, the leaflets 20–60, the blade linear to linear-oblong, 4–15 mm long, 1–2 mm wide, asymmetric, the apex acuminate, 2–4-nerved, the margin ciliate, the upper and lower surfaces glabrous; stipules peltate, appendiculate proximally, linear-lanceolate, 10–25 mm long, 1–4 mm wide. Flowers 1–7 in an axillary raceme; calyx bilabiate, 3–4 mm long, the upper lobe 2-fid, the lower lobe 3-fid, glabrous to sparsely hispidulous; corolla 5–10 mm long, yellowish orange, usually with red or purple lines, the banner glabrous. Fruit with (3)6–8(9) segments, the upper margin straight or slightly curved, the lower margin crenate, the segments semiorbicular, 3–5 mm long and wide, glabrous or puberulent, sometimes with glandular trichomes, usually verrucose, the gynophore 1–3 mm long; seeds elliptic, 2–3 mm long, dark brown.

Moist, disturbed sites. Frequent; nearly throughout. Georgia, Alabama, Florida, and Louisiana; West Indies, Mexico, Central America, and South America; Africa, Asia, Australia, and Pacific Islands. Native to North America, West Indies, Central America, and South America. Summer–fall.

Aeschynomene fluitans Peter [Floating.] GIANT WATER SENSITIVE PLANT.

Aeschynomene fluitans Peter, Abh. Königl. Ges. Wiss. Göttingen, Math.-Phys. Kl., ser 2. 13(2): 82, t. 2. 1928.

Floating perennial herb; stem hollow, to 4.5 m long, to 1 cm wide, spongy, glabrous, with numerous adventitious roots. Leaves 8–10 cm long, the leaflets 16–26 pairs, the blade oblong to linear-oblong, 1–2.5 cm long, 2–7 mm wide, the apex rounded, usually mucronulate, the base rounded to truncate, the margin entire or finely serrulate, the upper and lower surfaces glabrous; stipules lanceolate, appendiculate proximally, 13–20 mm long, 3–5 mm wide. Flowers solitary, the peduncle and pedicel together 3.5–10 cm long; calyx 12–14 mm long, bilabiate, the upper lobe 1, the lower lobe 3-fid, glabrous; corolla 1.5–2.5(3) cm long, yellow, the banner glabrous. Fruit with 1–4(5) segments, the upper margin straight or slightly curved, the lower margin crenate, the segments irregularly oblong, glabrous, tuberculate along the margin and slightly so medially, the gynophore 5–7 mm long; seeds reniform, 6–8 mm long, dark purplish brown.

Ponds and lakes. Rare; Hillsborough County. Escaped from cultivation. Florida; Africa. Native to Africa. Summer–fall.

Aeschynomene histrix Poir. var. **incana** Benth. [Hispid, in reference to the fruit; hoary, in reference to the stems and leaves.] PORCUPINE JOINTVETCH.

Aeschynomene incana Vogel, Linnaea 12: 90. 1838; non (Swartz) G. Meyer, 1818. *Aeschynomene histrix* Poiret var. *incana* Bentham, in Martius, Fl. Bras. 15(1): 69. 1859.

Prostrate perennial herb; stem to 1 m long, canescent. Leaves 1–6 cm long, the leaflets (8)15–30, the blade oblong-elliptic, 2–5(6) mm long, 1–2(5) mm wide, subsymmetrical, the apex obtuse to subacute, 1-nerved, the margin entire or sparsely denticulate-ciliate, the upper and lower surfaces canescent; stipules basally attached, linear-lanceolate, 4–15 mm long, 1–2 mm wide. Flowers 4–15 in a raceme; calyx campanulate, 2–3 mm long, the lobes subequal, glabrous or puberulent; corolla (4)5–6(7) mm long, yellow, sometimes with reddish veins, the banner puberulent abaxially. Fruit 2–3(5)-segmented, the upper margin straight, the lower margin deeply constricted, the segments semiorbicular, 2–3 mm long and wide, the surface appressed-pubescent, inconspicuously reticulate, the gynophore 1–2 mm long; seed 1–2 mm long and wide, black.

Disturbed sites. Rare; Escambia County. Florida; Mexico, Central America, and South America. Native to South America. Summer.

Aeschynomene indica L. [Of India.] INDIAN JOINTVETCH.

Aeschynomene indica Linnaeus, Sp. Pl. 2: 713. 1753. *Hedysarum neli-tali* Roxburgh, Fl. Ind., ed. 1832. 3: 365. 1832, nom. illegit.

Erect annual or perennial herb, to 2.5 m; stem glabrous or moderately glandular-hispid. Leaves 5–10 cm long, the leaflets (30)50–70, the blade elliptic-oblong, 2–10 mm long, 1–3 mm wide, the apex obtuse to subacute, nearly symmetrical, 1-nerved, the margin entire, the upper and lower surfaces glabrous; stipules peltate, appendiculate, linear-lanceolate, 10–15 mm long, 2–3 mm wide. Flowers solitary or 2–5 in a raceme; calyx bilabiate, 4–6 mm long, the upper lip 3-dentate, the lower lip 2-dentate, glabrous; corolla 7–10 mm long, yellow to purplish, the banner glabrous. Fruit with 5–8(13) segments, the upper margin nearly straight, the lower margin crenate, the segments subquadrate, 5–6 mm long, 4–6 mm wide, reticulate, becoming smooth at maturity, sometimes muricate or verrucose near the center at maturity, glabrous or sparsely hispid, the gynophore 3–10 mm long, glabrous; seeds 3–4 mm long, 2–3 mm wide, dark brown.

Lakes and pond margins, ditches, and wet, disturbed sites. Frequent; nearly throughout. North Carolina south to Florida, west to Missouri, Oklahoma, and Texas; West Indies and South America; Africa, Asia, Australia, and Pacific Islands. Native to Africa, Asia, and Australia. Spring–summer.

Aeschynomene pratensis Small [Of meadows.] MEADOW JOINTVETCH.

Aeschynomene pratensis Small, Bull. New York Bot. Gard. 3: 423. 1905. TYPE: FLORIDA: Miami-Dade Co.: near the slough between Camp Jackson and Long [Pine] Key, May 1904, *Small & Wilson 1960* (holotype: NY).

Erect or decumbent perennial herb, sometimes suffrutescent, to 3 m; stem glabrous or sparsely hispid. Leaves 4–7 cm long, the leaflets 15–35, the blade linear-oblong, 5–10 mm long, 1–3 mm wide, the apex obtuse, nearly symmetrical, 1-nerved, the margin entire, the upper and lower surfaces glabrous; stipules peltate, appendiculate, linear-lanceolate, 5–15 mm long, 1–2 mm wide. Flowers solitary or 2–5 in a raceme; calyx bilabiate, 5–6 mm long, the lips entire or nearly so, glabrous or the upper sometimes with a few cilia; corolla 10–12 mm long, yellowish with purple markings, the banner glabrous. Fruit with 3–9 segments, with margins crenate, the lower more deeply indented, the segments subelliptic, 7–8 mm long, 5–6 mm wide, verrucose, glabrous, the gynophore 10–15 mm long, glabrous or glandular-hirsute; seeds 4–5 mm long, 3–4 mm wide, dark brown.

Pinelands. Occasional; Sarasota and Palm Beach Counties, southern peninsula. Endemic. Winter–spring.

Our plants are var. *pratensis*. Specimens from the West Indies, Central America, and South America are named var. *caribeae* Rudd. This taxon is distinguished from the typical variety by its consistently smaller flowers and fruits (see Rudd, 1955).

Aeschynomene pratensis is listed as endangered in Florida (Florida Administrative Code, Chapter 5B-40).

Aeschynomene rudis Benth. [Of disturbed areas.] ZIGZAG JOINTVETCH.

Aeschynomene rudis Bentham, Pl. Hartw. 116. 1843.

Erect annual herb, to 2 m; stem glabrous or moderately glandular hispid. Leaves 4–10 cm long, the leaflets 30–40(50), the blade oblong to linear-oblong, 6–15 mm long, 2–3 mm wide, subsymmetrical, 1-nerved, the margin entire, sometimes ciliate, the upper and lower surfaces glabrous; stipules peltate, appendiculate, linear-lanceolate, 7–15 mm long, 2–3 mm wide. Flowers solitary or 2–7 in a raceme; calyx bilabiate, 5–8 mm long, the upper lip shallowly 2-lobed, the lower lip 3-dentate, glabrous, the margin ciliate; corolla (8)10–15 mm long, yellow, the banner glabrous, the margin glandular-ciliate. Fruit (3)6–12-segmented, the upper margin straight or nearly so, the lower margin crenate, the segments subquadrate, 4–6 mm long and wide, the surface inconspicuously to moderately reticulate-nerved, usually muricate or verrucose in the center at maturity, subglabrous or lightly glandular-hispidulous, the gynophore 3–6(10) mm long; seeds 3–4 mm long, 2–3 mm wide, dark brown to black.

Disturbed sites. Rare; Lake County. Pennsylvania south to Florida, west to Missouri, Arkansas, and Texas, also California; West Indies, Mexico, Central America, and South America. Native to tropical America. Spring–summer.

Aeschynomene viscidula Michx. [Sticky pubescent.] STICKY JOINTVETCH.

Aeschynomene viscidula Michaux, Fl. Bor.-Amer. 2: 74. 1803. Secula viscidula (Michaux) Small, Fl. Miami 90, 200. 1913. Aeschynomene prostrata Poiret, in Lamarck, Encycl., Suppl. 4: 76. 1816, nom. illegit. TYPE: FLORIDA/GEORGIA.

Prostrate perennial herb; stem to 1 m long, viscid, the trichomes glandular, also crisp-pubescent. Leaves 0.5–2.5 cm long, with 5–9 leaflets, the blade obovate, 4–10(18) mm long, 3–7(10) mm wide, slightly asymmetrical, 1-nerved, the apex obtuse, the margin ciliate-denticulate, the upper surface glabrous or subglabrous, the lower surface glabrous or slightly subappressed-puberulent; stipules attached at the base, ovate, 2–4 mm long, ca. 2 mm wide. Flowers solitary or 2–8 in a raceme; calyx campanulate, 3–4 mm long, the lobes subequal, hispidulous, the margin ciliate; corolla 5–7 mm long, yellow or orangish with reddish venation, the banner pubescent on the dorsal surface. Fruit 2–3(5)-segmented, the upper margin straight or nearly so, the lower margin deeply indented, the segments semiorbicular, 4–5 mm long and wide, slightly to moderately reticulate, densely white-tomentose, usually viscid-hispid, rarely the terminal segment glabrous, the gynophore 1–3(5) mm long; seed 2–3 mm long, ca. 2 mm wide, light brown.

Sandhills, scrub, and flatwoods. Frequent; northern counties, central peninsula, Miami-Dade County. Georgia south to Florida, west to Mississippi, also Texas; West Indies, Mexico, Central America, and South America. Summer–fall.

EXCLUDED TAXA

Aeschynomene histrix Poiret—Because infraspecific categories were not recognized, the type variety was reported for Florida by implication by Small (1933, as Secula histrix (Poiret) Small) and Wilhelm (1984). All Florida collections are of var. incana.

Aeschynomene virginica (L.) Britton et al.—This northeastern U.S. species was reported for Florida by Chapman (1860, as Aeschynomene hispida Willdenow) and Small (1903, 1913a, 1933), who misapplied the name to material of A. indica.

Albizia Durazz. 1772.

Trees. Leaves bipinnate, petiolate, the petiole with a gland, stipulate. Flowers in umbellate heads, axillary or fasciculate-racemose in corymbose compound inflorescences; sepals 5, connate, tubular, the lobes very short; petals 5, connate, the corolla tubular-funnelform, the lobes much shorter than the tube; stamens numerous, basally fused into a tube, strongly exserted. Fruit a legume, strongly compressed.

A genus of about 70 species; North America, West Indies, Mexico, Central America, South America, Africa, Asia, and Australia. [Commemorates Il Sig. Cavalier Filippo degl' Albrizzi, Italian nobleman who introduced *A. julibrissin* into Tuscany in 1749.]

Selected reference: Barneby and Grimes (1996).

1. Inflorescence an umbel 2.5–5 cm in diameter.
   2. Flowers pinkish; pinnae 5–12(15) pairs; leaflets ca. 20(30) pairs, 0.7–1.5 cm long .... **A. julibrissin**
   2. Flowers cream-colored; pinnae 2–4(5) pairs; leaflets 3–10 pairs, 2–4 cm long................**A. lebbeck**
1. Inflorescence a head 1–1.5 cm in diameter.
   3. Leaflets elliptic, 2–4 cm long ....**A. procera**
   3. Leaflets oblong, 1–2.5 cm long..**A. lebbekoides**

Albizia julibrissin Durazz. [Modification of Persian name.] SILKTREE; MIMOSA.

Albizia julibrissin Durazzini, Mag. Tosc. 3(4): 11. 1772. *Mimosa julibrissin* (Durazzini) Scopoli, Delic. Fl. Faun. Insubr. 1: 18. 1786. *Acacia julibrissin* (Durazzini) Willdenow, Enum. Pl. 1052. 1809. *Sericandra julibrissin* (Durazzini) Rafinesque, Sylva. Tellur. 119. 1838. *Feuilleea julibrissin* (Durazzini) Kuntze, Revis. Gen. Pl. 1: 188. 1891.

Tree, to 6 m; bark light, smooth, the branches glabrate. Leaves ca. 20(30) pairs, oblong, the pinnae 7–15 mm long, strongly asymmetric, the midvein submarginal or marginal and mucronate-excurrent, the secondary venation evident, the upper and lower surfaces glabrate, the pubescence primarily along the leaf rachis and sometimes the margins of the leaflets, sometimes with a small gland between the terminal pinnae, the petiole with a flat, proximal elliptic gland. Flowers in a head ca. 2.5 cm in diameter, in a terminal corymbose raceme, pink or rose; calyx 2–4 mm long, glabrous or sparsely villous; corolla 7–11 mm long, glabrous; stamens 24–34, 2.5–3.5 cm long, the tube 6–16 mm long. Fruit oblong, 12–20 cm long, 15–25 mm wide, the apex and base acuminate, laterally compressed, the valves membranous, the margins straight or irregularly constricted.

Disturbed sites. Frequent; northern counties south to Hillsborough, Polk, and Osceola Counties. Escaped from cultivation. Massachusetts and New York south to Florida, west to California; West Indies and South America; Europe, Africa, Asia, and Pacific Islands. Native to Asia. Spring.

Albizia julibrissin is listed as a Category I invasive species in Florida by the Florida Exotic Pest Plant Council (FLEPPC, 2015).

Albizia lebbeck (L.) Benth. [Arabic name.] WOMAN'S TONGUE.

Mimosa lebbeck Linnaeus, Sp. Pl. 1: 516. 1753. *Acacia lebbeck* (Linnaeus) Willdenow, Sp. Pl. 4: 1066. 1806. *Albizia lebbeck* (Linnaeus) Bentham, London J. Bot. 3: 87. 1844. *Albizia latifolia* Boivin ex Miquel, Fl. Ned. Ind. 1: 22. 1855, nom. illegit. *Feuilleea lebbeck* (Linnaeus) Kuntze, Revis. Gen. Pl. 1: 184. 1891.

Tree, to 12 m; branches glabrate. Leaves with 2–4(50) pairs of pinnae, the leaflets 3–10 pairs, oblong-elliptic, 2–4 cm long, asymmetric with displaced midvein, reticulate-nerved, glabrous or inconspicuously villous, the rachis sometimes with glands between some pinnae, the petiole with a large proximal gland. Flowers in an umbellate head 5–6 cm in diameter, in a terminal, corymbose raceme, cream-colored, the pedicel and calyx densely villosulous; calyx 4–6 mm long, glabrous; corolla 7–14 mm long, glabrous; stamens 26–52, 2–3.5(4) cm long, the tube 3–19 mm long. Fruit oblong, 15–25 cm long, 3–4 cm wide, subsessile, the apex acute, the base cuneate, laterally compressed, the margins widely expanded, splitting at dehiscence, glabrous.

Disturbed sites. Occasional; central and southern peninsula. Escaped from cultivation. Florida, Texas, and California. West Indies, Mexico, Central America, and South America; Africa, Asia, and Pacific Islands. Native to Africa and Asia. Spring.

Albizia lebbeck is listed as a Category I invasive species in Florida by the Florida Exotic Pest Plant Council (FLEPPC, 2015).

Albizia lebbeckoides (DC.) Benth. [Resembling *A. lebbeck*.] INDIAN ALBIZIA.

Acacia lebbeckoides de Candolle, Prodr. 2: 467. 1825. *Albizia lebbeckoides* (de Candolle) Bentham, London J. Bot. 3: 89. 1844. *Feuilleea lebbeckoides* (de Candolle) Kuntze, Revis. Gen. Pl. 1: 188. 1891.

Tree, to 15 m; branches glabrous. Leaves with 3–6 pairs of pinnae, the leaflets 5–10 pairs, oblong to narrowly oblong, 1–2.5 cm long, 3–6 mm wide, asymmetric, with a displaced midvein, pinnate-veined, the apex rounded to acute, mucronate, the base truncate, the proximal part widened, the upper and lower surfaces glabrous or glabrate, often with scattered trichomes along the margin, the petiole with a circular substipitate gland. Flowers in a head ca. 1 cm in diameter, in an axillary raceme, green; calyx narrowly campanulate, 1–2 mm, puberulous to tomentose; corolla tubular to campanulate, 4–5 mm long, puberulous to villous; stamens 6–8 mm long, the tube 5–6 mm long. Fruit oblong, 7–15 cm long, 1.5–2.5 cm wide, the apex acute, the base rounded, laterally compressed, glabrous.

Disturbed sites. Rare; Miami-Dade County. Escaped from cultivation. Florida; Asia. Native to Asia. Spring.

Albizia procera (Roxb.) Benth. [Tall.] TALL ALBIZIA.

Mimosa procera Roxburgh, Pl. Coramandel 2: 12, t. 121. 1799. *Acacia procera* (Roxburgh) Willdenow, Sp. Pl. 4: 1063. 1806. *Albizia procera* (Roxburgh) Bentham, J. London Bot. 3: 89. 1844. *Feuilleea procera* (Roxburgh) Kuntze, Revis. Gen. Pl. 1: 188. 1891.

Tree, to 20 m; bark smooth, pale gray. Leaves with 3–5 pairs of pinnae, the leaflets 6–7(10) pairs, elliptic to ovate, 2–3.5(4) cm long, 1–1.8(2) cm wide, slightly asymmetric, pinnate-nerved, the apex obtuse to rounded, often mucronate, the base broadly cuneate, the upper surface glabrous, lustrous, the lower surface strigose with minute trichomes to glabrate, the petiole with a proximal, elliptic gland, 2–5 mm long. Flowers in a head ca. 1.5 cm in diameter, in an axillary, fasciculate raceme or congested in a terminal compound raceme, greenish yellow to golden yellow; calyx campanulate, 3–4 mm long, glabrous; corolla 6–7 mm long, glabrous; stamens 30–40, 9–11 mm long, the tube 5–6 mm long. Fruit oblong, 10–16 cm long, 1.5–3 cm wide, the apex obtuse, the base cuneate, laterally compressed, glabrous.

Disturbed sites. Rare; Miami-Dade County. Escaped from cultivation. Florida; West Indies and Central America; Asia and Australia. Native to Asia and Australia. Spring.

Alysicarpus Neck. ex Desv., nom. Cons. 1813. MONEYWORT

Annual or perennial herbs. Leaves 1-foliolate, petiolate, stipellate and stipulate. Flowers in axillary or terminal pseudoracemes, bracteate; calyx lobes 4, the upper lobe 2-lobed, longer than the tube; corolla papilionaceous, the standard broad, the keel adherent to the wings; stamens diadelphous (9 + 1). Fruit an indehiscent loment.

A genus of about 30 species; North America, West Indies, Mexico, Central America, South America, Africa, Asia, Australia, and Pacific Islands. [From the Greek meaning "chain fruit."]
Selected reference: Verdcourt (1971).

1. Loment laterally compressed, constricted between segments; calyx lobes imbricate at the base ....
.... **A. rugosus**
1. Loment subterete, not constricted between segments; calyx lobes not imbricate at the base.
2. Inflorescence dense, the rachis internodes shorter than the flowers ....**A. vaginalis**
2. Inflorescence lax, the rachis internodes much longer than the flowers........................**A. ovalifolius**

Alysicarpus ovalifolius (Schumach. & Thonn.) J. Léonard [With ovate leaves.] FALSE MONEYWORT; ALYCE CLOVER.

Hedysarum ovalifolium Schumacher & Thonning, Beskr. Guin. Pl. 359. 1827. *Desmodium ovalifolium* (Schumacher & Thonning) Walpers, Report. Bot. Syst. 1: 737. 1842. *Alysicarpus ovalifolius* (Schumacher & Thonning) J. Léonard, Bull. Jard. Bot. État Bruxelles 24: 88. 1954.

Erect or spreading annual herb, to 6 dm; stem puberulent or glabrescent. Leaves 1-foliolate, the leaflets usually dimorphic, the lower ones with the blade elliptic or oblong, the upper ones lanceolate, 1–10 cm long, 0.6–3 cm wide, the upper surface glabrous, the lower surface sparsely puberulent to glabrescent, the petiole ca. 5 mm long; stipules lanceolate, 6–8 mm long, striate. Flowers paired in a terminal or axillary raceme; calyx bilabiate, with minute spreading hooked trichomes and sparsely ciliate with straight trichomes, the lobes subequal, 2–4 mm long; corolla subequaling the calyx, pink or reddish purple. Fruit with (1)4–6 segments, 1.5–2 cm long, ca. 2 mm wide, with minute hooked trichomes.

Disturbed sites. Frequent; nearly throughout. Escaped from cultivation. Florida; Africa and Asia. Native to Africa and Asia. Summer–fall.

Alysicarpus rugosus (Willd.) DC. [Wrinkled, in reference to the seed.] RED MONEYWORT.

Hedysarum rugosum Willdenow, Sp. Pl. 3: 1172. 1802. *Alysicarpus rugosus* (Willdenow) de Candolle, Prodr. 2: 353. 1825.

Erect annual or perennial herb, to 6 dm; stem pilose. Leaves 1-foliolate, the leaflets with the blade spatulate to elliptic, 3–4.5 cm long, 2–10 mm wide, the apex obtuse or emarginate, the base rounded, the margin ciliate, the upper surface glabrous, the lower surface sparsely subappressed-villous, the petiole 3–9 mm long; stipules lanceolate, 4–6 mm long, striate. Flowers 10–25 in a terminal or axillary raceme; calyx 5–7 mm long, glabrous; corolla 6–7 mm long, white. Fruit oval to oblong, 4–12 mm long, 2–3 mm wide, the segments 1–4, constricted on both margins, conspicuously cross-rugose, glabrate.

Disturbed sites. Rare; Miami-Dade County. Escaped from cultivation. Florida; Africa, Asia, and Australia. Native to Africa, Asia, and Australia. Summer–fall.

Alysicarpus vaginalis (L.) DC. [Sheathed, in reference to the stipules.] WHITE MONEYWORT.

Hedysarum vaginale Linnaeus, Sp. Pl. 2: 746. 1753. *Alysicarpus vaginalis* (Linnaeus) de Candolle, Prodr. 2: 353. 1825. *Alysicarpus vaginalis* (Linnaeus) de Candolle var. *typicus* King, Mat. Fl. Malay. Penins. 3: 133. 1897, nom. inadmiss.

Ascending or trailing perennial herb, to 1 m; stem glabrous. Leaves 1-foliolate, the blade suborbicular to lanceolate, to ca. 3.5 cm long, to 1.4 cm wide, the apex rounded or acute, the base broadly rounded to cordate, the margin minutely ciliolate, the upper and lower surfaces glabrous, the petiole slender, to 1 cm long; stipules 4–6 mm long, joined opposite the leaflet and enclosing 2 stipels, scarious. Flowers 6–12 in an axillary or terminal raceme 1–3 cm long, each subtended by a scarious deciduous bract; calyx 4–5 mm long, deeply cleft into 5 lanceolate lobes; corolla slightly exceeding the calyx, deep reddish purple, fading to violet or pink, the standard suborbicular, clawed, the wings obliquely oblong, adnate to the blunt incurved keel. Fruit subterete, 1–2 cm long, 2–3 mm wide, 4- to 7-jointed, the joints indehiscent.

Disturbed sites. Rare; Palm Beach and Lee Counties, southern peninsula. Escaped from cultivation. Virginia south to Florida, west to Texas; West Indies, Mexico, Central America, and South America; Africa, Asia, Australia, and Pacific Islands. Native to Africa and Asia. Summer–fall.

Amorpha L. 1753. FALSE INDIGO

Shrubs. Leaves alternate, odd-pinnately compound, petiolate, stipulate, the leaflets opposite or subopposite, pulvinate, stipellate. Flowers in terminal solitary or clustered racemes, or sometimes paniculiform, bracteate and bracteolate; sepals 5, basally connate; corolla consisting of only the banner petal; stamens 10, monadelphous basally at early anthesis, becoming free, the anthers dorsifixed, dehiscing longitudinally. Fruit a loment.

A genus of 16 species; North America, Mexico, South America, Europe, Africa, and Asia. [From the Greek *a*, without, and *morphos*, shape, in reference to the simple nonpapilionaceous flowers.]

Selected reference: Wilbur (1964, 1975).

1. Midvein of the leaflets exserted into a slender nonglandular mucro; lowest calyx lobe shorter than the tube ....**A. fruticosa**
1. Midvein of the leaflets exserted into a conspicuous swollen glandular mucro; lowest calyx lobe longer than the tube ....**A. herbacea**

Amorpha fruticosa L. [Shrubby.] BASTARD FALSE INDIGO.

Amorpha fruticosa Linnaeus, Sp. Pl. 2: 713. 1753. *Amorpha fruticosa* Linnaeus var. *vulgaris* Pursh, Fl. Amer. Sept. 466. 1814, nom. inadmiss. *Amorpha fruticosa* Linnaeus var. *typica* C. K. Schneider, Ill. Handb. Laubholzk. 2: 72. 1907, nom. inadmiss.

Amorpha croceolanata P. W. Watson, Dendrol. Brit. 2: t. 139. 1825. *Amorpha fruticosa* Linnaeus var. *croceolanata* (P. W. Watson) Mouillefert, Traite Arb. Arbriss. 577. 1894.

Amorpha caroliniana Croom, Amer. J. Sci. Arts 25: 74. 1834. *Amorpha fruticosa* Linnaeus var. *caroliniana* (Croom) S. Watson, Smithson. Misc. Collect. 258: 188. 1878.

Amorpha virgata Small, Bull. Torrey Bot. Club 21: 17, pl. 171. 1894.

Amorpha bushii Rydberg, in Britton, N. Amer. Fl. 24: 31. 1919.

Amorpha curtissii Rydberg, in Britton, N. Amer. Fl. 24: 30. 1919. TYPE: FLORIDA: Duval Co.: near river, Jacksonville, 6 May & 21 Aug 1894, *Curtiss 4703* (holotype: NY; isotypes: KANU, US).

Amorpha dewinkeleri Small, Man. S.E. Fl. 689. 1933. TYPE: FLORIDA: Lee Co.: near Ft. Shackleford, Big Cypress, 19 May 1917, *Small 8349* (lectotype: NY; isolectotypes: GH, MO, US). Lectotypified by Wilbur (1975: 400).

Shrub, to 3(4) m; stem puberulent to glabrate, rarely glabrous, eglandular or occasionally sparsely gland-dotted. Leaves 10–25(28) cm long, the leaflet with the blade elliptic to oblong, rarely ovate, (1)2–4(5) cm long, (0.5)1–2(2.5) cm wide, the apex acute to rounded or rarely emarginate, the midvein extended into a slender nonglandular mucro, the base acute to rounded, the margin entire, the upper and lower surfaces puberulent to glabrate or glabrous, the petiolules 2–4 mm long, the petiole 1–4 cm long; stipules linear, 2–4 mm long, pilosulose. Flowers in a terminal or subterminal spiciform raceme or a paniculiform inflorescence of 2–8(12) racemes, (5)10–20(25) cm long, the rachis puberulent to pilosulose or glabrate, sparsely glandular or eglandular; bracts and bracteoles linear to setaceous, 1–3 mm long, sparsely glandular or eglandular; calyx tube obconic to funnelform or campanulate, 2–3(4) mm long, puberulent to pilosulose or glabrous, gland-dotted on the upper ⅓ or sometimes eglandular, the calyx lobes triangular-dentate, the margin ciliate; banner petal broadly obovate to obcordate, 5–6 mm long, 3–4 mm wide, ca. 1 mm long, reddish purple, the margin entire or slightly erose, the apex emarginate; stamens 6–8 mm long, connate basally 1–2(3) mm. Fruit 5–9 mm long, (2)3–4 mm wide, the upper margin straight or curved, the lower margin curved, the surfaces glabrous or pubescent, eglandular or glandular-dotted; seeds 1–2, elliptic, 3–5 mm long, 1–2 mm wide, brown, smooth, lustrous.

Hammocks and stream banks. Frequent; nearly throughout. Quebec south to Florida, west to Manitoba, Washington, Oregon, and California; Mexico and South America; Europe, Africa, and Asia. Native to North America and Mexico. Summer.

Amorpha herbacea Walter [Herbaceous.] CLUSTERSPIKE FALSE INDIGO.

Shrub, to 1(3) m; stem sparsely to densely puberulent or glabrous, usually obscurely gland-dotted. Leaves (6)8–20(34) cm long, the leaflet with the blade elliptic to oblong or ovate to suborbicular, 1–2.5(4) cm long, 0.5–1(1.5) cm wide, the apex obtuse to rounded or emarginate, the base obtuse to rounded or truncate to subcordate, the margin entire or crenulate, sometimes inconspicuously so, the upper and lower surfaces pubescent or glabrous, the midvein exserted into a conspicuously swollen, glandular mucro, the petiolules 1–2 mm long, the petiole 1–10(18) mm long, puberulent or strigillose, rarely glabrous, gland-dotted; stipules acicular, 1–2(3) mm long, puberulent to glabrate. Flowers in a terminal or subterminal spiciform raceme or a paniculiform inflorescence of 4–12(20) branches, (3)10–18(40) cm long, the rachis puberulent, sparsely gland-dotted; bracts and bracteoles linear, 1–2(3) mm long, sparsely pubescent, sparsely gland-dotted; calyx tube turbinate to narrowly campanulate or cylindric, 2–3 mm long, puberulent or glabrous, gland dotted on the upper ⅔ to ½ or eglandular, the calyx lobes triangular; banner petal broadly obcordate, 4–7 mm long, 2–3 mm wide, white, lavender, or violet, the margin entire or slightly erose, the apex obtuse; stamens 6–8 mm long, free. Fruit 4–6 mm long, ca. 2 mm wide, the upper margin straight or slightly curved, the lower margin strongly curved upward, the surfaces pubescent or glabrous, gland dotted on the upper ⅔ to ½; seeds 1–2, elliptic, 3–5 mm long, 1–2 mm wide, brown, smooth, lustrous.

1. Leaflet margin entire or inconspicuously crenulate .... var. **herbacea**
1. Leaflet margin conspicuously crenulate...var. **crenulata**

Amorpha herbacea Walter var. **herbacea** CLUSTERSPIKE FALSE INDIGO.

> *Amorpha herbacea* Walter, Fl. Carol. 179. 1788. *Amorpha pubescens* Willdenow, Berlin Baumz. 17. 1796, nom. illegit. *Amorpha pumila* Michaux, Fl. Bor.-Amer. 2: 64. 1803, nom. illegit. *Amorpha herbacea* Walter var. *typica* C. K. Schneider, Ill. Handb. Lauholzk. 2: 69. 1907, nom. inadmiss.
> *Amorpha cyanostachya* M. A. Curtis, Boston J. Nat. Hist. 1: 140. 1835.
> *Amorpha herbacea* Walter var. *boyntonii* C. K. Schneider, Ill. Handb. Laubholzk. 2: 69. 1907. TYPE: FLORIDA: Pasco Co.: Richland, s.d., *Curtiss 6664* (lectotype: MO; isolectotypes: CU, GA, GH, ISC, MIN, NEB, NY, PH, UC, US). Lectotypified by Wilbur (1964: 55).
> *Amorpha floridana* Rydberg, in Britton, N. Amer. Fl. 24: 31. 1919. *Amorpha herbacea* Walter var. *floridana* (Rydberg) Wilbur, J. Elisha Mitchell Sci. Soc. 80: 55. 1964. TYPE: FLORIDA: without data, *Chapman s.n.* (holotype: NY).

Leaflet margin entire or inconspicuously crenulate.

Sandhills, flatwoods, and scrub. Occasional; peninsula west to central panhandle. North Carolina south to Florida, west to Alabama. Spring–summer.

Amorpha herbacea var. **crenulata** (Rydb.) Isely [Having small rounded teeth, in reference to the leaf margin.] MIAMI LEAD PLANT.

> *Amorpha crenulata* Rydberg, in Britton, N. Amer. Fl. 24: 30. 1919. *Amorpha herbacea* Walter var. *crenulata* (Rydberg) Isely, Sida 11: 433. 1986. TYPE: FLORIDA: Miami-Dade Co.: between Coconut Grove and Cutler, 9 May 1904, *Small & Wilson 1898* (holotype: NY; isotype: F).

Leaflet margin conspicuously crenulate.

Pine rocklands. Rare; Miami-Dade County. Endemic. Summer.

Amorpha herbacea var. *crenulata* is listed as endangered in Florida (Florida Administrative Code, Chapter 5B-40) and in the United States (U.S. Fish and Wildlife Service, 50 CFR 23).

EXCLUDED TAXON

> *Amorpha glabra* Michaux—Reported for Florida by Small (1903), the basis unknown. Excluded from Florida by Wilbur (1975).

Amphicarpaea Elliott ex Nutt., nom. cons. 1918. HOGPEANUT

Annual vines. Leaves alternate, 3-foliolate, stipellate, petiolate, stipulate. Flowers chasmogamous, papilionaceous in axillary pseudoracemes or sometimes branched, also with some cleistogamous flowers on runners from the lower stem nodes, bracteate; corolla papiliona-ceous; stamens 10, diadelphous (9 + 1). Fruit a dehiscent legume.

A genus of 3 species; North America, Mexico, Africa, and Asia.

Falcata J. F. Gmel., nom. rej., 1792.

Selected reference: Turner and Fearing (1964).

Amphicarpaea bracteata (L.) Fernald [With bracts.] AMERICAN HOGPEANUT.

Glycine bracteata Linnaeus, Sp. Pl. 2: 754. 1753. *Glycine monoica* Linnaeus, Sp. Pl., ed. 2. 1023. 1763, nom. illegit. *Savia volubilis* Rafinesque, Med. Repos., ser. 2. 5: 352. 1808, nom. illegit. *Amphicar-paea monoica* Elliott, J. Acad. Nat. Sci. Philadelphia 1: 372. 1818, nom. illegit. *Falcata bracteata* (Lin-naeus) Farwell, Pap. Michigan Acad. Sci. 3: 101. 1923. *Amphicarpaea bracteata* (Linnaeus) Fernald, Rhodora 35: 276. 1933. *Amphicarpa bracteata* (Linnaeus) Fernald var. *typica* B. Boivin & Raymond, Naturaliste Canad. 69: 225. 1943, nom. inadmiss.

Glycine comosa Linnaeus, Sp. Pl. 2: 754. 1753. *Amphicarphaea monoica* (Linnaeus) Elliott var. *co-mosa* (Linnaeus) Eaton, Man. Bot., ed. 3. 172. 1822. *Amphicarphaea comosa* (Linnaeus) G. Don ex Loudon, Hort. Brit. 314. 1830. *Falcata comosa* (L.) Kuntze, Revis. Gen. Pl. 1: 182. 1891. *Amphicar-paea bracteata* (Linnaeus) Fernald var. *comosa* (Linnaeus) Fernald, Rhodora 39: 318. 1937.

Procumbent or twining, annual, herbaceous vine, to 2 m; stem hirsute or strigose. Leaves with the leaflet blade broadly ovate to deltoid, 3–10 cm long, 2–8 mm wide, the apex acute to acu-minate, the base rounded to slightly cordate, the lateral leaflets with the base asymmetrical, the margin entire, the lower surface appressed pubescent or glabrate, the petiolule 1–10 mm long, minutely stipellate, the petiole 2–11 mm long; stipules lanceolate to ovate, 3–8 mm long, glabrous or puberulent. Chasmogamous flowers in a pedunculate raceme 2–9 cm long, the pedicel 2–4 mm long, the peduncle 1–4 cm long; bracts ovate, broadest above the middle, 2–4 mm long; calyx 5–6 mm long, the lobes shorter than the tube; corolla 1–1.5 cm long, white with lavender tips or light lavender purple; stamens 10, 11–12 mm long. Cleistogamous flowers on filiform stolons from the lower stem nodes, smaller than the chasmogamous; calyx 5-lobed ca. ½ its length; corolla with the petals generally lacking; stamens usually 1 or 2. Fruit of the chasmogamous flowers oblong, 1.5–3 cm long, 8–10 mm wide, straight or falcate, flattened, glabrous, the sutures villous; seeds 2–4. Fruit of cleistogamous flowers subterranean, ellipsoid or subglobose, 0.8–1.5 cm long; seed 1.

Mesic hammocks. Rare; northern peninsula south to Citrus County, west to the central pan-handle. Quebec south to Florida, west to Manitoba, Montana, Oklahoma, and Texas; Mexico. Summer.

Apios Fabr., nom. cons. 1759. GROUNDNUT

Perennial vines. Leaves alternate, odd-pinnately compound, pulvinate, stipellate, petiolate, stipulate. Flowers in axillary pseudoracemes, bracteate and bracteolate; sepals 4-toothed; co-rolla papilionaceous; stamens 10, diadelphous (9 + 1). Fruit a dehiscent legume.

A genus of 5 species; North America and Asia. [From the Greek *apios*, pear, alluding to the shape of the tubers.]

Selected reference: Woods (2005).

Apios americana Medik. [Of America.] GROUNDNUT.

> *Glycine apios* Linnaeus, Sp. Pl. 2: 753. 1753. *Apios americana* Medikus, Vorles. Churpfälz. Phys.-Okon. Ges. 2: 355. 1787. *Apios tuberosa* Moench, Methodus 165. 1794, nom. illegit. *Apios perennis* Vahl ex Hornemann, Hort. Bot. Hafn. 682. 1815, nom. illegit. *Gonancylis thyrsoidea* Rafinesque, First Cat. Gard. Transylv. Univ. 14. 1824, nom. illegit. *Apios apios* (Linnaeus) MacMillan, Bull. Torry Bot Club. 19: 15. 1892, nom. inadmiss.

Perennial herbaceous vine, to 3 m; stem glabrous or tomentose, with fleshy tubers, these oblong, oval, or globose, 4–12 cm long, 2–10 cm wide, moniliform. Leaves (3)5–7(9)-foliolate, 10–22 cm long, the leaflets with the blade ovate to ovate-lanceolate, 4–7(9) cm long, 2–4 cm wide, the apex acuminate to acute, apiculate, the base rounded, somewhat asymmetrical, the margin entire, the upper and lower surfaces glabrous to pubescent, stipels linear-triangular, ca. 1 mm long, the petiole 2–5 cm long, glabrous or velutinous; stipules linear-triangular, 4–6 mm long. Flowers 40–60 in a dense axillary pseudoraceme 3–14 cm long, the pedicels 2–3 mm long; bracts and bracteoles lanceolate, 2–3 mm long; calyx campanulate, ca. 3 mm long, the lobes ca. 1 mm long, green, red, or pinkish red, glabrous or puberulous; corolla 12–14 mm long, the petals subequal, deep maroon to pale maroon and white, the standard oblate, the wings obovate, falcate, auriculate basally, the keels incurved, united apically, auriculate basally. Fruit linear-oblong, 6–10(12) cm long, 6–7 mm wide, the apex aristate to acuminate; seeds 6–11, elliptic to oblong, 5–6 mm long, brown to reddish brown.

Wet hammocks. Frequent; nearly throughout. Quebec south to Florida, west to Ontario, South Dakota, Nebraska, Colorado, Oklahoma, and Texas. Summer–fall.

Arachis L. 1753. PEANUT

Annual or perennial herbs. Leaves alternate, even-pinnately compound or 3-foliolate, estipellate, petiolate, the stipules adnate to the petiole base. Flowers in axillary spikes or sometimes subpaniculate, bracteate and bracteolate; calyx lobes 5, bilabiate; corolla papilionaceous; stamens 10, monadelphous, the anthers dorsifixed. Fruit inarticulate loments, geocarpic.

A genus of 70–80 species; nearly cosmopolitan. [From the Greek *a*, without, and *rachis*, backbone, apparently in reference to the lack of a central upright branch.]

Selected reference: Krapovickas and Gregory (1994).

1. Stoloniferous-trailing perennial; petiole 2.5–3.5 cm long; corolla 1.5–2 cm long..................**A. glabrata**
1. Ascending or decumbent annual; petiole 5–10 cm long; corolla 1–1.5 cm long....................**A. hypogaea**

Arachis glabrata Benth [Glabrous.] GRASSNUT.

> *Arachis glabrata* Bentham., Trans. Linn. Soc. London 18: 159. 1841. *Arachis glabrata* Bentham forma *typica* Hoehne, Fl. Bras. 25(2): 19. 1940, nom. inadmiss.

Perennial herb; stem prostrate, to 40 cm long, puberulent or glabrescent. Leaves 4(3)-foliolate, the leaflets with the blade obovate or ovate to elliptic or oblong, 0.5–2.5 cm long, 0.5–1.5 cm wide, the apex obtuse to acute or retuse, the base rounded, the margin glabrous or villous, sometimes with scattered bristles, the upper surface minutely pubescent to glabrate, often with longer persistent trichomes on the midrib, the lower surface glabrous or glabrescent, the petiole 2.5–3.5 cm long; stipules adnate to the petiole base, linear-lanceolate, 8–15 mm long. Flowers in a spike or sometimes subpaniculate; calyx 0.5–1 cm long, bilabiate, bristly-villous; corolla 15–18 mm long, papilionaceous, yellow or yellow-orange. Fruit ovoid, ca. 1 cm long, ca. 0.5 cm wide, longitudinally striate; seeds ovoid, pale tan.

Disturbed sites. Occasional; peninsula, Calhoun County. Escaped from cultivation. Georgia, Alabama, and Florida; West Indies, Mexico, Central America, and South America; Africa, Asia, and Australia. Native to South America. Spring–fall.

Arachis hypogaea L. [From the Greek *hypo*, below, and *Gaea*, Greek goddess of earth, in reference to the plant producing fruits below ground.] PEANUT.

> *Arachis hypogaea* Linnaeus. Sp. Pl. 2: 741. 1753. *Arachidna hypogaea* (Linnaeus) Moench, Methodus 122. 1794. *Arachis hypogaea* Linnaeus subsp. *oleifera* A. Chevalier, Rev. Int. Bot. Appl. Agric. Trop. 13: 777. 1933, nom. inadmiss. *Arachis hypogaea* Linnaeus forma *typica* Hoehne, Fl. Bras. 25(2): 18. 1940.

Annual herb; stem ascending or decumbent, to 1.3 m, glabrous or villous. Leaves 4(3)-foliolate, the leaflets with the blade elliptic to ovate or obovate, 2–6 cm long, 1.5–3 cm wide, the apex rounded to subacute, sometimes emarginate, the base rounded to subacute, the margin entire, villous, sometimes with bristles, the upper surface glabrous, the lower surface glabrous or puberulent, the petiole 5–10 cm long; stipules adnate to the petiole base, lanceolate to subfalcate, 2–3(5) cm long. Flowers in a spike or sometimes subpaniculate; calyx 10–12 mm long, bilabiate, glabrous; corolla 1–1.5 cm long, papilionaceous, light yellow to orange, the banner often with reddish lines toward the base, glabrous; 8 stamens fertile, the 2 proximal ones sterile filaments. Fruit oblong, 2–6 cm long, 1–2 cm wide, sometimes constricted between the seeds, the surface reticulate; seeds 1–3(6), oblong, 1–2.5 cm wide, 1–1.3 cm wide, reddish brown.

Disturbed sites. Occasional; central and southern peninsula, panhandle. Escaped from cultivation. Massachusetts south to Florida, west to Utah and Texas; West Indies, Mexico, Central America, and South America; Europe, Africa, Asia, Australia, and Pacific Islands. Native to South America. Summer–fall.

EXCLUDED TAXON

> *Arachis prostrata* Bentham—Reported for Florida by Wunderlin (1998) and Wunderlin and Hansen (2003), who misapplied the name to material of *A. glabrata*.

Astragalus L. 1753. MILKVETCH

Perennial herbs. Leaves alternate, odd-pinnately compound, petiole, stipulate. Flowers in axillary or terminal racemes, bracteate, bracteolate or ebracteolate; calyx 5-lobed; corolla papilionaceous; stamens 10, diadelphous (9 + 1). Fruit a dehiscent legume.

A genus of about 3,000 species; North America, Mexico, Central America, and South America; Europe, Africa, and Asia. [From the Greek *astragalos*, ankle bone.]

Phaca L., 1753.

Selected references: Barneby (1964, 1977).

1. Plant glabrate or strigulose; fruit glabrous ....**A. obcordatus**
1. Plant villous or pilose; fruit villous ....**A. villosus**

Astragalus obcordatus Elliott [Obcordate in shape, in reference to the leaflets.] FLORIDA MILKVETCH.

Astragalus obcordatus Elliott, Sketch Bot. S. Carolina 2: 227. 1823. *Astragalus elliottii* D. Dietrich, Syn. Pl. 4: 1080. 1847, nom. illegit. *Tragacantha obcordata* (Elliott) Kuntze, Revis. Gen. Pl. 2: 946. 1891. *Tium obcordatum* (Elliott) Rydberg, in Small, Fl. S.E. U.S. 619, 1332. 1903. *Batidophaca obcordata* (Elliott) Rydberg, in Britton, N. Amer. Fl. 24: 321. 1929. *Phaca obcordata* (Elliott) Rydberg ex Small, Man. S.E. Fl. 710, 1933. TYPE: FLORIDA: Duval/Nassau Co.: S of the St. Marys (miscited by Elliott as "Southern Districts of Georgia"), s.d., *Baldwin s.n.* (lectotype: PH). Lectotypified by Barneby (1964: 991).

Perennial herb; stem prostrate or decumbent, to 40(50) cm, glabrate. Leaves 5–12(15) cm long, (9)15–27-pinnate, the leaflets with the blade oblong-obovate or cuneate-obcordate, 2–7(10) mm long, 1–4 mm wide, the apex retuse, the margin entire, the upper surface glabrous, the lower surface sparsely strigose on the midrib, the petiole 0.5–2 cm long; stipules free, ovate-triangular, 3–6 mm long. Flowers 5–15(18) in a raceme 4–6 cm long; bracts 2–3 mm long; bracteoles absent or 1–2; calyx campanulate, 5–6 mm long, strigulose, the lobes lanceolate to broadly subulate, 2–3 mm long; corolla 9–11 mm long, pink or bluish lilac, the banner obovate, ovate, or rhombic-cuneate, 8–10(11) mm long, recurved, the wings 8–9 mm long, the keel 6–8 mm long, the apex round. Fruit ascending or spreading, ellipsoid to obovoid, (13)15–25(17) mm long, 4–7 mm wide, sessile, glabrous; seeds 13–18.

Sandhills. Occasional; northern peninsula south to Hillsborough and Polk Counties, west to central panhandle. Florida and Mississippi. Spring.

Astragalus villosus Michx. [With shaggy trichomes.] BEARDED MILKVETCH.

Astragalis villosus Michaux, Fl. Bor.-Amer. 2: 67. 1803. *Phaca villosa* (Michaux) Nuttall, N. Amer. Fl. 2: 97. 1818. *Tragacantha villosa* (Michaux) Kuntze, Revis. Gen. Pl. 2: 949. 1891. *Astragalus intonsus* E. Sheldon, Minnesota Bot. Stud. 1: 23. 1894, nom. illegit. *Tium intonsum* Rydberg, in Small, Fl. S.E. U.S. 619, 1332. 1903, nom. illegit. *Batidophaca villosa* (Michaux) Rydberg, in Britton, N. Amer. Fl. 24: 319. 1929. *Phaca intonsa* Rydberg ex Small, Man. S.E. Fl. 710. 1933, nom. illegit.

Perennial herb; stem prostrate or decumbent, to 20 cm, hirsute. Leaves 3–10 cm long, (3)7–15-pinnate, the leaflets with the blade obovate, oblanceolate, or suborbicular, (2)5–22 mm long, 2–7 mm wide, the apex obtuse to emarginate, the margin entire, the upper surface glabrous, the lower surface villous-pilose, the petiole 0.5–1 cm long; stipules usually free, rarely basally connate, ovate-triangular, 2–8 mm long. Flowers (5)8–24 in a raceme 1–3(4) cm long; bracts 2–5 mm long; bracteoles absent; calyx campanulate, 5–8 mm long, pilose, the lobes lanceolate, 3–5 mm long; corolla 9–11 mm long, pale yellow to greenish gold, the banner ovate-cuneate, 9–11 mm long, recurved, the wings 9–11 mm long, the keel 7–8 mm long, the apex narrowly deltate, acute or subacute, often beak-like. Fruit ascending or spreading, lunate-ellipsoid, (15)17–25 mm long, 3–5 mm wide, hirsute; seeds 12–18.

Sandhills. Occasional; northern counties south to Hernando and Lake Counties. South Carolina south to Florida, west to Tennessee and Mississippi. Spring.

EXCLUDED TAXON

Astragalus michauxii (Kuntze) F. J. Hermann—Reported for Florida by Chapman 1860, as *Astragalus glaber* E. Sheldon), Small (1903, 1913, both as *Tium apilosum* Rydberg; 1933, as *Tium michauxii* (Kuntze) Rydberg), Radford et al. (1964, 1968), Wunderlin (1998), and Wunderlin and Hansen (2003). The basis of the reports by Chapman and Small is not known. Later reports are based on Barneby (1964), who included Polk County in his map of range of the species, which was apparently based on a North Carolina collection with bad label information. The species was excluded from Florida by Isely (1990).

Baptisia Vent. 1808. WILD INDIGO

Perennial herbs. Leaves 1- to 3-foliolate, estipellate, petiolate or sessile, stipulate. Flowers in terminal racemes or 1–3 in the leaf axils, bracteate, bracteolate or ebracteolate; calyx campanulate, 4-lobed, the upper lobe double; corolla papilionaceous, the wing petals auriculate, the keel petals slightly connate, auriculate; stamens 10, free. Fruit a dehiscent legume.

A genus of 15–17 species; North America. [From the Greek *bapto*, to dip or dye, in reference to some species used as a dye.]

Selected references: Isely (1981); Turner (2006).

1. Leaves 1-foliolate.
    2. Leaves perfoliolate ....**B. perfoliata**
    2. Leaves subsessile but not perfoliolate...**B. simplicifolia**
1. Leaves 3-foliolate (uppermost 2-foliolate or 1-foliolate).
    3. Pedicel bracteolate; corolla 11–14 mm long.
        4. Calyx lobes much longer than the tube..**B. calycosa**
        4. Calyx lobes subequaling the tube...**B. lecontei**
    3. Pedicel ebracteolate; corolla more than 14 mm long.
        5. Flowers white.
            6. Corolla 1.4–1.6(1.8) cm long; calyx 4.5–6.5 mm long; fruit cylindric to oblong-lanceolate, 0.7–0.9 cm in diameter, yellow-brown at maturity..**B. albescens**
            6. Corolla 2–2.5 cm long; calyx 7–8 mm long; fruit subspheroid, ellipsoid to ellipsoid-cylindric, 1–2.5(3) cm in diameter, black at maturity...**B. alba**
        5. Flowers cream or yellow.
            7. Keel petals strongly incurved; fruit woody, 1–2.5 cm wide, 1–1.2 cm wide ....**B. lanceolata**
            7. Keel petals not strongly incurved; fruit thinly coriaceous, 3–4 cm long, 1.8–2.5 cm wide......
            ....**B. megacarpa**

Baptisia alba (L.) Vent. [White, in reference to the flower color.] WHITE WILD INDIGO.

Crotalaria alba Linnaeus, Sp. Pl. 2: 716. 1753. *Sophora alba* (Linnaeus) Linnaeus, Syst. Nat., ed. 12. 2: 287. 1767. *Podalyria alba* (Linnaeus) Willdenow, Sp. Pl. 2: 503. 1799. *Baptisia alba* (Linnaeus) Ventenat, Dec. Gen. Nov. 9. 1808.

Baptisia pendula Larisey, Ann. Missouri Bot. Gard. 27: 170, t. 25(1). 1940. *Baptisia lactea* (Rafinesque) Thieret var. *pendula* (Larisey) B. L. Turner, Phytologia 88: 255. 2006.

Baptisia pendula Larisey var. *obovata* Larisey, Ann. Missouri Bot. Gard. 27: 172, t. 25(2). 1940. *Baptisia lactea* (Rafinesque) Thieret var. *obovata* (Larisey) Isely, Brittonia 30: 471. 1978.

Baptisia psammophila Larisey, Ann. Missouri Bot. Gard. 27: 180, t. 26(3). 1940. TYPE: FLORIDA: Leon Co.: Tallahassee, 8 Apr 1902, *Beadle* (?)*6046* (holotype: GH; isotypes: NY, US).

Erect perennial, to 2 m; stem glabrous. Leaves 3-foliolate, the leaflets with the blade elliptic-obovate to oblanceolate, 2–6 cm long, 1–2 cm wide, the apex acute to obtuse, the base cuneate, the margin entire, the upper and lower surfaces glabrous, the petiole 5–15 mm long; stipules small, deciduous. Flowers numerous in an ascending raceme; bracts lanceolate, 7–10 mm long, deciduous; bracteoles absent; calyx campanulate, 7–8 mm long, the lobes 1–2.5 mm long, the tube glabrous; corolla 2–2.5 cm long, papilionaceous, white, the standard sometimes purple-marked near the base, the keel nearly straight. Fruit subspheroid, ellipsoid to ellipsoid-cylindric, 2–4(5) cm long, 1–2.5(3) cm wide, coriaceous-woody, glabrous, smooth, black.

Flatwoods, hammocks, and riverbanks. Occasional; northern counties south to Citrus and Lake Counties. North Carolina south to Florida, west to Tennessee and Alabama. Spring.

Baptisia albescens Small [Turning white, in reference to the flower.] SPIKED WILD INDIGO.

Baptisia albescens Small, Fl. S.E. U.S. 600, 1331. 1903.

Erect perennial, to 1.5(2) m; stem glabrous or puberulent. Leaves 3-foliolate, the leaflets with the blade obovate to narrowly elliptic-lanceolate, 2–4(5) cm long, 1.5–2 cm wide, the apex acute, the base narrowly cuneate, the margin entire, the upper surface glabrous, the lower surface glabrous or puberulent, the petiole 0.5–1(2) cm long; stipules small, deciduous. Flowers (4)8–20 in an ascending raceme; bracts deciduous; bracteoles absent; calyx campanulate, 4.5–6.5 mm long, the lobes 1–2 mm long the tube minutely ciliate at the orifice, otherwise glabrous; corolla 1.4–1.6(1.8) cm long, papilionaceous, white, sometimes striate, the standard often with a purple spot near the base, the wing and keel petals straight, the keel only slightly incurved. Fruit cylindric to oblong-lanceolate, 2–3(3.5) cm long, 0.7–0.9 mm wide, coriaceous, glabrous, rugose-reticulate, yellow-brown.

Flatwoods and open hammocks. Rare; central panhandle. North Carolina south to Florida, west to Tennessee and Alabama. Spring.

Baptisia calycosa Canby [With evident calyx, in reference to the foliaceous calyx lobes.] FLORIDA WILD INDIGO.

Erect perennial, to 1 m; stem pubescent or glabrous. Leaves 3-foliolate, the leaflets with the blade obovate to spatulate, 1.5–2.5 cm long, 0.5–1.5 cm wide, the apex obtuse, the base cuneate, the margin entire, the upper and lower surfaces glabrous or pubescent, the petiole to 1–5 mm long; stipules elliptic, to 1.5 cm long, usually persistent. Flowers in a terminal, foliose raceme and appearing axillary; bracts foliaceous, lanceolate, 1–2 cm long, persistent; bracteoles paired above the middle of the pedicel; calyx 9–13 mm long, lobed to ca. ⅔ of the base, the lobes elliptic, slightly ciliate, sometimes initially pubescent; corolla 11–14 mm long, papilionaceous, yellow or yellowish green, the keel nearly straight. Fruit elliptic to subglobose, 1–1.2 cm long, 0.8–1 cm wide, coriaceous or woody, pubescent or glabrous, black.

1. Stem glabrous ....var. **calycosa**
1. Stem cinereous, villous-hirsute, or tomentose ....var. **villosa**

Baptisia calycosa Canby var. **calycosa** FLORIDA WILD INDIGO.

> *Baptisia calycosa* Canby, Bot. Gaz. 3: 65. 1878. TYPE: FLORIDA: St. Johns Co.: St. Augustine, Jul 1877, *Reynolds s.n.* (holotype: NY; isotypes: GH, NCU, NY).

Stem glabrous.

Flatwoods and sandhills. Rare; Clay and St. Johns Counties. Endemic. Spring.

Baptisia calycosa var. *calycosa* is listed as endangered in Florida (Florida Administrative Code, Chapter 5B-40).

Baptisia calycosa var. **villosa** Canby [Villous.] HAIRY FLORIDA WILD INDIGO.

> *Baptisia calycosa* Canby var. *villosa* Canby, Bot. Gaz. 12: 39. 1887. *Baptisia hirsuta* Small, Fl. S.E. U.S. 598, 1331. 1903. TYPE: FLORIDA: Walton Co.: DeFuniak Springs, May, *Curtiss 699* (holotype: NY; isotype: NCU).

Stem cinereous, villous-hirsute, or tomentose.

Hammocks. Rare; Holmes and Washington Counties; western panhandle. Endemic. Spring.

Baptisia calycosa var. *villosa* is listed as threatened in Florida (Florida Administrative Code, Chapter 5B-40).

Baptisia lanceolata (Walter) Elliott [Lanceolate, in reference to the leaflets.] GOPHERWEED.

> *Sophora lanceolata* Walter, Fl. Carol. 135. 1788. *Baptisia lanceolata* (Walter) Elliott, Sketch Bot. S. Carolina 1: 467. 1817.
> *Lasinia reticulata* Rafinesque, New. Fl. 2: 47. 1837.
> *Baptisia elliptica* Small, Fl. S.E. U.S. 599, 1331. 1903. *Baptisia lanceolata* (Walter) Elliott var. *elliptica* (Small) B. L. Turner, Phytologia 88: 256. 2006. TYPE: FLORIDA: Franklin Co.: near Apalachicola, Apr, *Curtiss 689* (holotype: NY).
> *Baptisia elliptica* Small var. *tomentosa* Larisey, Ann. Missouri Bot. Gard. 27: 150. 1940. *Baptisia lanceolata* (Walter) Elliott, var. *tomentosa* (Larisey) Isely, Brittonia 30: 471. 1978. TYPE: FLORIDA: Walton Co.: near DeFuniak Springs, 27 Apr 1898, *Curtis 6379* (holotype: MO; isotypes: GH, MU, NEB, NY).

Erect perennial, to 1 m; stem glabrate or appressed-villosulous. Leaves 3-foliolate, the leaflets broadly obovate, elliptic-lanceolate, or oblanceolate, 3.5–10 cm long, 1–4.5 cm wide, the apex obtuse, the base cuneate, the margin entire, the upper surface glabrate, the lower surface villosulous to glabrate, the petiole 0–4(8) mm long; stipules small, deciduous. Flowers solitary and axillary or 2–4(10) in a short raceme; bracts deciduous; bracteoles absent; calyx 8–10 mm long, pubescent or glabrous, the lobes 2–2.5(3) mm long; corolla 2–2.5 cm long, papilionaceous, cream or yellow, the standard often medially red-streaked or spotted, the keel strongly incurved. Fruit subspherical, ovoid-ellipsoid, or lanceoid, 1–2.5 cm long, 1–1.2 cm wide, woody, glabrate, black.

Sandhills, flatwoods, and oak hammocks. Frequent: northern and central peninsula south to Hillsborough and Polk Counties, central and western panhandle. South Carolina, Georgia, Florida, and Alabama. Spring.

Baptisia lecontei Torr. & A. Gray [Commemorates Louis LeConte (1782–1838).] PINELAND WILD INDIGO.

> *Baptisia lecontei* Torrey & A. Gray, Fl. N. Amer. 1: 386. 1840.
> *Baptisia lecontei* Torrey & A. Gray forma *robustior* Larisey, Ann. Missouri Bot. Gard. 27: 145. 1940. TYPE: FLORIDA: Jackson Co.: Marianna, 1838, *Chapman s.n.* (holotype: NY).

Erect perennial, to 1 m; stem puberulent or glabrate. Leaves 3-foliolate, the leaflets with the blade obovate to narrowly spatulate, (1.5)3–5(5.5) cm long, (0.5)1–3 cm wide, the apex rounded, the base cuneate, the margin entire, the upper and lower surfaces subappressed puberulent to glabrate, the petiole 1–5(10) mm long; stipules small, deciduous. Flowers well spaced in a terminal raceme; bracts foliaceous, lanceolate, 3–5 mm long; bracteoles paired above the middle of the pedicel; calyx 6–7 mm long, puberulent or glabrate, the lobes subequaling the tube; corolla 11–14 mm long, papilionaceous, yellow, the standard with black-maroon or brownish markings, the keel only slightly curved. Fruit ovoid or subglobose, ca. 1 cm long, ca. 1 cm wide, coriaceous or woody, puberulent, brown.

Sandhills and flatwoods. Frequent; northern and central peninsula west to central panhandle. Georgia and Florida. Spring–summer.

Baptisia megacarpa Chapm. ex Torr. & A. Gray [Large fruited.] APALACHICOLA WILD INDIGO.

> *Baptisia megacarpa* Chapman ex Torrey & A. Gray, Fl. N. Amer. 1: 386. 1840. TYPE: FLORIDA: Gadsden Co.: without data, *Chapman s.n.* (syntypes: MO, NY).
> *Baptisia riparia* Larisey, Ann. Missouri Bot. Gard. 27: 192, t. 26(1). 1940. TYPE: FLORIDA: Leon Co.: banks of the Ochlockonee River, 10 mi. west of Tallahassee, 20 Apr 1933, *Totten s.n.* (holotype: NCU).
> *Baptisia riparia* Larisey var. *minima* Larisey, Ann. Missouri Bot. Gard. 27: 193, t. 26(2). 1940. TYPE: FLORIDA: Gadsden Co.: near Ochlockonee River, Havana, 15 Apr 1934, *Griscom 21581* (holotype: GH).

Erect perennial, to 1.5 m; stem glabrous or glabrate. Leaves 3-foliolate, the leaflets with the blade elliptical, 4–6(9) cm long, 2–3(4) cm wide, the apex acute, the base cuneate, the margin entire, the upper and lower surfaces glabrous, the petiole 1.5–2 cm long; stipules small, deciduous. Flowers 2–10(15) in a terminal raceme, often appearing axillary due to overtopping lateral branches; bracts lanceolate, 2–3 mm long; bracteoles absent; calyx 8.5–10 mm long, the lobes 3–5 mm long, glabrous; corolla ca. 2 cm long, papilionaceous, cream or yellow, the keel only slightly curved. Fruit ellipsoid, 4–3.5 cm long, 2–3 cm wide, thinly coriaceous, conspicuously cross reticulate-striate, glabrous, tan or brownish.

Moist hammocks and floodplain forests. Rare; Central panhandle. Georgia, Alabama, and Florida. Spring–summer.

Baptisia megacarpa is listed as endangered in Florida (Florida Administrative Code, Chapter 5B-40).

Baptisia perfoliata (L.) R. Br. [Leaves appearing pierced by the stem.] CATBELLS.

Crotalaria perfoliata Linnaeus, Sp. Pl. 2: 714. 1753. *Sophora perfoliata* (Linnaeus) Walter, Fl. Carol. 135. 1788. *Rafnia perfoliata* (Linnaeus) Willdenow, Sp. Pl. 3: 949. 1802. *Podalyria perfoliata* (Linnaeus) Michaux, Fl. Bor.-Amer. 1: 263. 1803. *Baptisia perfoliata* (Linnaeus) R. Brown, in W. T. Aiton, Hortus Kew. 315. 1811. *Pericaulon perfoliatum* (Linnaeus) Rafinesque, New Fl. 2: 51. 1837 ("1836").

Erect perennial, to 8 dm; stem glabrous, glaucous. Leaves 1-foliolate, the blade broadly ovate-elliptic, 5–8 cm long, 4–6 cm wide, the apex rounded, the base perfoliolate, the margin entire, the upper and lower surfaces glabrous; stipules small, deciduous. Flowers solitary, axillary; bracts small; bracteoles absent; calyx 5–6.5 mm long, the lobes subequaling the tube or shorter, the inner surface matted pubescent, the outer surface glabrous; corolla 10–13 mm long, papilionaceous, yellow, the keel only slightly curved. Fruit ovoid to subglobose, 1–1.5 cm long, ca. 1 cm wide, woody, glabrous, smooth, brown.

Sandhills. Rare; Seminole, Orange, and Osceola Counties. South Carolina, Georgia, and Florida. Spring.

Baptisia simplicifolia Croom [Simple-leaved.] SCAREWEED.

Baptisia simplicifolia Croom, Amer. J. Sci. Arts 25: 74. 1834. *Eaplosia ovata* Rafinesque, New Fl. 2: 52. 1837 ("1836"), nom. illegit. TYPE: FLORIDA: Gadsden Co.: near Quincy, s.d., *Croom s.n.* (lectotype: NY). Lectotypified by Isely (1981: 225).

Erect perennial, to 1 m; stem glabrous. Leaves 1-foliolate, the blade ovate, 5–7(8) cm long, 4–6(7) cm wide, the apex obtuse, the base rounded, the margin entire, the upper and lower surfaces glabrous, sessile or subsessile. Flowers numerous in a raceme to 20 cm long; bracts lanceolate-spatulate, 3–4 mm long; calyx 5–6 mm long, the lobes subequaling the tube; corolla 1–1.5 cm long, papilionaceous, pale yellow or yellow-green, the keel only slightly incurved. Fruit ovoid, ca. 10 mm long, ca. 5 mm wide, woody, glabrous, smooth, brown.

Flatwoods. Rare; central panhandle. Endemic. Summer.

Baptisia simplicifolia is listed as endangered in Florida (Florida Administrative Code, Chapter 5B-40).

EXCLUDED TAXA

Baptisia alba var. *macrophylla* (Larisey) Isely—This more western taxon was reported for Florida by Chapman (1860, as *B. leucantha* Torrey & A. Gray), Small (1903, 1913a, and 1933, all as *B. leucantha* Torrey & A. Gray), Correll and Johnston (1970, as *B. leucantha* Torrey & A. Gray), Wunderlin (1982, as *B. lactea* (Rafinesesque) Thieret), and Clewell (1985, as *B. lactea* (Rafinesesque) Thieret), the name misapplied to *B. alba* var. *alba*, the state's only variety of this species.

Baptisia tinctoria (Linnaeus) Ventenat—Reported for Florida by Small (1903) and by Radford et al. (1964), based on misidentification of material of *B. lecontei* (fide Isely 1981).

Baptisia ×microphylla Nuttall—Reported for Florida by Chapman (1860), Small (1903, 1913a, and 1933), and Radford et al. (1964, 1968), but Isely (1981, 1990, 1998) relegates this name to the synonymy of the hybrid *B. perfoliata* × *B. tinctoria* (Linnaeus) Ventenat known only from South Carolina.

Baptisia ×serenae M. A. Curtis—Reported for Florida by Radford et al. (1964, 1968), but Isely (1990, 1998) relegates this name to the synonymy of the hybrid *B. albescens* × *B. tinctoria* (Linnaeus) Ventenat, which has not been found in the state.

Bauhinia L. 1753.

Trees or shrubs. Leaves alternate, 1-foliolate, petiolate, stipulate. Flowers in terminal, subterminal, or axillary racemes, bisexual, bracteate, bracteolate; sepals 5, connate, the calyx spathaceous at anthesis; corolla caesalpiniaceous; stamens 3, 5, or 10, monadelphous or diadelphous (9 + 1), the anthers dorsifixed. Fruit a legume, dehiscent.

A genus of 150–160 species; North America, West Indies, Mexico, Central America, Africa, Asia, Australia, and Pacific Islands. [Commemorates Caspar Bauhin (1560–1624), Swiss botanist and physician, and his elder brother, Jean Bauhin (1541–1612), Swiss botanist; the bilobed leaves symbolizing the brotherly relationship.]

1. Plant with intrastipular spines; stamens 10...**B. aculeata**
1. Plant lacking intrastipular spines; stamens 3 or 5.
    2. Flower buds narrowly club-shaped and slightly winged toward the apex; stamens 3 ....
    ....**B. purpurea**
    2. Flower buds spindle-shaped and tapering to the apex, not winged; stamens 5...............**B. variegata**

Bauhinia aculeata L. [Prickly, in reference to the intrastipular spines.] WHITE ORCHID TREE.

Bauhinia aculeata Linnaeus, Sp. Pl. 1: 374. 1753.

Tree or shrub, to 8 m; stem tomentose when young, soon glabrate, with 1 or 2 slightly curved intrastipular spines. Leaves 1-foliolate, the blade suborbicular to broadly lanceolate, 3–6 cm long, 2.5–5 cm wide, the apex emarginate or lobed ⅓–½ its length, the lobes parallel or slightly divergent, each rounded to acute, the base cordate to rounded, the upper and lower surfaces tomentose, strigose, or glabrate, the petiole 0.5–3 cm long, tomentose, strigose, or glabrate; stipules linear, ca. 4 mm long. Flowers solitary or 3–5 in a subterminal or axillary raceme; bract and bracteoles linear, 2–5 mm long; buds linear-lanceolate, 3–4 cm long, with free tips to 1–2 mm long; calyx with the tube 3–10 mm long, the lobes 2–3 cm long, remaining attached and spathaceous at anthesis; corolla white, the petals narrowly elliptic to obovate, 3–8(10) cm long, white, the claw sparsely pilose to glabrous; stamens 10, the 3 alternate ones slightly shorter, the filaments 2–3(7.5) cm long, short-connate basally, upward arcuate, sparsely pilose basally. Fruit linear to narrowly ellipsoid, 7–15 cm long, 1.5–2.5 cm wide, pubescent, the gynophore 1–2 mm long; seeds 5–10, oblong to suborbicular, 6–11 mm long, brown.

Disturbed sites. Rare; Miami-Dade County. Escaped from cultivation. Florida; West Indies, Central America, and South America. Native to tropical America. Summer.

Bauhinia purpurea L. [Purple, in reference to the petals.] PURPLE ORCHID TREE; BUTTERFLY TREE.

Bauhinia purpurea Linnaeus, Sp. Pl. 1: 375. 1753. *Phanera purpurea* (Linnaeus) Bentham, in Miquel, Pl. Jungh. 1: 262. 1852. *Bauhinia purpurea* Linnaeus var. *genuina* Kurz, J. Asiat. Soc. Bengal, Pt. 2, Nat. Hist. 45: 288. 1876, nom. inadmiss. *Caspareopsis purpurea* (Linnaeus) Pittier, in Pittier et al., Cat. Fl. Venez. 1: 363. 1945.

Tree, to 10 m; stem glabrescent. Leaves 1-foliolate, the blade suborbicular, 6–12 cm long, 6–12 cm wide, the apex emarginate or bilobate ⅓–½ its length, the base rounded to cordate, the upper surface glabrous, the lower surface sparsely pubescent, the petiole to 2 cm long; stipules broadly lanceolate, 1–2 mm long. Flowers 6–10 in a terminal or subterminal raceme; bracts and bracteoles ovate, 1–2 mm long; buds clavate, 3–4 cm long, 4- or 5-angled in the upper part; calyx tube 7–12 mm long, the lobes 2–3 cm long, remaining attached and spathaceous at anthesis; corolla pink to dark purple, the 2 lateral and 2 lower petals narrowly lanceolate, the posterior one oblanceolate, 3–5 cm long, glabrous; stamens 3, the filaments 3–4 cm long, short connate with the staminodes, upward arcuate, glabrous. Fruit linear, 20–25 cm long, 2.5–3.5 cm wide, glabrous, the stipe 1 cm long; seeds ca. 10, suborbicular, ca. 1.5 cm long, brown.

Disturbed sites. Rare; Indian River, St. Lucie, Broward, and Miami-Dade Counties. Escaped from cultivation. Florida and Texas; West Indies, Mexico, Central America, and South America; Africa, Asia, and Australia. Native to Asia. Fall.

Bauhinia variegata L. [Variegated, in reference to the petals.] ORCHID TREE; MOUNTAIN EBONY.

Bauhinia variegata Linnaeus, Sp. Pl. 1: 375. 1753. *Phanera variegata* (Linnaeus) Bentham, in Miquel, Pl. Jungh. 1: 262. 1852.

Tree, to 15 m; stem pubescent when young, soon glabrous. Leaves 1-foliolate, the blade ovate to suborbicular, 6–16 cm long, 6–16 cm wide, the apex emarginate or bilobate ¼–⅓ its length, the lobes rounded, the base cordate, the upper surface glabrous, the lower surface sparsely puberulous, the petiole 3–4 cm long, glabrous; stipules broadly lanceolate, 1–2 mm long. Flowers 3–8 in a subterminal raceme; bract and bracteoles triangular, minute; buds fusiform, 3–4 cm long; calyx tube ca. 1.5 cm long, the lobes 2–3 cm long, remaining attached and spathaceous at anthesis; corolla purple or white, the petals obovate, 4–5.5 cm long; stamens 5, the filaments 2–4 cm long, short connate with the staminodes, upward arcuate, glabrous. Fruit linear, 20–30 cm long, 1–2.5 cm wide, glabrous, the gynophore 1.5 cm long; seeds 10–25, suborbicular, 1.5 cm long, brown.

Disturbed sites. Occasional; central and southern peninsula. Escaped from cultivation. Florida, Louisiana, Texas, and California; West Indies, Mexico, Central America, and South America; Africa, Asia, and Australia. Native to Asia. Spring.

Bauhinia variegata is listed as a Category I invasive species in Florida by the Florida Exotic Pest Plant Council (FLEPPC, 2015).

Caesalpinia L. 1753. NICKER

Trees, shrubs, or woody or herbaceous vines; stems armed with prickles or unarmed. Leaves even- or odd-bipinnately compound, the pinnae even-pinnate, stipulate or estipulate. Flowers axillary or terminal in simple or paniculiform racemes, bisexual or unisexual; calyx 5-lobed; corolla caesalpiniaceous, the petals 5, the standard usually the smallest and distinctly clawed; stamens 10, free. Fruit a dehiscent or indehiscent legume.

A genus of about 150 species; nearly cosmopolitan. [Commemorates Andreas Caesalpini (1519–1603), Italian physician and botanist.]

Caesalpinia sens. lat. is considered polymorphic based on molecular data with eight segregate genera recognized by Lewis (2005). According to this interpretation, our six species fall into two genera, *Caesalpinia* sens. str. (3 spp.) and *Guilandina* (3 spp.). Since there is still considerable disagreement on this and pending further study, we follow the traditional approach and retain them in *Caesalpinia* sens. lat. following Isely (1990, 1998) and others

Guilandina L., 1753; *Ticanto* Adans., 1763.

1. Erect shrub or small tree.
   2. Petals 1.5–2 cm long; stamens to 5 cm long, long-exserted .... **C. pulcherrima**
   2. Petals 6–10 mm long; stamens to 1.5 cm long, only slightly exserted.
      3. Leaflets elliptic to obovate, 0.5–1.5 cm long, chartaceous; fruit 2–4 cm long .... **C. pauciflora**
      3. Leaflets round-obovate, 1–4 cm long, subcoriaceous; fruit 4–8 cm long .... **C. vesicaria**
1. Reclining, scrambling, or climbing, herbaceous or woody vine.
   4. Fruit unarmed, indehiscent .... **C. crista**
   4. Fruit spiny, dehiscent.
      5. Stipules present, conspicuous; seeds gray .... **C. bonduc**
      5. Stipules absent; seeds yellow .... **C. major**

Caesalpinia bonduc (L.) Roxb. [An old genus name.] GRAY NICKER.

Guilandina bonduc Linnaeus, Sp. Pl. 1: 381. 1753. *Guilandina bonducella* Linnaeus, Sp. Pl., ed. 2. 545. 1762, nom. illegit. *Bonduc minus* Medikus, Theodora 41: 1786. *Caesalpinia bonducella* Fleming, Asiat. Res. 11: 159. 1810, nom. illegit. *Guilandina bonduc* Linnaeus var. *minor* (Medikus) de Candolle, Prodr. 2: 489. 1825, nom. inadmiss. *Caesalpinia bonduc* (Linnaeus) Roxburgh, Fl. Ind., ed. 1832. 2: 362. 1832. *Caesalpinia jayabo* Gómez de la Maza y Jiménez var. *cyanosperma* Gómez de la Maza y Jiménez, Anales Soc. Esp. Hist. Nat. 19: 234. 1890, nom. illegit.

Reclining or scrambling woody or herbaceous vine; stem to 6 m long, armed with straight prickles or unarmed, puberulent. Leaves even-pinnately compound, to 4 dm long, with recurved, usually paired prickles, the pinnae 4–5 pairs, 4–12 cm long, usually with recurved prickles, the leaflets 4–8 pairs, the blade ovate to elliptic, 2–4(6) cm long, 1.5–3 cm wide, chartaceous, slightly asymmetrical, the apex mucronate, the base obtuse or rounded, the upper and lower surfaces glabrate or puberulent along the veins, short-petiolulate, the petiole 10–15 cm long; stipules conspicuous and persistent, 1–2 cm long, foliaceous, resembling reduced or incised leaflets. Flowers in a simple or compound axillary raceme to 20 cm long, bisexual or staminate; bracts conspicuous, recurved; calyx 5–8 mm long, pubescent; petals broadly oblong, 7–10 mm long, orange-yellow; stamens subequaling or shorter than the petals. Fruit suborbicular to oval,

3–9 cm long and wide, compressed, densely prickly, tardily dehiscent; seeds 1–3, suborbicular, compressed, 1.5–2 cm long and wide, gray.

Coastal strands. Frequent; central and southern peninsula. Florida, Louisiana, and Texas; West Indies, Mexico, Central America, and South America; Africa, Asia, Australia, and Pacific Islands. Summer.

Caesalpinia crista L. [Crest, terminal tuft, apparently in reference to the spine-crested fruits.] YELLOW NICKER.

> *Caesalpinia crista* Linnaeus, Sp. Pl. 1: 380. 1753. *Guilandina crista* (Linnaeus) Small, Fl. S.E. U.S. 591, 1331. 1903.
> *Guilandina nuga* Linnaeus, Sp. Pl., ed. 2. 546. 1762. *Ticanto nuga* (Linnaeus) Medikus, Theodora 52. 1786. *Caesalpinia nuga* (Linnaeus) R. Brown, in W. T. Aiton, Hortus Kew. 2: 32. 1811.

Climbing woody vine; stem to 10 m long, with a few recurved prickles. Leaves even-bipinnately compound, 2–3 dm long, the rachis with blackish recurved prickles, the pinnae 2–3(4) pairs, the leaflets 4–6 pairs, the blade ovate or elliptic, 3–6 cm long, 1.5–3 cm wide, chartaceous, the apex obtuse rounded, sometimes emarginate or acute, the base cuneate or obtuse, the upper and lower surfaces glabrous, short-petiolulate, the petiolule short, the petiole 10–15 cm long. Flowers numerous in a terminal paniculate raceme 10–20 cm long; calyx lobes lanceolate, ca. 6 mm long, glabrous; petals unequal, 4 ovate, short-clawed, yellow, glabrous, the upper one tinged with red stripes, attenuate to the claw, the inner surface pubescent; stamens with the filament inflated and pubescent at the base. Fruit obliquely ovoid, 3–4 cm long, 2–3 cm wide, reticulate; seed 1, yellow, compressed.

Disturbed sites. Rare; Palm Beach County, Monroe County keys. Escaped from cultivation. Florida; West Indies, Mexico, Central America, and South America; Africa, Asia, Australia, and Pacific Islands. Native to Asia. Summer–fall.

Caesalpinia major (Medik.) Dandy & Exell [Greater.] HAWAII PEARLS; YELLOW NICKER.

> *Guilandina bonduc* Linnaeus, Sp. Pl., ed. 2. 545. 1762; non Linnaeus, 1753. *Bonduc majus* Medikus, Theodora 43, t. 3. 1786. *Guilandina bonduc* Linnaeus var. *major* (Medikus) de Candolle, Prodr. 2: 480. 1825. *Caesalpinia jayabo* Gómez de la Maza y Jiménez, Anales Soc. Esp. Hist. Nat. 19: 234. 1890. nom. illegit. *Guilandina major* (Medikus) Small, Fl. S.E. U.S. 591, 1331. 1903. *Caesalpinia major* (Medikus) Dandy & Exell, J. Bot. 76: 180. 1938. *Caesalpinia globulorum* Bakhuizen van den Brink f. & P. Royen, Blumea 12: 62. 1963, nom. illegit.

Scrambling or reclining, herbaceous or suffrutescent vine; stem to 5 m long, usually armed with prickles. Leaves even-bipinnately compound, 1.5–5 dm long, the rachis with a few recurved prickles, the pinnae in 3–5 pairs, unarmed or with prickles, the leaflets 4–7 pairs, the blade suborbicular to ovate-oblong, 2–5.5 cm long, 1.5–4 cm wide, chartaceous, the apex acute to rounded or emarginate, the base broadly cuneate to rounded, the upper and lower surfaces glabrous, short-petiolulate. The petiole 10–15 cm long; estipulate. Flowers numerous in an axillary simple or paniculate raceme, unisexual; bracts ascending or spreading; calyx 8–9 mm long, glabrous; petals 8–12 mm long, orange-yellow; stamens shorter than the petals. Fruit oval to suborbicular, 4–7(9) cm long, 2–4(7) cm wide, prickly; seeds 1–2, 1.5–2 cm long, yellow.

Coastal strands. Rare; Martin, Palm Beach, and Miami-Dade Counties, Monroe County keys. Florida; West Indies, Central America, and South America; Africa, Asia, and Pacific Islands. All year.

Caesalpinia major is listed as endangered in Florida (Florida Administrative Code, Chapter 5B-40).

Caesalpinia pauciflora (Griseb.) C. Wright [Few-flowered.] FEWFLOWER HOLDBACK.

Libidibia pauciflora Grisebach, Cat. Pl. Cub. 78. 1866. *Caesalpinia pauciflora* (Grisebach) C. Wright, in Sauvalle, Anales Acad. Ci. Med. Habana 5: 404. 1869. *Poinciana pauciflora* (Grisebach) Small, Fl. S.E. U.S. 591, 1331. 1903.

Ascending or scrambling shrub, to 2 m; stem prickly, glabrate. Leaves even-bipinnately compound, 4–10 dm long, prickly, the pinnae 3–5 pairs, the leaflets 4–7 pairs, the blade elliptic or obovate, 5–14 mm long, 3–10 mm wide, chartaceous, lacking secondary venation, the apex rounded or slightly emarginate, the base asymmetrical, broadly cuneate, the upper and lower surfaces glabrate, short-petiolulate, the petiole 10–15 cm long; stipules paired prickles or inconspicuous. Flowers in a terminal raceme 5–20 cm long; bracts small, caducous; calyx irregular, the outer lobe 5–6 mm long, larger than the other 4; petals 6–9 mm long, yellow; stamens slightly exceeding the petals, the filaments short pilose at the base. Fruit asymmetric-lunate, 2–4 cm long, 0.8–1 cm wide, smooth; seeds few.

Pinelands and tropical hammocks. Rare; Monroe County keys. Florida; West Indies. All year.

Caesalpinia pauciflora is listed as endangered in Florida (Florida Administrative Code, Chapter 5B-40).

Caesalpinia pulcherrima (L.) Sw. [Beautiful and little.] PRIDE-OF-BARDADOS; DWARF POINCIANA.

Poinciana pulcherrima Linnaeus, Sp. Pl. 1: 380. 1753. *Caesalpinia pulcherrima* (Linnaeus) Swartz, Observ. Bot. 166. 1791.

Shrub or small tree, to 4 m; stem prickly or unarmed, glabrous. Leaves even-bipinnately compound, 1.5–3(4) dm long, the pinnae 5–8(10) pairs, the leaflets 6–10 pairs, the blade elliptic to obovate, 1–2.5 cm long, 0.5–1 cm wide, chartaceous, the apex rounded or slightly emarginate, usually mucronate, the base asymmetrical, broadly cuneate to rounded, the upper and lower surfaces glabrous, the lower surface pale and with evident venation, short-petiolulate, the petiole 4–8 mm long; stipules inconspicuous. Flowers in an elongate terminal raceme 10–20 cm long; bracts caducous; sepals unequal, the outer lobe 1–1.5 cm long, larger than the other 4, orange-red; petals obovate, 1.5–2 cm long, obovate, long-clawed, yellow-orange or yellow, the margin fringed; stamens long-exserted, to 5 cm long, red. Fruit obliquely oblong or oblanceolate, 6–10(12) cm long, 1.5–2 cm wide, dark brown, smooth.

Disturbed sites. Rare; Lee and Miami-Dade Counties, Monroe County keys. Escaped from cultivation. Florida, Texas, Arizona, and California; West Indies, Mexico, Central America, and South America; Africa, Asia, Australia, and Pacific Islands. Native to tropical America(?). All year.

Caesalpinia vesicaria L. [Bladder-like, inflated, in reference to the slightly inflated fruit.] INDIAN SAVIN TREE.

Caesalpinia vesicaria Linnaeus, Sp. Pl. 1: 381. 1753. *Nicarago vesicaria* (Linnaeus) Britton & Rose, in Britton, N. Amer. Fl. 23: 319. 1930.

Shrub or small tree, to 7 m; branchlets glabrous, armed with a few prickles or unarmed. Leaves even-bipinnately compound, 1–2.5 dm long, the pinnae 2–3 pairs, the leaflets 1–3 pairs, the blade round-obovate, 1–4 cm long and wide, subcoriaceous, the apex rounded, subtruncate, or emarginate, the base broadly cuneate, rounded, often asymmetrical, the upper and lower surfaces glabrous, short-petiolulate, the petiole 2–3 cm long; stipules inconspicuous. Flowers in a simple or paniculiform raceme 2–3 dm long, the pedicel 6–9 mm long; calyx 8–9 mm long, glabrous; corolla 8–10 mm long, slightly longer than the calyx, yellow; stamens subequaling the petals, the filaments densely villous. Fruit narrowly oblong, 4–8 cm long, 1–2 cm wide, slightly inflated, pubescent when young, becoming glabrous; seeds ca. 6, orbicular, slightly compressed, brown, lustrous.

Disturbed sites. Rare; Miami-Dade County. Escaped from cultivation. Florida; West Indies, Mexico, and Central America. Native to West Indies, Mexico, and Central America.

EXCLUDED TAXON

Caesalpinia ciliata (Bergius ex Wikstöm) Urban—Reported for Florida by Small (1913a, 1913d, 1913e, 1933, all as *Guilandina ovalifolia* (Urban) Britton), the name apparently misapplied to material of *C. major*.

Cajanus Adans., nom. cons. 1763.

Annual or perennial herbs. Leaves alternate, 3-foliolate, stipellate, petiolate, stipulate. Flowers in terminal or axillary racemes; calyx 4-lobed; corolla papilionaceous; stamens 10, diadelphous (9 + 1). Fruit a dehiscent legume.

A genus of 34 species; nearly cosmopolitan. [From the Malayan vernacular name *catjang*, meaning beans in general.]

Cajan Adans., 1763, orth. var.

Cajanus cajan (L.) Huth, nom. cons. [Malayan name for beans in general.] PIGEONPEA.

Cytisus cajan Linnaeus, Sp. Pl. 2: 739. 1753. *Cajanus flavus* de Candolle, Prodr. 2: 406. 1825, nom. illegit. *Cajanus indicus* Sprengel, Syst. Veg. 3: 248. 1826, nom. illegit. *Cajanus cajan* (Linnaeus) Huth, Helios 11: 133. 1893.

Erect annual or perennial, suffrutescent herb, to 3 m; stem strigulose. Leaves 3-foliolate, the leaflets with the blade elliptical, ovate, or lanceolate, 4–14 cm long, 1.5–4.5 cm wide, the apex acute to acuminate, the base cuneate, the upper and lower surfaces glabrous, the lower surface pale and sometimes obscurely glandular, short-petiolulate, the stipels setaceous, 1–4 mm long, caducous, the petiole 1–8 mm long; stipules triangular-lanceolate, 2–6 mm long. Flowers solitary or 2–5 in a short, aggregated, terminal or axillary raceme; calyx campanulate, the tube 3–6 mm long, the lobes 3–7 mm long, pubescent; corolla 1–1.5 cm long, yellow, the standard commonly suffused with orange, red, or purple; stamens 1–1.5 cm long. Legume oblong, 4–9 cm long, 1–1.5 cm wide, laterally compressed, straight or falcate, glandular-punctate, densely pubescent; seeds 2–9, globose to ellipsoid-reniform, 4–9 mm long, 3–8 mm wide, white, brown, purplish, to almost black, sometimes mottled.

Disturbed sites. Rare; Broward and Miami-Dade Counties; Monroe County keys. Escaped from cultivation. Florida; West Indies, Mexico, Central America, and South America; Africa, Asia, Australia, and the Pacific Islands. Native to Africa, Asia, Australia, and the Pacific Islands. All year.

Callerya Endl. 1843.

Woody vines. Leaves alternate, odd-pinnately compound, stipellate, petiolate, stipulate. Flowers numerous in terminal or axillary panicles, pedicellate, bracteate, bracteolate; sepals 5, connate, the calyx 5-lobed; corolla papilionaceous; stamens 10, diadelphous (9 + 1). Fruit a dehiscent legume.

A genus of about 20 species; North America, Asia, and Australia. [From the Greek *calli*, beautiful, and *eryo*, to guard, in reference to the flower buds protected by the bracts and bracteoles.]

Selected reference: Schot (1994).

Callerya reticulata (Benth.) Schot [Netlike, in reference to the secondary leaf venation.] EVERGREEN WISTERIA.

Millettia reticulata Bentham, in Miquel, Pl. Jungh. 249. 1852. *Phaseoloides reticulata* (Bentham) Kuntze, Revis. Gen. Pl. 1: 201. 1891. *Callerya reticulata* (Bentham) Schot, Blumea 39: 29. 1994.

Woody vine, to 13 m; bark gray to reddish brown, thinly ridged, the branchlets brown pubescent when young, glabrate in age. Leaves (5)7- to 9-foliolate, 1–2 cm long, the leaflet blade ovate, ovate-elliptic, to oblong, 1.5–4 cm wide, chartaceous, the apex obtuse, acute, or slightly retuse, the base rounded, the margin entire, the upper surface glabrous, lower surface glabrous or sparsely puberulent, the stipels linear, 2–3 mm long, the petiole 2–5 cm long; stipules narrowly triangular, 3–6 mm long. Flowers in a terminal or axillary erect or spreading panicle 1–2 dm long, the rachis reddish brown pubescent, the pedicel 3–5 mm long; bracts of the inflorescence narrowly triangular, 2–3 mm long, caducous, the bracts of the flowers narrowly triangular, 1–2 mm long, usually persistent; bracteoles subtending the flower bud ovate, 1–2 mm long, usually persistent; calyx 3–4 mm long, the tube glabrescent, the lobes short, slightly unequal, ciliate; corolla papilionaceous, purple, 1–1.3 cm long, glabrous. Fruit linear, ca. 15 cm long, 1–1.5 cm wide, flat; seeds 3–6, oblong, ca. 11 mm long, black.

Disturbed hammocks. Rare; Calhoun County. Escaped from cultivation. Florida; Asia. Native to Asia. Spring–summer.

Calliandra Benth., nom. cons. 1840.

Shrubs. Leaves alternate, even-bipinnately compound, petiolate, stipulate. Flowers in solitary or paired axillary heads; sepals connate, the calyx 5-lobed; petals connate, 5-lobed; stamens numerous, connate near the base. Fruit a dehiscent legume.

A genus of about 130 species; North America, West Indies, Mexico, and South America. [From the Greek *calli*, beautiful, and *andros*, male, in reference to the colorful stamens.]

Selected reference: Barneby (1998).

Calliandra haematocephala Hassk. [From the Greek *haemat*, blood-red, and *cephalus*, headed, in reference to the red-colored flowering heads.] POWDERPUFF TREE.

Calliandra haematocephala Hasskarl, Retzia 1: 216. 1855; Natuurk. Tijdschr. Ned.-Indië 10: 216. 1856. *Anneslia haematocephala* (Hasskarl) Britton & P. Wilson, Bot. Porto Rico 6: 348. 1926.

Shrub, to 3 m; branchlets glabrate to tawny-villous. Leaves with 1 pair of pinnae, the leaflets even-pinnate, 5–10 pairs, the blade falcate-lanceolate to half-elliptic, asymmetrical, 2–3(6) cm long, 0.5–1 cm wide, 2- to 3-nerved from the base, the apex acute to obtuse, mucronate, the base cuneate to rounded or subcordate, the upper and lower surfaces glabrous or pilosulous, the petiole 1–2 cm long. Flowers in a subterminal head 4–7 cm in diameter; calyx tubular, 2.5–4 mm long, the lobes short, red-pink; corolla tubular, 8–9 mm long, the lobes short, red-pink; stamens red-pink. Fruit oblanceolate-oblong, 6–10 cm long, 0.5–1 cm wide, compressed, glabrate; seeds 2–5, elliptic, 8–10 mm long, brown, minutely specked or mottled.

Disturbed sites. Rare; Escambia, Brevard, and Broward Counties. Escaped from cultivation. Florida; West Indies, Mexico, Central America, and South America; Africa, Asia, and Pacific Islands. Native to South America. Summer–fall.

Canavalia DC., nom. cons. 1825. JACKBEAN

Annual or perennial herbs or vines. Leaves alternate, 3-foliolate, petiolate, stipellate, stipulate. Flowers in axillary pseudoracemes, bracteate, bracteolate; calyx 2-lipped, the upper 2-toothed, the lower 3-toothed; corolla papilionaceous; stamens 10, monadelphous. Fruit a dehiscent legume.

A genus of about 60 species; North America, West Indies, Mexico, Central America, and South America; Africa, Asia, Australia, and Pacific Islands. [From the Malabar *kana-valli*, *kanum*, forest, and *valli*, climber.]

Selected references: Sauer (1964).

1. Seeds red to brown with darker marbling; plant of coastal strands and beaches......................**C. rosea**
1. Seeds white, off-white, olive, or brownish, not marbled; plant of disturbed upland sites.
    2. Plant usually erect; fruit 15–35 cm long, 3–3.5 cm wide, seeds white or off-white.......**C. ensiformis**
    2. Plant twining; fruit 6–20 cm long, 2–3 cm wide; seeds olive to brownish.................**C. brasiliensis**

Canavalia brasiliensis Mart. ex Benth. [Of Brazil.] BRAZILIAN JACKBEAN.

Canavalia brasiliensis Martius ex Bentham, Comm. Legum. Gen. 71. 1837.

Twining or prostrate, perennial, herbaceous vine; stem to 3 m long, sparsely pubescent to glabrate. Leaves with the leaflet blade 3–14 cm long, 3–9.5 cm wide, the apex obtuse, acute, or acuminate, the base cuneate, the upper and lower surfaces strigose to glabrate, the petiolules 7–8.5 mm long, moderately to densely pubescent, the petiole 4–11.5 cm long. Flowers in an axillary pseudoraceme to 25 cm long; bracteoles ovate, ca. 1 mm long and wide; calyx 6–12 mm long, the lower surface pubescent, the upper surface glabrous; corolla 2–2.5 cm long, papilionaceous, lavender to blue-violet. Fruit oblong, 6–20 cm long, 2–3 cm wide, slightly compressed, glabrate; seeds 4–12, oblong, 1.4–1.9 cm long, 1–1.2 cm wide, olive, the hilum ca. ½ the seed length.

Disturbed pinelands. Rare; Pasco, Broward, and Miami-Dade Counties. Escaped from cultivation. Florida; West Indies, Mexico, Central America, and South America. Native to tropical America. Fall.

Canavalia ensiformis (L.) DC. [Swordlike, in reference to the fruit.] WONDERBEAN; JACKBEAN.

Dolichos ensiformis Linnaeus, Sp. Pl. 2: 725. 1753. *Canavalia ensiformis* (Linnaeus) de Candolle, Prodr. 2: 404. 1825. *Canavalia ensiformis* (Linnaeus) de Candolle var. *normalis* Kuntze, Revis. Gen. Pl. 3(2): 55. 1898, nom. inadmiss.

Erect or twining, annual or perennial herbaceous vine; stem to 2(10) m long, glabrous or glabrate. Leaves with the leaflet blade ovate-elliptic, 6–15(20) cm long, 2.5–8 cm wide, the apex obtuse or subacute, the base cuneate, the upper and lower surfaces glabrate, the petiolules 4–8 mm long., the petiole 5–12 cm long. Flowers in an axillary pseudoraceme to 15 cm long; bracteoles ovate, ca. 2 m long; calyx 10–14 mm, the lower surface sparsely pubescent, the upper surface glabrous; corolla 2–2.8 cm long, papilionaceous, lavender to pink. Fruit narrowly oblong, 15–35 cm long, 3–3.5 cm wide, slightly compressed, glabrate; seeds 9–15, oblong, ca. 2 cm long, white or off-white, the hilum ca. ½ the seed length.

Disturbed sites. Rare; Lee County. Escaped from cultivation. Illinois south to Florida and Arizona; West Indies, Mexico, Central America, and South America; Africa, Asia, Australia, and Pacific Islands. Native to tropical America. Summer–fall.

Canavalia rosea (Sw.) DC. [Rose-colored, in reference to the corolla.] BAYBEAN; SEASIDE JACKBEAN.

Dolichos roseus Swartz, Prodr. 105. 1788. *Canavalia rosea* (Swartz) de Candolle, Prodr. 2: 404. 1825.
Dolichos maritimus Aublet, Hist. Pl. Guiane 765. 1775. *Dolichos obtusifolius* Lamarck, Encycl. 2: 295. 1786, nom. illegit.; non Jacquin, 1768. *Canavalia obtusifolia* de Candolle, Prodr. 2: 404. 1825, nom. illegit. *Canavalia maritima* (Aublet) Urban, Repert. Spec. Nov. Regni Veg. 15: 400. 1919; non Du Petit-Thouars, 1813.
Canavalia maritima Du Petit-Thouars, J. Bot. Agroc. 1: 80. 1813.

Prostrate-trailing or twining perennial vine; stem to 10 m long, pubescent to glabrate. Leaves with the leaflet blade suborbicular, elliptic, or oblong, 4–12 cm long, 2.5–6 cm wide, the apex obtuse or emarginate, the base cuneate to rounded, the upper surface moderately to densely pubescent or glabrous, the lower surface sparsely pubescent, the petiolules 3–8 mm long, the petiole 2–5.5 cm long. Flowers in an axillary pseudoraceme 9–21 cm long; bracteoles ovate, ca. 1 mm long; calyx 10–12 mm long, the lower surface sparsely to moderately densely pubescent, the upper surface glabrous; corolla 2.5–3 cm long, papilionaceous, lavender to reddish purple. Fruit oblong, 10–15 cm long, 2–3.5 cm wide, turgid to moderately compressed, glabrate; seeds (1)4–8, elliptic, ca. 1.5 cm long, red to brown with darker marbling, the hilum ca. ½ the seed length.

Coastal strands and beaches. Frequent; Dixie County, central and southern peninsula. Florida, west to Texas; West Indies, Mexico, Central America, and South America; Africa, Asia, Australia, and Pacific Islands. All year.

EXCLUDED TAXA

Canavalia altissima (Jacquin) Macfadyen—This synonym of the more tropical *Mucuna urens* (Linnaeus) Medikus was cited for Florida by Chapman (1897), who misapplied the name to material of *C. brasiliensis.*

Canavalia gladiata (Jacquin) de Candolle—Reported for Florida by Chapman (1897), Small (1903, 1913a, 1933), Long and Lakela (1971), and Isely (1990, 1998), but all material seen by Sauer (1964) is from cultivation. No specimen seen by us.

Canavalia lineata (Thunberg) de Candolle—This tropical species was reported by Small (1913a, 1913b, 1913d, 1933), who misapplied the name to material of *C. rosea.*

Cassia L., nom. cons. 1753.

Trees. Leaves alternate, even-pinnately compound, the leaflets opposite, petiolate, estipellate, stipulate. Flowers in terminal simple or paniculiform racemes, bracteate, bracteolate; sepals 5, zygomorphic; corolla caesalpiniaceous; stamens heterantherous (7 fertile, 3 long and 4 short); staminodes 3. Fruit an indehiscent legume.

A genus of about 30 species; Old and New World tropics. [From the Greek *kasia*, a kind of cinnamon.]

Selected reference: Irwin and Barneby (1982).

Cassia fistula L. [Hollow throughout, like a pipe, but closed at the ends, in reference to the seed pod.] GOLDEN SHOWER.

Cassia fistula Linnaeus, Sp. Pl. 1: 377. 1753. *Cathartocarpus fistulus* (Linnaeus) Persoon, Syn. Pl. 1: 459. 1805. *Bactyrilobium fistulum* (Linnaeus) Willdenow, Enum. Pl. 440. 1809.

Tree, to 20 m; stem glabrate or finely pubescent when young. Leaves even-pinnately compound, (15)20–55(60) cm long, the leaflets of (2)3–7 pairs, the blade ovate, (7)9–20 cm long, (4)5–8 cm wide, the apex acute, the base rounded, the upper and lower surfaces glabrate, the petiole 4–7(9) cm long. Flowers (7)15–75 in a simple or paniculiform raceme; sepals obovate to oblong-elliptic, 6–10 mm long, reflexed at anthesis; corolla caesalpiniaceous, yellow, the petals elliptic-oblanceolate or obovate, 2–3 cm long, subequal; 3 long fertile stamens (2)3–4 cm long, sigmoid, the anthers dehiscent by lateral slits and a basal pore, the 4 shorter fertile stamens in 2 pairs, 5–7 or 11–13 mm long, erect, the anthers dehiscent by a basal pore; staminodes 4–13 mm long. Fruit linear-cylindric, 30–60 cm long, 1.5–2.5 cm wide; seeds obovate-ellipsoid, 8–10 mm long, 6–7 mm wide, compressed, embedded in a glutinous blackish pulp.

Disturbed sites. Rare; Miami-Dade County. Escaped from cultivation. Florida; West Indies, Mexico, Central America, South America; Africa, Asia, and Australia. Native to Asia. Spring–summer.

EXCLUDED TAXON

Cassia javanica Linnaeus var. *indochinensis* Gagnepain—Reported for Florida by Wunderlin (1998) and Wunderlin and Hansen (2003) based on a Miami-Dade County specimen now considered to be from cultivated material.

Centrosema (DC) Benth., nom. cons. 1837. BUTTERFLY PEA

Perennial herbs. Leaves alternate, 1- or 3 foliolate, petiolate, stipellate, stipulate. Flowers in axillary pseudoracemes, bracteate, bracteolate; calyx 5-lobed; corolla papilionaceous, the banner gibbous above the claw; stamens 10, diadelphous (9 + 1). Fruit a dehiscent legume.

A genus of about 35 species; North America, West Indies, Mexico, Central America, Africa, and Asia. [From the Greek *kentron*, spur, and *sema*, vexillum (standard), in reference to the gibbous or short-spurred standard petal.]

Bradburya Raf., nom. rej. 1817.

1. Leaves 1-foliolate; petiole conspicuously winged... **C. sagittatum**
1. Leaves 3-foliolate; petiole not winged.
   2. Lower calyx lobe 5–8 mm long, subulate to lanceolate, the upper and lateral calyx lobes 3–4 mm long; leaflets chartaceous, the veins not prominent on the lower surface .... **C. arenicola**
   2. Lower calyx lobe 8–11 mm long, subulate, the upper and lateral calyx lobes 7–8 mm long; leaflets subcoriaceous, the veins prominent on the lower surface .... **C. virginianum**

Centrosema arenicola (Small) F. J. Herm. [*Arenus*, sandy place, and *cola*, dweller, in reference to growing in sandy areas.] PINELAND BUTTERFLY PEA; SAND BUTTERFLY PEA.

Bradburya arenicola Small, Fl. S.E. U.S. 561, 1332. 1903. *Centrosema arenicola* (Small) F. J. Hermann, J. Wash. Acad. Sci. 38: 237. 1948. TYPE: FLORIDA: Lake Co.: vicinity of Eustis, 16–31 Jul 1894, *Nash 1366* (holotype: NY).

Bradburya floridana Britton, Torreya 4: 142. 1904. *Centrosema floridanum* (Britton) Lakela, Sida 1: 182. 1963. TYPE: FLORIDA: Hillsborough Co.: Tampa, 25 Aug 1903, *Britton & Wilson 31* (holotype: NY).

Trailing or climbing, perennial herbaceous vine; stem to 3 m long, glabrate to sparsely strigose. Leaves 3-foliolate, the leaflets with the blade elliptic or elliptic-lanceolate to lanceolate, 1.5–5.5 cm long, 0.8–2.5 cm wide, the apex acute, the base rounded to truncate, the upper surface uncinate, the lower surface glabrate, the petiolule ca. 2 cm long, the stipels linear to subulate, 1–2 mm long, the petiole 1–4 cm long; stipules deltoid, ca. 1 mm long. Flowers solitary or 2(4) in an axillary raceme; bracts oval-ovate, 4–6 mm long; bracteoles ovate-lanceolate, 8–10 mm long; calyx tube 9–12 mm long, the upper lobes 3–4 mm long, subulate to lanceolate, the lower lobe 5–8 mm long; corolla papilionaceous, lavender to pinkish lavender, the banner 2–3 cm long, the claw 5–6 mm long, the spur ca. 2 mm long, the wings 2–2.5 cm long, the keel 2–2.5 mm long. Fruit (7)9–12.5 cm long, 5–6 mm wide, glabrate; seeds 8–12, 3–4 mm long, 4–5 mm wide, brownish black.

Sandhills. Occasional; northern and central peninsula, Dixie County. Endemic. Summer–fall.

Centrosema arenicola is listed as endangered in Florida (Florida Administrative Code, Chapter 5B-40).

Centrosema sagittatum (Humb. & Bonpl. ex willd.) Brandegee [Arrow-shaped, in reference to the leaves.] ARROWLEAF BUTTERFLY PEA.

Glycine sagittata Humboldt & Bonpland ex Willdenow, Enum. Pl. 757. 1809. *Rudolphia dubia* Kunth, in Humboldt et al., Nov. Gen. Sp. 6: 452. 1824, nom. illegit. *Centrosema hastatum* Bentham, Comm. Legum. Gen. 56. 1837, nom. illegit. *Centrosema dubium* Hemsley, Biol. Cent.-Amer., Bot. 1: 294. 1889, nom. illegit. *Bradburya sagittata* (Humboldt & Bonpland ex Willdenow) Rose, Contr. U.S. Natl. Herb. 8: 46. 1903. *Centrosema sagittatum* (Humboldt & Bonpland ex Willenow) Brandegee, Zoë 5: 202. 1905.

Perennial herbaceous vine; stems to 2 m long, glabrate or slightly strigose. Leaves 1-foliolate, the leaflets with the blade deltate, 8–15 cm long and wide, the apex acute or acuminate, the base deeply sagittate, the petiole 2–4 cm long, winged; stipules lanceolate, 4–5 mm long. Flowers solitary or paired at a node; bracts lanceolate, 2–3 mm long; bracteoles 5–8 mm long; calyx 10–12 mm long, the lobes deltate to lanceolate; corolla papilionaceous, 4–5 cm long, yellow, the banner purple-blotched. Fruit 10–18 cm long, 6–8 mm wide; seeds 10–12, 3–4 mm long and wide, brownish black.

Disturbed hammocks. Rare; Alachua County. Florida; West Indies, Mexico, Central America, and South America. Native to tropical America. Summer–fall.

Centrosema virginianum (L.) Benth. [Of Virginia.] SPURRED BUTTERFLY PEA.

Clitoria virginiana Linnaeus, Sp. Pl. 2: 753. 1753. *Centrosema virginianum* (Linnaeus) Bentham, Comm. Legum. Gen. 56: 1837. *Bradburya virginiana* (Linnaeus) Kuntze, Revis. Gen. Pl. 1: 164. 1891. *Cruminium virginianum* (Linnaeus) Britton, Bull. Torrey Bot. Club 18: 269. 1891. *Centrosema virginianum* (Linnaeus) Bentham var. *genuinum* Stehle et al., Fl. Guadeloupe 2: 108. 1948, nom. inadmiss.

Clitoria virginiana Linnaeus var. *angustifolia* de Candolle, Prodr. 2: 234. 1825. *Centrosema virginianum* (Linnaeus) Bentham var. *angustifolium* (de Candolle) Grisebach, Fl. Brit. W.I. 193. 1860. *Bradburya virginiana* (Linnaeus) Kuntze var. *angustifolia* (de Candolle) de Candolle ex Kuntze, Revis. Gen. Pl. 3(2): 53. 1898. *Centrosema virginianum* (Linnaeus) Bentham subsp. *angustifolium* (de Candolle) Hadac, Folia Geobot. Phytotax. 5: 432. 1970.

Clitoria virginiana Linnaeus var. *elliptica* de Candolle, Prodr. 2: 234. 1825. *Centrosema virginianum* (Linnaeus) Bentham var. *ellipticum* (de Candolle) Fernald, Rhodora 43: 587. 1941.

Trailing or twining, perennial herbaceous vine; stem to 3 m long, glabrate to strigose. Leaves 3-foliolate, the leaflets with the blade ovate, elliptic, lanceolate, or linear, (1.5)2–6(10) cm long, 0.8–4.5 cm wide, the apex acute, the base rounded, the lower surface glabrate, the upper surface minutely uncinate, the stipels linear, 3–5 mm long, the petiolules ca. 2 mm long, the petiole

1–5 cm long; stipules ovate-deltoid, 3–5 mm long. Flowers solitary or paired at a node, 1–3 cm long, 1–2(4)-flowered; bracts ovate, 4–6 mm long; bracteoles lance-ovate, 6–8 mm long; calyx 10–12 mm long, the upper lobes 7–8 mm long, the lower lobe 8–11 mm long; corolla papilionaceous, lavender to blue-violet, the banner 2.5–3.5(4) cm long, with a white medial stripe, the claw 2–4 mm long, the spur 1–2 mm long, the wings 14–19 mm long, the keel 16–20 mm long. Fruit (6)8–11 cm long, 3–5 mm wide, glabrate; seeds 14–18, 3–4 mm long, 4–6 mm wide, dark brown.

Moist to dry hammocks, sandhills, and coastal swales. Frequent; nearly throughout. New Jersey south to Florida, west to Illinois, Missouri, Oklahoma, and Texas; West Indies, Mexico, Central America, and South America; Africa and Asia. Summer–fall.

Cercis L. 1753. REDBUD

Trees or shrubs. Leaves alternate, simple, petiolate, stipulate. Flowers in fasciculate, cauliflorous clusters, bracteate; calyx 5-lobed; petals pseudopapilionaceous (appearing papilionaceous, but the standard smallest and in front of the other petals); stamens 10, free; fruit a dehiscent legume.

A genus of about 10 species; North America, Mexico, Europe, and Asia. [From the Greek *kerkis*, weaver's shuttle, apparently in reference to the shape of the fruit.]

Cercis canadensis L. [Of Canada.] EASTERN REDBUD.

> *Cercis canadensis* Linnaeus, Sp. Pl. 1: 374. 1753. *Cercis canadensis* Linnaeus var. *typica* Hopkins, Rhodora 44: 200. 1942, nom. inadmiss.
> *Cercis canadensis* Linnaeus forma *glabrifolia* Fernald, Rhodora 38: 234. 1936.

Shrub or tree, to 6(10) m; branchlets glabrate or puberulent. Leaves deciduous, with the blade broadly ovate, 5–10 cm long and wide, the apex acute to acuminate, the base cordate or truncate, the blade palmately veined, the upper surface glabrous, the lower surface glabrate or puberulent; stipules inconspicuous, caducous. Flowers in cauliflorous fascicles appearing before the leaves, the pedicels 6–10 mm long; calyx shallowly lobed; corolla pseudopapilionaceous, bright pink to lilac or rarely white. Fruit elliptic-falcate, 4–8(10) cm long, 9–15 mm wide, flat, glabrous, the ventral suture with a wing ca. 1–2 mm wide.

Mesic hammocks. Frequent; northern counties south to Hillsborough and Polk Counties. Massachusetts, New York, and Ontario south to Florida, west to Wisconsin, Iowa, Nebraska, Kansas, and New Mexico; Mexico. Spring.

Chamaecrista Moench 1794. SENSITIVE PEA

Annual or perennial herbs. Leaves alternate, even-pinnately compound, petiolate, stipulate. Flowers 1-several in reduced axillary racemes, bracteate, 2-bracteolate; caesalpinoid; sepals 5; petals 5; androecium with 5–10 functional stamens of various lengths, the anthers basifixed. Fruit a dehiscent, flat legume.

A genus of about 300 species; nearly cosmopolitan. [From the Greek *chamae*, low growing, and the Latin *crista*, crest, in reference to a low growing colorful plant.]

Selected reference: Irwin and Barneby (1982).

1. Leaflets 1 pair ....**C. rotundifolia**
1. Leaflets 2 or more pairs.
   2. Plant erect.
      3. Corolla 0.8–1 cm in diameter, the larger petals 4–7(8) mm long................................**C. nictitans**
      3. Corolla 2.5–3.5 cm in diameter, the larger petals 13–20 cm long............................**C. fasciculata**
   2. Plant prostrate or ascending.
      4. Plant a suffrutescent perennial; corolla ca. 2 cm in diameter, the larger petals 12–15 mm long ...
      ....**C. lineata**
      4. Plant annual; corolla to 1.5 cm in diameter, the larger petals 4–7 mm long.
         5. Plant pilose or hispid; petiolar gland small or obsolete, sessile ....**C. pilosa**
         5. Plant glabrate or puberulent; petiolar gland conspicuous, stipitate........................**C. serpens**

Chamaecrista fasciculata (Michx.) Greene [Fascicled.] PARTRIDGE PEA.

Cassia chamaecrista Linnaeus, Sp. Pl. 1: 379. 1753, nom. rej. *Cassia pulchella* Salisbury, Prodr. Stirp. Chap. Allerton 326. 1796, nom. illegit. *Grimaldia chamaecrista* (Linnaeus) Schrank ex Link, Handbuch 2: 141. 1831. *Xamacrista triflora* Rafinesque, Sylva Tellur. 127. 1838, nom. illegit. *Cassia chamaecrista* Linnaeus var. *normalis* Kuntze, Revis. Gen. Pl. 1: 169. 1891, nom. inadmiss. *Chamaecrista chamaecrista* (Linnaeus) Britton, Bull. Torrey Bot. Club 44: 12. 1917, nom. inadmiss.

Cassia fasciculata Michaux, Fl. Bor.-Amer. 1: 262. 1803. *Chamaecrista fasciculata* (Michaux) Greene, Pittonia 3: 242. 1897.

Cassia chamaecrista Linnaeus var. *robusta* Pollard, Bull. Torrey Bot. Club 21: 218. 1894. *Cassia robusta* (Pollard) Pollard, Bull. Torrey Bot. Club 24: 150. 1897. *Chamaecrista robusta* (Pollard) A. Heller, Cat. N. Amer. Pl., ed. 2. 5. 1900. *Cassia fasciculata* Michaux var. *robusta* (Pollard) J. F. Macbride, Contr. Gray Herb. 59: 24. 1919. *Chamaecrista fasciculata* (Michaux) Greene var. *robusta* (Pollard) Moldenke, Boissiera 7: 2. 1943.

Cassia depressa Pollard, Bull. Torrey Bot. Club 22: 515, pl. 251. 1895. *Chamaecrista depressa* (Pollard) Greene, Pittonia 3: 242. 1897. *Cassia fasciculata* Michaux var. *depressa* (Pollard) J. F. Macbride, Contr. Gray Herb. 59: 25. 1919. *Chamaecrista fasciculata* (Michaux) Greene var. *depressa* (Pollard) C. F. Reed, Phytologia 63: 410. 1987. TYPE: FLORIDA: Gadsden Co.: River Junction, 5 Sep 1895, *Nash 2571* (holotype: US; isotypes: LE, NY).

Chamaecrista bellula Pollard, Proc. Biol. Soc. Wash. 15: 20. 1902. TYPE: FLORIDA: Franklin Co.: St. Vincent, 9 Sep 1899, *Tracy 6326* (holotype: US; isotype: NY).

Chamaecrista brachiata Pollard, Proc. Biol. Soc. Wash. 15: 21. 1902. *Cassia brachiata* (Pollard) J. F. Macbride, Contr. Gray Herb. 59: 24. 1919. *Cassia fasciculata* Michaux var. *brachiata* (Pollard) Pullen ex Isely, Mem. New York Bot. Gard. 25(2): 87, 202. 1975. TYPE: FLORIDA: Miami-Dade Co.: Miami, 4–7 Apr 1898, *Pollard & Collins 245* (holotype: US; isotype: NY).

Chamaecrista littoralis Pollard, Proc. Biol. Soc. Wash. 15: 20. 1902. *Cassia fasciculata* Michaux var. *littoralis* (Pollard) J. F. Macbride, Contr. Gray Herb. 59: 25. 1919. *Cassia littoralis* (Pollard) Cory, Rhodora 38: 406. 1936.

Chamaecrista puberula Greene, Pittonia 5: 134. 1903. *Cassia fasciculata* Michaux var. *puberula* (Greene) J. F. Macbride, Contr. Gray Herb. 59: 25. 1919. *Cassia greenei* Standley, Publ. Field Mus. Nat. Hist., Bot. Ser. 11: 159. 1936.

Chamaecrista deeringiana Small & Pennell, in Pennell, Bull. Torrey Bot. Club 44: 345. 1917. *Cassia deeringiana* (Small & Pennell) J. F. Macbride, Contr. Gray Herb. 59: 24. 1919. TYPE: FLORIDA: Miami-Dade Co.: near Silver Palm, 22 Jun 1915, *Small et al. 6454* (holotype: NY).

Ascending to erect annual or perennial, to 30(10) cm; stem glabrate or puberulent. Leaves 3–7 cm long, the leaflets 5–18 pairs, the blade oblong, 0.5–1.5 cm long, 2–4 mm wide, the upper and lower surfaces strigulose or puberulent, the petiole with a sessile or stipitate gland; stipules lanceolate, striate. Flower axillary, 1–3, the pedicel 1–1.5 cm long; sepals 9–12 mm long; corolla yellow, red, or bicolored, 2.5–3.5 cm in diameter, the larger petals 15–20 mm long; functional stamens 10, the anthers red, yellow, or bicolored. Fruit oblong, 3–5.5(8) cm long, 3–5(8) mm wide, hirsute or glabrous.

Sandhills, open hammocks, flatwoods, dunes, and disturbed sites. Common; nearly throughout. Massachusetts and New York south to Florida, west to Minnesota, South Dakota, Nebraska, Kansas, and New Mexico; Europe. Native to North America. Spring–summer.

Chamaecrista lineata (Sw.) Greene var. keyensis (Pennell) H. S. Irwin & Barneby [Narrow, in reference to the fruits; from the Florida keys.] NARROWPOD SENSITIVE PEA; KEY CASSIA.

Chamaecrista keyensis Pennell, Bull. Torrey Bot. Club 44: 344. 1917. *Cassia keyensis* (Pennell) J. F. Macbride, Contr. Gray Herb. 59: 24. 1919. *Chamaecrista lineata* (Swartz) Greene var. *keyensis* (Pennell) H. S. Irwin & Barneby, Mem. New York Bot. Gard. 35: 756. 1982. TYPE: FLORIDA: Monroe Co.: Big Pine Key, 2 May 1917, *Pennell 9533* (holotype: NY?; isotype: US).

Prostrate or ascending perennial herb, to 8 dm; stem pilosulous. Leaves 1–3 cm long, the leaflets 4–7 pairs, the blade obovate-oblong, 8–12 mm long, 3–8 mm wide, the apex rounded, mucronate, the base rounded, the upper and lower surfaces pilosulous, the petiole with a sessile gland; stipules lanceolate, striate. Flower axillary, solitary, the pedicel 1–1.5 cm long; sepals 8–10 mm long; corolla yellow, ca. 2 cm in diameter, the larger petals 12–15 mm long; functional stamens 10, the anthers reddish or yellow and red. Fruit oblong, 2.5–4 cm long, 3.5–4.5 mm wide, pubescent.

Pinelands. Rare; Miami-Dade County and Monroe County keys. Endemic. Summer–spring.

Chamaecrista lineata var. *keyensis* is listed as endangered in Florida (Florida Administrative Code, Chapter 5B-40).

Chamaecrista nictitans (L.) Moench [*Nictare*, to wink, alluding to the sensitive reaction of the leaves to touch.] SENSITIVE PEA.

Spreading to erect annual herb, to 6(10) dm; stem pilose, puberulent, or glabrate. Leaves 2–5 cm long, leaflets 7–25 pairs, the blade oblong, 0.6–1.2 cm long, 2–3 mm wide, the apex mucronate, the base cuneate, the upper and lower surfaces glabrous or puberulent, the petiole with a stipitate gland; stipules lanceolate, striate. Flower axillary, 1–3, the pedicel 1–1.5 cm long; sepals 4–6 mm long; corolla yellow, 8–10 mm in diameter, the larger petals 4–7(8) mm long; functional stamens 5–9, reddish. Fruit oblong, 2–4(5) cm long, 4–5 mm wide, hirsute or glabrous.

1. Plant incurved-puberulent to glabrate...var. **nictitans**
1. Plant conspicuously pilose..var. **aspera**

Chamaecrista nictitans (L.) Moench var. **nictitans**

Cassia nictitans Linnaeus, Sp. Pl. 1: 380. 1753. *Chamaecrista nictitans* (Linnaeus) Moench, Methodus 272. 1794. *Nictitella amena* Rafinesque, Sylva Tellur. 128. 1838, nom. illegit. *Cassia chamaecrista* Linnaeus var. *nictitans* (Linnaeus) Kuntze, Revis. Gen. Pl. 1: 169. 1891.

Cassia procumbens Linnaeus, Sp. Pl. 380. 1753. *Chamaecrista procumbens* (Linnaeus) Greene, Pittonia 4: 28. 1899.

Cassia multipinnata Pollard, Bull. Torrey Bot. Club 22: 515, pl. 250. 1895. *Chamaecrista multipinnata* (Pollard) Greene, Pittonia 3: 243. 1897. *Cassia nictitans* Linnaeus var. *multipinnata* (Pollard) J. F. Macbride, Contr. Gray Herb. 59: 25. 1919. TYPE: FLORIDA: Duval Co.: near Jacksonville, 15 Sep & 27 Oct 1894, *Curtiss 5157* (lectotype: US). Lectotypified by Pennell (1917: 359).

Cassia multipinnata Pollard var. *nashii* Pollard, Bull. Torrey Bot. Club 22: 515. 1895. *Chamaecrista multipinnata* (Pollard) Greene var. *nashii* (Pollard) Greene ex Pollard, in Small, Fl. S.E. U.S. 588. 1903. TYPE: FLORIDA: Gadsden Co.: River Junction, 5 Sep 1895, *Nash 2577* (holotype: NY; isotypes: LE, US).

Cassia aspera Muhlenberg ex Elliott var. *mohrii* Pollard, Bull. Torrey Bot. Club 24: 150. 1897. *Chamaecrista aspera* (Muhlenberg ex Elliott) Greene var. *mohrii* (Pollard) Pollard ex A. Heller, Cat. N. Amer. Pl., ed. 2. 5. 1900. *Cassia nictitans* Linnaeus var. *mohrii* (Pollard) J. F. Macbride, Contr. Gray Herb. 59: 25. 1919. *Chamaecrista mohrii* (Pollard) Small ex Britton & Rose, in Britton, N. Amer. Fl. 23: 298. 1930.

Cassia nictitans Linnaeus var. *hebecarpa* Fernald, Rhodora 38: 423, pl. 448(1-3). 1936. *Chamaecrista nictitans* (Linnaeus) Moench var. *hebecarpa* (Fernald) C. F. Reed, Phytologia 63: 411. 1987.

Stem and leaves incurved-puberulent to glabrate, the petiolar gland cylindric to clavate. Flower with the functional stamens usually 5. Fruit incurved-puberulent to glabrate.

Sandhills, flatwoods, hammocks, and disturbed sites. Frequent; nearly throughout. Maine south to Florida, west to Wisconsin, Kansas, and Arizona; West Indies, Mexico, Central America, and South America; Asia and Pacific Islands. Native to North America and tropical America. Spring–fall.

Chamaecrista nictitans (L.) Moench var. **aspera** (Muhl. ex Elliott) H. S. Irwin & Barneby

Cassia aspera Muhlenberg ex Elliott, Sketch Bot. S. Carolina 1: 474. 1817. *Cassia nictitans* Linnaeus var. *aspera* (Muhlenberg ex Elliott) Torrey & A. Gray, Fl. N. Amer. 1: 396. 1840. *Chamaecrista aspera* (Muhlenberg ex Elliott) Greene, Pittonia 3: 243. 1897. *Chamaecrista nictitans* (Linnaeus) Moench var. *aspera* (Muhlenberg ex Elliott) H. S. Irwin & Barneby, Mem. New York Bot. Gard. 35: 838. 1982.

Cassia simpsonii Pollard, Bull. Torrey Bot. Club 21: 221. 1894. *Chamaecrista simpsonii* (Pollard) Pollard ex A. Heller, Cat. N. Amer. Pl., ed. 2. 5. 1900. *Cassia aspera* Muhlenberg ex Elliott var. *simpsonii* (Pollard) J. F. Macbride, Contr. Gray Herb. 59: 25. 1919. TYPE: FLORIDA: Monroe Co.: Big Pine Key, May 1891, *Simpson 174* (holotype: US).

Stem and leaves conspicuously pilose, the petiolar gland cylindric to clavate. Flowers with the functional stamens 5–8. Fruit pilose.

Sandhills, flatwoods, hammocks, and disturbed sites. Common; nearly throughout. South Carolina, Georgia, and Florida. Spring–fall.

Chamaecrista pilosa (L.) Greene [With short, weak, thin trichomes.] HAIRY SENSITIVE PEA.

Cassia pilosa Linnaeus, Syst. Nat., ed. 10. 1017. 1759. *Disterepta pilosa* (Linnaeus) Rafinesque, Sylva Tellur. 126. 1838. *Chamaecrista pilosa* (Linnaeus) Greene, Pittonia 4: 28. 1899.

Prostrate perennial herb; stem to 1 m, pilose or hispid. Leaves 1–5 cm long, the leaflets 2–5 pairs, the blade obovate-oblong or oblong-lanceolate, 0.5–2.5 cm long, 1–2 cm wide, the apex mucronate, the base unequally rounded, the margin ciliate, the upper and lower surfaces glabrous, the petiole with or without a small stipitate gland; stipules lanceolate, striate. Flower axillary, solitary, the pedicel 1.5–2 cm long; sepals ca. 4 mm long; corolla yellow, the petals 4–6 mm long; functional stamens 5, the anthers yellow or red. Fruit oblong, usually slightly falcate, 3–4.5 cm long, 3–4 mm wide, pubescent.

Disturbed sites. Rare; St Lucie, Highlands, Martin, and Palm Beach Counties. Florida; West Indies, Mexico, Central America, and South America. Native to West Indies and South America. All year.

Chamaecrista rotundifolia (Pers.) Greene [With round leaves.] ROUNDLEAF SENSITIVE PEA.

Cassia rotundifolia Persoon, Syn. Pl. 1: 456. 1805. *Chamaecrista rotundifolia* (Persoon) Greene, Pittonia 4: 31. 1899.

Prostrate annual or perennial herb; stem to 7 dm, pilose and/or puberulent. Leaves with 2 leaflets, the blade asymmetrically obovate, 0.5–2.5 cm long, 0.4–1.5 cm wide, the apex emarginate, mucronate or not, the base asymmetrical, cordate to cuneate, the upper surface glabrous, the lower surface sparsely pilose to glabrate, the petiole 2–8 mm long, eglandular; stipules lanceolate, striate. Flowers axillary, solitary, the pedicel 1–2 cm long; sepals 3–5 mm long; corolla yellow, 6–9 mm in diameter, the petals subequal, the longer 3–5 mm long; functional stamens 5, the anthers yellow. Fruit oblong, straight or slightly curved, 1.5–4 cm long, 3–5 cm wide, puberulent or glabrate.

Disturbed sites. Occasional; central peninsula. Florida; West Indies, Mexico, Central America, and South America; Africa. Native to tropical America. All year.

Chamaecrista serpens (L.) Greene [Creeping, snakelike.] SLENDER SENSITIVE PEA.

Cassia serpens Linnaeus, Syst. Nat., ed. 10. 1018. 1759. *Ophiocaulon serpens* (Linnaeus) Rafinesque, Sylva Tellur. 129. 1838. *Chamaecrista serpens* (Linnaeus) Greene, Pittonia 4: 29. 1899.

Prostrate annual herb; stem to 5 dm long, glabrous or puberulent. Leaves 1–2.5 cm long, the leaflets 4–6(7) pairs, the blade asymmetrically elliptic-obovate or -oblong, 3–6(8) mm long, 2–4 mm wide, the apex rounded, mucronate, the base cuneate, the upper surface glabrous, the lower surface puberulent or glabrous, the petiole ca. 2 mm long, with a stipitate, discoid gland, slightly pilose. Flower axillary, solitary, the pedicel 1–2 cm long; sepals 5–6 mm long; corolla yellow, 1–1.5 cm in diameter, the larger petals 5–7 mm long; functional stamens 10, the anthers yellow. Fruit oblong, 1.5–3(4) cm long, ca. 4 mm wide, sparsely pubescent or glabrate.

Disturbed sites. Rare; Lake, Polk, Hillsborough, and Highlands Counties. Florida, Arizona, and New Mexico; West Indies, Mexico, Central America, and South America. Native to the southwest United States, West Indies, Mexico, Central America, and South America. Summer–fall.

EXCLUDED TAXA

Chamaecrista chamaecristoides (Colladon) Greene—Reported for Florida by Small (1903, 1913a), the name of this Texas-Mexican species misapplied to material of *C. fasciculata*.

Chamaaecrista lineata var. *brachyloba* (Grisebach) H. S. Irwin & Barneby—This Cuban species was reported for Florida by Small (1903, 1913a, 1913d, all as *Chamaecrista grammica* (Sprengel) Pollard), the name misapplied to material of *C. lineata* var. *keyensis*.

Chapmannia Torr. & A. Gray 1838.

Perennial herbs. Leaves alternate, odd-pinnately compound, the leaflets alternate or subopposite, petiolate, stipulate. Flowers in spikes, bracteate, bracteolate; calyx 4–5-lobed; corolla papilionaceous; stamens 10, monadelphous, the anthers dorsifixed. Fruit a loment.

A genus of 7 species; North America, Mexico, Central America, South America, and Africa. [Commemorates Alvan Wentworth Chapman (1809–1899), American botanist.]

Selected reference: Thulin (1999).

Chapmannia floridana Torr. & A. Gray [Of Florida.] FLORIDA ALICIA.

Chapmannia floridana Torrey & A. Gray, Fl. N. Amer. 1: 355. 1838. TYPE: FLORIDA: East Florida, without data, *Leavenworth s.n.* (lectotype: NY). Lectotypified by Gunn et al. (1980: 179).

Erect perennial herb, to 1 m; stem appressed villous. Leaves 3–9-foliolate, the leaflets with the blade obovate to lanceolate, 5–20 mm long, 2–8 mm wide, the apex rounded to emarginate, apiculate, the base cuneate, the margin entire, the upper surface glabrous to sparsely pubescent, the lower surface villous, petiolules ca. 1 mm long, the petiole ca. 10 mm long; stipules subulate, 3–6 mm long. Flowers solitary or 2–4 in a spike subtended by reduced leaves and forming a paniculiform inflorescence; bracts and bracteoles lance-ovate, 1–2 mm long; calyx 5–8 mm long, glabrous or pubescent, the upper 2 lobes nearly connate, the lateral lobes triangular, ca. 2 mm long, the lower lobe narrowly triangular, ca. 4 mm long; corolla papilionaceous, yellow, 1–1.4 cm long, glabrous, the banner suborbicular, 10–14 mm long, the wings 12–18 mm long, the keel 8–12 mm long. Fruit subterete, 10–30 mm long, 3–4 mm wide, constricted between the seeds, the segments 1–4; seeds elliptic, 3–4 mm long, 2–3 mm wide, flattened, yellowish.

Sandhills and scrub. Frequent; peninsula. Endemic. Spring–summer.

Clitoria L. 1753. PIGEONWINGS

Perennial herbs or vines. Leaves alternate, odd-pinnately compound, petiolate, stipellate, stipulate. Flowers in axillary pseudoracemes, bracteate, bracteolate; calyx bilabiate; corolla papilionaceous, resupinate; stamens 10, diadelphous (9 + 1), the anthers dorsifixed. Fruit a dehiscent legume.

A genus of about 60 species; North America, West Indies, Mexico, Central America, South America, Africa, and Australia. [From the Greek *clitoris*, in reference to the closed flower resembling external human female genitalia.]

Martiusia Schult., 1822.

Selected reference: Fantz (1977).

1. Leaves 5- to 7-foliolate...**C. ternatea**
1. Leaves 3-foliolate.
   2. Leaflets of the upper leaves linear to linear-lanceolate; gynophore of the fruit conspicuously exserted from the persistent calyx..**C. fragrans**
   2. Leaflets of the upper leaves ovate to ovate-lanceolate; gynophore of the fruit included or only slightly exceeding the persistent calyx..**C. mariana**

Clitoria fragrans Small [Fragrant, in reference to the flowers.] SWEETSCENTED PIGEONWINGS.

Clitoria fragrans Small, Torreya 26: 57. 1926. *Martiusia fragrans* (Small) Small, Man. S.E. Fl. 722. 1933. TYPE: FLORIDA: Highlands Co.: near De Soto City, 20 May 1925, *Small s.n.* (holotype: NY).

Erect perennial herb, to 1.5 m; stem glabrous, usually glaucous. Leaves 3-foliolate, the leaflet blade linear, linear-lanceolate, to oblong-lanceolate, 2–4.5 cm long, 0.5–1.5 cm wide, the apex obtuse to retuse, mucronate, the base rounded, the upper surface pubescent, the lower surface glabrous, the petiolule ca. 2 mm long, the stipels linear to subulate, 1–3 mm long, petiole 1.5–3(3.5) cm long, glaucescent; stipules ovate to lance-ovate, 2–4 mm long. Flowers solitary or 2(4) in a pseudoraceme, with chasmogamous and/or cleistogamous flowers; bracts linear-lanceolate, 2–5 mm long; bracteoles linear-lanceolate, (3)4–5 mm long (cleistogamous flowers 2(3) mm). Chasmogamous flowers with the calyx tube 7–10 mm long, upper lobe 5–8 mm long, the lower lobes ovate, 5–6 mm long; corolla lilac, the banner 3.5–4.5 mm long, the wings 21–24 mm long, the keel 8–11 mm long. Cleistogamous flowers with the calyx tube 3–4 mm long, the lobes ca. 4 mm long. Fruit linear-elliptic 3–5.5 cm long, 6–8 mm wide (those from cleistogamous flowers 2–4 cm long), glabrous, glaucous, the stipe 15–21 mm long (those from cleistogamous fruits 9–14 mm long); seeds 2–5(8), cuboidal, ca. 4 mm long and wide, reddish brown.

Sandhills and scrub. Rare; Lake, Orange, Polk, and Highlands Counties. Endemic. Spring.

Clitoria fragrans is listed as endangered in Florida (Florida Administrative Code, Chapter 5B-40) and threatened in the United States (U.S. Fish and Wildlife Service, 50 CFR 23).

Clitoria mariana L. [Of Maryland.] ATLANTIC PIGEONWINGS.

Clitoria mariana Linnaeus, Sp. Pl. 2: 753. 1753. *Vexillaria mariana* (Linnaeus) Eaton, Man. Bot. 83. 1817. *Nauchea mariana* (Linnaeus) Descourtilz, Mém. Soc. Linn. Paris 4: 9. 1826. *Ternatea mariana* (Linnaeus) Kuntze, Revis. Gen. Pl. 1: 210. 1891. *Martiusia mariana* (L.) Small, Man. S.E. Fl. 722. 1933.

Clitoria mariana Linnaeus var. *pubescentia* Fantz, Sida 16: 727. 1995. TYPE: FLORIDA: Lake Co.: near edge of lake at Leesburg, 7 Jun 1967, *Baltzell 120* (holotype: FLAS).

Perennial vine; stem trailing, to 1.5 m, glabrous. Leaves 3-foliolate, the leaflet blade ovate to oblong or elliptic, 2–9(11.5) cm long, 1–4(6.5) cm wide, the apex acute to obtuse, sometimes mucronate, the base broadly cuneate to subcordate, the upper surface glabrate or uncinate-pubescent, glaucescent, the lower surface glabrate or sparsely pubescent on the major veins or moderately to densely pilose-sericeous, the petiolule 2–4 mm long, the stipels linear to subulate, 3–8 mm long, the petiole 2–10 cm long; stipules lanceolate to ovate-lanceolate, 4–8 mm long. Flowers 2(4) in a pseudoraceme, with chasmogamous and\or cleistogamous flowers; bracts lanceolate, 3–4 mm long; bracteoles lanceolate to ovate-lanceolate, 4–9 mm long (cleistogamous flowers 3–5 mm). Chasmogamous flowers with the calyx tube 10–14 mm long, the upper lobe 7–9 mm long, the lower lobes 5–9 mm long; corolla blue to pale purplish, the banner 4–6 mm long, the wings 7–11 mm long, the keel 8–13 mm long. Cleistogamous flowers with the calyx tube ca. 1 mm long, the lobes 2–3 mm long. Fruit linear-elliptic, 2.5–5.5(7) cm long, 6–9 mm wide, sparsely uncinate or glabrous, the stipe 12–17 mm long (those from cleistogamous flowers 5–10 mm); seeds cuboidal to globular, 3–5 mm long and wide, black.

Sandhills, flatwoods, scrub, and hammocks. Frequent; nearly throughout. New York south to Florida, west to Minnesota, Nebraska, Kansas, and Arizona; Mexico and Central America. Spring–summer.

Clitoria mariana is reported from Asia in various floras, but that material is best treated as *C. mariana* var. *orientalis* Fantz (see Fantz and Predeep, 1992).

Clitoria ternatea L. [A pre-Linnaean generic name, in threes, apparently in reference to the 3-foliolate leaves of some species.] ASIAN PIGEONWINGS.

Clitoria ternatea Linnaeus, Sp. Pl. 2: 753. 1753. *Clitoria spectabilis* Salisbury, Prodr. Stirp. Chap. Allerton 336. 1796, nom. illegit. *Ternatea vulgaris* Kunth, in Humboldt et al., Nov. Gen. Sp. 6: 415. 1824. *Nauchea ternatea* (Linnaeus) Descourtilz, Mém. Soc. Linn. Paris 4: 8. 1826.

Perennial vine; stem to 5 m, glabrous. Leaves 5- to 7-pinnate, the leaflet blade ovate, elliptic, or oblong, 1–5(7) cm long, (0.5)1–3 cm wide, the apex acute to obtuse or retuse, the base cuneate to rounded, the upper and lower surfaces glabrate, the petiolule 1–3 mm long, the stipels 1–3 mm long, the petiole 1–4 cm long; stipules linear, 4–10 mm long. Flowers solitary or paired in a pseudoraceme, chasmogamous; bracts ovate to lanceolate, 2–4 mm long; bracteoles broadly ovate to suborbicular, (4)6–11(15) mm long; calyx tube 8–14 mm long, the lobes 7–12 mm long; corolla pale blue to azure or blue-violaceous with a white to yellow medial strip, or white with a greenish white medial strip, the banner 4–5.5 cm long, the wings 6–10 mm long, the keel 7–10 mm long. Fruit linear, (5)7–11 cm long, 8–11 mm wide, sparsely strigose and uncinate pubescent to glabrate, subsessile; seeds subreniform, 4–6 mm long, 5–6 mm wide, brown or black.

Disturbed sites. Rare; Miami-Dade County and Monroe County keys. Escaped from cultivation. Georgia, Florida, Texas, and California; West Indies, Mexico, Central America, and South America; Africa, Asia, Australia, and Pacific Islands. Native to tropical America.

Coronilla L., nom. cons. 1753. CROWNVETCH

Perennial herbs. Leaves alternate, odd-pinnately compound, petiolate, estipellate, stipulate. Flowers in axillary, pedunculate heads, bracteate, ebracteolate; calyx slightly zygomorphic; corolla papilionaceous; stamens 10, diadelphous (9 + 1), the anthers basifixed. Fruit a loment.

A genus of about 22 species; North America, Africa, Asia, and Pacific Islands. [*Corona*, crown, and *illa*, diminutive, in refrence to the inflorescence.]

Securigea DC., 1805.

Coronilla varia L. [Variable, in reference to the varying flower color.] PURPLE CROWNVETCH.

Coronilla varia Linnaeus, Sp. Pl. 2: 743. 1753. *Ornithopus varius* (Linnaeus) Hornemann, Hort. Bot. Hafn. 697. 1815. *Securigera varia* (Linnaeus) Lassen, Svensk. Bot. Tidskr. 83: 86. 1989.

Sprawling or ascending perennial herb. Stem to 1.2 m long, rhizomatous, furrowed or ridged. Glabrate or sparsely villous. Leaves odd-pinnately compound, 5–15 cm long, the leaflets 11–25, the blade oblong or elliptic, 6–25 mm long, 2.5–12 mm wide, the apex rounded, mucronate, the base cuneate, the margin entire, narrowly scarious, the upper and lower surfaces glabrate or sparsely villous; stipules linear to oblong, 1–6 mm long, dark-tipped. Flowers 6–25 in a head, the peduncles 4–14 cm long, the pedicels 2–5 mm long; bract dark tipped; calyx campanulate, 2–4 mm long, the teeth ca. 1 mm long; corolla (8)10–15 mm long, white, pink, purple, or bicolored. Fruit 2–6(8) cm long, ca. 2 mm wide, the segments 2–10(12), oblong, 4–6 mm long, slightly constricted between the seeds, 4-angled in cross-section, the surface with weakly anastomosing longitudinal veins; seeds 2–12, oblong, 3–4 mm long, smooth.

Disturbed sites. Rare; Leon and Lee Counties. Escaped from cultivation. Nearly throughout North America; Europe, Africa, Asia, and Pacific Islands. Native to Europe, Africa, and Asia. Spring–summer.

Based on morphology, *Securigera* is sometimes considered a separate genus with *C. varia* placed in it (see Lassen, 1989). In our opinion, conclusive evidence is lacking and we here recognize *Coronilla* in the broad sense.

Crotalaria L., nom. cons. 1753. RATTLEBOX

Annual or perennial herbs. Leaves alternate, 3-foliolate or 1-foliolate, estipellate, petiolate, stipulate. Flowers terminal or axillary racemes, bracteate, bracteolate or ebracteolate; calyx lobes unequal or subequal; corolla papilionaceous; stamens monodelphous. Fruit an inflated or turgid, dehiscent legume; seeds numerous, arillate.

A genus of about 600 species; nearly cosmopolitan. [From the Greek *krotalon*, castanet, rattle, in reference to the ripe seeds rattling in the dry pods.]

Selected references: Polhill (1982); Windler (1974).

1. Leaves 3-foliolate.
    2. Fruit pilose ....**C. incana**
    2. Fruit strigose or pubescent.
        3. Fruit 0.8–1.5 cm long.
            4. Fruit puberulent; lower leaf surface with evident, loose trichomes ....**C. virgulata**
            4. Fruit short-pubescent; lower leaf surface with inconspicuous, short, appressed trichomes ...
            ....**C. pumila**

3. Fruit 2–5 cm long.
    5. Leaves linear or linear-lanceolate, the apex acute.
        6. Flowers 1.8–2 cm long; fruit ca. 15 mm in diameter, not upcurved at the tip ....
        ....**C. ochroleuca**
        6. Flowers 0.8–1 cm long; fruit 4–6 mm in diameter, upcurved at the tip .... **C. lanceolata**
    5. Leaves elliptic, ovate, or oblanceolate, the apex rounded.
        7. Calyx strigulose; bracts caducous; fruit 5–6 mm in diameter ....**C. pallida**
        7. Calyx glabrous or only slightly puberulent; bracts persistent; fruit 8–12 mm in diameter
        ....**C. trichotoma**
1. Leaves 1-foliolate.
    8. Flowers blue and white...**C. verrucosa**
    8. Flowers yellow.
        9. Corolla 1.7–2.5(3) cm long.
            10. Fruit pubescent ....**C. juncea**
            10. Fruit glabrous.
                11. Bracts 5–8 mm long, persistent; calyx glabrous or glabrate.........................**C. spectabilis**
                11. Bracts 2–3 mm long, caducous; calyx strigose...**C. retusa**
        9. Corolla 0.7–1.4 cm long.
            12. Leaves glabrous on the upper surface ....**C. purshii**
            12. Leaves strigose on the upper surface.
                13. Stipules absent; keel beak short, only slightly twisted distally, curved back toward the
                standard.. **C. avonensis**
                13. Stipules persistent; keel beak elongate, spirally twisted distally, projected upward and
                outward.
                    14. Style base geniculate..**C. rotundifolia**
                    14. Style base smoothly curved.. **C. alata**

Crotalaria alata Buch.-Ham. ex D. Don [Winged, in reference to the decurrent, winged stipules.] WINGED RATTLEBOX.

Crotalaria alata Buchanan-Hamilton ex D. Don, Prodr. Fl. Nepal. 241. 1825.

Erect or spreading annual or suffruticose perennial herb, to 1 m; stem zigzag, pubescent. Leaves 1-foliolate, the leaflets blade oblong-lanceolate to elliptic-oblong, 3–9 cm long, 1–5 cm wide, the apex obtuse, mucronate, the base cuneate, the upper and lower surfaces pubescent, the petiole 1–4 mm long; stipules triangular, ca. 5 mm long, with narrow, decurrent wings ca. 2–5 mm wide. Flowers 2–12 in a terminal or axillary raceme, the pedicel 3–5 mm long, the peduncle 5–15 cm long; bracts ovate-lanceolate, 3–8 mm long, subcordate on a short stalk; bracteoles arising from the calyx tube, similar to the bracts; calyx 6–10 mm long, the lobes lanceolate, pubescent; corolla yellow, the standard obovate-orbicular, 5–8(10) mm long, the wings oblong, shorter than the standard, the keel beak elongate, spirally twisted, projecting upward and outward. Fruit oblong, 3–4 cm long, sparsely pubescent or glabrous.

Disturbed sites. Rare; Alachua County. Escaped from cultivation. Florida; Africa, Asia, and Australia. Native to Africa, Asia, and Australia. Summer–fall.

Crotalaria alata is known in Florida only from a 1939 collection.

Crotalaria avonensis DeLaney & Wunderlin [In reference to Avon Park.] AVON PARK RATTLEBOX; AVON PARK HAREBELLS.

> *Crotalaria avonensis* DeLaney & Wunderlin, Sida 13: 315. 1989. TYPE: FLORIDA: Highlands Co.: E of Grassy Pond in the "Big Scrub," ca. 4 mi. SE of Avon Park, T33S, R29E, Sect. 34, S ½, 25 Apr 1988, *Delaney 1623* (holotype: USF; isotypes: USF).

Erect perennial herb, to 2 dm; stem loosely sericeous with white or yellowish white trichomes. Leaves 1-foliolate, the leaflet blade broadly elliptic to orbicular, (5)8–19 mm long, (4)7–16 mm wide, the apex obtuse to emarginate, apiculate, the base rounded to broadly cuneate, the upper and lower surfaces sericeous, the petiole 1–3 mm long, reddish brown or green; stipules absent. Flowers in a terminal raceme, the peduncles (1)10–17(33) mm long, the pedicels 2–3 mm long; bracts linear-elliptic, 2–3(4) mm long, short petiolate; bracteoles linear-elliptic, ca. 2 mm long; calyx 7–8 mm long, the tube 2–3 mm long, with ascending, loosely appressed white or yellowish white trichomes; corolla 8–9(10) mm long, yellow, the standard with brown-red lines, the keel ca. 4 mm long, smoothly incurved; style smoothly incurved below the middle. Fruit oblong, 1.5–2.5 cm long, 6–8 mm wide, glabrous or glabrate distally along the dorsal suture.

Scrub. Rare; Polk and Highlands Counties. Endemic. Spring–summer.

Crotalaria avonensis is listed as endangered in Florida (Florida Administrative Code, Chapter 5B-40) and endangered in the United States (U.S. Fish and Wildlife Service, 50 CFR 23).

Crotalaria incana L. [Hoary, white.] SHAKESHAKE.

> *Crotalaria incana* Linnaeus, Sp. Pl. 2: 716. 1753. *Crotalaria pubescens* Moench, Methodus 161. 1794, nom. illegit.

Erect or sprawling annual or perennial herb, to 2 m; stem hirsutulous or puberulous. Leaves 3-foliolate, the leaflet blade ovate, elliptic, or obovate, 1.5–5(6) cm long, 1–3 cm wide, the apex rounded, mucronate, the base broadly cuneate to rounded, the upper surface glabrate, the lower surface glabrate, sometimes puberulent along the veins; stipules filamentous, caducous. Flowers axillary or in an exserted raceme 4–20 cm long; bracts and bracteoles small; calyx 7–10 mm long, glabrous; corolla 10–13 mm long, yellow. Fruit short cylindric, 14–24 cm long, 1–1.5 cm wide, pilose.

Disturbed sites. Occasional; Alachua County, central and southern peninsula. South Carolina, Georgia, Florida, Alabama, Oklahoma, and Texas; West Indies, Mexico, Central America, and South America; Africa, Asia, Australia, and Pacific Islands. Native to Africa. Summer–fall.

Crotalaria juncea L. [Rushlike.] SUNN HEMP.

> *Crotalaria juncea* Linnaeus, Sp. Pl. 2: 714. 1753.

Erect annual herb, to 2.5 m; stem subappressed-pubescent. Leaves 1-foliolate, the leaflet blade elliptic to oblong, (2.5)3–10(13) cm long, 0.5–1.5(2.5) cm wide, the apex obtuse to subacute, mucronate, the base cuneate, the upper surface sparsely pubescent, the lower surface subsericeous; stipules filiform, ca. 2 mm long, caducous. Flowers in a terminal raceme to 25 cm long; bracts lanceolate-oblong, 3–5 mm long; bracteoles linear, 2–5 mm long, inserted at the base of the calyx; the pedicel ca. 5 mm long; calyx 1.5–2 cm long, the lobes lanceolate, 3–4 times the length of the tube, pubescent; corolla 1.5–2 cm long, yellow with dark reddish or brown stripes. Fruit oblong-cylindric, 2.5–3.5(5.5) cm long, 1–1.7 cm wide, tomentose, light brown.

Disturbed sites. Rare; Putnam and Miami-Dade Counties. Florida; West Indies and South America; Europe, Africa, Asia, Australia, and Pacific Islands. Native to Asia. Summer–fall.

Crotalaria lanceolata E. Mey. [Lance-shaped, in reference to the leaflets.] LANCELEAF RATTLEBOX.

Crotalaria lanceolata E. Meyer, in E. Meyer & Drège, Comm. Pl. Afr. Austr. 24. 1836.

Erect annual or short-lived perennial herb, to 1 m; stem strigulose to glabrescent. Leaves 3-foliolate, the leaflet blade lanceolate to linear, 4–10 cm long, 0.5–1.5 cm wide, the apex acuminate, mucronate, the base cuneate, the upper surface sparsely pilose or glabrous, the lower surface strigulose, the petiole 3–6 cm long; stipules obsolescent. Flowers in an exserted raceme 10–35 cm long, the pedicel 3–5 mm long; bracts subulate, 1–3 mm long; bracteoles ca. 1 mm long, at the calyx base; calyx 3–3.5 mm long, the lobes shorter than the tube, strigulose to glabrate; corolla 8–10 mm long, yellow. Fruit narrowly cylindric, 2–4 cm long, 4–6 mm wide, apically upcurved, strigose.

Disturbed sites. Frequent; nearly throughout. North Carolina south to Florida, west to Louisiana; South America; Africa, Asia, Australia, and Pacific Islands. Native to Africa, Asia, and Australia. Summer–fall.

Crotalaria ochroleuca G. Don [Pale yellow, in reference to the flower color.] SLENDERLEAF RATTLEBOX.

Crotalaria ochroleuca G. Don, Gen. Hist. 2: 138. 1832.

Erect annual or short-lived perennial herb, to 1.5 m; stem strigose. Leaves 3-foliolate, the leaflet blade linear to linear-lanceolate or elliptic-lanceolate, 5–15 cm long, 2–12 mm wide, the apex acute, mucronate, the base cuneate, the upper surface glabrous, the lower surface strigose, the petiole 3–5 cm long; stipules obsolescent. Flowers in an exserted raceme 1.5–3.5 dm long; bracts linear-caudate or subulate, 2–4 mm long, expanded at the base, semipersistent; bracteoles linear, 1–2 mm long, inserted at the base of the calyx; calyx 6–8 mm long, glabrous, contrasting with the strigulose pedicel; corolla 1.8–2 cm long, yellow, with reddish or maroon veins. Fruit oblong-cylindric, (4)5–7 cm long, (1)1.5–2 cm wide, strigulose.

Disturbed sites. Occasional; northern counties, central peninsula. North Carolina south to Florida, west to Louisiana; South America; Africa, Asia, and Australia. Native to Africa. Summer–fall.

Crotalaria pallida Aiton var. **obovata** (G. Don) Polhill [Pale, in reference to the flowers; obovate, in reference to the leaflets.] SMOOTH RATTLEBOX.

Crotalaria obovata G. Don, Gen. Hist. 2: 138. 1823. *Crotalaria pallida* Aiton var. *obovata* (G. Don) Polhill, Kew Bull. 22: 265. 1968.

Erect annual herb, to 2 m; stem strigulose. Leaves 3-foliolate, the leaflet blade elliptic obovate to obovate, 2.5(7) cm long, 3–4 cm wide, the apex rounded or retuse, mucronate, the base cuneate, the upper surface glabrous, the lower surface strigulose, the petiole about as long as the leaflets; stipules filiform, ca. 3 mm long, caducous. Flowers in a short or well-exserted raceme 10–25 cm long; bracts linear, to 5 mm long, caducous; bracteoles linear, to 4 mm long; calyx 6–8 mm long, the lobes subequal or longer than the tube, strigose; corolla 11–15 mm long, yellow, usually with reddish veins. Fruit narrowly oblong-oblanceolate, 3–4 cm long, 5–6 mm wide, minutely puberulous.

Disturbed sites. Frequent; nearly throughout. North Carolina south to Florida, west to Mississippi; West Indies, Mexico, Central America, and South America; Africa, Asia, Australia, and Pacific Islands. Native to Africa. All year.

Crotalaria pumila Ortega [Dwarf.] LOW RATTLEBOX.

Crotalaria pumila Ortega, Nov. Pl. Desc. Dec. 2: 23. 1797.

Decumbent or ascending annual or suffruticose perennial herb; stems to 1 m, strigulose. Leaves 3-foliolate, the leaflet blade obovate to elliptic-oblong, 0.7–1.2(3) cm long, 3–8(10) mm wide, the apex rounded or emarginate, mucronate, the base cuneate, the upper surface glabrous, the lower surface strigulose, the petiole 0.5–1 cm long; stipules bristlelike, caducous. Flowers 4–8(10) in an ascending axillary or rarely terminal raceme 1–6(10) cm long; bracts and bracteoles minute, caducous; calyx 3–4.5 mm long, strigose; corolla 7–9 mm long, yellow. Fruit ellipsoid, 1–1.5 cm long, ca. 4 mm wide, strigulose or puberulent.

Hammocks and coastal dunes. Frequent; Alachua County, central and southern peninsula. Maryland, Florida, Oklahoma and Texas west to Utah and Arizona; West Indies, Mexico, Central America, and South America; Pacific Islands. Native to North America and tropical America. All year.

Crotalaria purshii DC. [Frederick Traugott Pursh (1774–1820), German-American botanist.] PURSH'S RATTLEBOX.

Crotalaria laevigata Pursh, Fl. Amer. Sept. 369. 1814; non Lamarck, 1786. *Crotalaria purshii* de Candolle, Prodr. 2: 124. 1825. *Crotalaria cuneifolia* Rafinesque, New. Fl. 2: 55. 1837 ("1836"), nom. illegit; non (Forsskål) Schrank, 1828.

Erect or ascending perennial herb, to 8 dm; stem strigose. Leaves 1-foliolate, the leaflet blade elliptic or spatulate to obovate-oblong, 2–5 cm long, 0.5–10 mm wide, the medial and upper ones oblong-lanceolate to linear or filiform, 3–8(10) cm long, 2–5 mm wide, the apex acute, mucronate, the base cuneate, the upper surface glabrous, the lower surface strigulose, the petiole short or obsolete; stipules of the medial and upper leaves conspicuous, tapering, decurrent with ascending free lobes. Flowers 2–5 in an exserted raceme 6–25 cm long; calyx 9–12 mm long, strigose; corolla 8–12 mm long, yellow. Fruit ellipsoid, 1.5–2.5(3) cm long, 8–10 mm wide, glabrous.

Flatwoods and sandhills. Occasional; northern counties south to Lake County. Virginia south to Florida, west to Texas. Spring–summer.

Crotalaria retusa L. [With rounded, shallowly notched end, in reference to the leaflets.] RATTLEWEED.

Crotalaria retusa Linnaeus, Sp. Pl. 2: 715. 1753. *Crotalaria retusifolia* Stokes, Bot. Mat. Med. 3: 516. 1812, nom. illegit.

Crotalaria alatipes Rafinesque, New Fl. 2: 57. 1837 ("1836"). TYPE: FLORIDA.

Erect annual herb, to 8 dm; stem strigose. Leaves 1-foliolate, the leaflet blade obovate, spatulate to oblanceolate, 3.5–11 cm long, 1.5–4 cm wide, the apex rounded, emarginate, or obtuse, the base cuneate, the upper and lower surfaces strigose, the petiole 2–4 mm long; stipules linear to subulate, 1–5 mm long. Flowers 5–20 in a terminal raceme 1–30 cm long; bracts subulate-caudate to lanceolate-caudate, 2–3 mm long; bracteoles filiform, 1–2 mm long, on the pedicel; calyx 12–15 mm long, strigose; corolla 2–3 cm long, yellow, the standard sometimes finely purple-veined. Fruit ovoid-oblong or cylindric, 2.5–4 cm long, 1–1.5 cm wide, glabrous.

Disturbed sites. Occasional; Alachua and Leon Counties, central and southern peninsula. New Jersey, Kentucky, North Carolina south to Florida, west to Texas; West Indies, Mexico, Central America, and South America; Africa, Asia, Australia, and Pacific Islands. Native to Africa, Asia, and Australia. Summer–fall.

Crotalaria rotundifolia J. F. Gmel. [With round leaves.] RABBITBELLS.

Crotalaria rotundifolia J. F. Gmelin, Syst. Nat. 2: 1095. 1792. *Crotalaria rotundifolia* J. F. Gmelin var. *vulgaris* Windler, Phytologia 21: 264. 1971, nom. inadmiss.

Crotalaria sagittalis Linnaeus var. *ovalis* Michaux, Fl. Bor.-Amer. 2: 55. 1803. *Crotalaria ovalis* (Michaux) Pursh, Fl. Amer. Sept. 469. 1814.

Crotalaria maritima Chapman, Fl. South. U.S., ed. 2. 614. 1883. TYPE: FLORIDA: Monroe Co.: Middle Cape, Cape Sable, Everglades National Park, 18 Apr 1964, *Ward 3939* (neotype: FLAS). Neotypified by Ward (2009: 222).

Crotalaria linaria Small, Man. S.E. Fl. 679, 1505. 1933. *Crotalaria maritima* Chapman var. *linaria* (Small) H. Senn, Rhodora 41: 347. 1939. *Crotalaria rotundifolia* J. F. Gmel var. *linaria* (Small) Fernald & B. G. Schubert, Rhodora 50: 203. 1948. TYPE: FLORIDA: Monroe Co.: Big Pine Key, s.d., *Small & Mosier 6034* (holotype: NY?).

Prostrate, decumbent, or ascending perennial herb; stem to 4 dm long, pilose or strigose. Leaves 1-foliolate, the leaflet blade ovate to elliptic-oblong to linear, 1–5 cm long, 2–15 mm wide, the apex acute to rounded, mucronate, the base cuneate to rounded, the upper and lower surfaces strigose; stipules linear, ca. 5 mm long, with narrow decurrent wings. Flowers 2–5 in an erect or ascending raceme 4–20 cm long; bracts and bracteoles 2–4 mm long; calyx 8–12 mm long, villosulous or strigose; corolla 8–13 mm long, yellow. Fruit ellipsoid, (1)1.5–2.5(3) cm long, 7–12 mm wide, glabrous.

Flatwoods, sandhills, and dry, disturbed sites. Common; nearly throughout. Maryland south to Florida, west to Arkansas and Louisiana; West Indies, Mexico, Central America, and South America. All year.

Crotalaria spectabilis Roth [Spectacular, in reference to the showy flowers.] SHOWY RATTLEBOX.

Crotalaria spectabilis Roth, Nov. Pl. Sp. 341. 1821.

Crotalaria sericea Retzius, Observ. Bot. 5: 26. 1788; non Burman f. *Crotalaria retzii* Hitchcock, Rep. (Annual) Missouri Bot. Gard. 4: 74. 1893.

Erect annual herb, to 1.5 m; stem strigulose. Leaves 1-foliolate, the leaflet blade obovate to elliptic, 5–15 cm long, 3–8 cm wide, the apex rounded, mucronate, the base cuneate, the upper surface glabrous, the lower surface strigulose, the petiole 2–8 mm long; stipules obliquely oblong-ovate, 3–7 mm long. Flowers in a terminal raceme 1–5 cm long; bracts ovate-cordate, 5–8 mm long; bracteoles lanceolate, 1–2 mm long; calyx 1–1.5 cm long, glabrous; corolla 1.7–2.5 cm long, yellow, the standard lined with reddish purple. Fruit ovoid or ellipsoid, 3–4.5 cm long, 1–2 cm wide, glabrous.

Disturbed sites. Frequent; nearly throughout. Virginia south to Florida, west to Missouri, Oklahoma, and Texas; West Indies, Mexico, Central America, and South America; Africa, Asia, Australia, and Pacific Islands. Native to Asia. All year.

Crotalaria trichotoma Bojer [Three branched.] WEST INDIAN RATTLEBOX.

Crotalaria trichotoma Bojer, Ann. Sci. Nat., Bot., ser. 2. 4: 265. 1835.
Crotalaria zanzibarica Bentham, London J. Bot. 2: 584. 1843.
Crotalaria usaramoensis Baker f., J. Linn. Soc., Bot. 42: 346. 1914.

Erect annual or short-lived perennial herb, to 1.5 m; stem strigose. Leaves 3-foliolate, the leaflets with the blade elliptic to elliptic-oblong or lanceolate, 4–10 cm long, 1–4 cm wide, the apex obtuse to retuse, mucronate, the base broadly cuneate, the upper surface glabrous or rarely strigose, the lower surface strigose, the petiole shorter than the blade; stipules obsolescent. Flowers in a terminal raceme 1–3.5 cm long; bracts linear-caudate, 2–4 mm long; bracteoles slightly smaller than the bracts, usually at the calyx base; calyx 4–6 mm long, glabrous or slightly puberulent; corolla 13–15 mm long, yellow, the standard reddish purple tinged. Fruit oblong-cylindric, 3–4.5 cm long, 8–12 mm wide, minutely strigulose or glabrate.

Disturbed sites. Rare; Miami-Dade County. Florida; West Indies, Central America, and South America; Africa, Asia, Australia, and Pacific Islands. Native to Africa. All year.

Crotalaria verrucosa L. [Warty.] BLUE RATTLEBOX.

Crotalaria verrucosa Linnaeus, Sp. Pl. 2: 715. 1753. *Crotalaria angulosa* Lamarck, Encycl. 2: 197. 1786, nom. illegit. *Crotalaria flexuosa* Moench, Suppl. Meth. 55. 1802, nom. illegit. *Anisanthera versicolor* Rafinesque, Fl. Tellur. 2: 60. 1837 ("1836"), nom. illegit. *Crotalaria verrucosa* Linnaeus var. *genuina* Stehle, Bull. Mus. Hist. Nat. (Paris), ser. 2. 18: 101. 1946, nom. inadmiss.

Erect annual, to 1 m; stem pubescent. Leaves 1-foliolate, the leaflet blade ovate to elliptic, 5–13 cm long, 3–8 cm wide, the apex acute, mucronate, the base broadly cuneate, the upper and lower surfaces thinly pubescent, the petiole 3–9 mm long; stipules ovate-falcate, 7–21 mm long. Flowers 12–24 in a terminal raceme; bracts lanceolate-caudate, 1–5 mm long; bracteoles filiform, 1–2 mm long; calyx 7–11 mm long, thinly strigose; corolla 8–12 mm long, blue or purple. Fruit oblong-clavate, 3–5 cm long, 0.7–1.2 cm wide, strigose or pilose.

Disturbed sites. Rare; Martin and Miami-Dade Counties. Escaped from cultivation. Florida; West Indies, Mexico, Central America, and South America; Africa, Asia, Australia, and Pacific Islands. Native to Asia. All year.

Crotalaria virgulata Klotzsch subsp. **grantiana** (Harv.) Polhill [Long and slender; commemorates W. B. Grant, collector of the type specimen in Port Natal, South Africa.] GRANT'S RATTLEBOX.

Crotalaria grantiana Harvey, in Harvey & Sonder, Fl. Cap. 2: 43. 1862. *Crotalaria virgulata* Klorzsch subsp. *grantiana* (Harvey) Polhill, Crotalaria Afr. Madag. 293. 1982.

Decumbent to erect annual or short-lived perennial, to 1.8 m; stem subappressed to spreading hirsute. Leaves 3-foliolate, the leaflet blade oblanceolate to elliptic-obovate, 5–20 mm long, 2–8 mm wide, the apex rounded, mucronate, the base cuneate, the upper surface glabrous or pubescent, the lower surface glabrous, the petiole (2)4–12 mm long; stipules subulate to attenuate-triangular, 1–3 mm long. Flowers usually 3–8 in a primary leaf-opposed raceme, 2–6 cm long, the secondary inflorescences 1- to 2-flowered, on a filiform axis 1–2 cm long from near the base of short shoots and appearing axillary; bracts linear-subulate, 1–2(3) mm long; bracteoles setaceous; calyx 3.5–9 mm long, pubescent; corolla 6–12(14) mm long, yellow, the standard often lined with red. Fruit oblong-obovate, 8–15 mm long, 5–9 mm wide, pubescent.

Disturbed sites. Rare; Highlands County. Escaped from cultivation. Florida; South America; Africa and Australia. Native to Africa. Summer.

EXCLUDED TAXA

Crotalaria angulata Miller—Reported for Florida by Radford et al. (1964, 1968) and Long and Lakela (1971), the name misapplied to material of *C. rotundifolia*.

Crotalaria brevidens Bentham—Reported for Florida by Clewell (1985), who misapplied the name to material of *C. ochroleuca*.

Crotalaria brevidens Bentham var. *intermedia* (Kotschy) Polhill—Reported for Florida by Radford et al. (1964, 1968, both as *C. intermedia* Kotschy), Wunderlin (1982, as *C. intermedia* Kotschy), and Wilhelm (1984), the name misapplied to material of *C. ochroleuca*.

Crotalaria pallida Aiton—Reported for Florida by Small (1933, as *C. striata* de Candolle), Radford et al. (1964, 1968, both as *C. mucronata* Desvaux), Long and Lakela (1971, as *C. mucronata* Desvaux), and Wunderlin (1982, as *C. mucronata* Desvaux). Because infraspecific categories were not recognized, the typical variety was reported by implication for Florida by Wilhelm (1984) and Clewell (1985). All Florida material is of *C. pallida* var. *obovata*.

Crotalaria sagittalis Linnaeus—Reported for Florida by Chapman (1860), Small (1903, 1913a, 1933), Radford et al. (1964, 1968), Correll and Johnston (1970), and Wilhelm (1984). Excluded from Florida by Isely (1990, 1998).

Dalbergia L. f., nom. cons. 1782. INDIAN ROSEWOOD

Trees or shrubs. Leaves alternate, odd-pinnately compound or 1-foliolate, estipellate, petiolate, stipulate. Flowers in axillary racemose paniculiform or corymbiform inflorescences, bracteate, bracteolate; calyx zygomorphic, 5-lobed; corolla papilionaceous; stamens 9 or 10, monadelphous or diadelphous (9 + 1), the anthers basifixed. Fruit an indehiscent legume, laterally compressed.

A genus of about 150 species; North America, West Indies, Mexico, Central America, South America, Africa, Asia, and Australia. [Commemorates the Swedish brothers Nils Ericsson Dahlberg (1736–1820) and Carl Gustav Dahlberg (1721–1781), the former a student of Linnaeus and royal Swedish physician, and the latter who owned an estate in Surinam from where he sent specimens to Linnaeus.]

Amerimnon P. Browne, nom. rej. 1756; *Ecastaphyllum* P. Browne, nom. rej. 1756.

1. Tree; leaves 3- to 5(7)-foliolate ....**D. sissoo**
1. Scandent or trailing shrub; leaves 1-foliolate.
  2. Fruit samaroid, suborbicular, ovate, or reniform; leaflets coriaceous.................... **D. ecastaphyllum**
  2. Fruit flattened but not samaroid, oval or oblong; leaflets chartaceous.............................**D. brownei**

Dalbergia brownei (Jacq.) Schinz [Commemorates Patrick Browne (1720–1790), Irish physican and botanist.] BROWNE'S INDIAN ROSEWOOD.

Amerimnon brownei Jacquin, Enum. Syst. Pl. 27. 1760. *Dalbergia amerimnum* Bentham, J. Proc. Linn. Soc., Bot. 4(Suppl.): 36. 1860, nom. illegit. *Dalbergia brownei* (Jacquin) Schinz, Bull. Herb. Boissier 6: 731. 1898.

Shrubs or trees, to 10 m, sometimes somewhat scandent. Leaves 1-foliolate, the leaflet blade ovate to ovate-elliptic, 4–8 cm long, 2–4 cm wide, the apex acute, the base cordate to rounded, the upper and lower surfaces glabrous or slightly appressed-pubescent, lustrous, the petiolule 2–3 mm long, the petiole 0.5–1.5 cm long; stipules deltoid-ovoid, 1–2 mm long, caducous. Flowers in a corymbose-paniculiform inflorescence; bracts deltoid, 1–2 mm long, bracteoles elliptic-ovate, ca. 1 mm long; calyx ca. 4 mm long, the lobes subequal, the lower lobe ca. 1 mm longer than the 2 lateral lobes, the 2 upper lobes connate nearly to the apex and longer than the lateral lobes; corolla 8–10 mm long, white or pinkish; stamens 10, monadelphous. Fruit elliptic to oblong, 1.5–5 cm long, ca. 1 cm wide, glabrous, lustrous; seeds 1–4.

Tropical hammocks and mangrove margins. Rare; Miami-Dade County and Monroe County keys. Florida; West Indies, Central America, and South America. Spring–summer.

Dalbergia brownei is listed as endangered in Florida (Florida Administrative Code, Chapter 5B-40).

Dalbergia ecastaphyllum (L.) Taub. [From the Greek *ek*, out, *aster*, star, and *phyllum*, leaf, in reference to the unifoliolate leaves.] COINVINE.

Hedysarum ecastaphyllum Linnaeus, Syst. Nat., ed. 10. 1169. 1759. *Pterocarpus ecastaphyllum* (Linnaeus) Bergius, Kongl. Vetensk. Akad. Handl. 30: 116. 1769. *Ecastaphyllum brownei* Persoon, Syn. Pl. 2: 277. 1807. *Dalbergia ecastaphyllum* (Linnaeus) Taubert, in Engler & Prantl, Nat. Pflanzenfam. 3(3): 335. 1894. *Ecastaphyllum ecastaphyllum* (Linnaeus) Britton, Mem. Brooklyn Bot. Gard. 1: 55. 1918, nom. inadmiss. *Amerimnon ecastaphyllum* (Linnaeus) Standley, J. Wash. Acad. Sci. 15: 459. 1925. *Dalbergia brownei* (Persoon) Kuntze ex Small, Fl. S.E. U.S. 620. 1903, nom. illegit.

Ecastaphyllum brownei Persoon var. *psilocalyx* Radlkofer ex Koepff, Anat. Charakt. Dalberg. 40. 1892.

Shrubs or trees, to 4(6) m, sometimes scandent. Leaves 1-foliolate, the leaflet blade elliptic to ovate, (2.5)5–8 cm long and wide, the apex abruptly narrowed to an obtuse tip, the base rounded to subcordate, the upper and lower surfaces subsericeous when young, moderately appressed-pubescent or glabrous at maturity, the petiolules 3–5 mm long, the petiole 4–6 mm long; stipules lanceolate, ca. 10 mm long, caducous. Flowers in a racemose fascicle; bracts deltoid, ca. 1 mm long; bracteoles deltoid-ovoid, ca. 1 mm long; calyx 3–4 mm long, the lobes subequal; corolla 8–9 mm long, white or sometimes pinkish; stamens 10, monadelphous or diadelphous (9 + 1). Fruit suborbicular to subreniform, 2–3.5 cm long, 1.5–2 cm wide, appressed-pubescent to glabrescent; seed 1.

Hammocks, coastal strands, and shell middens. Frequent; central and southern peninsula. Florida; West Indies, Mexico, Central America, and South America; Africa and Asia. Spring–summer.

Dalbergia sisso Roxb. ex DC. [From the Hindi vernacular name śiśam.] INDIAN ROSEWOOD.

Dalbergia sissoo Roxburgh ex de Candolle, Prodr. 2: 416. 1825. *Amerimnon sissoo* (Roxburgh ex de Candolle) Kuntze, Revis. Gen. Pl. 1: 159. 1891.

Trees, to 25 m. Leaves 3–5-foliolate, ca. 15 cm long, the leaflets with the blade suborbicular, 2–6.5 cm long and wide, the apex abruptly acuminate, the base rounded, the upper and lower surfaces puberulent or glabrous, the petiolules 2–6 mm long, the petiole (2)3–6 cm long; stipules lanceolate, ca. 4 mm long. Flowers in a subcymose, paniculiform inflorescence; bracts obovate, ca. 2 mm long, caducous; bracteoles elliptic, ca. 2 mm long, caducous; calyx 3–5 mm long, puberulent, subequal, the lower lobe ca. 2 mm longer than the lateral, the upper 2 lobes fused nearly to the apex; corolla 8–10 mm long, creamy white to yellowish; stamens 9, monadelphous. Fruit narrowly elliptic to oblong, 4–10 cm long, 0.5–1.5 cm wide, usually reticulate, glabrous; seeds 1–2(4).

Disturbed sites. Occasional; central and southern peninsula. Escaped from cultivation. Florida; West Indies; Africa, Asia, and Australia. Native to Asia. Spring–summer.

Dalbergia sissoo is listed as a Category II invasive species in Florida by the Florida Exotic Pest Plant Council (FLEPPC, 2015).

Dalea L., nom. cons. 1758. PRAIRIECLOVER

Perennial herbs. Leaves alternate, odd-pinnately compound, the leaflets opposite or subopposite, estipellate, petiolate, stipulate. Flowers in spikes, bracteate; calyx 5-lobed; corolla with 5 petals, subequal, the upper 1 (banner petal) reflexed, the wing and keel petals similar, epistemonous; stamens 5 or 10, monadelphous. Fruit a dehiscent legume; seed 1.

A genus of about 170 species; North America, Mexico, Central America, and South America. [Commemorates Samuel Dale (1659–1739), English botanist, apothecary, and physician.]

Kuhnistera Lam., 1792; *Parosela* Cav., 1802; *Petalostemon* Michx., nom. cons. 1803.

Selected reference: Barneby (1977).

1. Flower spike subtended by conspicuous involucral bracts (head resembling that of the Asteraceae); calyx segments plumose..**D. pinnata**
1. Flower spike lacking conspicuous involucral bracts (head not resembling that of the Asteraceae); calyx segments foliose.
   2. Stamens 10; wing and keel petals borne near the middle of the staminal tube ....**D. carthagenensis**
   2. Stamens 5; wing and keel petals borne near the top of the staminal tube.
      3. Flower spike ellipsoid or cylindric; bracts as long as or longer than the calyx ....**D. carnea**
      3. Flower spike globose; bracts much shorter than the calyx ....**D. feayi**

Dalea carnea (Michx.) Poir. [*Carneus*, flesh-colored, in reference to the flowers.] WHITETASSELS.

Erect to ascending or spreading to decumbent, perennial herb, 1 m; stem glabrous below the inflorescence, gland-dotted distally. Leaves 1.5–4 cm long, odd-pinnately compound, the leaflets 5–9(11), the leaflet blade elliptic to oblanceolate, (5)6–17 mm long, the apex acute, mucronate, the base cuneate, the margin entire, the lower surface gland-dotted, the petiolule short, the petiole 4–6 mm long. Flowers in a spike 0.5–3(3.5) cm long, dense, not involucrate; bracts 2–5 mm long; calyx 3–4 mm long, asymmetrical, deeply recessed opposite the banner, the teeth subulate, ciliolate, hyaline areas between the ribs eglandular or with 1–3 small glands; corolla rose or deep pink or white, the banner petal 4–6 mm long, the 4 epistemonous petals 2–3 mm long, attached at the separation of the filaments; stamens 5. Fruit ca. 3 mm long, glabrous or distally pilosulose, somewhat glandular.

1. Corolla rose or deep pink, rarely white...var. **carnea**
1. Corolla white.
    2. Plant spreading or decumbent; leaves shorter than the internodes; bracts inconspicuous, equaling the calyx.. var. **gracilis**
    2. Plant erect or ascending; leaves equaling the internodes; bracts conspicuous, exceeding the calyx.
    .... var. **albida**

Dalea carnea (Michx.) Poir. var. **carnea**

> *Petalostemon carneum* Michaux, Fl. Bor.-Amer. 2: 49. 1803. *Psoralea carnea* (Michaux) Poiret, in Lamarck, Encycl. 5: 694. 1804. *Dalea carnea* (Michaux) Poiret, in Cuvier, Dict. Sci. Nat. 12: 462. 1819. *Kuhnistera carnea* (Michaux) Kuntze, Revis. Gen. Pl. 1: 192. 1891. TYPE: FLORIDA: "Liux humides pres Nord West riv. en Floride," s.d., *Michaux s.n.* (holotype: P).
> *Petalostemon roseum* Nuttall, Amer. J. Sci. Arts. 5: 298. 1822. *Kuhnistera rosea* (Nuttall) Kuntze, Revis. Gen. Pl. 1: 192. 1891. TYPE: FLORIDA: "East-Florida," Oct–Nov 1821, *Ware s.n.* (holotype: GH).

Stem erect to ascending. Leaves with (5)7–9(11) leaflets. Corolla rose or deep pink, rarely white.

Flatwoods. Frequent; peninsula and eastern panhandle. Georgia south to Florida, west to Louisiana. Summer.

Dalea carnea var. **albida** (Torr. & A. Gray) Barneby [*Albidus*, somewhat white, in reference to the flowers.]

> *Petalostemon carneum* Michaux var. *albidum* Torrey & A. Gray, Fl. N. Amer. 1: 311. 1838. *Petalostemon albidum* (Torrey & A. Gray) Small, Fl. S.E. U.S. 630. 1902. *Dalea carnea* (Michaux) Poiret var. *albida* (Torrey & A. Gray) Barneby, Mem. New York Bot. Gard. 27: 255. 1977. *Dalea albida* (Torrey & A. Gray) D. B. Ward, Novon 14: 369. 2004.

Stems erect to ascending. Leaves usually with 5 leaflets, equaling the internodes. Floral bracts conspicuous, exceeding the calyx; corolla white; stamens 6–9 mm long.

Sandhills. Occasional; northern peninsula south to Polk County, west to central panhandle. Georgia, Florida, and Alabama. Summer.

Dalea carnea var. **gracilis** (Nutt.) Barneby [Thin, slender.]

> *Petalostemon gracile* Nuttall, J. Acad. Nat. Sci. Philadelphia 7: 92. 1834. *Kuhnistera gracilis* (Nuttall)
> Kuntze, Revis. Gen. Pl. 1: 192. 1891. *Dalea carnea* (Michaux) Poiret var. *gracilis* (Nuttall) Barneby,
> Mem. New York Bot. Gard. 27: 256. 1977. *Dalea gracilis* (Nuttall) D. B. Ward, Novon 14: 369. 2004;
> non Kunth, 1819. *Dalea montjoyae* M. Woods, Phytoneuron 2013–23: 1. 2013. TYPE: FLORIDA:
> without data, *Ware s.n.* (lectotype: PH). Lectotypified by Ward (2004: 369).

Stems spreading to procumbent. Leaves usually with 7 leaflets, shorter than the internodes.
Floral bracts inconspicuous, equaling the calyx; corolla white; stamens 5–6 mm long.

Wet flatwoods and coastal swales. Occasional; Jefferson County, central and western panhandle. Georgia south to Florida, west to Louisiana. Summer.

Dalea carthagenensis (Jacq.) J. F. Macbr. var. **floridana** (Rydb.) Barneby [Of
Carthagena; of Florida.] FLORIDA PRAIRIECOVER.

> *Parosela floridana* Rydberg, in Britton, N. Amer. Fl. 24: 114. 1920. *Dalea carthagenensis* (Jacquin) J.
> F. Macbride var. *floridana* (Rydberg) Barneby, Mem. New York Bot. Gard. 27: 519. 1977. TYPE:
> FLORIDA: Miami-Dade Co.: Miami, Nov 1878, *Garber s.n.* (holotype: NY; isotypes: NY).

Erect perennial, suffruticose herb, to 2 m; stem erect, to 20 dm, sparsely glandular-verruculose
distally. Leaves odd-pinnately compound, 3–5.5(7) cm long, the leaflets 15–23(25), the leaflet
blade obovate to oblong-elliptic, 4–18(20) mm long, the apex obtuse to rounded, mucronate,
the base cuneate, the margin entire, the upper surface glabrous, the lower surface glabrous,
gland-dotted, the petiolule short, the petiole 2–3 cm long. Flowers in a spike(2)4–15(20) mm
long, loose, not involucrate; bracts 3–5 mm long; calyx 5–7 mm long, asymmetrical, the teeth
triangular-aristate, gland-spurred laterally, becoming plumose, the hyaline space between the
ribs with 2–8 small glands; corolla greenish white or cream, maroon in age, the banner petal
4–5 mm long, the 4 epistemonous petals 2–5 mm long, attached below the middle of the staminal tube; stamens 10. Fruit 2–3 mm long, villosulous and gland-dotted distally.

Pinelands and hammocks. Rare; Palm Beach County, southern peninsula. Endemic. All
year.

Dalea carthagenensis var. *floridana* is listed as endangered in Florida (Florida Administrative Code, Chapter 5B-40).

Dalea feayi (Chapm.) Barneby [Commemorates William Féay (1803–1879), American
physician and botanist.] FEAY'S PRAIRIECLOVER.

> *Petalostemon feayi* Chapman, Fl. South. U.S., ed. 2. 615. 1883. *Kuhnistera feayi* (Chapman) Nash, Bull.
> Torrey Bot. Club 22: 149. 1895. *Dalea feayi* (Chapman) Barneby, Mem. New York Bot. Gard. 27:
> 257. 1977. TYPE: FLORIDA: Polk Co.: Bartow, s.d., *Féay s.n.* (lectotype: US; isolectotypes: GH,
> NY). Lectotypified by Barneby (1977: 257).

Erect perennial, suffrutescent herb, to 7 dm; stem glabrous below the inflorescence, with scattered small glands. Leaves odd-pinnately compound, 1.5–3.5 cm long, the leaflets 7–9, the leaflet blade linear, (6)7–12(14) mm long, the apex acute to obtuse, mucronate, the base cuneate,
the margin subentire, the upper surface glabrous, the lower surface glabrous, gland-dotted,
the petiolule short, the petiole 4–6 mm long. Flowers in a spike 6–11 mm long, dense, not
involucrate; bracts 1–2 mm long; calyx 3–4 mm long, asymmetrical, deeply recessed opposite

the banner, the teeth subulate, glabrous, the space between the ribs with 1–3 glands; corolla pink to lavender, rarely white, the banner petal 4–5 mm long, the 4 epistemonous petals ca. 3 mm long, attached at the separation of the filaments; stamens 5. Fruit ca. 3 mm long, glabrous, somewhat glandular.

Scrub and sandhills. Frequent; peninsula, central panhandle. Georgia and Florida. Summer–fall.

Dalea pinnata (J. F. Gmel.) Barneby [Featherlike, in reference to the compound leaves.] SUMMER FAREWELL.

Erect perennial herb, to 9(10) dm; stem distally finely to coarsely tuberculate. Leaves odd-pinnately compound, 1–2.5 cm long, the leaflets 3–11(13), the leaflet blade linear to elliptic-oblanceolate, 5–11 mm long, the apex acute to obtuse, mucronate, the base cuneate, the margin subentire, the upper surface glabrous, the lower surface gland-dotted, the petiolule short or obsolete, the petiole 4–6 mm long. Flowers in a spike 6–12 mm long, dense, conspicuously involucrate with several whorls of clearly differentiated sterile bracts below the spike, the basal bracts 5–8 mm long; calyx 5–8 mm long, slightly asymmetrical, the teeth linear, becoming plumose, the hyaline space between the ribs eglandular; corolla white, the banner petals 5–8 mm long, the 4 epistemonous petals 3–5 mm long, attached at the separation of the filaments; stamens 5. Fruit 2–3 mm long, distally pilosulous.

1. Leaflets elliptic-oblanceolate, often folded, but not involute, 1–2 mm wide .... var. **adenopoda**
1. Leaflets filiform to linear, usually involute, 0.5(1) mm wide.
    2. Leaflets 5–9(15); epistemonous petals 3–4 mm long .... var. **pinnata**
    2. Leaflets 3; epistemonous petals 5–6 mm long .... var. **trifoliata**

Dalea pinnata (J. F. Gmel.) Barneby var. **pinnata**

> *Kuhnia pinnata* J. F. Gmelin, Syst. Nat. 2: 375. 1791. *Kuhnistera caroliniensis* Lamarck, Encycl. 3(2): 370. 1792, nom. illegit. *Dalea kuhnistera* Willdenow, Sp. Pl. 3: 1337. 1802, nom. illegit. *Petaloste-mon corymbosum* Michaux, Fl. Bor.-Amer. 2: 50. 1803, nom. illegit. *Psoralea corymbosa* Poiret, in Lamarck, Encycl. 5: 694. 1804, nom. illegit. *Dalea corymbosa* Poiret, in Cuvier, Dict. Sci. Nat. 12: 462. 1819, nom. illegit. *Kuhnistera pinnata* (J. F. Gmelin) Kuntze, Revis. Gen. Pl. 1: 192. 1891. *Petalostemon pinnatum* (J. F. Gmelin) S. F. Blake, Rhodora 17: 131. 1915. *Petalostemon caroliniensis* Sprague, Bull. Misc. Inform. Kew 1939: 331. 1939. *Dalea pinnata* (J. F. Gmelin) Barneby, Mem. New York Bot. Gard. 27: 278. 1977.

Stem finely glandular-tuberculate below the flower spike. Leaflets 5–9(15), the blade filiform to linear, 0.5(1) mm wide, usually involute. Involucre 6–9 mm wide. Epistemonous petals (including the claw) 3–4 mm long.

Sandhills and scrub. Frequent; northern counties, central peninsula. North Carolina south to Florida, west to Louisiana. Summer–fall.

Dalea pinnata var. **adenopoda** (Rydb.) Barneby [From the Greek *adenos*, gland, and *podion*, foot, in reference to the basally glandular stem.]

> *Kuhnistera adenopoda* Rydberg, in Britton, N. Amer. Fl. 24: 136. 1920. *Petalostemon adenopodum*

(Rydberg) Wemple, Iowa State J. Sci. 45(1): 23. 1970. *Dalea pinnata* (J. F. Gmelin) Barneby var. *adenopoda* (Rydberg) Barneby, Mem. New York Bot. Gard. 27: 279. 1977. *Dalea adenopoda* (Rydberg) Isely, Brittonia 38: 353. 1986. TYPE: FLORIDA: Hillsborough Co.: Tampa, Oct 1877, *Garber s.n.* (holotype: GH; isotype: NY).

 Kuhnistera truncata Small, Bull. Torrey Bot. Club 51: 380. 1924. TYPE: Palm Beach Co.: Earman, 2 May 1921, *Rane s.n.* (holotype: NY).

Stem coarsely glandular-tuberculate below the flower spike. Leaflets 5–11(13), 1–2 mm wide, not involute. Involucre 10–13 mm wide. Epistemonous petals (including the claw) 4–6 mm long.

 Sandhills and scrub. Occasional; central and southern peninsula. Endemic. Summer–fall.

Dalea pinnata var. **trifoliata** (Chapm.) Barneby [With three leaflets.]

 Petalostemon corymbosum Michaux var. *trifoliatum* Chapman, Fl. South. U.S., ed. 3. 101. 1897. *Petalostemon pinnatum* (J. F. Gmelin) S. F. Blake subsp. *trifoliatum* (Chapman) Wemple, Iowa State J. Sci. 45(1): 27. 1970. *Dalea pinnata* (J. F. Gmelin) Barneby var. *trifoliata* (Chapman) Barneby, Mem. New York Bot. Gard. 27: 279. 1977. TYPE: FLORIDA: Franklin Co.: Apalachicola, s.d., *Chapman s.n.* (lectotype: US). Lectotypified by Wemple (1970: 28).

Stem finely glandular-tuberculate below the flower spike. Leaflets 3, the blade filiform to linear, 0.5(1) mm wide, involute. Involucre 6–9 mm wide. Epistemonous petals (including the claw) 5–6 mm long.

 Sandhills. Occasional; central and western panhandle. Georgia south to Florida, west to Louisiana. Summer–fall.

EXCLUDED TAXA

 Dalea carthagenensis (Jacqin) J. F. Macbride—Because infraspecific categories were not recoginzed, the typical variety was reported by implication by Wunderlin (1982). All Florida material is var. *floridana*.

 Dalea carthagenensis subsp. *domingensis* (de Candolle) R. T. Clausen—Reported for Florida by Chapman (1897, as *D. domingensis* de Candolle), Small (1903, 1913a, 1913b, 1913d, all as *Parosela domingensis* (de Candolle) Millspaugh), and Long and Lakela (1971, the name misapplied to material of *D. carthagenensis* var. *floridana*.

 Dalea multiflora (Nuttall) Shinners—Reported for Florida by Chapman (1897, as *Petalostemon multiflorum* Nuttall), based on a misidentification of material of *D. feayi*.

Delonix Raf. 1837. POINCIANA

Trees. Leaves alternate, even-pinnately compound, petiolate, stipulate. Flowers in terminal or axillary corymbose racemes; calyx 5-lobed; petals 5, caesalpiniaceous; stamens 10, free, the anthers longitudinally dehiscent. Fruit a dehiscent legume.

 A genus of about 11 species; North America, West Indies, Mexico, Central America, South America, Africa, Asia, Australia, and Pacific Islands. [From the Greek *delos*, evident, and *onyx*, claw, in reference to the long-clawed petals.]

Delonix regia (Bojer ex Hook.) Raf. [Royal, in reference to its majestic appearance.]
ROYAL POINCIANA.

Poinciana regia Bojer ex Hooker, Bot. Mag. 56, pl. 2884. 1829. *Delonix regia* (Bojer ex Hooker) Rafinesque, Fl. Tellur. 2: 92. 1837 ("1836"). *Delonix regia* (Bojer ex Hooker) Rafinesque var. *genuina* Stehle, Bull. Mus. Hist. Nat. (Paris), ser. 2. 18: 186. 1946, nom. inadmiss.

Tree, to 12 m; bark thin, grayish-brown, the branchlets pubescent. Leaves even-pinnately compound, 30–50 cm long, the pinnae 10–25 pairs, the leaflets in 20–40 pairs, oblong, 4–10 mm long, the apex and base rounded, the surfaces strigillose to glabrate, the petioles 7–12 cm long, reddish or yellow; stipules pinnate. Flowers in a large and showy corymbose raceme; calyx ca. 2 cm long, deeply 5-lobed, glabrous, the lobes subequal, valvate; corolla caesalpiniaceous, the petals 5–7 cm long, long-clawed, spreading and often reflexed, bright red, often mottled with orange, the standard larger than the other 4; stamens declined, shorter than the petals. Fruit broadly linear, 4–6 dm long, 5–7 cm wide, flattened, nearly solid between the oblong transverse seeds, dark brown or blackish, hard and woody.

Disturbed sites. Rare; Lee County, southern peninsula. Escaped from cultivation. Florida, West Indies, Mexico, Central America, and South America; Africa, Asia, Australia, and Pacific Islands. Native to Africa. Spring–summer.

Desmanthus Willd., nom. cons. 1806 BUNDLEFLOWER

Perennial herbs or shrubs. Leaves even-bipinnately compound, the leaflets opposite, petiolate, stipulate. Flowers in axillary heads, variously sterile, staminate, and/or bisexual; sepals 5, connate, 5-lobed; petals 5, free, equal; stamens 5 or 10. Fruit a dehiscent legume, crowded apically on the peduncle.

A genus of 24 species; nearly cosmopolitan. [From the Greek *desme*, bundle, and *anthos*, flower, in reference to the dense inflorescences.]

Acuan Medik., nom. rej. 1786.

Selected reference: Luckow (1993).

1. Fruit broadly oblong, falcate, 1.5–2.5 cm long; pinnae 5–18 pairs; stamens 5 ....**D. illinoensis**
1. Fruit oblong to linear, straight or only moderately curved, 4–6(8) cm long; pinnae 2–4 pairs; stamens 10 (rarely 5 in *D. virgatus*).
   2. Petiole of mature leaves 1–5(6) mm long ....**D. virgatus**
   2. Petiole of mature leaves (3)5–16 mm long ....**D. leptophyllus**

Desmanthus illinoensis (Michx.) MacMill. ex B. L. Rob. & Fernald [Of Illinois.]
PRAIRIE BUNDLEFLOWER.

Mimosa illinoensis Michaux, Fl. Bor.-Amer. 2: 254. 1803. *Acacia brachyloba* Willdenow, Sp. Pl. 4: 1071. 1806, nom. illegit. *Darlingtonia brachyloba* de Candolle, Ann. Sci. Nat. (Paris) 4: 97. 1824. *Darlingtonia illinoensis* (Michaux) de Candolle ex Torrey, Ann. Lyceum Nat. Hist. New York 2: 191. 1828. *Darlingtonia brachyloba* de Candolle var. *illinoensis* (Michaux) Torrey & A. Gray, Fl. N. Amer. 1: 401. 1840. *Desmanthus brachylobus* Bentham, J. Bot. (Hooker) 4: 358. 1842. *Acuan illinoensis* (Michaux) Kuntze, Revis. Gen. Pl. 1: 158. 1891. *Desmanthus illinoensis* (Michaux) MacMillan ex B. L. Robinson & Fernald, in A. Gray, Manual, ed. 7. 503. 1908.

Mimosa virgata W. Bartram, Travels Carolina 421. 1791; non Linnaeus, 1753. *Acacia virgata* Rafinesque, Fl. Ludov. 136. 1817; non (Linnaeus) Gaertner. *Darlingtonia virgata* Rafinesque, New Fl. 1: 43. 1836. *Neptunia virgata* (Rafinesque) Trelease ex Branner & Coville, Rep. (Annual) Geol. Surv. Arkansas 1888(4): 178. 1891.

Perennial herb, to 1.5 m; stem glabrous or pubescent along the ridges. Leaves even-bipinnately compound, 3.5–12 cm long, the pinnae 5–18 pairs, 15–35 mm long, the lowest pair with a sessile, orbicular, crateriform nectary, the leaflets 15–35 pairs, oblong, falcate, 2–6 mm long, ca. 1 mm wide, the apex acute to mucronate, the base rounded-oblique, the margin sparsely ciliate, the upper and lower surfaces glabrous, the petiole 2–10 mm long; stipules setiform, 4–12 mm long, with a small winged margin at the base, glabrous or pubescent. Flowers numerous in a solitary or paired, axillary, pedunculate head, usually all bisexual, occasionally with a few staminate and sterile at the base, the peduncle 1–6 cm long; bract subtending each flower 1–3 mm long, peltate, subulate; calyx obconic, 1–2 mm long, green or white, glabrous; petals oblanceolate, 2–3 mm long, pale green or white, glabrous; stamens 5, 4–8 mm long. Fruit oblong, 1.5–2.5 cm long, 4.5–7 mm wide, slightly to strongly falcate, dark brown, glabrous; seeds 2–5, ovate, 3–4 mm long, red-brown.

Coastal lagoon margins. Rare; Leon, Wakulla, and Escambia Counties. Pennsylvania south to Florida, west to North Dakota, Nevada, and New Mexico; Europe. Native to North America. Summer.

Desmanthus leptophyllus Kunth [From the Greek *leptos*, slender, and *phyllon*, leaf, in reference to the leaves.] SLENDERLOBE BUNDLEFLOWER.

Desmanthus leptophyllus Kunth, in Humboldt et al., Nov. Gen. Sp. 6: 264. 1823.

Shrub, to 3 m; stem glabrous. Leaves even-bipinnately compound, 4–11 cm long, the pinnae 4–8 pairs, 1.5–3.5 cm long, the lowest pair with a sessile, elliptic or orbicular, crateriform or flattened nectary, the leaflets 16–36 pairs, narrowly linear, straight or sometimes falcate, 2–6 mm long, ca. 1 mm wide, the apex acute to attenuate, the base truncate-oblique, the margin finely ciliate, the upper and lower surface glabrous, the petiole 3–16 mm long; stipules setiform, 3–10 mm long, with very small auricles at the base, glabrous or pubescent. Flowers numerous in a solitary, axillary, pedunculate head, these staminate or bisexual, rarely with 1–2 sterile; bract subtending each flower deltate, ca. 1 mm long, peltate; calyx obconic, 2–3 mm long, pale green with a white margin, glabrous; petals oblanceolate, 3–5 mm long, red with a green or purple tip, glabrous; stamens 10, 5–8 mm long. Fruit linear, 5.5–8.5 cm long, 2–5 mm wide, falcate, reddish brown to black, glabrous or sparsely pubescent; seeds 11–26, ovate, 2–3 mm long, red or yellow-brown.

Pine rocklands and dry, disturbed sites. Rare; Miami-Dade County. Florida; West Indies, Mexico, Central America, and South America. Native to tropical America. Summer.

Desmanthus virgatus (L.) Willd. [Long and slender.] WILD TANTAN.

Mimosa virgata Linnaeus, Sp. Pl. 1: 519. 1753. *Mimosa angustisiliqua* Lamarck, Encycl. 1: 10. 1783, nom. illegit. *Acuan virgatum* (Linnaeus) Medikus, Theodora 62: 1786. *Acacia virgata* (Linnaeus) Gaertner, Fruct. Sem. Pl. 2: 317. 1791. *Desmanthus virgatus* (Linnaeus) Willdenow, Sp. Pl. 4: 1047. 1806. *Acacia angustisiliqua* Desfontaines, Tabl. Ecole Bot., ed. 3. 300. 1829, nom. illegit.

Desmanthus depressus Humboldt & Bonpland ex Willdenow, Sp. Pl. 4: 1046. 1806. *Mimosa depressa* (Humboldt & Bonpland ex Willdenow) Poiret, in Lamarck, Encycl., Suppl. 1: 58. 1810. *Acuan depressum* (Humboldt & Bonpland ex Willdenow) Kuntze, Revis. Gen. Pl. 1: 158. 1891. *Desmanthus virgatus* (Linnaeus) Willdenow var. *depressus* (Humboldt & Bonpland ex Willdenow) B. L. Turner, Field & Lab. 18: 61. 1950.

Prostrate, decumbent, or erect perennial herb, to 1.5 m; stem glabrous or sparsely pubescent. Leaves even-bipinnately compound, 2.5–8 cm long, the pinnae 2–5 pairs, 1–3 cm long, the lowest pair with an orbicular, sessile, crateriform nectary, the leaflets 11–23 pairs, linear-oblong, 3–7 mm long, ca. 1 mm wide, the apex rounded to acute, the base square-oblique, the margin finely ciliate, the upper and lower surfaces glabrous, the petiole 1–5(6) mm long; stipules setiform, 2–9 mm long, auriculate. Flowers numerous in a solitary, axillary, pedunculate head, these sterile, staminate, or bisexual; bract subtending each flower deltate, 1–2 mm long, pale green with a green tip, peltate; calyx obconic, 2–3 mm long, pale green with a white margin, glabrous; petals oblanceolate, 3–4 mm long, green with a red or purple tip, glabrous; stamens 10, rarely 5, 4–7 mm long. Fruit linear, 2–9 cm long, 3–4 mm wide, straight or slightly falcate, brown to nearly black, glabrous; seeds 9–27, ovate, 2–3 mm long, red or yellow-brown.

Sandhills, shell middens, hammocks, and disturbed sites. Occasional; central and southern peninsula. Florida and Texas; West Indies, Mexico, Central America, and South America; Africa, Asia, Australia, and Pacific Islands. Native to Texas and tropical America. Summer–fall or all year in southern peninsula.

EXCLUDED TAXON

Desmanthus luteus (Leavenworth) Bentham ex Chapman—Reported for Florida by Chapman (1860, 1883, 1897), who misapplied the name to material of *Neptunia pubescens*.

Desmanthus pernambucanus (Linnaeus) Thellung—Reported for Florida by Chapman (1860, as *Desmanthus diffusus* Willdenow), and as naturalized in Florida by Luckow (1993). No specimens cited by Luckow (1993) and no voucher specimens seen.

Desmodium Desv., nom. cons. 1813. TICKTREFOIL

Perennial or annual herbs. Leaves 3-foliolate, sometimes also 1-foliolate below, the leaflets entire, stipellate or estipellate, petiolate, stipulate. Flowers in compound or simple terminal or axillary pseudoracemes, bracteate, ebracteolate; sepals connate, 5-lobed; corolla papilionaceous; stamens 10, diadelphous (9 + 1). Fruit a stipitate loment dividing into 1-seeded segments.

A genus of about 275 species; nearly cosmopolitan. [From the Greek *desmos*, chain, and *odion*, to resemble, in reference to the jointed pod of some species resembling a chain.]

Meibomia Heist. ex Fabr., nom. rej. 1759; *Sagotia* Duchass. & Walp., nom. rej. 1851.

1. Plant prostrate or procumbent, vinelike.
    2. Flowers in an axillary cluster; leaflets broadly obovate, 0.5–1 cm long............................**D. triflorum**
    2. Flowers in a raceme; leaflets various but not obovate, more than 1 cm long.
        3. Leaves with an evident, lighter colored area on the upper surface surrounding the midrib.........
        ....**D. incanum**

3. Leaves solid green.
    4. Loment uncinate-puberulent only along the sutures.................................**D. ochroleucum**
    4. Loment evenly uncinate-puberulent over the entire surface.
        5. Stipules caducous... **D. lineatum**
        5. Stipules persistent or semipersistent.
            6. Leaflets suborbicular to broadly ovate; loment crenate on the upper suture, notched or deeply crenate on the lower suture......................................**D. rotundifolium**
            6. Leaflets ovate-elliptic; loment linear, equally crenate on both sutures ... **D. scorpiurus**
1. Plant erect or ascending, never vinelike.
    7. Flowers in fascicles of (2)3–5(7).. **D. tortuosum**
    7. Flowers in pairs.
        8. Inflorescence axis pilose or villous as well as uncinate-puberulent........................**D. canescens**
        8. Inflorescence axis glabrous or uncinate-puberulent only.
            9. Leaves with an evident, lighter colored area on the upper surface surrounding the midrib...
            ....**D. incanum**
            9. Leaves solid green.
                10. Stipules persistent, most or all present (or sometimes all caducous in *D. cuspidatum*).
                    11. Loment segments 2–4, 5–7 mm long; corolla 6–7 mm long; lowermost leaves (if present) 1-foliolate, the upper ones 3-foliolate..............................**D. floridanum**
                    11. Loment segments 4–6, 9–11 mm long; corolla 8–12 mm long; leaves all 3-foliolate .....
                    .... **D. cuspidatum**
                10. Stipules caducous or only a few present.
                    12. Loment segments 2–3, rounded on the lower suture.
                        13. Leaflets linear-lanceolate.
                          14. Stem conspicuously uncinate-villous or -puberulent; leaves usually sessile or subsessile.. **D. sessilifolium**
                        14. Stem glabrate or inconspicuously puberulent (appearing glabrous without magnification).
                            15. Loment convex on the upper surface ....**D. tenuifolium**
                            15. Loment straight or barely convex on the upper surface.............. **D. strictum**
                        13. Leaflets oblong-ovate to suborbicular.
                          16. Leaflets elliptic to ovate-lanceolate, the terminal leaflet 5–7 cm long, distinctly longer and narrower than the others ....**D. obtusum**
                        16. Leaflets ovate, oblong, or suborbicular, the terminal leaflet 1–3 cm long, similar to the lateral ones.
                            17. Petiole 1–3(5) mm long; stem and petiole usually pilose; pedicel 3–8 mm long .... **D. ciliare**
                            17. Petiole 2.5–10 mm long; stem and petiole glabrous or sparsely uncinate-puberulent; pedicel 8–15 mm long.....................................**D. marilandicum**
                    12. Loment segments 3–5, obtusely angled on the lower suture, or if rounded, then the plants moderately villous.
                    18. Plants conspicuously villous or cinereous.
                        19. Loment straight, the segments (3)4–5(6); leaflets densely villous on the lower surface, the reticulum mostly obscured..................................... **D. viridiflorum**
                        19. Loment usually upwardly curved (upper suture convex), the segments 2–4; leaflets moderately villous on the lower surface, the reticulum usually evident
                        .... **D. nuttallii**

    18. Plant neither conspicuously villous nor cinereous except sometimes on the lower surface of the leaflets.

        20. Stem and leaves glabrous .... **D. laevigatum**

        20. Stems and leaves evidently pubescent or glabrate, but with at least some pubescence on the leaves.

            21. Leaflets conspicuously uncinate-pubescent on the veins on the lower surface, but otherwise glabrous .... **D. fernaldii**

            21. Leaflets slightly strigose to conspicuously subappressed villous on the lower surface, some also uncinate-pubescent.

                22. Leaflet pubescence usually scant, of short (to 0.5 mm long), straight appressed trichomes or of longer, curving, somewhat spreading trichomes; medial stem glabrous or glabrate, if with a few trichomes, these usually minute and uncinate .... **D. paniculatum**

                22. Leaflet pubescence usually evident, of longer (more than 0.5 mm), either pilose or with uncinate trichomes or both (if lacking, then with this type of indumentum on the petioles).

                    23. Stem and petiole pilose .... **D. perplexum**

                    23. Stem and petiole uncinate-pubescent .... **D. glabellum**

Desmodium canescens (L.) DC. [Becoming gray.] HOARY TICKTREFOIL.

Hedysarum canescens Linnaeus, Sp. Pl. 2: 748. 1753. *Desmodium canescens* (Linnaeus) de Candolle, Prodr. 2: 328. 1825. *Meibomia canescens* (Linnaeus) Kuntze, Revis. Gen. Pl. 1: 195. 1891. *Pleurolobus canescens* (Linnaeus) MacMillan, Metasp. Minnesota Valley 320. 1892.

Erect or ascending perennial herb, to 1.5(2) m; stem pilose, usually also with uncinate pubescence. Leaves 3-foliolate, the leaflet blade ovate, 5–10 cm long, 2–8 cm wide, the apex acute or acuminate, the base rounded to subcordate, the upper surface uncinulate on the veins, otherwise glabrate, the lower surface uncinulate-puberulent with a few villous trichomes, the petiole 4–10 cm long; stipules deltate to ovate-lanceolate, amplexicaul, commonly reflexed. Flowers in a terminal, branched raceme, the pedicel 8–13 mm long; calyx 3–5 long; corolla 9–13 mm long, purple. Fruit sinuate above, notched below, the isthmi narrow, the segments 4–6, each 6–10 mm long, convex above, obtusely angled below.

Flatwoods. Occasional; northern peninsula south to Hernando County, west to central panhandle. Massachusetts and New York south to Florida, west to Ontario, Minnesota, Nebraska, Kansas, Oklahoma, and Texas. Summer–fall.

Desmodium ciliare (Muhl. ex Willd.) DC. [With cilia.] HAIRY SMALL-LEAF TICKTREFOIL.

Hedysarum ciliare Muhlenberg ex Willdenow, Sp. Pl. 3: 1196. 1802. *Desmodium ciliare* (Muhlenberg ex Willdenow) de Candolle, Prodr. 2: 327. 1825. *Meibomia ciliaris* (Muhlenberg ex Willdenow) S. F. Blake, Bot. Gaz. 78: 275. 1924.

Ascending to erect perennial herb, to 1.5 m; stem conspicuously or thinly pilose or sometimes uncinate-pubescent. Leaves 3-foliolate, the leaflet blade broadly ovate or elliptic-ovate to lanceolate, 1.5–3 cm long, 0.5–2 cm wide, the apex obtuse to rounded, the base rounded to subcordate, the upper and lower surfaces inconspicuously subappressed-pubescent to glabrate, the petiole 1–3(5) mm long; stipules subulate, caducous. Flowers in a terminal, branched raceme, the pedicel 3–8 mm long; calyx 1.5–2 mm long; corolla 3.5–5 mm long, lavender-purple. Fruit sinuate above, notched below, the isthmi narrow, the segments 1–2(3), each 4–5 mm long, convex above, rounded below.

Sandhills and dry open hammocks. Occasional; nearly throughout. Massachusetts and New York south to Florida, west to Ontario, Michigan, Illinois, Kansas, Oklahoma, and Texas; West Indies, Mexico, and Central America. Fall.

Desmodium cuspidatum (Muhl. ex Willd.) DC. ex Loudon [With a cusp.] LARGEBRACT TICKTREFOIL.

Hedysarum cuspidatum Muhlenberg ex Willdenow, Sp. Pl. 3: 1198. 1802. *Desmodium cuspidatum* (Muhlenberg ex Willdenow) de Candolle ex Loudon, Hort. Brit. 309. 1830. *Meibomia cuspidata* (Muhlenberg ex Willdenow) Schindler, Repert. Spec. Nov. Regni Veg. 20: 140. 1924. *Desmodium bracteosum* (Michaux) de Candolle var. *cuspidatum* (Muhlenberg ex Willdenow) de Candolle, Prodr. 2: 329. 1825.

Hedysarum grandiflorum Walter, Fl. Carol. 185. 1788; non Pallas, 1773. *Desmodium grandiflorum* de Candolle, Prodr. 2: 338. 1825. *Meibomia grandiflora* (de Candolle) Kuntze, Revis. Gen. Pl. 1: 196. 1891. *Pleurolobus grandiflorus* (de Candolle) MacMillan, Metasp. Minnesota Valley 321. 1892.

Hedysarum bracteosum Michaux, Fl. Bor.-Amer. 2: 73. 1803. *Desmodium bracteosum* (Michaux) de Candolle, Prodr. 2: 329. 1825. *Meibomia bracteosa* (Michaux) Kuntze, Revis. Gen. Pl. 1: 195. 1891.

Erect or ascending perennial herb, to 1.5 m; stem glabrous or slightly pilose. Leaves 3-foliolate, the leaflet blade ovate to ovate-lanceolate, 5–10 cm long, 2–4 mm wide, the apex acuminate, the base rounded, the upper surface glabrate, the lower surface glabrate or spreading villosulous, the petiole 3–6 cm long; stipules lanceolate to subulate, semipersistent. Flowers in a terminal, branched raceme, the pedicel 4–8 mm long; calyx 3–4 mm long, corolla purple, 8–12 mm long. Fruit sinuate above, notched below, the isthmi narrow, the segments 4–6, each 9–11 mm long, slightly convex above, obtusely angled below.

Moist to dry hammocks. Rare; Jefferson, Leon, and Jackson Counties. New Hampshire south to Florida, west to Ontario, Minnesota, Nebraska, Kansas, Oklahoma, and Texas. Summer–fall.

Desmodium fernaldii B. G. Schub. [Commemorates Merritt Lyndon Fernald (1873–1950), American botanist.] FERNALD'S TICKTREFOIL.

Desmodium fernaldii B. G. Schubert, Rhodora 52: 147. 1950.

Ascending to erect perennial herbs, to 1 m; stem glabrous or uncinate-pubescent. Leaves 3-foliolate, the leaflet blade ovate, 3–9 cm long, 1.5–5 cm wide, the apex obtuse, the base broadly cuneate to rounded, the upper surface glabrate or slightly uncinulate-puberulent, the lower surface uncinulate along the veins, the petiole 2–4 cm long; stipules deltate, caducous. Flowers in a terminal branched or unbranched raceme, the pedicel 4–8(10) mm long; calyx 2–3 mm long; corolla 7–8 mm long, purplish. Fruit sinuate above, notched below, the isthmi narrow, the segments (2)3–5, each 4–8 mm long, convex above, obtusely angled below.

Sandhills. Rare; Alachua, Leon, and Escambia Counties. Maryland south to Florida, west to Texas. Fall.

Desmodium floridanum Chapm. [Of Florida.] FLORIDA TICKTREFOIL.

Desmodium floridanum Chapman, Fl. South. U.S. 102. 1860. *Meibomia floridana* (Chapman) Kuntze, Revis. Gen. Pl. 1: 198. 1891. TYPE: FLORIDA: Franklin Co.: Apalachicola, s.d., *Chapman s.n.* (lectotype: US). Lectotypified by Schubert (1950: 147).

Ascending to erect perennial herb, to 1 m; stem sparsely to densely villous-uncinate or uncinulate-puberulent. Leaves 1- or 3-foliolate below, 3-foliolate above, the leaflet blade rhombic or broadly ovate, 3–7 cm long, 1.5–3 mm wide, the apex obtuse, the base rounded, the upper surface glabrate to slightly villous, the veins uncinulate, the lower surface closely and spreading villous, the petiole 2–4 cm long; stipules lanceolate, striate, sometimes reflexed. Flowers a simple or branched raceme, the pedicel 5–8 mm long; calyx 3–4 mm long; corolla 6–7 mm long, purple. Fruit straight or curved, sinuate above, notched below, the isthmi narrow, the segments 2–4, each 5–7 mm long, convex above, rounded or obtusely angled below.

Sandhills, flatwoods, and dry hammocks. Frequent; nearly throughout. South Carolina, Florida, Georgia, and Alabama. Summer–fall.

Desmodium glabellum (Michx.) DC. [Smooth.] DILLENIUS' TICKTREFOIL.

Hedysarum glabellum Michaux, Fl. Bor.-Amer. 2: 73. 1803. *Desmodium glabellum* (Michaux) de Candolle, Prodr. 2: 329. 1825. *Hedysarum paniculatum* Linnaeus var. *obtusum* Desvaux, Mém. Soc. Linn. Paris 4: 316. 1825. *Meibomia glabella* (Michaux) Kuntze, Revis. Gen. Pl. 1: 198. 1891. *Meibomia paniculata* (Linnaeus) Kuntze var. *obtusa* (Desvaux) Schindler, Repert. Spec. Nov. Regni Veg. 22: 282. 1926.

Ascending or erect perennial herb, 1(1.5) m; stem uncinate-pubescent. Leaves 3-foliolate, the leaflet blade broadly ovate or rhombic to narrowly ovate or ovate-lanceolate, 2–8(10) cm long, 0.5–2 cm wide, the apex obtuse or rounded, the base rounded, the upper surface glabrate, the veins uncinulate, the lower surface spreading villous, the petiole (1)2–5 cm long; stipules subulate, usually caducous. Flowers in a terminal, branched raceme, the pedicel 5–10(12) mm long; calyx 2–3 mm long; corolla 6–7(8) mm long, lilac to purple. Fruit sinuate above, notched below, the isthmi narrow, the segments (2)3–5, each 4–8 mm long, convex above, obtusely angled below.

Hammocks, often along stream banks. Occasional; northern counties south to Levy County. New York south to Florida, west to Michigan, Iowa, Kansas, Oklahoma, and Texas. Summer–fall.

Desmodium incanum (Se.) DC. [Hoary, white.] ZARZABACOA COMUN.

Hedysarum incanum Swartz, Prodr. 107. 1788, nom. cons. *Aeschynomene incana* (Swartz) G. Meyer, Prim. Fl. Esseq. 245. 1818. *Desmodium incanum* (Swartz) de Candolle, Prodr. 2: 332. 1825. *Meibomia adscendens* (Swartz) Kuntze var. *incana* (Swartz) Kuntze, Revis. Gen. Pl. 1: 195. 1891. *Meibomia incana* (Swartz) Vail, Bull. Torrey Bot. Club 19: 118. 1892.

Hedysarum racemosum Aublet, Hist. Pl. Guiane 774. 1775. *Hedysarum canum* J. F. Gmelin, Syst. Nat. 2: 1124. 1792, nom. illegit. *Hedysarum racemiferum* J. F. Gmelin, Syst. Nat. 2: 1125. 1792, nom. illegit. *Desmodium canum* Schinz & Thellung, Mem. Soc. Sci. Nat. Neuchatel 5: 371. 1913. *Meibomia cana* (Schinz & Thellung) S. F. Blake, Bot. Gaz. 78: 276. 1924. *Desmodium frutescens* Schindler, Repert. Spec. Nov. Regni Veg. 21: 9. 1925, nom. illegit.

Hedysarum canescens Miller, Gard. Dict., ed. 8. 1768; non Linnaeus, 1753. *Hedysarum supinum* Swartz,

Prodr. 106. 1788; non Chaix ex Villars, 1779. *Desmodium supinum* de Candolle, Prodr. 2: 332. 1825. *Desmodium incanum* (Swartz) de Candolle var. *supinum* (de Candolle) Hooker & Arnott, Bot. Beechey Voy. 417. 1841. *Meibomia supina* Britton, in Morong & Britton, Ann. New York Acad. Sci. 7: 83. 1892, nom. illegit.

Ascending to erect perennial herb, to 1 m; stem villous to glabrate or partly uncinuate-pubescent. Leaves 3-foliolate, the leaflet blade elliptic to ovate, 2–4(6) cm long, 0.8–1.2 mm wide, the apex and base rounded, the upper surface glabrate, the lower surface villous to substrigose, with or without uncinate pubescence, the petiole 1–2 cm long; stipules lanceolate, striate, usually persistent. Flowers in a simple terminal raceme, the pedicel 5–9 mm long; calyx ca. 2 mm long; corolla 5–8 mm long, purple. Fruit oblong, straight or slightly upcurved, slightly sinuate or linear above, crenate below, the isthmi broad to narrow, the segments 4–7(9), each 4–5 mm long, straight or slightly convex above, broadly rounded below.

Open hammocks and disturbed sites. Frequent; nearly throughout. Georgia, Florida, and Texas; West Indies, Mexico, Central America, and South America; Africa, Asia, Australia, and Pacific Islands. Native to tropical America. Spring–fall, all year in southern counties.

Desmodium laevigatum (Nutt.) DC. [Smooth.] SMOOTH TICKTREFOIL.

Hedysarum laevigatum Nuttall, Gen. N. Amer. Pl. 1: 109. 1818. *Desmodium laevigatum* (Nuttall) de Candolle, Prodr. 329. 1825. *Meibomia laevigata* (Nuttall) Kuntze, Revis. Gen. Pl. 1: 198. 1891.

Ascending to erect perennial herb, to 1(1.5) m; stem glabrous. Leaves 3-foliolate, the leaflet blade ovate to ovate-lanceolate 3.5–8 cm long, 1–2(4) cm wide, the apex acute to obtuse, the base cuneate to rounded, the upper and lower surfaces glabrous, the petiole 2–6 cm long; stipules ovate-lanceolate, caducous. Flowers in a terminal, branched raceme, the pedicel 1–1.5(2) cm long; calyx 3–4 mm long; corolla 8–10 mm long, lavender to purple. Fruit straight or sinuate above, notched below, the isthmi narrow, the segments (2)3–5, each 5–8 mm long, convex or straight above, obtusely angled or rounded below.

Open, mesic hammocks. Occasional; northern counties. New York south to Florida, west to Illinois, Missouri, Oklahoma, and Texas. Summer–fall.

Desmodium lineatum DC. [Marked with lines, in reference to the stem.] SAND TICKTREFOIL.

Hedysarum lineatum Michaux, Fl. Bor.-Amer. 2: 72. 1803; non Linnaeus, 1759. *Desmodium lineatum* de Candolle, 2: 330. 1825. *Meibomia lineata* (de Candolle) Kuntze, Revis. Gen. Pl. 1: 196. 1891. *Meibomia arenicola* Vail, Bull. Torrey Bot. Club 23: 139. 1896, nom. illegit. *Desmodium arenicola* F. J. Hermann, J. Wash. Acad. Sci. 38: 237. 1948, nom. illegit.
Meibomia lineata (de Candolle) Kuntze var. *polymorpha* Vail, Bull. Torrey Bot. Club 19: 109. 1892. *Meibomia arenicola* Vail var. *polymorpha* (Vail) Vail, in Small, Fl. S.E. U.S. 636. 1903. *Meibomia polymorpha* (Vail) Small, Man. S.E. Fl. 733. 1933. TYPE: FLORIDA/LOUISIANA.

Procumbent perennial herb; stem to 1 m long, uncinate-puberulent, closely pilose. Leaves 3-foliolate, the leaflet blade suborbicular or broadly ovate, the terminal one 1.5–2.3 cm long, 1.5–2.3 cm wide, the lateral ones slightly smaller, the apex obtuse, the base rounded to truncate, the upper surface glabrate, the lower surface glabrate or inconspicuously spreading villous, the petiole 0.5–1.5 cm long; stipules lanceolate, caducous or semipersistent. Flowers in an ascending terminal, branched raceme, and also in an axillary, simple raceme, the pedicel 6–12 mm long; calyx ca. 2 mm long; corolla 3–5 mm long, purple or rarely white. Fruit sinuate above, notched below, the isthmi narrow, the segments 2–3(4), each 4–5 mm long, convex above, rounded below.

Sandhills, flatwoods, and mesic hammocks. Occasional; northern counties south to Lake, Sumter, and Hernando Counties, also Miami-Dade County. Maryland and West Virginia south to Florida, west to Texas. Summer–fall.

Desmodium marilandicum (L.) DC. [Of Maryland.] SMOOTH TICKTREFOIL.

> *Hedysarum marilandicum* Linnaeus, Sp. Pl. 2: 748. 1753. *Desmodium marilandicum* (Linnaeus) de Candolle, Prodr. 2: 328, 1825. *Meibomia marilandica* (Linnaeus) Kuntze, Revis. Gen. Pl. 1: 198. 1891.

Ascending or erect perennial herb, to 1 m; stem glabrous or sparsely uncinate-puberulent. Leaves 3-foliolate, the leaflet blade broadly ovate or elliptic-ovate, 1.5–2.5(4) cm long, 1–2(3) cm wide, the apex and base rounded, the upper and lower surfaces glabrous or with a few trichomes, the petiole 1–2.5 cm long; stipules subulate, caducous. Flowers in a terminal, branched raceme, the pedicel 8–15 mm long; calyx ca. 2 mm long; corolla ca. 5 mm long, lavender to red-violet. Fruit sinuate above, notched below, the isthmi narrow, the segments 2(3), each 3–4 mm long, convex above, rounded below.

Hammocks. Occasional; peninsula west to central panhandle. New Hampshire south to Florida, west to Ontario, Michigan, Illinois, Kansas, Oklahoma, and Texas. Summer–fall.

Desmodium nuttallii (Schindl.) B. G. Schub. [Commemorates Thomas Nuttall (1786–1859), English botanist who worked in America (1808–1841).] NUTTALL'S TICKTREFOIL.

> *Meibomia nuttallii* Schindler, Repert. Spec. Nov. Regni Veg. 23: 354. 1927. *Desmodium nuttallii* (Schindler) B. G. Schubert, Rhodora 52: 142. 1950.

Ascending or erect perennial herb, to 1 m; stem villous with weakly uncinate trichomes and short uncinulate pubescence. Leaves 3-foliolate, the leaflet blade ovate to narrowly ovate, 2–8 cm long, 1–4 cm wide, the terminal one usually the largest, the apex acute to obtuse, the base rounded, the upper surface slightly uncinulate on the veins, the lower surface villous and often velvety, the petiole 1–3 cm long; stipules subulate, caducous. Flowers in a terminal, branched raceme, the pedicel 4–10 mm long; calyx 2–3 mm long; corolla 6–7 mm long, purple or pink. Fruit incurved or straight, sinuate above, notched below, the isthmi narrow, the segments 2–4, each 4–5 mm long, convex above, rounded or obtusely angled below.

Hammocks. Rare; central and western panhandle, Alachua County. New York south to Florida, west to Illinois, Missouri, Oklahoma, and Texas. Summer.

Desmodium obtusum (Muhl. ex Willd.) DC. [Obtuse shaped, in reference to the apex of the Leaflets.] STIFF TICKTREFOIL.

> *Hedysarum obtusum* Muhlenberg ex Willdenow, Sp. Pl. 3: 1190. 1802. *Desmodium obtusum* (Muhlenberg ex Willdenow) de Candolle, Prodr. 2: 329. 1825. *Meibomia obtusa* (Muhlenberg ex Willdenow) Vail, Bull. Torrey Bot. Club 19: 115. 1892.

Hedysarum rigidum Elliott, Sketch Bot. S. Carolina 2: 215. 1823. *Desmodium rigidum* (Elliott) de Candolle, Prodr. 2: 330. 1825. *Meibomia rigida* (Elliott) Kuntze, Revis. Gen. Pl. 1: 198. 1891.

Ascending to erect perennial herb, to 1.5 m; stem sparsely to densely uncinate-pubescent. Leaves 3-foliolate, the leaflet blade elliptic-ovate to lanceolate, (2.5)4–6(7.5) cm long, 2–4 cm wide, the apex rounded to obtuse, the base obtuse, the upper surface glabrate or uncinulate-puberulent, the lower surface appressed villous to glabrate, the petiole 3–12 mm long; stipules deltate, caducous. Flowers in a terminal, branched raceme, the pedicel 4–10 mm long; calyx 1.5–2.5 mm long; corolla 4.5–6 mm long, pink-purple, rarely white. Fruit sinuate above, notched below, the isthmi narrow, the segments 2–3, each 3–5 mm long, convex above, rounded below.

Sandhills and open hammocks. Occasional; northern counties. New Hampshire south to Florida, west to Michigan, Illinois, Colorado, Oklahoma, and Texas; West Indies, Mexico, and Central America. Summer–fall.

Desmodium ochroleucum M. A. Curtis ex Canby [From the Greek *ochros*, pale yellow, and *leucron*, white, in reference to the flower color.] CREAM TICKTREFOIL.

Desmodium ochroleucum M. A. Curtis ex Canby, Proc. Acad. Nat. Sci. Philadelphia 16: 17. 1864. *Meibomia ochroleuca* (M. A. Curtis ex Canby) Kuntze, Revis. Gen. Pl. 1: 198. 1891.

Decumbent perennial herb; stem to 1 m long, pilose and uncinate-puberulent. Leaves 3-foliolate, the leaflet blade ovate, 2.5–6 cm long, 2–5 mm wide, the apex and base rounded, the upper surface glabrate, the lower surface glabrate or uncinulate on the veins, the petiole 1.5–5 cm long; stipules deltate, aplexicaul. Flowers in an ascending to erect simple, axillary raceme, sometimes also in a terminal compound raceme, the pedicel 10–15 mm long; calyx 3–4 mm long; corolla 7–8 mm long, white to pale yellow. Fruit crenate above, notched below, the isthmi narrow, the segments 3–5, each 7–10 mm long, convex above, rounded below, usually uncinate only on the margin.

Open hammocks and disturbed sites. Rare; Jackson County. New Jersey south to Florida, west to Missouri, Tennessee, and Mississippi. Fall.

Desmodium ochroleucum is listed as endangered in Florida (Florida Administrative Code, Chapter 5B-40).

Desmodium paniculatum (L.) DC. [Branched, in reference to the inflorescence.] PANICLED TICKTREFOIL.

Hedyarum paniculatum Linnaeus, Sp. Pl. 2: 749. 1753. *Desmodium paniculatum* (Linnaeus) de Candolle, Prodr. 2: 329. 1825. *Meibomia paniculata* (Linnaeus) Kuntze, Revis. Gen. Pl. 1: 198. 1891. *Pleurolobus paniculatus* (Linnaeus) MacMillan, Metasp. Minnesota Valley 320. 1892. *Desmodium paniculatum* (Linnaeus) de Candolle var. *typicum* B. G. Schubert, Rhodora 52: 152. 1950, nom. inadmiss.

Desmodium dillenii Darlington, Fl. Cestr., ed. 2. 414. 1837. *Meibomia dillenii* (Darlington) Kuntze, Revis. Gen. Pl. 1: 195. 1891. *Pleurolobus dillenii* (Darlington) MacMillan, Metasp. Minnesota Valley 320. 1892. *Desmodium paniculatum* (Linnaeus) de Candolle var. *dillenii* (Darlington) Isely, Amer. Midl. Naturalist 49: 927. 1953.

Desmodium paniculatum (Linnaeus) de Candolle var. *angustifolium* Torrey & A. Gray, Fl. N. Amer. 1: 364. 1840. *Meibomia paniculata* (L.) Kuntze var. *angustifolia* (Torrey & A. Gray) Vail, Bull. Torrey

Bot. Club 19: 112. 1892. *Meibomia angustifolia* (Torrey & A. Gray) Kearney, Bull. Torrey Bot. Club 20: 481. 1893; non (Kunth) Kuntze, 1891. *Meibomia paniculata* (Linnaeus) Kuntze var. *chapmanii* Britton, Mem. Torrey Bot. Club 5: 204. 1894, nom. illegit.

Desmodium paniculatum (Linnaeus) de Candolle var. *pubens* Torrey & A. Gray, Fl. N. Amer. 1: 364. 1840. *Desmodium pubens* (Torrey & A. Gray) M. J. Young, Fl. Texas 233. 1873. *Meibomia paniculata* (Linnaeus) Kuntze var. *pubens* (Torrey & A Gray) Vail, Bull. Torrey Bot. Club 19: 112. 1892. *Meibomia pubens* (Torrey & A. Gray) Rydberg, Brittonia 1: 92. 1931. TYPE: FLORIDA: Hillsborough Co.: Tampa Bay, s.d., *Burrows s.n.* (syntype: NY).

Ascending or spreading to erect perennial herb, to 1(1.5) m; stem glabrate or sparsely strigose or uncinate. Leaves 3-foliolate, the leaflet blade ovate to lanceolate or narrowly oblong, 2–5(6) cm long, 1.5–2 cm wide, the apex acute to obtuse or rounded, the base rounded, the upper surface glabrate, the lower surface strigulose or subappressed villous, the petiole (1)2–5 cm long; stipules subulate, caducous. Flowers in a terminal, branched raceme, the pedicel 5–12(20) mm; calyx 2–3 mm long; corolla 6–7(8) mm long, lilac to purple. Fruit sinuate above, notched below, the isthmi narrow, the segments 3–5, each 4–7 mm long, convex above, obtusely angled below.

Sandhills and open hammocks. Frequent; nearly throughout. Quebec south to Florida, west to Ontario, Nebraska, Kansas, Oklahoma, and Texas. Summer–fall.

Desmodium perplexum B. G. Schub. [Confused, intricate, in reference to its taxonomy.] PERPLEXED TICKTREFOIL.

Desmodium perplexum B. G. Schubert, Rhodora 52: 154. 1950.

Ascending or erect perennial herb, to 1(1.5) m; stem sparsely or conspicuously pilose, some uncinate trichomes also present. Leaves 3-foliolate, the leaflets broadly to narrowly ovate, 3–9 cm long, 2–3 cm wide, the apex obtuse, the base rounded, the upper surface sparsely pubescent, occasionally uncinulate on the veins, the lower surface subappressed or spreading villous, the petiole (1)2–5 cm long; stipules subulate, caducous. Flowers in a terminal, branched raceme, the pedicel 3–7 mm long; calyx 2–3 mm long; corolla 6–8(9) mm long, lavender to purple. Fruit sinuate above, notched below, the isthmi narrow, the segments (2)3–5, each (4)5–8 mm long, convex above, obtusely angled below.

Hammocks. Rare; Franklin, Gadsden, Leon, and Marion Counties. Maine south to Florida, west to Wisconsin, Nebraska, Oklahoma, and Texas. Summer.

Desmodium rotundifolium DC. [With round leaves, in reference to the leaflets.] PROSTRATE TICKTREFOIL.

Hedysarum rotundifolium Michaux, Fl. Bor.-Amer. 2: 72. 1803; non Vahl, 1791. *Desmodium rotundifolium* de Candolle, Prodr. 2: 330. 1825. *Meibomia rotundifolia* (de Candolle) Kuntze, Revis. Gen. Pl. 1: 197. 1891. *Meibomia michauxii* Vail, Bull. Torrey Bot. Club 23: 140. 1896, nom. illegit.

Prostrate perennial herb; stem to 2(3) m long, pilose. Leaves 3-foliolate, the leaflet blade suborbicular or broadly ovate, the terminal one (2.5)3–3.5 cm long, 2–3.5 cm wide, the apex and base rounded, the upper surface slightly villous, the lower surface appressed-villous, the petiole 2–5 cm long; stipules ovate-lanceolate, subamplexicaul, usually persistent. Flowers in an erect, axillary, usually simple raceme, the pedicels (5)10–15 mm long; calyx 2–4 mm long; corolla 9–11 mm long, blue-purple. Fruit crenate above, notched or deeply crenate below, the isthmi narrow, the segments 3–6(7), each 4–6 mm long, convex above and below, thinly uncinate-puberulent.

Hammocks. Rare; Jackson, Liberty, Gadsden, and Columbia Counties. New Hampshire south to Florida, west to Ontario, Michigan, Illinois, Kansas, Oklahoma, and Texas. Summer.

Desmodium scorpiurus (Sw.) Desv. [Scorpion-like, apparently in reference to its sprawling, contorted stems.] SCORPION TICKTREFOIL.

Hedysarum scorpiurus Swartz, Prodr. 107. 1788. *Desmodium scorpiurus* (Swartz) Desvaux, J. Bot. Ag-ric. 1: 122. 1813. *Meibomia scopiurus* (Swartz) Kuntze, Revis. Gen. Pl. 1: 198. 1891.

Trailing to ascending perennial herb; stem to 6 dm long, uncinate-puberulent and/or incon-spicuously pilose. Leaves 3-foliolate, the leaflet blade ovate-elliptic, 1–2.5 cm long, 0.5–1 cm wide, the apex acute to obtuse, the base rounded, the upper and lower surfaces subglabrate or variously pubescent, the petiole 1–2.5 cm long; stipules deltate, clasping, semipersistent. Flow-ers in a terminal and axillary, usually simple raceme, the pedicel 3–10 mm long; calyx ca. 2 mm long; corolla 4–5 mm long, pink or pale pink. Fruit linear, crenate above and below, the isthmi broad, the segments 5–10, each 1–2 mm wide, the upper and lower margins convex.

Disturbed sites. Rare; Hillsborough and Lee Counties, southern peninsula. Florida; West In-dies, Mexico, Central America, and South America; Africa, Asia, Australia, and Pacific Islands. Native to tropical America. All year.

Desmodium sessilifolium Torr. & A. Gray [With sessile leaflets.] SESSILELEAF TICKTREFOIL.

Desmodium sessilifolium Torrey & A. Gray, Fl. N. Amer. 1: 363. 1840. *Meibomia sessilifolia* (Torrey & A. Gray) Kuntze, Revis. Gen. Pl. 1: 98. 1891.

Ascending to erect perennial herb, to 1(1.5) m; stem uncinate-villous or -puberulent. Leaves 3-foliolate, the leaflet blade narrowly elliptic to linear-lanceolate (3)4–5 cm long, 5–10 mm wide, the apex acute, the base cuneate, the upper surface glabrate or uncinulate-puberulent, sometimes sparsely subappressed-puberulent, the lower surface subappressed- or spreading-villosulous, sessile or the petiole 1–3 mm long; stipules subulate, often awn-tipped, caducous. Flowers in a terminal, branched raceme, the pedicel 2–5 mm long; calyx 2–3 mm long; corolla ca. 5 mm long, pale lavender to reddish-purple. Fruit sinuate above, notched below, the isthmi narrow, the segments 2(3), each 4–6 mm long, nearly straight or convex above, symmetrically rounded below.

Dry, open hammocks. Rare; Jefferson and Wakulla Counties. Massachusetts south to Flor-ida, west to Ontario, Michigan, Nebraska, Kansas, Oklahoma, and Texas. Summer.

Desmodium strictum (Pursh) DC. [Very upright.] PINEBARREN TICKTREFOIL.

Hedysarum strictum Pursh, Fl. Amer. Sept. 483. 1814. *Desmodium strictum* (Pursh) de Candolle, Prodr. 2: 329. 1825. *Meibomia stricta* (Pursh) Kuntze, Revis. Gen. Pl. 1: 198. 1891.

Ascending or erect perennial herb, to 1 m; stem inconspicuously uncinate-puberulent. Leaves 3-foliolate, the leaflet blade oblong to linear, 3–6(8) cm long, 4–5 mm wide, the apex obtuse, the base cuneate, the upper surface glabrous, the lower surface glabrous or sparsely puberulent, the petiole 0.5–1.5 cm long; stipules linear, caducous. Flowers in a terminal, branched raceme, the pedicel 5–10 mm long; calyx ca. 2 mm long; corolla ca. 4 mm long, pink or purple. Fruit slightly sinuate above, notched below, the isthmi narrow, the segments 1–2(3), each 4–6 mm long, concave above, rounded below.

Dry hammocks, sandhills, and flatwoods. Occasional; nearly throughout. New Jersey south to Florida, west to Missouri and Texas. Summer–fall.

Desmodium tenuifolium Torr. & A. Gray [With narrow leaves, in reference to the leaflets.] SLIMLEAF TICKTREFOIL.

Desmodium tenuifolium Torrey & A. Gray, Fl. N. Amer. 1: 363. 1840. *Meibomia tenuifolia* (Torrey & A. Gray) Kuntze, Revis. Gen. Pl. 1: 198. 1891.

Ascending to erect perennial herb, to 1 m; stem glabrate or inconspicuously uncinate-puberulent. Leaves 3-foliolate, the leaflet blade oblong to linear, 3–6(8) cm long, 4–5 mm wide, the apex obtuse to rounded, the base cuneate, the upper surface glabrous, the lower surface glabrous or sparsely puberulent below, the petiole 3–10 mm long; stipules linear, caducous. Flowers in a terminal, branched raceme, the pedicel 5–10 mm long; calyx ca. 2 mm long; corolla ca. 4 mm long, pink. Fruit sinuate above, notched below, the isthmi narrow, the segments 1–2(3), each 4–5 mm long, convex above, rounded below.

Wet flatwoods and hammocks. Occasional; northern counties south to Charlotte County. Maryland south to Florida, west to Louisiana. Summer–fall.

Desmodium tortuosum (Sw.) DC. [Twisted, in reference to the fruit.] DIXIE TICKTREFOIL.

Hedysarum tortuosum Swartz, Prodr. 107. 1788. *Desmodium tortuosum* (Swartz) de Candolle, Prodr. 2: 332. 1825. *Meibomia tortuosa* (Swartz) Kuntze, Revis. Gen. Pl. 1: 198. 1891.

Hedysarum purpureum Miller, Gard. Dict., ed. 8. 1768. *Meibomia purpurea* (Miller) Vail, in Small, Fl. S.E. U.S. 639. 1903. *Desmodium purpureum* (Miller) Fawcett & Rendle, Fl. Jamaica 4: 36. 1920; non Hooker & Arnott, 1832.

Erect annual herb, to 2 m; stem uncinulate-puberulent and pilose. Leaves 3-foliolate, the leaflet blade ovate, elliptic, or ovate-lanceolate, (2)4–10 cm long, 1.5–4 cm wide, the apex obtuse to rounded, the base broadly cuneate, the upper surface glabrate or uncinate, the lower surface subappressed-villous, also uncinulate-puberulent, the petiole (1)2–4 cm long; stipules ovate-acuminate to lanceolate-aristate, semipersistent, the lower ones amplexicaul. Flowers in a terminal simple or sometimes branched raceme, the pedicel 10–15 mm long; calyx 2–3 mm long; corolla ca. 6 mm long, lavender. Fruit crenate above and below, the isthmi narrow to moderately broad, the segments (3)4–6(7), each 3–4 mm long, rounded above and below.

Disturbed sites. Common; nearly throughout. North Carolina south to Florida, west to Texas; West Indies, Mexico, Central America, and South America; Africa, Asia, Australia, and Pacific Islands. Native to tropical America. Spring–fall.

Desmodium triflorum (L.) DC. [Three-flowered.] THREEFLOWER TICKTREFOIL.

Hedysarum triflorum Linnaeus, Sp. Pl. 2: 749. 1753. *Aeschynomene triflora* (Linnaeus) Poiret, in Lamarck, Encycl. 4: 451. 1798. *Desmodium triflorum* (Linnaeus) de Candolle, Prodr. 2: 334. 1825. *Sagotia triflora* (Linnaeus) Duchassaing de Fontbressin & Walpers, Linnaea 23: 738. 1850. *Nicolsonia triflora* (Linnaeus) Grisebach, Abh. Königl. Ges. Wiss. Göttingen 7: 202. 1857. *Meibomia triflora* (Linnaeus) Kuntze, Revis. Gen. Pl. 1: 197. 1891.

Prostrate annual or perennial stoloniferous herb; stem to 8 dm long, ascending-pilose or stri-gose. Leaves 3-foliolate, the leaflet blade broadly obovate or cuneate-obovate, 5–10 mm long, 4–12 mm wide, the apex emarginate, the base cuneate, the upper surface glabrate, the lower surface subappressed-pilose along the medial vein or slightly uncinulate, the petiole 3–8 mm long; stipules lanceolate, persistent. Flowers 1–4 in an axillary fascicle, the pedicel 5–8 mm long; calyx 3–4 mm long; corolla 4–6 mm long, pale pink to purplish. Fruit slightly upcurved to falcate, slightly sinuate above, weakly crenate below, the isthmi broad, the segments 3–5, each 3–4 mm long, nearly square, slightly concave above, convex below, inconspicuously uncinate, becoming glabrate.

Disturbed sites. Frequent; nearly throughout. Florida and Louisiana; West Indies, Mexico, Central America, and South America; Africa, Asia, Australia, and Pacific Islands. Native to tropical America, Africa, Asia, Australia, and Pacific Islands. All year.

Desmodium viridiflorum (L.) DC. [With green flowers.] VELVETLEAF TICKTREFOIL.

> *Hedysarum viridiflorum* Linnaeus, Sp. Pl. 2: 748. 1753. *Desmodium viridiflorum* (Linnaeus) de Can-
> dolle, Prodr. 2: 329. 1825. *Meibomia viridiflora* (Linnaeus) Kuntze, Revis. Gen. Pl. 1: 197. 1891.

Erect or ascending perennial herb, to 1 m; stem uncinate-villosulous. Leaves 3-foliolate, the leaflet blade broadly ovate or rhombic, 5–12(15) cm long, 2–4 cm wide, the apex rounded, sometimes slightly emarginate, the base broadly cuneate to subcordate, the upper surface ob-scurely strigose, uncinulate along the veins, the lower surface villous-pubescent, the petiole 2–5 cm long; stipules deltate, caducous. Flowers in a terminal, branched raceme, the pedicel 3–9 mm long; calyx 2–3 mm long; corolla 7–8 mm long, purple to pink or light lavender. Fruit sinuate above, notched below, the isthmi narrow, the segments (3)4–5(6), each (4)5–8(9) mm long, straight or convex above, obtusely angled below.

Dry hammocks. Occsional; northern and central peninsula, central and western panhan-dle. New Jersey and Pennsylvania south to Florida, west to Missouri, Oklahoma, and Texas. Summer–fall.

EXCLUDED TAXA

> *Desmodium glabrum* (Miller) de Candolle—Reported for Florida by Chapman (1860, as *D. molle* (Vahl) de Candolle), who misapplied the name to material of *D. tortuosum*.
>
> *Desmodium rhombifolium* (Elliott) de Candolle—This taxon has not been properly typified, but seems to be a synonym of a species in the *D. ciliare* group, fide Schubert (1950). Reports for Florida by Small (1903, 1913a, 1933, all as *Meibomia rhombifolia* (Elliott) Vail) are misapplications of the name to material of *D. floridanum* or possibly *D. fernaldii*).

Dichrostachys (DC.) Wight & Arn., nom. cons. 1834.

Shrubs; branches terminating as thorns. Leaves even-bipinnately compound, petiolate, stipu-late. Flowers in axillary spikes, bisexual or sterile; sepals 5, connate; petals 5, connate, subequal; stamens 10, free. Fruit an indehiscent legume.

A genus of 14 species; North America, West Indies, Africa, Asia, and Australia. [From the Greek *di*, two, *chroma*, colored, and *stachys*, spike, in reference to the flowers of two colors in a spike.]

Cailliea Guill. & Perr., nom. rej. 1832.

Selected reference: Brenan and Brummitt (1965).

Dichrostachys cinerea (L.) Wight & Arn. subsp. africana Brenan & Brummitt [Ash-gray; of Africa.] AROMA.

Dichrostachys cinerea (Linnaeus) Wight & Arnott subsp. *africana* Brenan & Brummitt, Bol. Soc. Brot., ser. 2. 39: 77. 1965.

Mimosa glomerata Forsskål, Fl. Aegypt.-Arab. 177. 1775. *Dichrostachys glomerata* (Forsskål) Chiovenda, Ann. Bot. (Rome) 13: 409. 1915. *Cailliea glomerata* (Forsskål) J. F. Macbride, Contr. Gray Herb. 59: 16. 1919.

Shrub, to 2 m; branches with spurs terminating as thorns, pubescent. Leaves even-bipinnately compound, the rachis with a stalked gland between the lowermost leaflet or between several pairs of leaflets, the pinnae 6–12 pairs, the leaflets 16–22 pairs, the blade oblong, (1)3–5(10) mm long, the apex acute to obtuse, the base obtuse to truncate, asymmetrical, the upper and lower surfaces sparsely pubescent, the petiole ca. 1 cm long. Flowers numerous in an axillary, solitary or clustered pendent spike, the upper flowers fertile, yellowish, the lower ones sterile, pinkish, fading to white; calyx 1–2 mm long, shallowly 5-lobed; corolla 4–5 mm long, 5-lobed; stamens of the fertile flowers 5–6 mm long; staminodes of the sterile flowers ca. 1 cm long. Fruit oblong, 4–6 cm long, ca. 1 cm wide, laterally compressed, the margin undulate, indehiscent, contorted at maturity; seeds several.

Disturbed sites. Rare; Lake, Polk, Palm Beach, and Miami-Dade Counties, Monroe County keys. Florida; West Indies; Africa. Native to Africa. Spring–summer.

EXCLUDED TAXON

Dichrostachys cinerea (Linnaeus) Wight & Arnott—The typical subspecies was reported for Florida by implication by Wunderlin (1982, 1998) and by Isely (1990, 1998). Our material is subsp. *africana*.

Enterolobium Mart. 1837.

Trees. Leaves even-bipinnately compound, stipellate, petiolate, stipulate. Flowers in axillary, pedunculate, solitary or fasciculate heads; sepals 5, connate, subequal; petals 5, connate; stamens ca. 80, basally connate. Fruit an indehiscent legume.

A genus of 11 species; North America, West Indies, Mexico, Central America, South America, and Africa. [From the Greek *enteron*, intestine, and *lobion*, pod or capsule, in reference to the curved fruit that resembles an intestine.]

Selected reference: Barneby and Grimes (1996).

Enterolobium contortisiliquum (Vell.) Morong [Twisted, flattened, in reference to the fruit.] EARPOD TREE.

Mimosa contortisiliqua Vellozo, Fl. Flumin. 436. 1881. *Feuilleea contortisiliqua* (Vellozo) Kuntze, Revis. Gen. Pl. 1: 185. 1891. *Enterolobium contortisiliquum* (Vellozo) Morong, in Morong & Britton, Ann. New York Acad. Sci. 7: 102. 1892.

Tree, to 30 m; bark smooth, the branches glabrous or sparsely pubescent. Leaves even-pinnately compound, the pinnae 4–10(15) pairs, the leaflet blade (12)15–30 pairs, opposite, asymmetrically linear-oblong to subfalcate, 8–15 mm long, 3–4 mm wide, the apex obliquely acute to obtuse, mucronate, the base obliquely rounded, the upper and lower surfaces glabrous or sparsely pubescent, the stipels linear-subulate, ca. 1 mm long, short-petiolulate, the petiole 3–5(8) cm long, glabrous or sparsely pubescent, with a sessile, elliptic gland above the middle; stipules linear-subulate, 2–3 mm long, caducous. Flowers in a short, fasciculate raceme, the head 1–2 cm in diameter, the peduncle 0.5–2 mm long, pedicel ca. 1 mm long; calyx ca. 2 mm long, white or cream, pubescent on the outer surface, 5-ribbed, the lobes triangular; corolla funnelform, 5–6 mm long, white or cream, pubescent on the outer surface, the lobes ca. 2 mm long; stamens ca. 80, 8–10 mm long, white or cream, glabrous, connate ca. 4 mm proximally. Fruit deeply reniform, (5.5)6–7(8) cm long, (4)5–6 cm wide, blackish, dull, glaucous; seeds (5)8–9(14), ellipsoid, ca. 1 cm long, flattened, dark brown, with an evident pleurogram.

Disturbed sites. Occasional; central peninsula, Broward County. Escaped from cultivation. Florida; West Indies and South America; Africa. Native to South America. Spring-summer.

EXCLUDED TAXON

Enterolobium cyclocarpum (Jacquin) Grisebach—The name was misapplied by Lakela and Wunderlin (1980) to Florida material of *E. contortisiliquum*.

Erythrina L. 1753.

Perennial suffrutescent herbs, shrubs, or trees. Leaves 3-foliolate, petiolate, stipellate, stipulate. Flowers in terminal or intercalary pseudoracemes, bracteate, bracteolate; sepals 5, connate; corolla papilionaceous; stamens diadelphous (9 + 1). Fruit a dehiscent legume.

A genus of about 120 species; North America, West Indies, Mexico, Central America, South America, Africa, Asia, and Australia. [*Erythro*, red, in reference to the flower color of some species.]

Erythrina herbacea L. [Herbaceous.] CORALBEAN; CHEROKEE BEAN.

Erythrina herbacea Linnaeus, Sp. Pl. 2: 706. 1753. *Erythrina humilis* Salisbury, Prodr. Stirp. Chap. Allerton 335. 1796, nom. illegit. *Corallodendron herbaceum* (Linnaeus) Kuntze, Revis. Gen. Pl. 1: 172. 1891.

Erythrina herbacea Linnaeus var. *arborea* Chapman, Fl. South. U.S., ed. 3. 117. 1897. *Erythrina arborea* (Chapman) Small, S.E. U.S. 647, 1332. 1903. TYPE: FLORIDA.

Erythrina herbacea Linnaeus forma *albiflora* Moffler & Crewz, Phytologia 52: 288. 1983. TYPE: FLORIDA: Pinellas Co.: Fort DeSoto Park, on St. Jean Key W of FL 693, T33S, R16E, Sec. 8, 24 April 1981, *Crewz 2132* (holotype: USF).

Erect perennial suffrutescent herb or shrub, to 1.5 m or small tree in Southern Florida to 5 m; stem glabrous, usually with prickles. Leaves 3-foliolate, the leaflet blade broadly ovate, (2)3–8(10) cm long and wide, often 3-lobed, the apex acute to obtuse, the lateral lobes (if present) rounded, the base broadly cuneate to subcordate, the upper and lower surfaces glabrous, often with prickles on the veins, petiolules, and petioles, the petiole 4–8 cm long. Flowers few to many in a long-pedunculate, terminal or intercalary axillary pseudoraceme; calyx tube short-cylindric or obconic, truncate, 5–8 mm long, scarcely 2-lipped, the lobes abortive; corolla elongate, 3–5 cm long, red, the standard remaining folded, the wing and keels much shorter; stamens subequaling the standard, red. Fruit oblong, 6–15(20) cm long, 1–1.5 cm wide, moniliform, torulose after dehiscence; seeds red.

Hammocks. Frequent; nearly throughout. North Carolina south to Florida, west to Oklahoma and Texas; Mexico. Winter–spring.

EXCLUDED TAXON

Erythrina crista-galli Linnaeus—Reported for Florida by Small (1933, as *Micropteryx crista-galli* (Linnaeus) Walpers). No vouchering specimens seen.

Galactia P. Browne 1756. MILKPEA

Perennial herbs. Leaves odd-pinnately compound, stipellate, petiolate, stipulate. Flowers in axillary pseudoracemes, bracteate, bracteolate; sepals 5, 4-lobate; corolla papilionaceous; stamens diadelphous (9 + 1). Fruit a dehiscent legume.

A genus of 70 species; North America, West Indies, Mexico, Central America, South America, Africa, Asia, and Australia. [From the Greek *galacto-*, milky, from Patrick Browne, who stated in the original description that it had "milky branches."]

Although *Galactia* has been much studied recently by several workers, there is still considerable disagreement on the generic and species delimitations (see Isely, 1990, 1998; Schrire, 2005; Ward and Hall, 2004). Therefore, this treatment, like all past ones, must be considered tentative pending still further study.

Selected reference: Ward and Hall (2004).

1. Leaves (5)7(9)-foliolate... G. elliottii
1. Leaves 3-foliolate.
  2. Plant erect, rarely sprawling ....G. erecta
  2. Plant prostrate or twining.
    3. Fruit 6–8(9) mm wide... G. striata
    3. Fruit 3–5.5(6) mm wide.
      4. Fruit sericeous or villous; corolla persistent on the fruit and drying dark .... G. mollis
      4. Fruit glabrous or pubescent; corolla soon deciduous.
        5. Inflorescence usually much exserted, bearing flowers ½ to most of the axis length from well-spaced nodes in which the distance between the flowers or flower clusters evidently exceeds the length of the flower and pedicel; corolla 8–12(14) mm long........G. regularis

5. Inflorescences usually short, if exserted, then flowering above the middle of the axis from crowded nodes in which the distance between the flowers or flower clusters is less than the length of the flower and pedicel; corolla (11)12–18 mm long .... **G. volubilis**

Galactia elliottii Nutt. [Commemorates Stephen Elliott (1771–1830), American legislator, banker, educator, and botanist.] ELLIOTT'S MILKPEA.

Galactia elliottii Nuttall, Gen. N. Amer. Pl. 2: 117. 1818. *Tephrosia elliottii* (Nuttall) Bentham, Ann. Wiener Mus. Naturgesch. 2: 127. 1838.

Galactia elliottii Nuttall var. *leavenworthii* Torrey & A. Gray, Fl. N. Amer. 1: 687. 1840. TYPE: FLORIDA: "East Florida," without data, *Leavenworth s.n.* (Holotype: NY?).

Twining perennial herb; stem to 1.5 m, glabrous. Leaves (5)7(9)-foliolate, the leaflet blade elliptic or oblong, 2–5 cm long, 0.5–1.5 cm wide, the apex rounded or emarginate, the base rounded, the upper and lower surfaces glabrous, the petiolule 1–2 mm long, the petiole 0.5–2 cm long, stipules setaceous, 2–4 mm long. Flowers in an axillary pseudoraceme 4–15 cm long or reduced to 1–2 flowers in the leaf axils; calyx 7–11 mm long, strigulose or glabrate; corolla 10–15(17) mm long, white with red striations. Fruit linear, 3–6 cm long, 5–9 mm wide, strigose.

Sandhills, scrub, and flatwoods. Common; peninsula, Taylor and Bay Counties. South Carolina, Georgia, and Florida. Spring–fall.

Galactia erecta (Walter) Vail [Upright.] ERECT MILKPEA.

Ervum erectum Walter, Fl. Carol. 187. 1788. *Galactia erecta* (Walter) Vail, Bull. Torrey Bot. Club 22: 502. 1895.

Galactia brachypoda Torrey & A. Gray, Fl. N. Amer. 1: 288. 1838. TYPE: FLORIDA: without data, *Chapman s.n.* (holotype: NY).

Galactia sessiliflora Torrey & A. Gray, Fl. N. Amer. 1: 288. 1838. TYPE: GEORGIA/FLORIDA.

Erect (rarely sprawling) perennial herb, to 3.5 dm; stem glabrate. Leaves 4–6(8), 3-foliolate, the leaflet blade elliptic or elliptic-oblong, 2–4(5) cm long, 5–10 mm wide, the apex rounded or emarginate, the base rounded, the upper and lower surfaces glabrate, the petiole 4–6 cm long; the stipules subulate, 3–4 mm long. Flowers 2–4(8) in a cluster in the leaf axil; calyx 4–6 mm long, slightly puberulent; corolla 7–9 mm long, white or lavender. Fruit linear, 2–3 cm long, 4–5 mm wide, sparsely villous.

Sandhills and flatwoods. Occasional; Clay and Suwannee Counties; central and western panhandle. North Carolina south to Florida, west to Texas. Spring–summer.

Galactia mollis Michx. [Soft.] SOFT MILKPEA.

Galactia mollis Michaux, Fl. Bor.-Amer. 2: 61. 1803.

Trailing or twining perennial herbs; stem to 1 m, villous. Leaves 3-foliolate, the leaflet blade ovate or elliptic, 2–4(4.5) cm long, 1–2 cm wide, the apex acute to rounded, the base broadly cuneate to rounded, the upper surface subappressed villous to glabrate, the lower surface subappressed villous, the petiole ca. 2 cm long; stipules subulate, 2–3 mm long. Flowers 6–10 on the upper ¼–½ of the axis, the inflorescence (3)5–12(15) cm long; calyx 5–8 mm long, villous; corolla 8–10 mm long, red-purple or white. Fruit linear, 2.5–3.5(4) cm long, 4.5–5 mm wide, sericeous or subappressed villous, the corolla persistent and drying dark.

Sandhills and scrub. Occasional; northern counties south to Hillsborough and Polk Counties. Georgia south to Florida, west to Mississippi. Summer.

Galactia regularis (L.) Britton et al. [Having all parts similar, apparently in reference to the leaflet blade shape.] DOWNY MILKPEA.

> *Dolichos regularis* Linnaeus, Sp. Pl. 2: 726. 1753. *Galactia regularis* (Linnaeus) Britton et al., Prelim. Cat. 14. 1888.
> *Galactia pilosa* Nuttall, Gen. N. Amer. Pl. 2: 116. 1818.
> *Galactia macreei* M. A. Curtis, Boston J. Nat. Hist. 1: 120. 1837. *Galactia pilosa* Nuttall var. *macreei* (M. A. Curtis) Torrey & A. Gray, Fl. N. Amer. 1: 287. 1838.
> *Galactia parvifolia* A. Richard, in Sagra, Hist. Fis. Cuba, Bot. 10: 176. 1845. *Galactia parvifolia* A. Richard var. *triphylla* Urban, Symb. Antill. 2: 314. 1900, nom. inadmiss.

Twining or trailing perennial herb; stem to 2 m, villous or loosely subappressed, strigose, or strigulose. Leaves 3-foliolate, the leaflet blade ovate to oblong-lanceolate, 1.5–4.5 cm long, 1–2.5 cm wide, the apex obtuse to rounded, the base cordate, the upper surface glabrate, the lower surface villous, subappressed, strigose or strigulose, the petiole 1–2 cm long; stipules lanceolate, 2–4 mm long. Flowers in a pseudoraceme (2)4–15(20) cm long, the inflorescence usually much exserted, bearing flowers ½ to nearly the axis length from well-spaced nodes in which the distance between the flowers or flower clusters exceeds the length of the flower and pedicel; calyx 4–7(8) mm long; corolla 8–12(14) mm long, pink to purple. Fruit linear, 2–6 cm long, 3–5(6) mm wide, strigose or puberulent.

Sandhills and open hammocks. Common; nearly throughout. New Jersey and Pennsylvania south to Florida, west to Kansas, Oklahoma, and Texas; West Indies. Summer–fall.

Galactia striata (Jacq.) Urb. [Striped, in reference to the corolla.] FLORIDA HAMMOCK MILKPEA.

> *Glycine striata* Jacquin, Hort. Bot. Vindob. 1: 32, t. 76. 1771. *Galactia striata* (Jacquin) Urban, Symb. Antill. 2: 320. 1900.
> *Galactia cubensis* Kunth, in Humboldt et al., Nov. Gen. Sp. 6: 429. 1824. *Galactia filiformis* Bentham var. *cubensis* (Kunth) Gómez de la Maza y Jiménez, Anales Soc. Esp. Hist. Nat. 23: 296. 1894; non Grisebach, 1866. *Galactia striata* (Jacquin) Urban var. *cubensis* (Kunth) Urban, Symb. Antill. 2: 322. 1900.
> *Galactia spiciformis* Torrey & A. Gray, Fl. N. Amer. 1: 288. 1838. TYPE: FLORIDA: Monroe Co.: Key West, s.d., *Bennett 283* (holotype: NY).

Twining perennial herb; stem to 4 m long, villous to retrorsely appressed-pubescent, sometimes glabrate in age. Leaves 3-foliolate, the leaflet blade ovate or ovate-elliptic, 2–6 cm long, 1–3 cm wide, the apex obtuse to emarginate, the base cuneate to rounded, the upper surface glabrescent, the lower surface subappressed-villous, the petiole 2–5 cm long; stipules subulate, 2–4 mm long. Flowers in a pseudoraceme (2)5–15(20) cm long; calyx 7–8 mm long, strigulose; corolla ca. 1 cm long, red-purple to blue, the standard sometimes white-striate. Fruit linear, 3–7 cm long, 6–8(9) mm wide, strigose or subappressed-villous.

Coastal hammocks and pinelands; occasional; Manatee, Sarasota, Charlotte, and Lee Counties, southern peninsula. Florida; West Indies, Mexico, Central America, and South America. All year.

Galactia volubilis(L.) Britton [Twining.] EASTERN MILKPEA.

Hedysarum volubile Linnaeus, Sp. Pl. 2: 750. 1753. *Galactia volubilis* (Linnaeus) Britton, Mem. Torrey Bot. Club 5: 208. 1894.

Galactia glabella Michaux, Fl. Bor.-Amer. 2: 62. 1803.

Galactia floridana Torrey & A. Gray, Fl. N. Amer. 1: 288. 1838. TYPE: FLORIDA: Hillsborough Co.: Tampa Bay, s.d., *Burrows s.n.* (holotype: NY).

Galactia pilosa Nuttall var. *angustifolia* Torrey & A. Gray, Fl. N. Amer. 1: 287. 1838. *Galactia volubilis* (Linnaeus) Britton var. *intermedia* Vail, Bull. Torrey Bot. Club 22: 508. 1895, nom. illegit.

Galactia floridana Torrey & A. Gray var. *microphylla* Chapman, Fl. South. U.S. 108. 1860. *Galactia microphylla* (Chapman) D. W. Hall & D. B. Ward ex Isely, Brittonia 38: 354. 1986. TYPE: FLORIDA: without data, *Chapman s.n.* (lectotype: NY). Lectotypified by Duncan (1977: 61).

Galactia mollis Michaux var. *microphylla* A. W. Wood, Amer. Bot. Fl. 98. 1870. TYPE: FLORIDA/GEORGIA.

Galactia fasciculata Vail, Bull. Torrey Bot. Club 22: 505. 1895. *Galactia volubilis* (Linnaeus) Britton var. *fasciculata* (Vail) D. B. Ward & D. W. Hall, Phytologia 86: 69. 2004. TYPE: FLORIDA: Hillsborough Co.: Tampa, 24 Aug 1895, *Nash 2480* (holotype: NY).

Galactia floridana Torrey & A. Gray var. *longeracemosa* Vail, Bull. Torrey Bot. Club 22: 505. 1895. TYPE: FLORIDA: without data, 1889, *Simpson s.n.* (holotype: ?).

Galactia pinetorum Small, Fl. Miami 93, 200. 1913. TYPE: FLORIDA: Miami-Dade Co.: between Coconut Grove and Cutler, 9 May 1904, *Small & Wilson 1592* (holotype: NY).

Galactia prostrata Small, Man. S.E. Fl. 719, 1505. 1933; non Bentham, 1838. *Galactia smallii* H. J. Rogers ex A. Herndon, Rhodora 83: 471. 1981. TYPE: FLORIDA: Miami-Dade Co.

Galactia minor W. H. Duncan, Phytologia 37: 59. 1977.

Galactia volubilis (Linnaeus) Britton var. *baltzelliana* D. B. Ward & D. W. Hall, Phytologia 86: 68. 2004. TYPE: Lake Co.: 1 mi. N of Leesburg, 25 Sep 1975, *Hall & Baltzell 413* (holotype: FLAS; isotypes: FLAS).

Trailing perennial herb; stem 1.5(3) m, strigulose. Leaves 3-foliolate, the leaflet blade ovate, elliptic to ovate-lanceolate, 1.5–5(7) cm long, 1–3(5) cm wide, the apex rounded or emarginate, the base rounded to cordate, the upper surface glabrate, the lower surface strigulose, the petioles 5–10 mm long; stipules lanceolate, 2–3 mm long. Flowers in a pseudoraceme (2)3–15 cm long, the inflorescence usually short, if exserted, then flowering above the middle of the axis from crowded nodes in which the distance between the flowers or flower clusters is less than the length of the flower and pedicel; calyx 6–9 mm long, strigose or glabrate; corolla (11)12–18 mm long, lavender to pink-purple or bicolored, rarely white. Fruit linear, 3–5 cm long, 4.5–5.5 mm wide, strigose.

Sandhills, scrub, and dry hammocks. Frequent; nearly throughout. New York south to Florida, west to Kansas, Oklahoma, and Texas. All year.

EXCLUDED TAXA

Galactia filiformis (Jacquin) Bentham—This species, an endemic of Hispaniola, was cited for Florida by Chapman (1897), who misapplied the name probably to material of *G. striata*.

Galactia mohlenbrockii R. H. Maxwell—Reported for Florida by Correll and Johnston (1970, as *Dioclea multiflora* (Torrey & A. Gray) C. Mohr), by Clewell (1985, as *Dioclea multiflora* (Torrey & A. Gray) C. Mohr), Isely (1990, as *Dioclea multiflora* (Torrey & A. Gray) C. Mohr), Wunderlin (1998), and Wunderlin and Hansen (2003). At least some of these reports were apparently based on misidentifications of *Phaseolus polystachios*). No voucher specimens seen.

Gleditsia J. Clayton 1753. LOCUST

Trees. Leaves alternate, evenly 1- or 2-pinnately compound, petiolate, the stipules obsolete. Flowers radially symmetrical or nearly so (caesalpinioid), borne in spikes or racemes, unisexual or bisexual; sepals, petals, and stamens borne on the hypanthium lip; sepals 3–5, free; petals 3–5, free; stamens 6–8. Fruit an indehiscent legume.

A genus of about 15 species; nearly cosmopolitan. [Commemorates Johann Gottlieb Gleditsch (1714–1786), German botanist and director of the Berlin Botanical Garden.]

1. Fruit ovate, 3–5(8) cm long; seeds 1(3)... **G. aquatica**
1. Fruit oblong, 20–40 cm long; seeds numerous..**G. triacanthos**

Gleditsia aquatica Marshall [Of water.] WATER LOCUST.

Gleditsia aquatica Marshall, Arbust. Amer. 54. 1785. *Gleditsia triacanthos* Linnaeus var. *aquatica* (Marshall) Castiglioni, Viagg. Staati Uniti 2: 249. 1790.

Gleditsia monosperma Walter, Fl. Carol. 254. 1788. *Asacara aquatica* Rafinesque, Sylva Tellur. 121. 1838, nom. illegit. *Caesalpiniodes monosperma* (Walter) Kuntze, Revis. Gen. Pl. 1: 166. 1891.

Tree, to 25 m; bark grayish brown to blackish, narrowly furrowed or warty, the stems with simple or branched thorns 7–14 cm long. Leaves 1- or 2-pinnately compound, 10–20 cm long, the pinnae 2–6(8) pairs, the leaflet blade 7–15 pairs, oblong-lanceolate, 2.5–5 cm long, 1.2–2.1 cm wide, the apex obtuse to rounded, the base asymmetrically cuneate, the upper and lower surfaces glabrous. Staminate flowers numerous, in elongate clusters 5–7 cm long; bisexual flowers few, in slender clusters 7–9 cm long; perianth of 2 similar 3- to 5-merous whorls, yellowish green; stamens 6–8. Fruit elliptic-, oval-, or ovate-oblique, 2.5–5 cm long, 2–3.5 cm wide, dark brown; seeds 1–3, suborbicular, 1–1.5 cm long.

Swamps, sloughs, and floodplain forests. Frequent; northern counties, central peninsula. New York south to Florida, west to Texas. Spring.

Gleditsia triacanthos L. [From the Greek *tri-*, three, and *acanthos*, spines, apparently in reference to the sometimes 3-branched spines.] HONEY LOCUST.

Gleditsia triacanthos Linnaeus, Sp. Pl. 2: 1056. 1753. *Gleditsia elegans* Salisbury, Prodr. Stirp. Chap. Allerton 233. 1796, nom. illegit. *Gleditsia triacanthos* Linnaeus var. *macrocarpos* Michaux, Fl. Bor.-Amer. 2: 257. 1803, nom. inadmiss. *Gleditsia polysperma* Stokes, Bot. Mat. Med. 1: 228. 1812, nom. illegit. *Caesalpiniodes triacanthos* (Linnaeus) Kuntze, Revis. Gen. Pl. 1: 166. 1891.

Tree, to 30(45) m; bark with deep fissures, and long, narrow scaly ridges between them, the stem with simple or branched thorns 6–15(40) cm long. Leaves 1- or 2-pinnately compound, 10–20 cm long, the pinnae 2–6(8) pairs, the leaflets 7–15 pairs, the blade oblong-lanceolate, 2.5–5 cm long, 1.2–2.1 cm wide, the apex obtuse to rounded, the base asymmetrically cuneate, the upper surface glabrous, the lower surface pubescent along the veins. Staminate flowers numerous, in elongate clusters 5–7 cm long; bisexual flowers few, in slender clusters 7–9 cm long; perianth of 2 similar 3- to 5-merous whorls, yellowish green; stamens 6–8. Fruit linear-oblong, 15–45 cm long, 2–3.5 cm wide, twisting at maturity, dark brown; seeds numerous, each surrounded by a sugary pulp, dark brown, ca. 8 mm long.

Moist hammocks. Occasional; panhandle, Levy County. Maine south to Florida, west to Ontario, Idaho, and California; South America; Europe, Africa, Asia, and Australia. Native to North America. Spring.

HYBRID

Gleditsia ×texana Sargent (*G. aquatica* × *G. triacanthos*) [Of Texas.]

Gleditsia texana Sargent, Bot. Gaz. 31: 1. 1901, pro sp.

Disturbed sites. Rare; Wakulla County. Spring.

Gliricidia Kunth 1824. QUICKSTICK

Trees or shrubs. Leaves alternate to subopposite, odd-pinnately compound, estipellate, petiolate, stipulate. Flowers in axillary clustered racemes, bracteate; calyx zygomorphic, 5-lobed; corolla papilionaceous; stamens 10, diadelphous (9 + 1), the anthers basifixed. Fruit a dehiscent legume.

A genus of 5 species; North America, West Indies, Mexico, Central America, South America, Asia, Africa, and Pacific Islands. [*Glis*, dormouse, and *caedo*, kill, in reference to the powdered bark and seeds being used as a rodent poison in the tropics.]

Selected references: Lavin and Sousa (1995); Lavin et al. (2003).

Gliricidia sepium (Jacq.) Kunth ex Walp. [*Sepes*, hedge, used for hedges.] QUICKSTICK.

Robinia sepium Jacquin, Enum. Syst. Pl. 28. 1760. *Lonchocarpus sepium* (Jacquin) de Candolle, Prodr. 2: 260. 1825. *Gliricidia sepium* (Jacquin) Kunth ex Walpers, Repert. Bot. Syst. 1: 679. 1842.

Trees or shrubs, to 15 m. Leaves (15)19–30(35) cm long, with (7)13–21(25) leaflets, the leaflet blade 4.5–8 cm long, 2–4 cm wide, the apex obtuse, the base narrowly cuneate to rounded, the margin entire, the upper and lower surfaces glabrate to strigose, the petiole 1.5–3 cm long; stipules triangular, 1–2 mm long. Flowers 20–100 in an axillary, usually clustered raceme; bracts triangular, ca. 1 mm long, the pedicels 5–10(15) mm long; calyx (5)6–9 mm long, glabrous to sparsely strigose, the lobes equal, minute; corolla papilionaceous, 15–22 mm long, glabrous. Fruit 10–17(23) cm long, 1.5–2 cm wide, brown, glabrous; seeds 3–10, lenticular.

Disturbed sites. Rare; Monroe County keys. Escaped from cultivation. Florida; West Indies, Mexico, Central America, and South America; Africa, Asia, Australia, and Pacific Islands. Native to tropical America. Spring.

Hylodesmum Ohashi & R. R. Mill 2000. TICKTREFOIL

Perennial herbs, monomorphic or dimorphic (leaves and flowers born on same stem or separate stems). Leaves alternate, 3-foliolate, estipellate, petiolate, stipulate. Flowers in terminal branched pseudoracemes, bracteate, ebracteolate; sepals connate, 5-lobed; corolla papilionaceous; stamens 10, monadelphous. Fruit a stipitate loment dividing into 1-seeded segments.

A genus of 14 species; North America, Mexico, Africa, and Asia. [From the Greek *hyle*, forest, and *desmos*, chain, in reference to an abbreviated form of *Desmodium*.]

Selected reference: Ohashi and Mill (2000).

1. Plant dimorphic, the flowering stems leafless or rarely with 1–3 leaves, scapose, distinct from the shorter foliar stems .... **H. nudiflorum**
1. Plants monomorphic, the foliar stems bearing the inflorescence.
  2. Leaves alternate; inflorescence axillary and terminal; flowers white .... **H. pauciflorum**
  2. Leaves usually subverticillate; inflorescence terminal; flowers rose-purple.............. **H. glutinosum**

Hylodesmum glutinosum (Muhl. ex Willd.) Ohashi & R. R. Mill [Glutinous, viscid.] POINTEDLEAF TICKTREFOIL.

Hedysarum glutinosum Muhlenberg ex Willdenow, Sp. Pl. 3: 1198. 1802. *Desmodium glutinosum* (Muhlenberg ex Willdenow) A. W. Wood, Class-Book Bot, 120, 1845. *Hylodesmum glutinosum* (Muhlenberg ex Willdenow) Ohashi & R. R. Mill, Edinburgh J. Bot. 57: 177. 2000.

Hedysarum acuminatum Michaux, Fl. Bor.-Amer. 2: 72. 1803. *Desmodium acuminatum* (Michaux) de Candolle Prodr. 2: 329. 1825. *Meibomia acuminata* (Michaux) S. F. Blake, Bot. Gaz. 78: 277. 1924.

Erect perennial herb, to 8 dm; stem monomorphic (foliar stem bearing the inflorescence), sparsely pilose or glabrate. Leaves in a subverticillate cluster of ca. 6, the leaflet blade broadly ovate-acuminate, 5–10 cm long, 2–10 cm wide, the apex acuminate, the base rounded to subcordate, the upper surface glabrate, the lower surface inconspicuously spreading-villous, the petiole 3–8 cm long; stipules linear-subulate, caducous. Flowers in a terminal branched pseudoraceme, the pedicel 3.5–5.5 mm long; calyx ca. 2 mm long, the lobes shorter than the tube; corolla pink to pink-purple, rarely white, 5–7 mm long. Fruit 7–9 mm long, sinuate or straight above, incised to the ventral suture below, the isthmi very narrow, the segments 2–3, each 8–9 mm long, usually narrowly concave above, rounded below.

Mesic to dry, open hammocks. Rare; Leon and Jackson Counties. Quebec south to Florida, west to North Dakota, South Dakota, Nebraska, Kansas, Oklahoma, and Texas; Mexico. Summer.

Hylodesmum nudiflorum (L.) Ohashi & R. R. Mill [Naked flowered, in reference to the leafless or nearly leafless flowering stems.] NAKEDFLOWER TICKTREFOIL.

Hedysarum nudiflorum Linnaeus, Sp. Pl. 2: 749. 1753. *Desmodium nudiflorum* (Linnaeus) de Candolle, Prodr. 2: 330. 1825. *Meibomia nudiflora* (Linnaeus) Kuntze, Revis. Gen. Pl. 1: 197. 1891. *Pleurolobus nudiflorus* (Linnaeus) MacMillan, Metasp. Minnesota Valley 321. 1892. *Hylodesmum nudiflorum* (Linnaeus) Ohashi & R. R. Mill, Edinburgh J. Bot. 57: 180. 2000.

Erect perennial herb, to 7 dm; stem dimorphic (sterile, foliose stem simple, bearing 5–6 apically subverticillate or slightly spaced leaves; fertile stem arising slightly below ground level from the base of the foliar stem, usually without leaves, taller than the sterile stem), glabrate or sparsely pilose. Leaves with the leaflet blade ovate, 4.5–8.5 cm long, 3–6 cm wide, the apex acute to obtuse, the base broadly cuneate to rounded, the upper surface glabrate, the lower surface sparsely pilose, the petiole 5–8(10) cm long; stipules linear-subulate, caducous. Flowers in a terminal branched pseudoraceme, the pedicel 12–25 mm long; calyx 2–3 mm long, the lobes shorter than the tube; corolla pink, rarely white, 6–8 mm long. Fruit 12–15 mm long, sinuate above, incised to the ventral suture below, the isthmi very narrow, the segments 2–3(4), each 7.5–8 mm long, concave above, asymmetrically rounded below.

Mesic hammocks. Occasional; northern counties. Quebec south to Florida, west to Ontario, Minnesota, Iowa, Kansas, Oklahoma, and Texas. Summer.

Hylodesmum pauciflorum (Nutt.) Ohashi & R. R. Mill [Few-flowered.] FEWFLOWERED TICKTREFOIL.

Hedysarum pauciflorum Nuttall, Gen. N. Amer. Pl. 2: 109. 1818. *Desmodium pauciflorum* (Nuttall) de Candolle, Prodr. 2: 330. 1825. *Meibomia pauciflora* (Nuttall) Kuntze, Revis. Gen. Pl. 1: 198. 1891. *Hylodesmum pauciflorum* (Nuttall) Ohashi & R. R. Mill, Edinburgh J. Bot. 57: 181. 2000.

Ascending or spreading perennial herb, to 6 dm; stem monomorphic (foliar stem bearing the inflorescence), sparsely pilose to glabrate. Leaves usually 4–6, separated on the stem, the leaflet blade ovate to rhombic, 3–7 cm long, 2–5 cm wide, the apex acute to subacuminate, the upper surface glabrous, the lower surface villous to puberulent, the margin ciliate, the petiole 3–7 cm long; stipules linear-subulate, caducous. Flowers usually in terminal simple pseudoracemes, sometimes also with secondary pseudoracemes from the upper leaf axils, the pedicel 2–7 mm long; calyx ca. 2 mm long, the lobes shorter than the tube; corolla 5–6 mm long, white. Fruit 6–9 mm long, sinuate above, incised to the ventral suture below, the isthmi very narrow, the segments 1–2(3), each 9–10 mm long, concave above, asymmetrically rounded below.

Moist hammocks. Rare; central panhandle. New York south to Florida, west to Kansas, Oklahoma, and Texas. Summer–fall.

Indigofera L. 1753. INDIGO

Annual or perennial herbs or shrubs. Leaves alternate, odd-pinnately compound or 3-foliolate, the leaflets opposite or alternate, entire, estipellate, petiolate, stipulate. Flowers in axillary racemes, bracteate, ebracteolate; sepals 5, connate, 5-lobed; corolla papilionaceous; stamens diadelphous (9 + 1). Fruit a dehiscent or indehiscent legume, septate.

A genus of about 700 species; nearly cosmopolitan. [Indigo and -*fer*, bearing, in reference to the plant being the source of indigo dye.]

Selected references: de Kort and Thijsse (1984); Lievens (1992).

1. Leaves 3-foliolate...**I. pilosa**
1. Leaves odd-pinnately compound.
   2. Fruit ovoid or shortly oblong, less than 1 cm long, indehiscent, 2- to 3-seeded...........**I. caroliniana**
   2. Fruit linear, more than 1 cm long, dehiscent, several-seeded.
      3. Stem hirsute or pilose.
         4. Plant with nonglandular trichomes; flowers in a much exserted raceme; fruits deflexed ......
         ....**I. hirsuta**
         4. Plant with glandular trichomes; flowers in a raceme equaling or shorter than the leaves; fruits spreading ....**I. colutea**
      3. Stem strigose, strigulose, or glabrate.

　　5. Erect or ascending suffrutescent herb or shrub.
　　　6. Fruit stout, 1.5–2 cm long, distinctly falcate...**I. suffruticosa**
　　　6. Fruit slender, 2.8–3.5 cm long, straight or nearly so...**I. tinctoria**
　　5. Prostrate, decumbent, or scrambling herb.
　　　7. Fruit 3–4.5 cm long; plant usually decumbent or scrambling....................................**I. trita**
　　　7. Fruit 1–2.5 cm long; plant prostrate.
　　　　8. Stipules deltoid or lanceolate; fruits closely imbricate, deflexed......................**I. spicata**
　　　　8. Stipules subulate; fruits loosely disposed, divergent to deflexed...................**I. miniata**

Indigofera caroliniana Mill. [Of Carolina.] CAROLINA INDIGO.

Indigofera caroliniana Miller, Gard. Dict., ed. 8. 1768.
Indigofera caroliniana Walter, Fl. Carol. 187. 1788; non Miller, 1768. *Anila caroliniana* Kuntze, Revis.
　　Gen. Pl. 2: 939. 1891.

Erect or ascending herbaceous or suffruticose perennial herbs, to 1.5 m; stem slightly strigose. Leaves odd-pinnately compound, 2.5–7 cm long, the leaflet blade (7)9–13, obovate to oblanceolate, 0.8–2(2.5) cm long, 4–9 mm long, the apex rounded, mucronate, the base cuneate, the upper and lower surfaces strigulose, the petiole 0.5–1 cm long; stipules obsolete. Flowers numerous in a slender and lax raceme, the pedicel 1–2 mm long; calyx ca. 2 mm long, the lobes deltate, much shorter than the tube; corolla 6–9 mm long, flesh-colored to pale yellowish white. Fruit ovoid or short-oblong, 7–9 mm long, ca. 3 mm wide, indehiscent, reflexed, woody, glabrate or inconspicuously strigulose; seeds 2–3, cubic.

　　Sandhills and scrub. Frequent; nearly throughout. North Carolina south to Florida, west to Louisiana. Spring–fall.

Indigofera colutea (Burm. f.) Merr. [Indian vernacular name.] RUSTY INDIGO.

Galega colutea Burman f., Fl. Indica 172. 1768. *Tephrosia colutea* (Burman f.) Persoon, Syn. Pl. 2: 329.
　　1807. *Indigofera colutea* (Burman f.) Merrill, Philipp. J. Sci. 19: 355. 1921.

Procumbent or prostrate annual or perennial herb; stem to 5 dm, with brown gland-tipped trichomes and appressed white medifixed, 2-branched trichomes. Leaves odd-pinnately compound, 1.5–4 cm long, the leaflets (5)9–11, opposite, the blade elliptic to narrowly elliptic, 4–14(18) cm long, 1.4–4(8) mm wide, the apex obtuse and mucronate, the base cuneate, the upper and lower surfaces with white medifixed and gland-tipped trichomes, the petiole 7–10 mm long; stipules linear, 3–4 mm long. Flowers in a (1)2–6(8) cm long, lax raceme equaling or shorter than the leaves; bracts linear, ca. 1 mm long, the pedicel ca. 1 mm long; calyx 1–2 mm long, with white medifixed and gland-tipped trichomes, the lobes longer than the tube; corolla ca. 4 mm long, red. Fruit cylindric, 1–2 cm long, 1–2 mm wide, dehiscent, divergent, with medifixed and gland-tipped trichomes; seeds 9–12, cubic.

　　Dry, disturbed sites. Rare; Hillsborough, Polk, and Manatee Counties. Florida; Africa, Asia, Australia, and Pacific Islands. Native to Asia, Australia, and Pacific Islands. Spring–fall.

Indigofera hirsuta L. [With coarse, stiff trichomes.] HAIRY INDIGO.

Indigofera hirsuta Linnaeus, Sp. Pl. 2: 751. 1753. *Anila hirsuta* (Linnaeus) Kunze, Revis. Gen. Pl. 2:
　　939. 1891.

Erect or sprawling, annual or suffrutescent biennial herb, to 1 m; stem brownish hirsute or pilose. Leaves odd-pinnately compound, 3–8 cm long, the leaflets 5–9, the blade elliptic to obovate, 2–4 cm long, 5–10 mm wide, the apex rounded, mucronate, the base cuneate, the upper and lower surfaces with subappressed trichomes, the petiole 0.5–1 cm long; stipules setaceous, ca. 1 cm long. Flowers in a raceme 6–20 cm long, much exserted beyond the leaves, the pedicel ca. 1 mm long; calyx 3.5–5 mm long, the lobes setaceous, bristly plumose, longer than the tube; corolla 6–7 mm long, salmon to maroon. Fruit oblong, 1.5–2 cm long, 2–3 mm long, reflexed, dehiscent, hispid; seeds 5–8, cubic.

Disturbed sites. Frequent; nearly throughout. Escaped from cultivation. South Carolina, Georgia, Alabama, and Florida; West Indies, Mexico, Central America, and South America; Africa, Asia, and Australia. Native to Africa, Asia, and Australia. All year.

Indigofera miniata Ortega [Flame-red, in reference to the flower color.] COASTAL INDIGO.

Indigofera miniata Ortega, Nov. Pl. Descr. Dec. 98. 1798.

Indigofera leptosepala Nuttall in Torrey & A. Gray, Fl. N. Amer. 1: 198. 1838. *Anila leptosepala* (Nuttall) Kuntze, Revis. Gen. Pl. 2: 939. 1891. *Indigofera miniata* Ortega var. *leptosepala* (Nuttall) B. L. Turner, Field & Lab. 24: 104. 1956.

Indigofera miniata Ortega var. *florida* Isley, Brittonia 34: 339. 1982. TYPE: FLORIDA: Miami-Dade Co.: between Peter's Prairie and Homestead, 10 Nov 1906, *Small & Carter 2571* (holotype: NY).

Prostrate or ascending perennial herb; stem to 5 dm, strigulose. Leaves odd-pinnately compound, 0.6–3 cm long, the leaflets (3)5–9, the blade obovate to narrowly oblanceolate, 0.5–2.5 cm long, 1–6 mm wide, the apex acute, mucronate, the base cuneate, the upper and lower surfaces strigulose, the petiole 1–3 mm long; stipules subulate, 2–6 mm long. Flowers 3–10 in a loose raceme, exceeding the leaves, the pedicel ca. 1 mm long, the peduncle 1–2 cm long; calyx 4–6 mm long, the lobes longer than the tube; corolla 7–12 mm long, salmon-red, rarely pink or orange. Fruit oblong, 1–2.5 cm long, 2–3 mm wide, subterete, divergent or deflexed, dehiscent; seeds 8–10, cubic.

Pinelands, coastal dunes and hammocks, and dry disturbed sites. Occasional; peninsula. Georgia, Alabama, Florida, Arkansas, Louisiana, Oklahoma and Texas; West Indies, Mexico, and Central America.

Although the Florida material has been placed in two fairly well-marked varieties (var. *leptosepala* and var. *florida*), We follow Lievens (1992) here who presents convincing evidence that *Indigofera miniata* consists of a single, wide-ranging, polymorphic species.

Indigofera pilosa Poir. [Soft hairy.] SOFTHAIRY INDIGO.

Indigofera pilosa Poiret, in Lamarck, Encycl., Suppl. 3: 151. 1813. *Anila pilosa* (Poiret) Kuntze, Revis. Gen. Pl. 2: 939. 1891.

Ascending or spreading annual herb; stem to 7 dm, pilose. Leaves 5–15 mm long, the leaflets 3, the blade elliptic, 2–3 cm long, 1–1.5 cm wide, the apex acute, the base cuneate, the upper and lower surfaces pilose. Flowers 1–2(3) in a short terminal or axillary raceme, the pedicel 1–2 mm long; calyx 2–3 mm long, the lobes linear-subulate, longer than the tube; ca. 5 mm long, reddish. Fruit straight, subterete, dehiscent, ascending or erect, 1.5–1.8 cm long, 2–3 mm wide, papery, hirsute; seeds 8–12, cubic.

Disturbed sites. Occasional; central peninsula. Florida; Africa. Native to Africa. Summer.

Indigofera spicata Forssk. [In spikes, in reference to the flowers and fruits.] TRAILING INDIGO.

> *Indigofera spicata* Forsskål, Fl Aegypt.-Arab. 138. 1775. *Anila spicata* (Forsskål) Kuntze, Revis. Gen. Pl. 2: 939. 1891.

Prostrate or ascending annual or perennial herb; stem to 3 dm, glabrate or strigulose. Leaves 1–3.5 cm long, the leaflets (3)5–8(11), the blade obovate to oblanceolate, (0.5)1–2.5 cm long, 3–15 mm wide, the apex rounded, the base cuneate, the upper and lower surfaces inconspicuously strigulose, the stipules deltate to lanceolate, 4–7 mm long. Flowers in an axillary raceme, the pedicel ca. 1 mm long; calyx 2–4 mm long, the lobes subulate, longer than the tube; corolla 6–8 mm long, salmon or pink to pale carmine. Fruit oblong, 1.5–2 cm long, ca. 2 mm wide, straight, dehiscent, deflexed, strigose; seeds 8–12, cubic.

Dry, disturbed sites. Frequent; nearly throughout. Florida; Africa, Asia, and Pacific Islands. Native to Africa. Summer–fall.

Indigofera suffruticosa Mill. [Somewhat woody.] ANIL DE PASTO.

> *Indigofera suffruticosa* Miller, Gard. Dict., ed. 8. 1768.
> *Indigofera anil* Linnaeus, Mant. Pl. 272. 1771. *Anila tinctoria* (Linnaeus) Kuntze var. *vera* Kuntze, Revis. Gen. Pl. 1: 160. 1891.

Erect or ascending suffrutescent perennial herb, to 1(2) m; stem strigose. Leaves 4–10 cm long, the leaflets (9)11–15, the blade elliptic to oblanceolate, 1–3 cm long, 0.5–1 cm wide, the apex obtuse, mucronate, the base cuneate, the upper surface strigose or glabrate, the lower surface strigose, the petiole 0.5–1 cm long; the stipules subulate, ca. 2 mm long. Flowers in a short axillary raceme, the pedicel ca. 1 mm long; calyx 1–2 mm long, the lobes deltate, shorter than or subequaling the tube; corolla 5–6 mm long, greenish yellow, orange, or purple-pink. Fruit oblong-falcate, 1.5–2 cm long, ca. 3 mm wide, reflexed, the tip upwardly curved, dehiscent, coriaceous, strigose to glabrate; seeds 4–8, cubic.

Disturbed sandhills and dry, open hammocks. Occasional; peninsula. North Carolina south to Florida, west to Texas; West Indies, Mexico, Central America, and South America; Africa, Asia, Australia, and Pacific Islands. Native to Africa, Asia, Australia, and Pacific Islands. Summer–fall.

Indigofera tinctoria L. [Tinctorius, used in dying.] TRUE INDIGO.

> *Indigofera tinctoria* Linnaeus, Sp. Pl. 2: 751. 1753. *Indigofera indica* Lamarck, Encycl. 3: 245. 1789, nom. illegit; non Miller, 1768.
> *Anila tinctoria* (Linnaeus) Kuntze, Revis. Gen. Pl. 1: 160. 1891. *Anila tinctoria* (Linnaeus) Kuntze var. *normalis* Kuntze, Revis. Gen. Pl. 1: 160. 1891, nom. inadmiss.

Erect or ascending perennial herb or shrub, to 1(2) m; stem strigulose with medifixed trichomes. Leaves 3–11 cm long, the leaflets 9–15, the blade ovate or elliptic, 1–2.5 cm long, 0.5–1 cm wide, the apex round to truncate, mucronate, the upper surface strigose with medifixed trichomes or glabrous, the lower surface strigose with medifixed trichomes, the petiole 1–2.5 cm long; stipules triangular, ca. 2 mm long. Flowers in a subsessile axillary raceme 2.5–5(9) cm long, the pedicel 1–2 mm long; bracts bristlelike, 1–2 mm long; calyx ca. 1.5 mm long, the lobes subequaling the tube, strigose with medifixed trichomes; corolla 5–6 mm long, reddish orange. Fruit linear, slightly falcate, 2.8–3.5 cm long, ca. 2 mm wide, subterete, usually moniliform, deflexed, strigose; seeds 5–12, ca. 2 mm long, cubic.

Disturbed sites. Occasional; Hillsborough County, southern peninsula. Escaped from cultivation. North Carolina south to Florida, west to Tennessee and Alabama; West Indies, Mexico, Central America, and South America; Africa, Asia, and Australia. Native to Africa, Asia, and Australia. Summer.

Indigofera tinctoria was widely grown in the southeast as the source for indigo dye.

Indigofera trita L. f. subsp. **scabra** (Roth) de Kort & Thijsse. [Rubbed or bruised; rough.] FLORIDA KEYS INDIGO.

Indigofera scabra Roth, Nov. Pl. Sp. 359. 1821. *Indigofera subulata* Poiret var. *scabra* (Roth) Meikle, Kew Bull. 5: 352. 1951. *Indigofera trita* Linnaeus f. var. *scabra* (Roth) Ali, Bot. Not. 3: 558. 1958. *Indigofera trita* Linnaeus f. subsp. *scabra* (Roth) de Kort & Thijsse, Blumea 30: 140. 1940.

Indigofera mucronata Sprengel ex de Candolle, Prodr. 2: 227. 1825; non Lamarck, 1789. *Indigofera jamaicensis* Sprengel, Syst. Veg. 3: 277. 1826.

Indigofera keyensis Small, Fl. Florida Keys 63, 155. 1913. *Indigofera mucronata* Sprengel ex de Candolle var. *keyensis* (Small) Isely, Brittonia 34: 340. 1982. *Indigofera trita* Linnaeus f. var. *keyensis* (Small) Kartesz & Gandhi, Phytologia 68: 423. 1990. TYPE: FLORIDA: Monroe Co.: Lower Matecumbe Key, Aug 1907, *Small 2570* (holotype: NY).

Decumbent or prostrate annual or short-lived perennial; stem to 1 m, strigulose. Leaves 2–5 cm long, the leaflets 5, the blade elliptic, 0.7–1.8 cm long, 3–6 mm wide, the apex obtuse to rounded, mucronate, the base broadly cuneate, the upper and lower surface strigulose, the petiole ca. 1 cm long; stipules setaceous, 5–8 mm long. Flowers in an axillary, short pedunculate raceme, the pedicel 1–2 mm long; calyx 2–3 mm long, the lobes lanceolate, longer than the tube; corolla 6–7 mm long, pink to salmon. Fruit oblong, 3–4.5 cm long, 2–3 mm wide, reflexed, dehiscent, strigulose; seeds 8–12, cubic.

Dry hammocks and disturbed sites. Rare; Collier and Miami-Dade Counties, Monroe County keys. Florida; West Indies, Mexico, Central America, and South America; Africa and Asia. All year.

EXCLUDED TAXA

Indigofera hendecaphylla Jacquin—Reported for Florida by Long and Lakela (1971) who misapplied the name to *I. spicata*.

Indigofera trita Linnaeus f. var. *subulata* (Vahl ex Poiret) Ali—This African taxon was reported for Florida by Chapman (1897, as *I. subulata* Vahl ex Poiret) and Small (1903, as *I. subulata* Vahl ex Poiret), who misapplied the name to material of *I. trita* subsp. *scabra*.

Kummerowia Schindl. 1912.

Annual herbs. Leaves 3-foliolate, estipellate, petiolate, stipulate. Flowers solitary, axillary, chasmogamous and cleistogamous, bracteolate, pedicellate; sepals 5, connate; corolla (chasmogamous flowers) papilionaceous, aborted in cleistogamous flowers; stamens 10, diadelphous (9 + 1). Fruit an indehiscent legume; seed 1.

A genus of 2 species; North America, Asia, and Australia. [Commemorates J. Kummerow (eighteenth century), Polish botanist.]

Kummerowia striata (Thunberg) Schindler [*Stria*, linear markings, in reference to the parallel veined stipules.] JAPANESE CLOVER.

> *Hedysarum striatum* Thunberg, in Murray, Syst. Veg., ed. 14. 675. 1784. *Desmodium striatum* (Thunberg) de Candolle, Prodr. 2: 337. 1825. *Lespedeza striata* (Thunberg) Hook & Arnott, Bot. Beechey Voy. 262. 1838. *Kummerowia striata* (Thunberg) Schindler, Repert. Spec. Nov. Regni Veg. 10: 403. 1912. *Microlespedeza striata* (Thunberg) Makino, Bot. Mag. (Tokyo) 28: 182. 1914. *Microlespedeza makinoi* Tanaka, Bult. Sci. Fak. Kjusu Imp. Univ. 1: 204, 209. 1924, nom. illegit.

Prostrate to ascending or erect annual herb; stem to 20 cm, retrorsely strigose. Leaves 3-foliolate, the leaflet blade obovate to elliptic-oblong, 0.5–1.4 cm long, 2–4 mm wide, transversely striate with parallel, unbranched secondary nerves, the apex rounded, mucronate, the base cuneate to rounded, the margin sparsely ciliate, the upper and lower surfaces glabrous, the petiole 1–2(4) mm long; stipules lanceolate, 2–3 mm long. Flowers solitary and axillary along the stems; bracteoles 2, striate. Chasmogamous flowers with the calyx 2–3 mm long; corolla ca. 4 mm long, pink. Fruit elliptic, ca. 3 mm long, subtended and covered ½–¾ by the persistent papery, reticulate calyx, sessile.

Disturbed sites. Frequent; nearly throughout. Escaped from cultivation. New York and Connecticut south to Florida, west to Iowa, Kansas, and New Mexico; Asia. Native to Asia. Summer–fall.

EXCLUDED TAXON

> *Kummerowia stipulacea* (Maximowicz) Makino—Reported for Florida by Radford et al. (1964, 1968, as *Lespedeza stipulacea* Maximowicz); this taxon excluded from the state by Isely (1990).

Lablab Adans. 1763.

Annual or perennial herbs. Leaves 3-foliolate, stipellate, petiolate, stipulate. Flowers in axillary pseudoracemes, bracteate and bracteolate; sepals 5, the calyx 4-lobed; corolla papilionaceous; stamens diadelphous (9 + 1). Fruit a dehiscent legume.

A monotypic genus; nearly cosmopolitan. [Arabic name for the plant, in reference to a description of the dull rattle of the seeds in the pod.]

Lablab purpureus (L.) Sweet [Purple, in reference to color of various plant parts (e.g., corolla, fruit, and leaves).] HYACINTHBEAN.

Dolichos purpureus Linnaeus, Sp. Pl. 2: 725. 1753. *Lablab purpureus* (Linnaeus) Sweet, Hort. Brit. 481. 1826.

Dolichos lablab Linnaeus, Sp. Pl. 2: 725. 1753. *Lablab niger* Medikus, Vorles Churpfälz. Phys.-Ocon. Ges. 2: 354. 1787. *Lablab vulgaris* Savi, Nuovo Giorn. Lett., ser. 3: 116. 1824, nom. illegit.

Trailing, twining, or erect annual or rarely perennial herb; stem to 3 m, pubescent or glabrate. Leaves 3-foliolate, the leaflet blade broadly ovate, 4–10 cm long, 4–10 cm wide, the apex acute or acuminate, the base subtruncate, the upper and lower surfaces pubescent or glabrate, stipels lanceolate; the stipules lanceolate, reflexed, caducous. Flowers in an axillary, erect raceme 1.5–2.5 cm long, 2–5 clustered at each node; bracts lanceolate, caducous; bracteoles broadly ovate, calycine. persistent; calyx campanulate, 6–8 mm long, the lobes shorter than the tube; corolla ca. 2 cm long, white or purple, the standard with a pair of basal or medial callosities, the keel scythe-shaped. Fruit broadly oblong, 5–10 cm long, 1.5–3 cm wide, straight or slightly curved, coriaceous, glabrescent, the ventral (upper) suture bumpy-nodose, shallowly septate between the seeds; seeds 3–5, oblong, compressed, white, purple, or purple-black, with a white aril.

Disturbed sites. Rare; Miami-Dade County, Monroe County keys. Escaped from cultivation. New York and Massachusetts south to Florida, west to Ontario and Ohio; West Indies, Mexico, Central America, and South America; Europe, Africa, Asia, Australia, and Pacific Islands. Native to Africa. Fall.

Lathyrus L., nom. cons. 1753. PEAVINE

Annual herbs. Leaves pinnately 2-foliolate, terminated with a tendril, estipellate, petiolate, stipulate. Flowers in axillary racemes, bracteate, ebracteolate; calyx 5-lobed; corolla papilionaceous; stamens 10, diadelphous (9 + 1); style flattened, laterally bearded on the inner side. Fruit a dehiscent legume.

A genus of 50 species; North America, Africa, Europe, and Asia. [*Lathyros*, a leguminous plant of Theophrastus, the name said to be composed of *la*, very, and *thuros*, passionate, the original plant reputed to be an aphrodisiac.]

1. Fruit conspicuously pustulate-hirsute; corolla 10–14 mm long...**L. hirsutus**
1. Fruit glabrous; corolla 6–9 mm long.. **L. pusillus**

Lathyrus hirsutus L. [With erect stiff trichomes.] CALEY PEA; SINGLETARY PEAVINE.

Lathyrus hirsutus Linnaeus, Sp. Pl. 2: 732. 1753. *Lastila hirsuta* (Linnaeus) Alefield, Bonplandia 9: 146. 1861.

Ascending or sprawling annual herb; stem to 1 m, narrowly winged, glabrous or hirsute. Leaves with 2 leaflets, the blade narrowly elliptic to oblong-lanceolate or oblanceolate, 2–7 cm long, 3–8 mm wide, the apex acute to rounded, the base cuneate, the upper and lower surfaces glabrate, with several longitudinal nerves, the tendrils branched, the petiole 2–3 cm long; stipules sagittate, 5–10 mm long, foliaceous. Flowers 1–3(4), well spaced in a raceme; calyx 5–7 mm long, the lobes subequal, narrowly deltate to short-lanceolate, the apex acuminate to aristate, subequaling the tube; corolla 10–14 mm long, pink or blue-purple or bicolored. Fruit oblong, 2.5–4.5 cm long, 6–8 mm wide, pustulate-hirsute.

Disturbed sites. Rare; Levy and Leon Counties. Escaped from cultivation. Maryland and Michigan south to Florida, west to New Mexico, also Oregon and California; Europe, Africa, and Asia. Native to Europe, Africa, and Asia. Spring–fall.

Lathyrus pusillus Elliott [Small.] TINY PEAVINE.

Lathyrus pusillus Elliott, Sketch Bot. S. Carolina 2: 223. 1824.

Sprawling or low twining annual herb; stem to 4(6) dm long, angled or narrowly winged, glabrate or sparsely villous. Leaves with 2 leaflets, the blade narrowly elliptic to oblong-lanceolate, 2.5–5(6) cm long, 2–5 mm wide, the apex acute, the base cuneate, the upper and lower surfaces glabrate, with several longitudinal nerves, the tendrils simple or branched, the petiole 1.5–2 cm long; stipules sagittate, 5–8 mm long, foliaceous. Flowers 1–2(3), apical; calyx 5–8, the lobes subequal, longer than the tube; corolla 6–9 mm long, pale lavender-blue. Fruit oblong, 2.5–4 cm long, 3–5 mm wide, slightly curved, glabrous.

Disturbed sites. Rare; Leon, Madison, and Marion Counties. Virginia south to Florida, west to Kansas, Oklahoma, and Texas, also Oregon; South America. Native to North America. Spring.

Leptospron (Benth.) A. Delgado 2011.

Perennial herbs. Leaves alternate, 3-foliolate, stipellate, petiolate, stipulate. Flowers in pedunculate, axillary pseudoracemes, bracteate, bracteolate; sepals 5, connate into a tube, unequal; petals 5, the corolla papilionaceous, the keel petals spiraled and the distal end projected downward; stamens 10, diadelphous (9 + 1). Fruit a dehiscent legume.

A genus of 2 species; North America, West Indies, Mexico, Central America, South America, Africa, Asia, Australia, and Pacific Islands. [From the Greek *leptos*, thin, weak, and the Latin *pronus*, projected downward, in reference to the keel petals.]

Leptospron adenanthum (G. Mey.) A. Delgado [From the Greek *adenos*, gland, and *anthos*, flower, in reference to the glands on the inflorescence.] WILD PEA.

Phaseolus adenanthus G. Meyer, Prim. Fl. Esseq. 239. 1818. *Vigna adenantha* (G. Meyer) Meréchal et al., Taxon 27: 202. 1978. *Leptospron adenanthum* (G. Meyer) A. Delgado, Amer. J. Bot. 98: 1710. 2011.

Twining, perennial herb; stem to 3 m, brown pubescent or glabrous. Leaves 3-foliolate, the leaflet blade ovate to lanceolate, 4–8 cm long, 2–3.5 cm wide, the apex obtuse to acute, the base subtruncate or rounded, the lateral ones oblique, the margin entire, the upper and lower surfaces at first pubescent, sparsely so to glabrate at maturity, the petiole 3–6 cm long; stipules 3–5 mm long, entire. Flowers several in a pedunculate axillary pseudoraceme, 3–15 cm long, the rachis brown strigose, with minute sessile glands; bracteoles 3–4 mm long; calyx 7–8 mm long, the tube campanulate, sparsely pubescent, with minute sessile glands, the lower lobe lanceolate, slightly shorter than the tube, the lateral ones turned sideways; corolla 2.5–3 cm long, purple-blue with white, fading yellow, the keel petals spiraled 2–3 turns, the distal end turned down. Fruit oblong, 7–10 cm long, 8–12 mm wide, falcate, laterally compressed, glabrous.

Disturbed sites. Rare; Pinellas, Palm Beach, and Miami-Dade Counties. Escaped from cultivation. Florida; West Indies, Mexico, Central America, and South America; Africa, Asia, Australia, and Pacific Islands. Native to South America. All year.

Lespedeza Michx. 1803.

Perennial herbs or shrubs. Leaves 3-foliolate, estipellate, petiolate, stipulate. Flowers in axillary or terminal pseudoracemes, chasmogamous and often cleistogamous, the latter usually in separate racemes or solitary in the leaf axils; bracts 2–3, subtending a pair of flowers; bracteoles calycine; calyx in chasmogamous flowers campanulate, 4- to 5-lobed, reduced in the cleistogamous; corolla papilionaceous in chasmogamous flowers, reduced in the cleistogamous; stamens diadelphous (9 + 1). Fruit an indehiscent legume, papery.

A genus of about 35 species; North America, Mexico, Central America, South America, Africa, Asia, Australia, and Pacific Islands. [Commemorates the Spanish governor of Cuba, Vicente Manuel de Céspedez y Velasco (fl. 1784–1790).]

Selected reference: Clewell (1966).

1. Plant suffrutescent; corolla (8)12–18 mm long.
    2. Calyx lobes longer than the tube, the apex acuminate; seeds purplish black; corolla (10)12–15 mm long...**L. thunbergii**
    2. Calyx lobes subequal or shorter than the tube, the apex rounded to subacute; seeds purple and green mottled; corolla 8–11 mm long...**L. bicolor**
1. Plant herbaceous; corolla 5–9(10) mm long.
    3. Calyx nearly equaling or exceeding the mature fruit; corolla white or cream-colored with a purple throat.
        4. Leaflets narrowly cuneate. widest at the apex; racemes 2- to 5-flowered, much shorter than the subtending leaves ....**L. cuneata**
        4. Leaflets widest at or below the middle; racemes 10- or more-flowered, equaling or exceeding the subtending leaves.
            5. Leaflets linear to narrowly oblong, the upper surface glabrous........................**L. angustifolia**
            5. Leaflets broadly elliptic to suborbicular, the upper surface strigose.
                6. Racemes subequaling the subtending leaves ....**L. capitata**
                6. Racemes usually well exserted beyond the subtending leaves ....**L. hirta**
    3. Calyx to ½ as long as the mature fruit; corolla blue-purple (nearly white in *L. repens*).
        7. Stem prostrate or trailing.
            8. Plant with spreading, curved trichomes ....**L. procumbens**
            8. Plant with strigose trichomes.. **L. repens**
        7. Stem erect or ascending.
            9. Leaflets linear...**L. virginica**
            9. Leaflets oblong to elliptic.
                10. Leaflets pubescent on the upper surface; stem pubescent...................**L. stuevei**
                10. Leaflets glabrous on the upper surface or sometimes strigose along the midrib; stem strigose...**L. violacea**

Lespedeza angustifolia (Pursh) Elliott [With narrow leaves.] NARROWLEAF
LESPEDEZA.

> *Lespedeza capitata* Michaux var. *angustifolia* Pursh, Fl. Amer. Sept. 480. 1814. *Lespedeza angustifolia*
> (Pursh) Elliott, Sketch. Bot. S. Carolina 2: 206. 1824. *Lespedeza hirta* (Linnaeus) Hornemann var.
> *angustifolia* (Pursh) Maximowicz, Trudy Imp. S.-Peterburgsk. Bot. Sada 2: 379. 1873. *Despeleza*
> *angustifolia* (Pursh) Nieuwland, Amer. Midl. Naturalist 3: 176. 1914.
>
> *Lespedeza angustifolia* (Pursh) Elliott var. *brevifolia* Britton, New York Acad. Sci. 12: 68. 1893. TYPE:
> FLORIDA: without data, *Chapman s.n.* (holotype: NY).

Erect or ascending perennial herb, to 1(1.5) m; stem strigose or pilosulous. Leaves with the leaflet blade elliptic to narrowly oblong, 2–4 cm long, 2–5 mm wide, the apex rounded, mucronate,
the base rounded to broadly cuneate, the upper surface glabrate, the lower surface strigose,
the petiole 1–3 mm long; stipules setaceous. Flowers 10–20+ in a compact, broadly ovoid to
subglobose raceme; calyx 5–8 mm long, the lobes much longer than the tube, exceeding the
corolla; corolla 5–7 mm long, white to cream. Fruit strigose, the chasmogamous 4–5 mm long,
the cleistogamous 3–4 mm long.

Sandhills, flatwoods, and open hammocks. Frequent; northern peninsula south to Hillsborough County, west to central panhandle. New York and Massachusetts south to Florida, west
to Tennessee and Mississippi. Summer–fall.

Lespedeza bicolor Turcz. [Two-colored, in reference to the flower.] SHRUBBY
LESPEDEZA.

> *Lespedeza bicolor* Turczaninow, Bull. Soc. Imp. Naturalistes Moscou 13: 69. 1840. *Lespedeza bicolor*
> Turczaninow var. *typica* Maximowitz, Trudy Imp. S.-Peterburgsk. Bot. Sada 2: 355. 1873, nom.
> inadmiss.

Ascending shrub, to 2(3) m; stem strigose or puberulous. Leaves (medial) with the leaflet blade
oval, 2–3 cm long, 1–3 cm wide, the apex emarginate, mucronate, the base broadly cuneate to
rounded, the upper and lower surfaces inconspicuously strigose or glabrate, the petiole 2–4 cm
long; stipules setaceous. Flowers 5–15 in a pedunculate raceme in the upper leaf axil; calyx ca. 3
mm long, the lobes subequaling the tube or shorter; corolla 8–11 mm long, purple or magenta,
rarely white. Fruit 7–8 mm long, strigose.

Disturbed sites. Rare; central and western panhandle, Jefferson and Hernando Counties.
Escaped from cultivation. New York and Massachusetts south to Florida, west to Ontario,
Michigan, Iowa, Kansas, and Texas; Asia. Native to Asia. Spring–fall.

Lespedeza capitata Michx. [In heads, in reference to the inflorescence.]
ROUNDHEAD LESPEDEZA; DUSTY CLOVER.

> *Lespedeza capitata* Michaux, Fl. Bor.-Amer. 2: 71. 1803. *Hedysarum conglomeratum* Poiret, in Lamarck,
> Encycl. 6: 416. 1804. *Lespedeza capitata* Michaux var. *vulgaris* Torrey & A. Gray, Fl. N. Amer. 1: 368.
> 1840, nom. inadmiss. *Despeleza capitata* (Michaux) Nieuwland, Amer. Midl. Naturalist 3: 176. 1914.
> *Lespedeza capitata* Michaux var. *typica* Fernald, Rhodora 43: 576. 1941, nom. inadmiss.

Ascending to erect perennial herb, to 1.5 m; stem silvery or cinereous, strigose or villous.
Leaves with the leaflet blade elliptic to oblong, 2–4 cm long, 1–2 cm wide, the apex rounded to
emarginate, the base broadly cuneate to rounded, the upper surface glabrate, the lower surface

gray-strigose or silver-sericeous, the petiole 1–3 mm long; stipules setaceous. Flowers 15–25 in a subcapitate, subterminal raceme; calyx 8–12 mm long, the lobes much exceeding the tube and the corolla; corolla 7–10 mm long, yellow-white. Fruit pubescent, the chasmogamous 5–7 mm long, the cleistogamous 4–5 mm long.

Flatwoods and wet savannas. Frequent; central and western panhandle. Maine south to Florida, west to Ontario, Minnesota, South Dakota, Nebraska, Kansas, Oklahoma, and Texas. Summer–fall.

Lespedeza cuneata (Dum. Cours.) G. Don [Wedge-shaped, in reference to the leaflet base.] SERICEA LESPEDEZA; CHINESE LESPEDEZA.

Anthyllis cuneata Dumort de Courset, Bot. Cult., ed. 2. 6: 100. 1811. *Aspalathus cuneatus* (Dumort de Courset) D, Don, Prodr. Fl. Nepal. 246. 1825. *Lespedeza cuneata* (Dumort de Courset) G. Don, Gen. Hist. 2: 307. 1832.

Erect, short-lived perennial herb, to 2 m; stem strigose or subpuberulous. Leaves with the leaflet blade oblanceolate or oblong, 1–2.5 cm long, 2–5 mm wide, the apex rounded to truncate or retuse, the base cuneate, the upper surface strigose or glabrate, the lower surface strigose, the petiole 1–2 cm long; stipules setaceous. Flowers (1)2–3 in a raceme shorter than the subtending leaves; calyx 4–5 mm long, lobes longer than the tube; corolla 7–8 mm long, whitish with purple markings. Fruit sparsely strigose to glabrate, the chasmogamous 4–5 mm long, the cleistogamous 3–4 mm long.

Open hammocks and disturbed sites. Occasional; northern counties, Brevard and Hardee Counties. Escaped from cultivation. New York and Massachusetts south to Florida, west to Ontario, Wisconsin, Nebraska, Kansas, Oklahoma, and Texas; Mexico, Central America, and South America; Africa, Asia, Australia, and Pacific Islands. Native to Asia and Australia. Summer–fall.

Lespedeza hirta (L.) Hornem. [Hairy.] HAIRY LESPEDEZA.

Hedysarum hirtum Linnaeus, Sp. Pl. 2: 748. 1753. *Lespedeza polystachya* Michaux, Fl. Bor.-Amer. 2: 71. 1803, nom. illegit. *Lespedeza hirta* (Linnaeus) Hornemann, Hort. Bot. Hafn. 699. 1815. *Lespedeza hirta* (Linnaeus) Hornemann var. *typica* Schindler, Bot. Jahrb. Syst. 49: 623. 1836, nom. inadmiss. *Despeleza hirta* (Linnaeus) Nieuwland, Amer. Midl. Naturalist 3: 176. 1914.
Lespedeza hirta (Linnaeus) Hornemann var. *calycina* Schindler, Bot. Jahrb. Syst. 49: 624. 1913. *Lespedeza capitata* Michaux var. *calycina* (Schindler) Fernald, Rhodora 43: 578. 1941.
Lespedeza hirta (Linnaeus) Horneman subsp. *curtissii* Clewell, Brittonia 16: 75. 1964. *Lespedeza hirta* (Linnaeus) Hornemann var. *curtissii* (Clewell) Isely, Brittonia 34: 354. 1986.

Ascending or erect perennial herb, to 1.5(2) m; stem villous to finely puberulent. Leaves with the leaflet blade ovate-elliptic or obovate, 1–4(5) cm long, 0.5–2 cm wide, the apex rounded to obtuse, mucronate, the base rounded to shallowly subcordate, the upper surface pubescent or glabrous, the lower surface pubescent, the petiole 1–1.5(2) cm long; the stipules setaceous. Flowers 10–20 in upper axillary, pedunculate, ovoid to short-oblong clusters, sometimes aggregated in a loose compound inflorescence; calyx 7–10 mm long, the lobes much longer than the tube, exceeding the corolla; corolla 6–9 mm long, cream yellow. Fruit strigose, the chasmogamous 6–8 mm long, the cleistogamous 5–7 mm long.

Sandhills, flatwoods, and open, dry hammocks. Frequent; northern counties, central peninsula. Maine south to Florida, west to Ontario, Michigan, Kansas, Oklahoma, and Texas. Fall.

Lespedeza procumbens Michx. [Trailing.] TRAILING LESPEDEZA.

Lespedeza procumbens Michaux, Fl. Bor.-Amer. 2: 70., t. 29. 1803.

Procumbent perennial herb; stem to 2 m, villous or villosulous. Leaves with the leaflet blade ovate or elliptic, 1–2.5 cm long, 0.8–2 cm wide, the apex rounded or emarginate, mucronate, the base rounded, the upper surface loosely strigose or glabrous, the lower surface subappressed villous, the petiole 0.5–1.5 cm long; stipules setaceous. Flowers 6–10(12) in an axillary or terminal pseudoraceme, the cleistogamous subsessile in the leaf axils; calyx 2–3 mm long, the lobes subequaling the tube; corolla 6–7 mm long, pink to purple. Fruit glabrate or sparsely strigose, the chasmogamous 4–5 mm long, the cleistogamous 3.5–4.5 mm long.

Hammocks. Occasional; central panhandle. New Hampshire south to Florida, west to Wisconsin, Illinois, Kansas, Oklahoma, and Texas. Fall.

Lespedeza repens (L.) W. P. C. Barton [Creeping.] CREEPING LESPEDEZA.

Hedysarum repens Linnaeus, Sp. Pl. 2: 749. 1753. *Lespedeza repens* (Linnaeus) W. P. C. Barton, Comp. Fl. Philadelph. 2: 77. 1818.

Prostrate perennial herb; stem to 1 m, glabrate or strigose, rarely pilosulous. Leaves with the leaflet blade elliptic to rarely ovate or obovate, 1–2(2.5) cm long, 0.5–1(2) cm wide, the apex rounded or retuse, mucronate, the base cuneate, the upper surface glabrate or finely strigulose, the lower surface strigulose; the petiole 0.7–1(1.5) cm long; the stipules setaceous, 2–4(5) mm long. Flowers 2–6(10), ascending, in a raceme, the peduncle 2–6 cm long; calyx 2.5–4 mm long, the lobes subequaling the tube, much shorter than the corolla; corolla 5–7 mm long, pink-lavender to violet, rarely whitish, red-violet striate. Legume strigulose, chasmogamous 4–6(7) mm long, cleistogamous 3–4 mm long.

Sandhills, flatwoods, and dry, open hammocks. Occasional; northern counties, Lake County. New York south to Florida, west to Wisconsin, Iowa, Kansas, Oklahoma, and Texas. Spring–fall.

Lespedeza stuevei Nutt. [Commemorates Herman Heinrich Theodore Wilhelm Stüwe (b. 1775), Prussian botanist and pharmacist.] TALL LESPEDEZA.

Lepedeza stuevei Nuttall, Gen. N. Amer. Pl. 2: 107. 1818.

Erect to ascending perennial herb, to 1.5 m; stem villosulous or strigose. Leaves with the leaflets broadly elliptic to elliptic oblong, 1–2.5(3) cm long, 3–6 mm wide, the apex rounded or retuse, mucronate, the base cuneate to rounded, the upper surface strigose, the lower surface subappressed villous or strigose, the petiole (0.5)1–1.5 cm long; stipules setaceous. Flowers 5–15 in an axillary glomerulate raceme, short pedunculate; calyx 3–4 mm long, much shorter than the corolla, the lobes subequaling the tube, strigose or villous; corolla 5–7 mm long, purple. Fruit strigose or villous, the chasmogamous 5–6 mm long, the cleistogamous 4–6 mm long.

Sandhills and dry, open hammocks. Occasional; northern counties south to Citrus County. Vermont and New York south to Florida, west to Illinois, Kansas, Oklahoma, and Texas. Summer–fall.

Lespedeza thunbergii (DC.) Nakai [Commemorates Carl Peter Thunberg (1743–1828), Swedish botanist and student of Linnaeus.] THUNBERG'S LESPEDEZA.

Desmodium thunbergii de Candolle, Prodr. 2: 337. 1825. *Lespedeza thunbergii* (de Candolle) Nakai, Lespedeza Japan 15. 1927.

Ascending shrub, to 2(3) m; stem strigose or puberulous. Leaves with the leaflet blade narrowly elliptic to oval, 2–3 cm long, 1–1.5 cm wide, the apex round, the base cuneate, the upper and lower surfaces strigose or glabrate, the petiole 2–4 mm long; stipules setaceous. Flowers 5–15 in a pedunculate raceme in the upper leaf axil; calyx ca. 3 mm long, the lobes longer than the tube; corolla (10)12–15 mm long, purple or magenta, rarely white. Fruit 7–8 mm long, strigose.

Disturbed sites. Rare; Jefferson, Walton, Leon, and Escambia Counties. Escaped from cultivation. Massachusetts and New York south to Florida, west to Ontario, Wisconsin, Illinois, Missouri, Arkansas, and Louisiana; Asia. Native to Asia. Spring.

Lespedeza violacea (L.) Pers. [Violet colored.] VIOLET LESPEDEZA.

Hedyarum violaceum Linnaeus, sp. Pl. 2: 749. 1753. *Lespedeza violacea* (Linnaeus) Persoon, Syn. Pl. 2: 318. 1807.

Hedysarum divergens Muhlenberg ex Willdenow, Sp. Pl. 3: 1196. 1802. *Lespedeza divergens* (Muhlenberg ex Willdenow) Pursh, Fl. Amer. Sept. 481. 1814. *Lespedeza violacea* (Linnaeus) Persoon var. *divergens* (Muhlenberg ex Willdenow) Torrey & A. Gray, Fl. N. Amer. 1: 367. 1940.

Lespedeza stuevei Nuttall var. *intermedia* S. Watson, in A. Gray, Manual, ed. 6. 141. 1889.

Erect or ascending perennial herb, to 8(15) dm; stem strigose or puberulent. The leaves with the leaflet blade obovate to elliptic, 1–3(4) cm long, 0.5–1 cm wide, the apex rounded or retuse, mucronate, the base cuneate to rounded, the upper surface glabrous, rarely strigose along the midrib, the lower surface strigose, the petiole 0.7–2.5 cm long; stipules setaceous. Flowers 4–15 in a congested short-pedunculate axillary raceme; calyx 3–5 mm long, much shorter than the corolla, the lobes equaling the tube or slightly longer; corolla 6–7 mm long, purple. Fruit strigose, the chasmogamous 5–7 mm long, the cleistogamous 3–5 mm long.

Hammocks. Occasional; Nassau County, central and western panhandle. New Hampshire and Vermont south to Florida, west to Ontario, Minnesota, Nebraska, Kansas, Oklahoma, and Texas. Summer–fall.

Lespedeza virginica (L.) Britton [Of Virginia.] SLENDER LESPEDEZA.

Medicago virginica Linnaeus, Sp. Pl. 2: 778. 1753. *Lespedeza sessiliflora* Michaux, Fl. Bor.-Amer. 2: 70. 1803, nom. illegit. *Hedysarum sessiliflorum* Poiret, in Lamarck, Encycl. 6: 414. 1804. *Lespedeza violacea* (Linnaeus) Persoon var. *sessiliflora* (Poiret) G. Don, Gen. Hist. 2: 307. 1832. *Lespedeza reticulata* (Muhlenberg ex Willdenow) Persoon var. *sessiliflora* (Poiret) Maximowicz, Trudy Imp. S.-Peterburgsk. Bot. Sada 2: 365. 1873. *Lespedeza reticulata* (Muhlenberg ex Willdenow) Persoon var. *virginica* (Linnaeus) MacMillan, Metasp. Minnesota Valley 318. 1892, nom. illegit. *Lespedeza virginica* (Linnaeus) Britton, Trans. New York Acad. Sci. 12: 64. 1893. *Lespedeza virginica* (Linnaeus) Britton, var. *typica* Schindler, Bot. Jahrb. Syst. 49: 614. 1913, nom. inadmiss.

Hedysarum reticulatum Muhlenberg ex Willdenow, Sp. Pl. 3: 1194. 1802. *Lespedeza reticulata* (Muhlenberg ex Willdenow) Persoon, Syn. Pl. 2: 318. 1807. *Lespedeza violacea* (Linnaeus) Persoon var. *reticulata* (Muhlenberg ex Willdenow) G. Don, Gen. Hist. 2: 307. 1832.

Lespedeza angustifolia Darlington, Fl. Cestr. 81: 1826; non (Pursh) Elliott, 1824. *Lespedeza violacea*

(Linnaeus) Persoon var. *angustifolia* Torrey & A. Gray, Fl. N. Amer. 1: 367. 1840. *Lespedeza reticulata* (Muhlenberg ex Willldenow) Persoon var. *angustifolia* (Torrey & A. Gray) Maximowicz, Trudy Imp. S.-Peterburgsk. Bot. Sada 2: 366. 1873.

Lespedeza virginica (Linnaeus) Britton, var. *sessiliflora* Schindler, Bot. Jahrb. Syst. 49: 616. 1913.

Ascending to erect perennial herb, to 8 dm; stem strigose or puberulent. Leaves with the leaflet blade oblong to linear, 1.5–3 cm long, 2–4 mm wide, the apex acute to obtuse, mucronate, the base cuneate, the upper and lower surfaces strigulose or glabrate, the petiole ca. 1 cm long; the stipules setaceous. Flowers 4–10 in a short-pedunculate congested raceme in the upper axils; calyx 3–5 mm long, strigulose, shorter than the corolla, the lobes subequal or longer than the tube; corolla 4–7 mm long, pinkish to purple, rarely white. Fruit strigose, the chasmogamous 4–7 mm long, the cleistogamous 4–5 mm long.

Sandhills and open, dry hammocks. Occasional; panhandle. New Hampshire south to Florida, west to Ontario, Minnesota, Iowa, Kansas, Oklahoma, and Texas; Mexico. Fall.

HYBRID

Lespedeza ×oblongifolia (Britton) W. Stone (*L. angustifolia* × *L. hirta*).

Lespedeza hirta (Linnaeus) Hornemann var. *oblongifolia* Britton, Trans. New York Acad. Sci. 12: 66. 1893. *Lespedeza oblongifolia* (Britton) W. Stone, Fl. New Jersey 509. 1912, pro. sp.

Lespedeza hirta (Linnaeus) Hornemann var. *appressipilis* S. F. Blake, Rhodora 26: 32. 1924. TYPE: FLORIDA: Duval Co.: without data, Oct, *Curtiss 639* (holotype: US).

Disturbed sites. Rare; Duval, Jefferson, Leon, and Gadsden Counties. Fall.

EXCLUDED TAXA

Lespedeza frutescens (Linnaeus) Hornemann—This northern taxon previously known as *L. violacea* (see Reveal & Barrie, 1991) was reported for Florida by Small (1903, 1913a, 1933), who misapplied the name to material of *L. violacea*. Also reported for Florida by Radford et al. (1964, 1968, both as *L. intermedia* Britton), Wilhelm (1984, as *L. intermedia* Britton), and Clewell (1985, as *L. intermedia* Britton), the name misapplied to material of *L. violacea*. Reveal and Barrie (1991) have shown this name to be homotypic with *L. frutescens*.

Lespedeza ×longifolia de Candolle—Reported for Florida by Fernald (1941, as *L. hirta* (Linnaeus) Hornemann var. *longifolia* (de Candolle) Fernald). Clewell (1968) excludes this nothotaxon from the state.

Lespedeza ×nuttallii Darlington—Reported for Florida by Small (1903, 1913a, 1933), but no vouchering specimens have been seen. Isely (1955) also reports this nothotaxon for Florida. Clewell (1966) excludes Florida from its range.

Lespedeza virgata (Thunberg) de Candolle—Reported for Florida by Anderson (1988) based on a misidentification of *L. repens*.

Leucaena Benth. nom. cons. 1842. LEADTREE

Shrubs or trees. Leaves alternte, even-bipinnately compound, petiolate, stipulate. Flowers in axillary, fasciculate heads, bracteate, bracteolate; sepals 5, connate, subequal; petals 5, free, subequal; stamens 10, free, dorsifixed. Fruit a dehiscent legume.

A genus of about 22 species; nearly cosmopolitan. [From the Greek *leucano*, becoming white, in reference to the white flowers.]

Selected reference: Hughes (1998).

Leucaena leucocephala (Lam.) de Wit [From the Greek *leucon*, white, and *cephalon*, head, in reference to the heads of white flowers.] WHITE LEADTREE.

Mimosa leucocephala Lamarck, Encycl. 1: 12. 1783. *Acacia leucocephala* (Lamarck) Link, Enum. Hort. Berol. Alt. 2: 444. 1822. *Leucaena leucocephala* (Lamarck) de Wit, Taxon 10: 54. 1961.

Leucaena glabrata Rose, Contr. U.S. Natl. Herb. 5: 140. 1897. *Leucaena leucocephala* (Lamarck) de Wit subsp. *glabrata* (Rose) Zárate, Phytologia 63: 305. 1987.

Trees or shrubs, to 10(20) m; branchlets puberulent. Leaves 10–20 cm long, the pinnae of 4–9 pairs, the pinnules 11–17 pairs, the pinnule blade oblong, 7–12 mm long, 2–4 mm wide, the apex acute, the base rounded to truncate, asymmetrical, the margin ciliate or glabrous, the upper and lower surfaces sparsely pubescent to glabrate, the petiole 3–6 cm long, with a shallow nectary gland near the lowest pinnae. Flowers in a head 2–3 cm in diameter; bracts and bractioles inconspicuous; calyx tube campanulate, ca. 1 mm long, the lobes subequal, strigose; corolla with the petals linear-spatulate, 4–5 mm long, white, puberulent. Fruit linear, 10–15 cm long, 1–1.5 cm wide, laterally compressed, velutinous to glabrescent or glabrous, the stipe 7–20 mm long; seeds 18–25, ovate to obovate.

Hammocks, coastal strands, and disturbed sites. Occasional; Alachua County, central and southern peninsula. Escaped from cultivation. Florida, Texas, Arizona, and California; Mexico, Central America, and South America; Europe, Africa, Asia, Australia, and Pacific Islands. Native to Mexico and Central America. All year.

Leucaena leucocephala is listed as a Category II invasive species in Florida by the Florida Exotic Pest Plant Council (FLEPPC, 2015).

EXCLUDED TAXON

Leucaena glauca (Linnaeus) Bentham—Reported for Florida by Small (1903, 1913a, 1913b, 1913c, 1913d, 1913e, 1933) and by West and Arnold (1946), the name misapplied to material of *L. leucocephala*.

Lonchocarpus Kunth, nom. cons. 1824. LANCEPOD

Trees. Leaves alternate, odd-pinnately compound, petiolulate, estipellate, petiolate, stipulate. Flowers in terminal and axillary racemes; bracteate and bracteolate; sepals 5, connate, obscurely 5-toothed; petals 5, the corolla papilionaceous; stamens 10, monadelphous. Fruit an indehiscent legume.

A genus of about 120 species; North America, Central America, West Indies, and Africa. [From the Greek *lonchos*, lance, and *carpos*, fruit, in reference to the lance-shaped fruit.]

Lonchocarpus punctatus Kunth [Dotted, in reference to the gland-dotted leaflets.] DOTTED LANCEPOD.

Lonchocarpus punctatus Kunth, in Humboldt et al., Nov. Gen. Sp. 6: 383. 1824.

Tree, to 15 m. Leaves with 3–4 pairs of leaflets, the blade ovate or elliptic-ovate, 4–10 cm long, 2.5–4 cm wide, the apex bluntly acuminate, the base rounded, the margin entire, the upper and lower surfaces glabrous, punctate with translucent glands, the petiole 4–6 cm long. Flowers few to many, paired, in a raceme to 20 cm long; calyx ca. 5 mm long; corolla papilionaceous, ca. 1.5 cm long, pink-purple, the standard mottled with yellow and white. Fruit linear-elliptic, 6–8 cm long, 3–4 cm wide, flattened, pale straw-colored; seeds 1–3.

Disturbed tropical hammocks. Rare; Monroe County keys. Escaped from cultivation. Florida; West Indies; South America. Native to West Indies and South America. All year.

EXCLUDED TAXON

Lonchocarpus violaceus (Jacquin) de Candolle—Reported for Florida by Long and Lakela (1971), the name misapplied to material of *L. punctatus*.

Lotononis (DC.) Eckl. & Zeyh., nom. cons. 1836.

Perennial herbs. Leaves 3-foliolate, petiolate, stipulate. Flowers in axillary umbellate heads; sepals 5, connate, 5-toothed; petals 5, the corolla papilionaceous; stamens 10, monadelphous, connate into a dorsally split sheath, the anthers dimorphic, 4 basifixed, 6 shorter and dorsifixed. Fruit a dehiscent legume.

A genus of about 150 species; North America, Europe, Africa, and Asia. [From *Lotus* and *Ononis*, suggesting a similarity to both genera.]

Selected reference: van Wyk (1991).

Lotononis bainesii Baker [Commemorates John Thomas Baines (1820–1875), artist and explorer in Africa, one of the collectors of the type specimen.] LOTONONIS.

Lotononis bainesii Baker, in Oliver, Fl. Trop. Afr. 2: 6. 1871.

Decumbent perennial herb; stem to 5 dm long, branching and rooting at the nodes, glabrous or sparsely puberulent. Leaves 3-foliolate, the leaflet blade linear-elliptic to lanceolate, 1.5–4 cm long, 0.6–1 cm wide, the apex rounded, mucronate, the base cuneate, the upper and lower surfaces glabrous or sparsely puberulent, the petiole 6–7 mm long; stipules ovate, 4–10 mm long, one smaller than the other, the base auriculate. Flowers 8–12 in an umbellate head, sessile, the peduncle to 25 cm long; bracts and bracteoles minute; calyx 3–4 mm long, cupulate, the upper tooth narrowly triangular and the longest, the other 4 teeth connate in 2 pairs; corolla papilionaceous, ca. 1 cm long, yellow, the standard orbicular to oblong, long-clawed, the wings obliquely ovate to oblong, the keel 8–9 mm long, incurved, obtuse, longer than the standard. Fruit linear, 7–12 mm long, 2–3 mm wide, slightly inflated, white-villous; seeds numerous, ca. 1 mm long.

Disturbed sites. Rare; Alachua County. Escaped from cultivation. Florida; Africa, Asia, and Australia. Native to Africa. Spring–summer.

Lotus L. 1753. TREFOIL

Perennial herbs. Leaves alternate, odd-pinnately compound, 5-foliolate, subsessile, stipulate. Flowers axillary, umbellate, bracteate; sepals 5, connate; petals 5, the corolla papilionaceous; stamens 10, diadelphous (9 + 1). Fruit a dehiscent legume.

A genus of about 125 species; North America, Europe, Africa, Asia, Australia, and Pacific Islands. [Originally applied to *Ziziphus lotus* (Rhamnaceae) and later to diverse leguminous plants.]

Lotus pedunculatus Cav. [With a peduncle.] BIRDSFOOT TREFOIL.

Lotus pedunculatus Cavanilles, Icon. 2: 52. 1793.
Lotus uliginosus Schkuhr, Bot. Handb. 2: 412, pl. 211. 1796. *Mullaghera uliginosa* (Schkuhr) Bubani, Fl. Pyren. 2: 509. 1899.

Erect or ascending perennial herb, to 12 dm; stem glabrate to sparsely pilose, rhizomatous. Leaves 10–34 mm long, the leaflet blade ovate-elliptic to elliptic, 8–25 mm long, 3–15 mm wide, the apex rounded, often mucronate, the upper and lower surfaces glabrate, subsessile; stipules gland-like. Flowers (4)5–15 in an umbel; bracts with (1)3 leaflets; calyx 5–8 mm long, the lobes triangular, 2–3 mm long, subequaling or slightly longer than the tube, the tube glabrate to pilose; corolla papilionaceous, 8–13(18) mm long, yellow, often mottled with red. Fruit cylindric, (10)15–35 mm long, 2–3 mm wide; seeds 15–35, globose to round-oblong, ca. 1 mm long, yellowish to yellowish brown, sometimes mottled, smooth.

Disturbed sites. Rare; Alachua and Osceola Counties. Escaped from cultivation. Canada, Florida, and Illinois, Washington and Idaho south to California; South America; Europe, Africa, Asia, and Pacific Islands. Native to Europe, Africa, and Asia. Spring–summer.

EXCLUDED TAXA

Lotus corniculatus Linnaeus—Reported for Florida by Small (1933) who cited this species from "Coastal Plain, Atlantic and Gulf States," which would presumably include Florida, apparently based on misapplication of the name to specimens of *L. pedunculatus*. No Florida specimens known.

Lupinus L. 1753. LUPINE

Annual or perennial herbs. Leaves alternate, palmately compound or 1-foliolate, estipellate, petiolate, stipulate. Flowers in terminal racemes, bracteate, bracteolate; sepals 5, the lobes 2–4, bilabiate; petals 5, the corolla papilionaceous; stamens monadelphous. Fruit a dehiscent legume.

A genus of about 250 species; North America, Mexico, Central America, South America, Europe, Africa, Asia, Australia, and Pacific Islands. [*Lupus*, wolf, in reference to the belief that the plants destroy the soil fertility.]

Selected reference: Dunn (1971).

1. Leaves 1-foliolate.
  2. Stipules obsolete..**L. westianus**
  2. Stipules evident.
    3. Corolla blue, the standard blue with a conspicuous white to cream spot....................**L. diffusus**
    3. Corolla pink, the standard pink with a conspicuous deep reddish purple spot............**L. villosus**
1. Leaves 5- to 11-foliolate.
  4. Corolla yellow..**L. luteus**
  4. Corolla lavender to blue-purple or 2-colored.
    5. Racemes included to only slightly exserted ....**L. angustifolius**
    5. Racemes moderately to strongly exserted.
      6. Plant a rhizomatous perennial; leaves 7- to 11-foliolate; corolla blue-violet or rarely white or
      pink..**L. perennis**
      6. Plant an annual; leaves 6(7)-foliolate; corolla dark blue..**L. texensis**

Lupinus angustifolius L. [Narrow leaves, in reference to the leaflets of this species.] NARROWLEAF LUPINE.

Lupinus angustifolius Linnaeus, Sp. Pl. 2: 721. 1753.

Erect annual herb, to 1 m; stem pubescent. Leaves (5)7- to 8(11)-foliolate, the leaflet blade oblong-lanceolate to linear, 2–3 cm long, 2–4 mm wide, the apex rounded, the base narrowly cuneate, the upper and lower surface glabrous, the petiole 3–5 mm long; stipules filiform. Flowers usually whorled in a short-pedunculate raceme with included or slightly exserted flowers; calyx with the upper lip bifurcate; corolla papilionaceous, 10–14 mm long, pinkish lavender, withering blue. Fruit 4–6 cm long, 1 cm wide, villous.

Disturbed sites. Rare; Alachua County. Escaped from cultivation. Maine, Georgia, Florida, and British Columbia; Europe, Africa, Asia, Australia, and Pacific Islands. Native to Europe, Africa, and Asia. Spring.

Lupinus diffusus Nutt. [Spreading.] SKYBLUE LUPINE.

Lupinus diffusus Nuttall, Gen. N. Amer. Pl. 2: 93. 1818. *Lupinus villosus* Willdenow var. *diffusus* (Nuttall) Torrey & A. Gray, Fl. N. Amer. 1: 382. 1840. *Lupinus villosus* Willdenow subsp. *diffusus* (Nuttall) L. Ll. Phillips, Res. Stud. State Coll. Wash. 23: 201. 1955.
Lupinus cumulicola Small, Man. S.E. Fl. 681. 1933. TYPE: Highlands Co.: sandhills E of Sebring, 1 May 1919, *Small & DeWinkeler 2081* (holotype: NY?).

Sprawling or ascending, annual or perennial herb; stem to 8 dm, silvery pubescent. Leaves 1-foliolate, the leaflet blade narrowly elliptic-lanceolate, 4–10 cm long, (1.5)2–3 cm wide, the apex rounded, mucronate, the base cuneate, the upper and lower surfaces densely sericeous or strigulose, the petiole 1–3 cm long; stipules with a setaceous free portion to 2 cm long. Flowers numerous, mostly whorled in a strongly exserted raceme; calyx symmetric, 7–10 mm long; corolla papilionaceous, 11–14 mm long, purple-blue, the standard with a conspicuous white or cream eyespot. Fruit oblong, 3–5 cm long, 8–10 mm wide, densely villous to sericeous.

Sandhills and dry, open hammocks. Common; peninsula, central and western panhandle. North Carolina south to Florida, west to Mississippi. Spring.

Lupinus luteus L. [Deep yellow, in reference to the corolla color.] YELLOW LUPINE.

Lupinus luteus Linnaeus, Sp. Pl. 2: 722. 1753.

Erect annual herb, to 8 dm; stem pubescent. Leaves (5)7–9(11)-foliolate, the leaflet blade ovate-oblong or oblanceolate, 2–4 cm long, ca. 1 cm wide, the apex rounded, mucronate, the base cuneate, the upper and lower surfaces pubescent, the petiole 6–7 cm long; stipules setaceous. Flowers whorled in a pedunculate, exserted raceme; calyx with the upper lip divided; corolla papilionaceous, 13–16 mm long, yellow. Fruit 4–5 cm long, ca. 1 cm wide, villous.

Disturbed sites. Rare; Jefferson County. Escaped from cultivation. Florida; Europe and Africa. Native to Europe and Africa. Spring.

Lupinus perennis L. [Perennial.] SUNDIAL LUPINE.

Lupinus perennis Linnaeus, Sp. Pl. 2: 721. 1753.

Lupinus gracilis Nuttall, J. Acad. Nat. Sci. Philadelphia 7: 115. 1834; non Kunth, 1824. *Lupinus perennis* Linnaeus var. *gracilis* Chapman, Fl. South. U.S. 89. 1860. *Lupinus nuttallii* S. Watson, Proc. Amer. Acad. Arts 8: 526. 1873. *Lupinus perennis* Linnaeus subsp. *gracilis* (Chapman) D. Dunn, Leafl. W. Bot. 10: 154. 1965.

Erect perennial herb, to 8 dm; stem glabrate or pubescent. Leaves 7- to 11-foliolate, the leaflet blade obovate to oblanceolate, 1–2.5 cm long, 3–5 mm wide, the apex obtuse to acute, mucronate, the base narrowly cuneate, the upper and lower surface glabrous or pubescent. Flowers subverticillate or alternate in a pedunculate, moderately to much exserted raceme; calyx symmetric, 4–8 mm long; corolla papilionaceous, 8–13 mm long, blue-violet, rarely pink or white. Fruit oblong, 3–5 cm long, 8–10 mm wide, villous.

Sandhills and dry, open hammocks. Frequent; northern counties. Maine south to Florida, west to Ontario, Minnesota, Iowa, and Texas. Spring.

Lupinus texensis Hook. [Of Texas.] TEXAS BLUEBONNET.

Lupinus texensis Hooker, Bot. Mag. 63: t. 3492. 1836.

Ascending or spreading annual herb; stem to 4 dm, pubescent. Leaves 5- to 6(7)-foliolate, the leaflet blade obovate or oblanceolate, 1–2.5 cm long, 5–8 mm wide, the apex obtuse to rounded, mucronate, the base cuneate, the upper and lower surfaces pubescent, the petiole 5–7 cm long; stipules setaceous. Flowers verticillate or alternate in a pedunculate, exserted raceme; calyx 6–8 mm long, symmetric, the upper lip toothed; corolla papilionaceous, 10–13 mm long, dark blue. Fruit 2.5–3.5 cm long, 8 mm wide, silky-villous.

Roadsides. Rare; St. Johns, Alachua, and Pinellas Counties. Florida, Louisiana, Oklahoma, and Texas. Native to Louisiana and Texas. Spring.

Lupinus villosus Willd. [With long weak trichomes.] LADY LUPINE.

Lupinus villosus Willdenow, Sp. Pl. 3: 1029. 1802.

Sprawling or ascending annual or biennial herb; stem to 6 dm, silvery or tawny villous. Leaves 1-foliolate, the leaflet blade elliptic to elliptic-oblong, 6–15 cm long, 2–3 cm wide, the apex rounded, mucronate, the base cuneate, the upper and lower surfaces villous, the petiole 4–5 cm long; stipules with the filamentous free portion 2(3) cm long. Flowers alternate or subverticillate in a short pedunculate, exserted raceme; calyx 7–11 mm long, symmetric; corolla papilionaceous, 10–14 mm long, lilac to lavender or pink, the standard with a red or purple maroon eyespot. Fruit elliptic or oblong, 2.5–4 cm long, ca. 1 cm wide, densely villous.

Sandhills and dry, open hammocks. Frequent; northern counties south to Polk County. North Carolina south to Florida, west to Louisiana. Spring.

Lupinus westianus Small [Commemorates George Mortimer West (1845–1926), St. Andrews, Florida, entrepreneur, founder of Panama City, Florida, horticulturalist, publisher, who provided material of the species to J. K. Small for the protologue.]

Spreading or ascending biennial herb; stem to 1.5 m, silvery subappressed-sericeous to shaggy pubescent. Leaves 1-foliolate, the leaflet blade ovate or elliptic, 3–7 cm long, 2–3 cm wide, the apex rounded to obtuse, mucronate, the base broadly cuneate, the upper and lower surfaces silvery subappressed-sericeous to shaggy pubescent, the petiole 1–2.5 cm long; stipules obsolete. Flowers mostly whorled in a pedunculate, exserted raceme; calyx 7–9 mm long, symmetric; corolla papilionaceous, 11–14 cm long, blue or pink to rose, the standard with a conspicuous red eyespot. Fruit ovate to ovate-oblong, ca. 1.5–2 cm long, ca. 6 mm wide, villous.

1. Flowers blue; central and western panhandle .... var. **westianus**
1. Flowers pink to rose; central peninsula .... var. **aridorum**

Lupinus westianus Small var. **westianus** GULF COAST LUPINE.

> *Lupinus westianus* Small, Torreya 26: 91. 1926. TYPE: FLORIDA: Bay Co.: St. Andrews, 4 May 1926, *Small et al. s.n.* (syntypes: NY).

Corolla blue.

Coastal dunes and sandhills. Occasional; central and western panhandle. Endemic. Spring.

Lupinus westianus is listed as Threatened in Florida (Florida Administrative Code, Chapter 5B-40).

Lupinus westianus Small var. **aridorum** (McFarlin ex Beckner) Isely [Living in dry areas.] BECKNER'S LUPINE; McFARLIN'S LUPINE.

> *Lupinus aridorum* McFarlin ex Beckner, Phytologia 50: 209. 1982. *Lupinus westianus* Small var. *aridorum* (McFarlin ex Beckner) Isely, Brittonia 38: 356. 1986. TYPE: FLORIDA: Orange Co.: bank of drainage in back of factories, on FL 437 just S of US 441 in Plymouth, 13 Apr 1970, *Beckner et al. 2375* (holotype: FLAS; isotypes: FLAS, FSU, GH, NCU, NY, USF).

Corolla pink to rose.

Scrub. Rare; Orange, Osceola, and Polk Counties. Endemic. Spring.

Lupinus westianus var. *aridorum* is listed as endangered in Florida (Florida Administrative Code, Chapter 5B-40) and endangered in the United States (U.S. Fish and Wildlife Service, 50 CFR 23).

EXCLUDED TAXA

> *Lupinus albus* Linnaeus—Reported for Florida by Isely (1990, 1998), Wunderlin (1998), and Wunderlin and Hansen (2003). No Florida specimens have been seen.

Lupinus luteolus Kellogg—Reported for Florida by Clewell (1985), the name misapplied to material of *L. luteus.*

Lupinus mexicanus Cervantes ex Lagasca y Segura—Reported for Florida by Clewell (1985, as *L. hartwegii* Lindley), based on cultivated material.

Lysiloma Benth. 1844. FALSE TAMARIND

Trees. Leaves alternate, even-bipinnately compound, petiolate, stipulate. Flowers in globose heads; sepals 5, connate, subequal; petals 5, connate, subequal; stamens 12–30, monadelphous. Fruit a laterally compressed, indehiscent legume.

A genus of about 9 species; North America, West Indies, Mexico, and Central America. [From the Greek *lyse*, to split, and *loma*, border, in reference to the fruit valves of some species separating at the margin.]

Selected references: Barneby and Grimes (1996); Thompson (1980).

1. Leaflets 10–30 pairs, oblong to oblong-lanceolate, 0.8–1.5 cm long .... **L. latisiliquum**
1. Leaflets 3–7 pairs, ovate to obovate, 1–2 cm long .... **L. sabicu**

Lysiloma latisiliquum (L.) Benth. [With wide siliqua, in Roman times, the word *siliqua* was mostly used for the fruit of Fabaceae.] FALSE TAMARIND.

Mimosa latisiliqua Linnaeus, Sp. Pl. 1: 519. 1753. *Acacia latisiliqua* (Linnaeus) Willdenow, Sp. Pl. 4: 1067. 1806. *Lysiloma latisiliquum* (Linnaeus) Bentham, Trans. Linn. Soc. London 30: 534. 1875. *Leucaena latisiliqua* (Linnaeus) Gillis, in Gillis & Stearn, Taxon 23: 190. 1974.

Lysiloma bahamensis Bentham, London J. Bot. 3: 82. 1844. *Acacia bahamensis* (Bentham) Grisebach, Fl. Brit. W.I. 221. 1860.

Tree, to 16 m; bark smooth, gray, splitting into scales, the branchlets glabrous. Leaves 8–14 cm long, the pinnae 2–5 pairs, 3–8 cm long, the leaflets 10–30 pairs, the blade oblong to oblong-lanceolate, 0.8–1.5 cm long, the apex obtuse, the base rounded, sessile, the upper and lower surfaces glabrous, the petiole ca. 6 cm long, with a large gland near the lowest pair of pinnae; stipules ovate, the apex acuminate. Flowers in a globose head 1.5–2 cm in diameter, racemose, the peduncle 2–4 cm long; calyx campanulate, ca. 1 mm long, subequal, white, glabrous; corolla 2–3 mm long, white, the lobes reflexed, subequal, villous; stamens ca. 20, much exserted. Fruit linear-elliptic to linear-oblong, ca. 20 cm long, 2–3 cm wide, nearly straight, indehiscent; seeds suborbicular, 1–2 cm long, laterally compressed, dark brown, lustrous.

Pinelands, hammocks, and coastal strands. Occasional; southern peninsula. Florida; West Indies, Mexico, and Central America. Spring–fall.

Lysiloma sabicu Benth. [Vernacular name for the species.] HORSEFLESH MAHOGANY.

Lysiloma sabicu Bentham, Hooker's J. Bot. Kew Gard. Misc. 6: 236. 1854.

Tree, to 10 m; bark gray, somewhat scaly, the branchlets glabrous. Leaves 10–20 cm long, the pinnae in 2–4 pairs, the leaflets 3–7 pairs, the blade ovate to obovate, 1–2.5 cm long, the apex rounded, the base rounded or cuneate, short petiolulate, the upper and lower surfaces glabrous, the petiole ca. 4 cm long, with a small circular gland; stipules obovate. Flowers in a long-pedunculate, solitary head about 1.5 cm in diameter; calyx ca. 1 mm long, subequal, greenish white; corolla 2–3 mm long, greenish white, subequal, the lobes villous; stamens 15–20, 4–6 mm long. Fruit oblong, 7–15 cm long, 2–3 mm wide, indehiscent; seeds suborbicular, 6–10 mm long, brown.

Disturbed hammocks. Rare; Miami-Dade County. Florida; West Indies. Native to West Indies. Spring–fall.

Macroptilium (Benth.) Urb. 1928. BUSHBEAN

Annual or perennial herbs. Leaves alternate, 3-foliolate, stipellate, petiolate, stipulate. Flowers in racemes, bracteate; sepals 5, connate; corolla papilionaceous, the standard obovate or orbicular, reflexed, with 2 reflexed basal auricles, the standard long-clawed, the claw partly adnate to the staminal tube, the keel twisted, long-clawed; stamens diadelphous (9 + 1). Fruit a dehiscent legume.

A genus of about 20 species; North America, West Indies, Mexico, Central America, South America, Africa, Asia, and Pacific Islands. [From the Greek *macro*, large, and *pter*, wing, in reference to the long wing petals.]

1. Leaflets velvety-pubescent or sericeous on the lower surface .... **M. atropurpureum**
1. Leaflets glabrate or sparsely strigose on the lower surface ....**M. lathyroides**

Macroptilium atropurpureum (Moç. & Sessé ex DC.) Urb. [Dark purple, in reference to the corolla.] PURPLE BUSHBEAN.

Phaseolus atropurpureus Moçiño & Sessé y Lacasta ex de Candolle, Prodr. 2: 395. 1825. *Phaseolus semi-erectus* Linnaeus var. *atropurpureus* (Moçiño & Sessé y Lacasta ex de Candolle) Gómez de la Maza y Jiménez, Anales Soc. Esp. Hist. Nat. 23: 295. 1894. *Phaseolus atropurpureus* Moçiño & Sessé y Lacasta ex de Candolle var. *genuinus* Hassler, Candollea 1: 457. 1923, nom. inadmiss. *Macroptilium atropurpureum* (Moçiño & Sessé y Lacasta ex de Candolle) Urban, Symb. Antill. 9: 457. 1928.

Sprawling perennial herb; stem 1–2 m, pilose. Leaves with the leaflet blade suborbicular or ovate to rhombic or oblong, 2–8 cm long, 1.5–3.5 cm wide, sometimes lobed, the apex obtuse or acute, the base rounded, the lateral ones oblique, usually with a lobe on the outer side, the upper and lower surfaces densely pilose, the petiole 3–4 cm long; stipules ovate, 4–5 mm long, pilose. Flowers few on a long pedunculate raceme; bracts linear, equaling or longer than the buds; calyx campanulate-tubular, 5–6 mm long, white pilose, the 2 upper teeth triangular, the lower 3 teeth subulate, subequaling or shorter than the tube; corolla papilionaceous, blackish purple, the standard 1.5–2 cm long, long clawed, the wing petals somewhat dimorphic and exceeding the standard in length, obliquely directed, the keel forming a spiral. Fruit linear, 5–9 cm long, ca. 3 mm wide; seeds 12–15, oblong-elliptic, ca. 4 mm long, strigose, black and brown marbled.

Disturbed sites. Occasional; Alachua County, central and southern peninsula. Escaped from cultivation. Florida, Texas, and Arizona; West Indies, Mexico, Central America, and South

America; Africa, Asia, and Pacific Islands. Native to southwestern United States and tropical America. Spring–fall.

Macroptilium lathyroides (L.) Urb. [Resembling *Lathyrus*.] WILD BUSHBEAN.

Phaseolus lathyroides Linnaeus, Sp. Pl., ed. 2. 1018. 1763. *Phaseolus lathyroides* Linnaeus var. *genuinus* Hassler, Candollea 1: 446. 1923, nom. inadmiss. *Phaseolus lathyroides* Linnaeus forma *typicus* Hassler, Candollea 1: 447. 1923. *Macroptilium lathyroides* (Linnaeus) Urban, Symb. Antill. 9: 457. 1928.

Erect or sometimes prostrate or twining annual or biennial herb; stem to 1.5 m, pilose or glabrous. Leaves with the leaflet blade linear-oblong to elliptic or subrotund, 3–7 cm long, 1–2.5 cm wide, the apex obtuse to acute, the base cuneate to rounded, the upper and lower surfaces glabrous to sparsely pilose, the petiole 0.5–5 cm long; stipules lanceolate, 5–10 mm long, subulate-tipped. Flowers in a long pedunculate raceme; bracts and bracteoles subulate; calyx tubular-campanulate, 4–6 mm long, the teeth short-triangular; corolla papilionaceous, purple-red, the standard ca. 1.5 cm long, clawed, the wing petals somewhat dimorphic and exceeding the standard in length, obliquely directed, the keel forming 1 spiral. Fruit linear, 8–10 cm long, 2–3 mm wide, strigose; seeds 18–30, oval, ca. 3 mm long, brownish gray mottled with white.

Disturbed sites. Frequent; peninsula, central and western panhandle. South Carolina, Georgia, Florida, Louisiana, and Texas; West Indies, Mexico, Central America, and South America; Africa, Asia, and Pacific Islands. Native to tropical America. Spring–fall.

Macroptilium lathyroides is listed as a Category II invasive species by the Florida Exotic Pest Plant Council (FLEPPC, 2015).

Medicago L., nom. cons. 1753. MEDICK

Annual or perennial herbs. Leaves 3-foliolate, estipellate, petiolate, stipulate, the stipules adnate to the petiole at the base. Flowers in axillary short racemose or subcapitate racemes; calyx 5-lobed, the lobes subequal; corolla papilionaceous, the wing and keel petals involved with an explosive tripping mechanism for pollination; stamens diadelphous (9 + 1). Fruit an indehiscent legume, falcate or spiral.

A genus of about 80 species; nearly cosmopolitan. [*Medice*, an ancient crop plant, probably alfalfa (*Medicago sativa*), this in turn from the Greek "from media" (Persia) with reference to its introduction from there to Europe, and the female suffix -*ago*.]

Selected reference: Small and Jomphe (1989).

1. Fruit reniform, 1-seeded; corolla 2–3 mm long..**M. lupulina**
1. Fruit spirally coiled, several-seeded; corolla 3–11 mm long.
    2. Corolla various shades of purple or varicolored (rarely yellow), 8–11 mm long; plant perennial......
    ....**M. sativa**
    2. Corolla yellow, 3–6 mm long; plant annual.
        3. Stipules entire or only slightly dentate...**M. minima**
        3. Stipules lacerate.
            4. Fruit 1–1.5 cm in diameter, flat, papery, unarmed ....**M. orbicularis**

4. Fruit 5–8 mm in diameter (excluding spines), burlike, firm, spiny (if unarmed, then with small tubercules).

    5. Upper and lower leaf surfaces pubescent .... **M. littoralis**

    5. Upper leaf surface glabrous, the lower sparsely pubescent to glabrate.

        6. Stipules shallowly lacerate (sinus less than ½ their length); margin of the fruit usually with a distinct spine-bearing ridge or nerve on each side of the suture; leaflets usually with a conspicuous central dark red spot..**M. arabica**

        6. Stipules strongly lacerate (sinus more than ½ their length); margin of the fruit lacking a distinct spine-bearing ridge or nerve on each side of the suture, the row of spines arising from a ridge lacking a central furrow; leaflets lacking a central dark red spot .... **M. polymorpha**

Medicago arabica (L.) Huds. [Of Arabia.] SPOTTED MEDICK.

Medicago polymorpha Linnaeus var. *arabica* Linnaeus, Sp. Pl. 2: 780. 1753. *Medicago arabica* (Linnaeus) Hudson, Fl Angl. 288. 1762. *Medica echinula* Lamarck, Fl. Franc. 2: 587. 1779 ("1778"), nom. illegit. *Medica arabica* (Linnaeus) Medikus, Vorles. Churpfälz. Phys.-Okon. Ges. 2: 386. 1787. *Medicago cordata* Desrousseaux, in Lamarck, Encycl. 3: 636. 1792, nom. illegit. *Medicago maculata* Sibthorp, Fl. Oxon. 232. 1794, nom. illegit.

Decumbent or erect annual herb; stem to 5 dm, glabrate or sparsely pubescent. Leaves with the leaflet blade broadly obovate or obcordate, 0.8–3 cm long, 1–3 cm wide, the apex rounded, retuse, the base cuneate, the margin denticulate, usually with a medial darker patch, the upper surface glabrous, the lower surface sparsely pubescent, the petiole 2–8(12) cm long; stipules ovate-lanceolate, 5–12 mm long, shallowly lacerate. Flowers 2–4 in a subcapitate raceme; calyx campanulate, 2–2.5 mm long; corolla papilionaceous, 4–5(6) mm long, yellow. Fruit coiled 4–6 times, subglobose or short cylndric, 9–14 mm, armed with sulcate spines, 9–14 mm in diameter (including spines), coriaceous, glabrous, brown, the submarginal nerves fused with the suture margin at maturity; seeds several, oblong-elliptic, yellowish brown.

Disturbed sites. Occasional; northern counties. Escaped from cultivation. Maine south to Florida, west to Illinois, Missouri, Oklahoma, and Texas, also Montana, British Columbia, south to California; South America; Europe, Africa, Asia, Australia, and Pacific Islands. Native to Europe, Asia, and Africa. Spring.

Medicago littoralis Rohde ex Loisel. [Of the seashore.] COASTAL MEDIC; WATER MEDIC.

Medicago littoralis Rohde ex Loiseleur-Deslongchamps, Not. Pl. Fl. France 118. 1810. *Medicago truncatula* Gaertner subsp. *littoralis* (Rohde ex Loiseleur-Deslongchamps) Ponert, Feddes Repert. 83: 639. 1973 ("1972").

Decumbent, annual or biennial herb; stem to 3 dm, sparsely villous. Leaves with the leaflet blade obovate, 5–8 mm long, 4–7 mm wide, the apex rounded, truncate, or retuse, the base cuneate, the margin denticulate distally, the upper and lower surfaces sparsely villous, the petiole 1–2 cm long; stipules ovate, dentate-lacerate. Flowers solitary or 2–3 in a subcapitate raceme, the peduncle 5–10 mm long, the pedicel ca. 1 mm long; calyx campanulate, ca. 3 mm long; corolla papilionaceous, 5–6 mm long, yellow or orange. Fruit tightly coiled 3–6 turns,

discoid to cylindric, 4–6 mm in diameter, glabrous, the dorsal suture spinose, the spines to ca. ½ the fruit diameter; seeds several, reniform, pale brown.

Disturbed sites near coastal beaches. Rare; Franklin, Walton, Okaloosa, and Santa Rosa Counties. Florida; Europe, Africa, and Asia. Native to Europe, Africa, and Asia. Summer–fall.

Medicago lupulina L. [Hoplike.] BLACK MEDICK.

Medicago lupulina Linnaeus, Sp. Pl. 2: 779. 1753. *Medica lupulina* (Linnaeus) Scopoli, Fl. Carniol., ed. 2. 2: 88. 1772. *Medicula lupulina* (Linnaeus) Medikus, Vorles. Churpfälz. Phys.-Oecon. Ges. 2: 386. 1787. *Medicago lupulina* Linnaeus var. *vulgaris* W.D.J. Koch, Syn. Fl. Germ. Helv. 161. 1835, nom. inadmiss. *Lupulina aurata* Noulet, Fl. Bass. Sous-Pyren. 157. 1837. *Melilotus lupulinus* (Linnaeus) Trautvetter, Bull. Sci. Acad. Imp. Sci. Saint-Petersbourg 8: 271. 1841; non Lamarck, 1779. *Lupularia parviflora* Opiz, Seznam 61: 1852, nom. illegit. *Medicago lupulina* Linnaeus var. *typica* Urban, Verh. Bot. Vereins Prov. Brandenburg 15: 52. 1873, nom. inadmiss.

Prostrate or ascending annual or short-lived perennial herb; stem 4(10) cm, pubescent, sometimes glandular. Leaves with the leaflet blade obovate, 0.8–1.5 cm long, 4–6 mm wide, the apex rounded to slightly retuse, the base cuneate, the margin denticulate in the upper half, the upper surface glabrous, the lower surface sparsely pubescent, the petiole 1–2 cm long; stipules ovate-lanceolate, ca. 10 mm long, entire or toothed. Flowers 10–20+ in an ovoid raceme, the peduncle 1.5–2 cm long, the pedicel ca. 1 mm long; calyx campanulate, 1–2 mm long; corolla papilionaceous, 2–3 mm long, yellow. Fruit reniform-incurved, 2–3 mm long, 1–2 mm wide, with strong concentric nerves, brown, becoming black at maturity, glabrous, rarely pubescent; seed 1, ovoid, brown.

Disturbed sites. Frequent; nearly throughout. Nearly throughout North America; West Indies, Mexico, Central America, and South America; Europe, Africa, Asia, and Pacific Islands. Native to Europe, Africa, and Asia. Spring–fall.

Medicago minima (L.) Bartal. [Very small.] BURR MEDICK.

Medicago polymorpha Linnaeus var. *minima* Linnaeus, Sp. Pl. 2: 780. 1753. *Medicago minima* (Linnaeus) Bartalini, Cat. Piante Siena 61. 1776. *Medicago minima* (Linnaeus) Bartalini var. *brevispina* Bentham, in Smith, Engl. Bot., Suppl. 1: sub t. 2635. 1831, nom. inadmiss. *Spirocarpus minimus* (Linnaeus) Opiz, Seznam 93. 1852. *Medicago minima* (Linnaeus) Bartalini var. *vulgaris* Urban, Verh. Bot. Vereins Prov. Brandenburg 15: 78. 1873, nom. inadmiss. *Medicago minima* (Linnaeus) subsp. *brevispina* Ponert, Feddes Repert. 83: 639. 1973 ("1972"), nom. inadmiss.

Procumbent or ascending annual herb; stem to 4(6) dm, villous. Leaves with the leaflet blade obovate to elliptic, 4–12 mm long, 2–5 mm wide, the apex rounded or retuse, the base cuneate, the margin denticulate toward the apex, the upper and lower surfaces villous, the petiole 5–10(20) mm long; stipules ovate, entire or slightly dentate. Flowers (1)3–6 in a subcapitate, short pedunculate raceme; calyx campanulate, 2–3 mm long; corolla papilionaceous, 3–5 mm long, yellow. Fruit coiled 3–4 turns, discoid to short cylindric, 6–8(10) mm in diameter (including spines), coriaceous, slightly pubescent, brown, the submarginal nerves sulcate; seeds several, reniform, brown.

Disturbed sites. Rare; Calhoun, Walton, and Escambia Counties. Escaped from cultivation. Massachusetts and New York south to Florida, west to Michigan, Kansas, Oklahoma, and

Texas, also Washington, Oregon, California, and Arizona; South America; Europe, Africa, Asia, Australia, and Pacific Islands. Native to Europe, Africa, and Asia. Spring.

Medicago orbicularis (L.) Bartal. [Rounded, in reference to the flowers in heads.] BLACKDISK MEDICK; BUTTON CLOVER.

Medicago polymorpha Linnaeus var. *orbicularis* Linnaeus, Sp. Pl. 2: 779. 1753. *Medicago orbicularis* (Linnaeus) Bartalini, Cat. Piante Siena 60. 1776. *Medica inermis* Lamarck, Fl. Franc. 2: 586. 1779 ("1778"), nom. illegit. *Medica orbicularis* (Linnaeus) Medikus, Vorles. Churpfälz. Phys.-Okon. Ges. 2: 286. 1787.

Prostrate annual herb; stem to 4 dm, glabrate. Leaves with the leaflet blade obovate, 0.7–1.5 cm long, 5–10 mm wide, the apex rounded or retuse, the base cuneate, the margin denticulate toward the apex, the upper and lower surfaces glabrate, the petiole 1–3 cm long; the stipules ovate-oblong, deeply lacerate. Flowers 1–2(4) in a loosely subcapitate raceme; calyx campanulate, 2–3 mm long; corolla papilionaceous, 3–5 mm long, yellow. Fruit coiled 3–5 turns, discoid, 1–1.5 cm in diameter, papery, brown, the submarginal nerves absent; seeds several.

Disturbed sites. Rare; Escambia County. Escaped from cultivation. New Jersey south to Florida, west to Illinois, Oklahoma, and Texas, also California; Europe, Africa, Asia, and Australia. Native to Europe, Africa, and Asia. Spring.

Medicago polymorpha L. [From the Greek *poly* and *morphos*, variable in form.] BURR CLOVER.

Medicago polymorpha Linnaeus, Sp. Pl. 2: 779. 1753. *Medica polymorpha* (Linnaeus) Scopoli, Fl. Carniol., ed. 2. 2: 89. 1772.

Medicago polymorpha Linnaeus var. *nigra* Linnaeus, Mant. Pl. 454. 1771. *Medicago nigra* (Linnaeus) Krocker, Fl. Siles. 2(2): 244. 1790. *Medicago hispida* Gaertner, Fruct. Sem. Pl. 2: 349. 1791, nom. illegit. *Medicago lappacea* Desrousseaux, in Lamarck, Encycl. 3: 637. 1792, nom. illegit. *Medicago denticulata* Willdenow var. *lappacea* Bentham, Cat. Pl. Pyrenees 103. 1826, nom. illegit. *Medicago denticulata* Willdenow var. *macracantha* Webb & Berthelot, Hist. Nat. Iles Canaries 3(2(2)): 64. 1842, nom. illegit. *Medicago lappacea* Desrousseaux var. *macracantha* Lowe, Man. Fl. Madeira 1: 158. 1862, nom. illegit. *Medicago hispida* Gaertner var. *lappacea* Burnat, Fl. Alpes Marit. 2: 107. 1896, nom. inadmiss.

Medicago denticulata Willdenow, Sp. Pl. 3: 1414. 1802. *Medicago denticulata* Willdenow var. *vulgaris* Bentham, Cat. Pl. Pyrenees 103. 1826, nom. inadmiss. *Medica denticulata* (Willdenow) Greene, Man. Bot. San Francisco 102. 1894. *Medicago hispida* Gaertner var. *denticulata* (Willdenow) Burnat, Fl. Alpes Marit. 2: 107. 1896. *Medicago polymorpha* Linnaeus var. *vulgaris* (Bentham) Shinners, Rhodora 58: 310. 1956, nom. illegit.

Prostrate or ascending annual or biennial herb; stem to 5 dm long, glabrate or glabrescent. Leaves with the leaflet blade obovate to obcordate, 0.8–2 cm long, 5–15 mm wide, the apex emarginate, the base cuneate, denticulate toward the apex, the upper surface glabrous, the lower surface pubescent to glabrate, the petiole 1–5 cm long; stipules ovate-oblong, 4–7 mm long, the base auriculate, deeply laciniate. Flowers (1)2–5 in a subcapitate raceme; calyx campanulate, ca. 3 mm long; corolla papilionaceous, 4–6 mm long, yellow. Fruit coiled 2–6 turns, the spines sulcate, straight, or curved, discoid, spheroid to short cylindric, 7–12 mm in diameter (including spines), coriaceous, glabrous, brown or black, the submarginal nerves not evident; seeds several, reniform, brown, smooth.

Disturbed sites. Occasional; nearly throughout. Escaped from cultivation. Nearly throughout North America; Mexico, Central America, and South America; Europe, Africa, Asia, Australia, and Pacific Islands. Native to Europe, Africa, and Asia. Spring.

Medicago sativa L. [Planted, cultivated.] ALFALFA.

Medicago sativa Linnaeus, Sp. Pl. 2: 778. 1753. *Medica sativa* (Linnaeus) Lamarck, Fl. Franc. 2: 584. 1779 ("1778"). *Medicago sativa* Linnaeus var. *vulgaris* Alefeld, Landw. Fl. 75. 1866, nom. inadmiss. *Medicago sativa* Linnaeus subsp. *macrocarpa* Urban, Verh. Bot. Vereins Prov. Brandenburg 15: 57. 1873, nom. inadmiss. *Medicago sativa* Linnaeus var. *falcata* Urban, Verh. Bot. Vereins Prov. Brandenburg 15: 57. 1873, nom. inadmiss. *Medica legitima* Greene, Man. Bot. San Francisco 101. 1894, nom. illegit.

Erect or ascending perennial herb; stem to 8(10) dm, strigose-puberulent. The leaves with the leaflet blade obovate to oblong-oblanceolate or narrowly lanceolate, 1–2.5(4) cm long, 3–10 mm wide, the apex rounded or retuse, the base cuneate, the margin denticulate toward the apex; stipules ovate-lanceolate, entire or denticulate near the base. Flowers 8–25 in an ovoid to oblong raceme; calyx campanulate, 4–6 mm long; corolla papilionaceous, 8–11 mm long, greenish blue, purple-black, greenish yellow, or yellow. Fruit coiled (1)2–3 turns or rarely falcate, the bur 4–6(8) mm in diameter, the valves dark brown, pubescent or glabrous; seeds 10–20, ovoid, yellow or brown, smooth.

Disturbed sites. Occasional; central and southern peninsula, central and western panhandle. Escaped from cultivation. Nearly throughout North America; Mexico, Central America, and South America; Europe, Africa, Asia, Australia, and Pacific Islands. Native to Europe, Africa, and Asia. Spring–fall.

Melilotus Mill. 1754. SWEETCLOVER

Annual or biennial herbs. Leaves 3-foliolate, estipellate, petiolate, stipulate, the stipules basally adnate to the petiole. Flowers in axillary racemes, bracteate, ebracteolate; calyx 5-lobed, subequal; corolla papilionaceous; stamens diadelphous (9 + 1). Fruit an indehiscent legume; seeds 1(2).

A genus of about 20 species; nearly cosmopolitan. [From the Greek *meli*, honey, and *lotos*, a leguminous fodder plant.]

Selected reference: Stevenson (1969).

1. Corolla white .... **M. albus**
1. Corolla yellow.
    2. Corolla 2–3 mm long; fruit obscurely reticulate or irregularly tuberculate .... **M. indicus**
    2. Corolla 5–7 mm long; fruit cross-rugose .... **M. officinalis**

Melilotus albus Medik. [White, in reference to the corolla.] WHITE SWEETCLOVER.

Melilotus albus Medikus, Vorles. Churpfälz. Phys.-Okon. Ges. 2: 382. 1787. *Trifolium album* (Medikus) Loiseleur-Deslongchamps, Fl. Gall. 479. 1807; non Crantz, 1769; nec Lamarck, 1779. *Sertula alba* (Medikus) Kuntze, Revis. Gen. Pl. 1: 205. 1891. *Medicago alba* (Medikus) Krause, in Sturm, Deutschl. Fl., ed. 2. 9: 128. 1901.

Erect annual or biennial herb, to 2 m; stem sparsely strigose to glabrate. Leaves with the leaflet blade obovate to elliptic-oblong, 1.5–3 cm long, (4)6–12 mm wide, the apex rounded, the base cuneate, the margin denticulate nearly to the base, the upper surface glabrous, the lower surface puberulent, the petiole 1–2 cm long; stipules subulate, 6–10 mm long, entire. Flowers 40–100 in a raceme 4–12 cm long, the pedicel 1–2 mm long; calyx ca. 2 mm long; corolla papilionaceous, 4–5 mm long, white. Fruit elliptic to oblong, 3–4 mm long, ca. 2 mm wide, reticulate; seeds 1(2), ovoid, papillate.

Disturbed sites. Frequent; nearly throughout. Escaped from cultivation. Nearly throughout North America; West Indies, Mexico, Central America, and South America; Europe, Africa, Asia, Australia, and Pacific Islands. Native to Europe and Asia. Spring–summer.

Melilotus indicus (L.) All. [Of India.] INDIAN SWEETCLOVER; ANNUAL YELLOW SWEETCLOVER.

> *Trifolium indicum* Linnaeus, Sp. Pl. 2: 765. 1753. *Melilotus indicus* (Linnaeus) Allioni, Fl. Pedem. 1: 308. 1785. *Melilotus levis* Moench, Methodus 110. 1794, nom. illegit. *Sertula indica* (Linnaeus) Kuntze, Revis. Gen. Pl. 1: 205. 1891.

Erect or spreading annual herb, to 60 dm; stem glabrous. Leaves with the leaflet blade obovate to oblanceolate, 1–2 cm long, 8–10 mm long, the apex round or retuse, the base cuneate, the margin denticulate to the apex, the upper surface glabrous, the lower surface strigose, the petiole 1–3 cm long; stipules lanceolate, 4–6 mm long, the base auriculate, the margin entire or few-toothed proximally. Flowers 15–25 in a compact, pedunculate raceme 1–2 cm long, the pedicel ca. 1 mm long; bracts filiform; calyx 1–2 mm long; corolla papilionaceous, 2–3 mm long, yellow. Fruit ovoid, 2–3 mm long, ca. 2 mm wide, reticulate, glabrous; seed 1, ovoid, ca. 2 mm long, dark brown, papillate.

Disturbed sites. Occasional; nearly throughout. Escaped from cultivation. Nearly throughout North America; West Indies, Mexico, and South America; Europe, Africa, Asia, Australia, and Pacific Islands. Native to Europe, Africa, and Asia. Spring–summer.

Melilotus officinalis Lam. [*Opificina*, shortened to *officina*, originally a workshop or shop, later a monastic storeroom, then an herb-store, pharmacy, or drug-shop.] YELLOW SWEETCLOVER.

> *Melilotus officinalis* Lamarck, Fl. Franc. 2: 594. 1779 ("1778"). *Brachylobus officinalis* (Linnaeus) Dulac, Fl. Hautes-Pyrénées 279. 1867.
>
> *Trifolium officinale* Linnaeus, Sp. Pl. 2: 765. 1753. *Melilotus officinalis* (Linnaeus) Desrousseaux, in Lamarck, Encycl. 4: 63. 1797; non Lamarck, 1779. *Sertula officinalis* (Linnaeus) Kuntze, Revis. Gen. Pl. 1: 205. 1891. *Medicago officinalis* (Linnaeus) Krause, in Sturm, Deutschl. Fl., ed. 2. 9: 127. 1901.

Erect biennial herb, to 2 m; stem strigose or glabrate. Leaves with the leaflet blade broadly obovate to ovate, 1–1.5(2) cm long, 4–8 mm wide, the apex rounded or retuse, the base cuneate, the margin denticulate, the petiole 1–2 cm long; stipules subulate to acicular, 3–5(7) mm long, the margin entire or with a few basal teeth. Flowers 30–70 in a raceme 4–12 cm long, the pedicel 1–2 mm long; bracts 1–2 mm long; calyx 2–3 mm long; corolla papilionaceous, 5–7 mm long, yellow. Fruit ovoid, 3–5 mm long, ca. 2 mm wide, dark brown, irregularly cross-rugose; seeds 1(2), ovoid, 2–3 mm long, yellow, smooth.

Disturbed sites. Rare; Escambia County. Escaped from cultivation. Nearly throughout North America; West Indies, Mexico, and South America; Africa, Europe, Asia, Australia, and Pacific Islands. Native to Europe and Asia. Spring–summer.

Millettia Wight & Arn., nom. cons. 1834.

Trees. Leaves alternate, odd-pinnately compound, the leaflets opposite, estipellate, petiolate, stipulate. Flowers in axillary pseudoracemes, bracteate, bracteolate; sepals 5, connate; Petals 5, the corolla papilionaceous; stamens 10, diadelphous (9 + 1), the anthers dorsifixed. Fruit an indehiscent legume.

A genus of about 150 species; North America, West Indies, Central America, Africa, Asia, Australia, and Pacific Islands. [Commemorates Charles Millett (fl. 1825–1834) British employee of the East India Company and plant collector in China.]

Pongamia Adans., nom. rej. 1763.

Millettia pinnata (L.) Panigrahi [With pinnate leaves.] KARUM TREE; PONGAME OILTREE, POONGA OIL-TREE.

Cytissus pinnatus Linnaeus, Sp. Pl. 2: 741. 1753. *Robinia mitis* Linnaeus, Sp. Pl., ed 2. 1044. 1763, nom. illegit. *Pongamia glabra* Ventenat, Jard. Malmaison t. 28. 1803, nom. illegit. *Pongamia mitis* Kurz, J. Asiat. Soc. Bengal, Pt. 2, Nat. Hist. 45(2): 128. 1876, nom. illegit. *Cajum pinnatum* (Linnaeus) Kuntze, Revis. Gen. Pl. 1: 167. 1891. *Galedupa pinnata* (Linnaeus) Taubert, in Engler & Prantl, Nat. Pflanzenfam. 3(3): 344. 1894. *Pongamia pinnata* (Linnaeus) Pierre, Fl. Forest. Cochinch. 4: sub pl. 385. 1899. *Millettia pinnata* (Linnaeus) Panigrahi, in Panigrahi & Murti, Fl. Bilaspur Distr. 1: 210. 1989.

Tree, to 30 m; branchlets glabrous. Leaves 6–17 cm long, the leaflets 5–9, the blade ovate to elliptic-oblong, 3–10 cm long, 2–5 cm wide, the apex acuminate, the base rounded, the petiolules 0.5–1 cm long, the petiole 2.5–5 cm long; stipules caducous. Flowers mostly paired in a pseudoraceme, the rachis sparsely pubescent or glabrous; bracts caducous; bracteoles in pairs at top of the pedicel; calyx 3–4 mm long, broadly campanulate, truncate; corolla 9–10 mm long, white to pink or lavender, the banner sericeous dorsally. Fruit narrowly elliptic, 5–8 cm long, 2.5–3 cm wide, flat, woody, glabrous; seed 1, reniform, ca. 2 cm long, brown.

Disturbed sites. Rare; Sarasota, Palm Beach, and Broward Counties. Escaped from cultivation. Florida; West Indies and Central America; Africa, Asia, Australia, and Pacific Islands. Native to Asia, Australia, and Pacific Islands. Summer.

Mimosa L. 1753. SENSITIVE PLANT

Annual or perennial herbs or shrubs. Leaves alternate, even-bipinnately compound, petiolate, stipulate. Flowers in heads; sepals 5, connate, subequal; petals 5, connate, subequal; stamens 4–10, free. Fruit a dehiscent or indehiscent legume.

A genus of about 500 species; nearly cosmopolitan. [From the Greek *mime*, for the apparent mimicry of animal reactions in reference to the sensitive leaves of some species.]

Leptoglottis DC., 1827; *Schrankia* Willd., nom. cons. 1806.

Selected reference: Barneby (1991).

1. Fruit tetragonal, the sutures as wide as the valves, each valve separating from the sutures entire .... **M. quadrivalvis**
1. Fruit laterally compressed, the sutures narrow, the valves breaking into 1-seeded segments.
  2. Pinnae (1)2 pairs...**M. pudica**
  2. Pinnae 4–12 pairs.
    3. Trailing or ascending perennial herb or shrub; stem and leaves with hooked prickles; fruit 7- to 15-segmented .... **M. pigra**
    3. Sprawling suffrutescent herb; stem and leaves unarmed; fruit usually 3-segmented.................... .... **M. strigillosa**

Mimosa pigra L. [Lazy, apparently in reference to its reclining habit.] BLACK MIMOSA.

Mimosa pigra Linnaeus, Cent. Pl. 1: 13. 1755, nom. cons.

Trailing or ascending perennial herb or shrub; stem to 1.5 m, with spreading or ascending trichomes and hooked prickles. Leaves with 5–12 pairs of pinnae, the leaflets 15–30+ pairs, the rachis with spreading or ascending trichomes and recurved prickles, the blade 5–10 mm long, 1–3 mm wide, the apex obtuse to acute, the base cuneate to rounded, asymmetric, the margin often ciliate near the base, the upper and lower surfaces strigose, the petiole ca. 1 cm long; stipules ovate-lanceolate, striate. Flowers numerous in an ovoid to capitate head ca. 1 cm in diameter, pink fading white; stamens 8, exserted. Fruit oblong, 4–8 cm long, 0.8–1 cm wide, laterally compressed, straight or slightly falcate, bristly pubescent, 7- to 15-segmented, breaking into 1-seeded segments and separating from the persistent sutures (replum) as a unit, the replum remaining intact.

Wet, disturbed sites. Rare; Highlands, Okeechobee, Martin, Palm Beach, and Broward Counties. Florida and Texas; West Indies, Mexico, Central America, and South America; Africa, Asia, and Australia. Native to South America. Spring–summer.

Mimosa pigra is listed as a Category I invasive species in Florida by the Florida Exotic Pest Plant Council (FLEPPC, 2015).

Mimosa pudica L. [Bashful, in reference to the sensitive leaves.] SENSITIVE PLANT; SHAMEPLANT.

Mimosa pudica Linnaeus, Sp. Pl. 1: 518. 1753.

Procumbent herbaceous or suffrutescent herb; stem to 1 m, with bristly trichomes and recurved prickles. Leaves with (1)2 pairs of pinnae, the leaflets 10–25 pairs, the blade linear-lanceolate, 6–15 mm long, 2–3 mm wide, the apex acute, apiculate, the base rounded to subcordate, asymmetric, the upper surface glabrous, the lower surface slightly hispid, the petiole 3–4 cm long; stipules lanceolate, 5–10 mm long, bristly. Flowers numerous in 1–2 globose long pedunculate heads ca. 1.5 cm in diameter, pink; bracts linear; calyx minute; corolla campanulate, the lobes pubescent on the outer surface; stamens 4, exserted. Fruit oblong, 1–2 cm long, 3–5 mm wide, the margin setose, the valves glabrate, 3–5 segmented, breaking into 1-seeded segments and separating from the persistent bristly sutures (replum) as a unit, the replum remaining intact; seeds ovoid, 3–4 mm long, light brown.

Disturbed sites. Rare; Baker, Hillsborough, Broward, and Miami-Dade Counties. Maryland, Virginia, and Florida; West Indies, Mexico, Central America, and South America; Africa, Asia, Australia, and Pacific Islands. Native to tropical America. All year.

The leaves of *Mimosa pudica* are sensitive to the touch.

Mimosa quadrivalvis L. [Four-valved, in reference to fruit splitting along four sutures.] SENSITIVE BRIER.

Trailing scrambling perennial herb; stem to 3 m, with recurved prickles, otherwise glabrous. Leaves with 3–8(11) pairs of pinnae, the leaflets 9–12 pairs, the blade, oblong, 2–4(6) mm long, 1–2 mm wide, the apex acute to rounded, the base rounded, asymmetric, the upper and lower surfaces glabrate, the rachis and petiole with recurved prickles. Flowers in a long-pedunculate, globose head (0.7)1–1.5(2) cm in diameter, pink to nearly white, the peduncle (1)2–4(6) cm long, with recurved prickles; stamens 6. Fruit oblong to linear (3)5–10)12 cm long, tetragonal, the 2 sutures about as wide as the 2 valves and separating from them at maturity entire, sparsely to densely prickly.

1. Leaflets with only the median vein evident on the lower surface...var. **angustata**
1. Leaflets with evident lateral venation on the lower surface..var. **floridana**

Mimosa quadrivalvis var. angustata (Torr. & A. Gray) Barneby [Narrow, in reference to the narrow leaflets and fruits.] SENSITIVE BRIER.

Schrankia angustata Torrey & A. Gray, Fl. N. Amer. 1: 400. 1840. *Morongia angustata* (Torrey & A. Gray) Britton, Mem. Torrey Bot. Club 5: 191. 1894. *Mimosa quadrivalvis* Linnaeus var. *angustata* (Torrey & A. Gray) Barneby, Mem. New York Bot. Gard. 65: 299. 1991.

Mimosa microphylla Dryander ex Smith, Nat. Hist. Lepidopt. Georgia 2: 123, pl. 62. 1797. *Morongia microphylla* (Dryander ex Smith) Britton, in Britton & A. Brown, Ill. Fl. N. U.S., ed. 2. 2: 334. 1913. *Schrankia microphylla* (Dryander ex Smith) J. F. Macbride, Contr. Gray Herb. 59: 9. 1919. *Leptoglottis microphylla* (Dryander ex Smith) Britton & Rose, in Britton, N. Amer. Fl. 23: 142. 1928.

Mimosa horridula Michaux, Fl. Bor.-Amer. 1: 254. 1803. *Schrankia uncinata* Willdenow, Sp. Pl. 4: 1043. 1806, nom. illegit. *Schrankia horridula* (Michaux) Chapman, Fl. South. U.S., ed. 2, suppl. 2. 683. 1892. *Morongia uncinata* Britton, Mem. Torrey Bot. Club 5: 191. 1894, nom. illegit. *Leptoglottis uncinata* Rydberg, Fl. Nebraska, Rosales 31. 1895, nom. illegit. *Morongia horridula* (Michaux) A. Heller, Cat. N. Amer. Pl. 2. 1898. TYPE: "Virginia ad Floridam," without data, *Michaux s.n.* (holotype: P).

Schrankia angustata Torrey & A. Gray var. *brachycarpa* Chapman, Fl. South U.S. 116. 1860. *Leptoglottis chapmanii* Small ex Britton & Rose, in Britton, N. Amer. Fl. 23: 141. 1928. *Schrankia chapmanii* (Small ex Britton & Rose) F. J. Hermann, J. Wash. Acad. Sci. 38: 237. 1948. TYPE: "Florida to North Carolina."

Leptoglottis angustisiliqua Britton & Rose, in Britton, N. Amer. Fl. 23: 143. 1928. *Schrankia angustisiliqua* (Britton & Rose) F. J. Hermann, J. Wash. Acad. Sci. 38: 237. 1948. TYPE: FLORIDA: Miami-Dade Co.: Brogdon Hammock, 19 Jun 1915, *Small & Mosier 6349* (holotype: NY).

Leaflets with only the median vein evident on the lower surface.

Sandhills, flatwoods, and mesic hammocks. Frequent; nearly throughout. Virginia south to Florida, west to Illinois and Texas. Summer–fall.

Mimosa quadrivalvis var. **floridana** (Chapm.) Barneby [Of Florida.] FLORIDA SENSITIVE PLANT.

> *Schrankia floridana* Chapman, Fl. South. U.S., ed. 2, suppl. 2, 683. 1892. *Morongia floridana* (Chapman) A. Heller, Cat. N. Amer. Pl. 4: 1898. *Leptoglottis floridana* (Chapman) Small ex Britton & Rose, in Britton, N. Amer. Fl. 23: 139. 1928. *Schrankia microphylla* (Dryander ex Smith) J. F. Macbride var. *floridana* (Chapman) Isely, Castanea 51: 204. 1986. *Mimosa quadrivalvis Linnaeus* var. *floridana* (Chapman) Barneby, Mem. New York Bot. Gard. 65: 300. 1991. TYPE: FLORIDA: Manatee Co.: Manatee, s.d. *Simpson s.n.* (isotype: NY).

Leaflets with evident lateral venation on the lower surface.

Sandhills and scrub. Frequent; northern and central peninsula. Georgia and Florida. Summer fall.

Mimosa strigillosa Torr. & A. Gray [Finely covered with straight, adpressed trichomes.] POWDERPUFF.

> *Mimosa strigillosa* Torrey & A. Gray, Fl. N. Amer. 1: 399. 1840.

Prostrate perennial herb; stem to 1.5 m, strigose. Leaves with 4–5(6) pinnae pairs, the leaflets 6–15 pairs, the blade linear, 4–8 mm long, ca. 1 mm wide, the apex obtuse to rounded, the base cuneate to rounded, asymmetric, the upper surface glabrous, the lower surface strigulose; stipules deltate, striate. Flowers in a globose to short-cylindric head ca. 2.5 cm in diameter, pinkish or lavender; stamens 8. Fruit oblong, ca. 2.5 cm long, ca. 1 cm wide, the segments 3, setulose, breaking into 1-seeded segments and separating from the persistent bristly sutures (replum) as a unit, the replum remaining intact.

Open, usually disturbed sites. Occasional; peninsula west to central panhandle. Georgia and Florida west to Texas; Mexico and South America. Spring–summer.

Mucuna Adans., nom. cons. 1763.

Annual or perennial woody or herbaceous vines. Leaves alternate, 3-foliolate, stipellate, petiolate, stipulate. Flowers in an axillary racemose or umbelliform inflorescence, bracteate; calyx 2-lipped; corolla papilionaceous; stamens 10, diadelphous (9 + 1). Fruit a dehiscent legume.

A genus of about 100 species; North America, West Indies, Mexico, Central America, South America, Asia, Africa, Australia, and Pacific Islands. [Tupi-Guarani vernacular name *mucuná*.]

> *Stizolobium* P. Browne, nom. rej. 1756.

Selected reference: Moura et al. (2013).

1. Corolla purple or rarely white, 3–4 cm long; flowers in a raceme; fruit 5–9 cm long, 1–1.5 cm wide; seed hilum ¼ the circumference of the seed .... **M. pruriens**
1. Corolla yellow, 6–6.5 cm long; flowers in an umbel; fruit 9–12(16) cm long, 4–6 cm wide; seed hilum nearly completely the circumference of the seed .... **M. sloanei**

Mucuna pruriens (L.) DC. [Causing itching or stinging.] COWITCH; VELVETBEAN.

Dolichos pruriens Linnaeus, Herb. Amb. 23. 1754. *Stizolobium pruriens* (Linnaeus) Medikus, Vorles. Churpfälz. Phys.-Okon. Ges. 2: 399. 1787. *Mucuna pruriens* (Linnaeus) de Candolle, Prodr. 2: 405. 1825. *Mucuna prurita* Hook, in Wight, Bot. Misc. 2: 348. 1831, nom. illegit. *Carpopogon pruriens* (Linnaeus) Roxburgh, Fl. Ind., ed. 1832. 3: 383. 1832. *Stizolobium pruritum* Piper & Tracy, U.S.D.A. Bur. Pl. Industr. Bull. 179: 10. 1910, nom. illegit.

Mucuna utilis Wallich ex Wight, Icon. Pl. Ind. Orient. 1: t. 280. 1840. *Mucuna pruriens* (Linnaeus) de Candolle var. *utilis* (Wallich ex Wight) Baker ex Burck, Ann. Jard. Bot. Buitenzorg 11: 187. 1893. *Stizolobium utile* (Wallich ex Wight) Piper & Tracy, U.S.D.A. Bur. Pl. Industr. Bull. 179: 14. 1910.

Stizolobium deeringianum Bort, U.S.D.A. Bur. Pl. Industr. Bull. 141: 31. 1909. *Mucuna deeringiana* (Bort) Merrill, Philipp. J. Sci. 5: 118. 1910. *Stizolobium utile* (Wallich ex Wight) Piper & Tracy subsp. *deeringianum* (Bort) Ditmann, Fl. Cult. Pl. 4: 403. 1937. TYPE: FLORIDA.

Annual herbaceous or woody vines; stem to 15 m, pubescent or strigose. Leaves with the terminal leaflet blade rhombic-ovate, (5)9–22 cm long, the apex acuminate to cuspidate, the base rounded, the margin entire, the upper leaf surface sparsely pubescent, the lower leaf surface strigose, the lateral leaflets with the base oblique, the petiole (3)6–20 cm long; stipules lanceolate, 3–5 mm long. Flowers in axillary racemes; calyx 12–15 mm long, the upper lip broad, acuminate, the lower lip 3-lobed, the lobes lanceolate; corolla papilionaceous, 3–4 cm long, purple or rarely white. Fruit 5–9 cm long, 1–15 mm wide, brown, lacking transverse ridges, velutinous, often with irritating trichomes; seeds (1)3–6, spherical to oblong, 1–2 cm long, black-brown to white with dark mottling, the hilum extending to ¼ the circumference of the seed, cream or black, elevated.

Disturbed sites. Rare; Lee and Palm Beach Counties, southern peninsula. Escaped from cultivation. North Carolina, South Carolina, Florida, and Alabama; West Indies, Mexico, Central America, and South America; Africa, Asia, and Australia. Native to Africa and Asia. Summer.

Mucuna sloanei Fawc. [Commemorates Sir Hans Sloan (1660–1753), British botanist.] HORSEEYE BEAN.

Mucuna sloanei Fawcett & Rendle, J. Bot. 55: 36. 1917.

Perennial herbaceous or woody vines; stem to 10 m, strigose. Leaves with the terminal leaflet blade ovate, 7–15 cm long, the apex acuminate, the base rounded, the margin entire, the upper leaf surface pubescent, the lower leaf surface strigose, the lateral leaflets with the base oblique, the petiole 3–11 cm long; stipules lanceolate, 1–3 mm long. Flowers in an axillary umbelliform inflorescence; calyx 15–20 mm long, the upper lip broad, rounded, the lower lip 3-lobed, the lobes lanceolate; corolla papilionaceous, 4.5–6.6 cm long, yellow. Fruit 9–12(16) cm long, 4–6 cm wide, brown, with transverse ridges, setose, with irritating trichomes; seeds 1–3, spherical, 2–3 cm long, brown to black, the hilum extending nearly completely the circumference of the seed, black, not elevated.

Disturbed sites. Rare; Broward and Miami-Dade Counties. Florida; West Indies, Central America, and South America; Africa and Pacific Islands. All year.

EXCLUDED TAXON

Mucuna urens (Linnaeus) Medikus—Reported for Florida by Chapman (1897, as *Canavalia altissima* (Jacquin) de Candolle), misapplied to material of *Canavalia brasiliensis*.

Neptunia Lour. 1790. PUFF

Perennial herbs. Leaves alternate, even-bipinnately compound, petiolate, stipulate. Flowers in a terminal spike, bracteate, dimorphic, the proximal flowers staminate, the distal ones carpellate; sepals 4, basally connate, subequal; petals 5, subequal; stamens 10, free. Fruit a dehiscent legume.

A genus of ca. 12 species; North America, West Indies, Mexico, Central America, South America, Africa, Asia, Australia, and Pacific Islands. [For Neptune, Roman god of the sea and other waters, in reference to the aquatic habitat of the type species.]

Selected reference. Windler (1966).

Neptunia pubescens Benth. [Hairy.] TROPICAL PUFF.

Neptunia pubescens Bentham, J. Bot. (Hooker) 4: 357. 1841.
Neptunia floridana Small, Bull. Torrey Bot. Club 25: 138. 1898. *Neptunia pubescens* var. *floridana* (Small) B. L. Turner, Amer. Midl. Naturalist 46: 89. 1951. TYPE: FLORIDA: without data, *Chapman s.n.* (lectotype: NY). Lectotypified by Turner (1951: 89).

Erect, ascending, decumbent, or prostrate, perennial herb; stem to 2 m, glabrous or pubescent. Leaves even-pinnately compound, with (2)3–6 pairs of pinnae, the pinnules (16)28–60(80), opposite, the pinnule blade oblong to oblong-lanceolate, 2–8 mm long, 1–2 mm wide, the margin ciliate, the upper and lower surfaces glabrous or pubescent, the petiole 1–3 cm long; stipules lanceolate. Flowers 15–30 in a spherical to ovoid congested spike, sessile or subsessile; bracts 0–2, subulate, 1–3 mm long, the proximal flowers sterile, flattened, with petaloid staminodes, the distal ones fertile; calyx campanulate, 2–3 mm long, the lobes equal; corolla yellow-green to white, 4–6 mm long, the lobes free nearly to the base. Fruit oblong to suborbicular, 5–13 mm long, the stipe 2–4 mm long; seeds 4–11, oblong-ellipsoid to obovoid, brown.

Riverbanks, flatwoods, salt marsh margins, and disturbed sites. Occasional; peninsula west to central panhandle. Florida west to Texas; West Indies, Mexico, Central America, and South America. Spring–fall.

EXCLUDED TAXON

Neptunia lutea (Leavenworth) Bentham—Reported for Florida by Chapman (1860, 1883, 1897, all as *Desmanthus luteus* (Leavenworth) Bentham ex Chapman), who misapplied the name to material of *N. pubescens*. Also reported by Small (1903, 1913a, 1933), the basis unknown. Excluded from Florida by all later authors; no specimens known.

Orbexilum Raf. 1832. LEATHERROOT

Perennial herbs. Leaves unifoliate, 3-foliolate, or palmately 5- to 7-foliolate, petiolate, stipulate. Flowers in spikes, bracteate; sepals 5, connate; corolla papilionaceous; stamens 10, monadelphous, dimorphic, the upper series basifixed, the lower series dorsifixed. Fruit a dehiscent legume.

A genus of 10 species; North America and Mexico. [*Orbicular*, circular, and *vexillum*, standard petal, in reference to the rounded vexillum characteristic of the genus.]

Rhytidomene Rydb., 1919.

Selected reference: Turner (2008).

1. Leaves 1-foliolate ....**O. virgatum**
1. Leaves 3- to 7-foliolate.
    2. Leaves (3)5- to 7-foliolate; leaflets linear-filiform..**O. lupinellus**
    2. Leaves 3-foliolate; leaflets elliptic-lanceolate to lanceolate....................................**O. pedunculatum**

Orbexilum lupinellus (Michx.) Isely [*Lupinus*, and *-ellus*, diminutive, in reference to its resemblance to a little *Lupinus*.] PIEDMONT LEATHERROOT.

Psoralea lupinellus Michaux, Fl. Bor.-Amer. 2: 58. 1803. *Lotodes lupinellus* (Michaux) Kuntze, Revis. Gen. Pl. 1: 194. 1891. *Rhytidomene lupinellus* (Michaux) Rydberg, in Britton, N. Amer. Fl. 24: 12. 1919. *Orbexilum lupinellus* (Michaux) Isely, Sida 11: 432. 1986.

Erect, perennial herb; stem to 75 cm, glabrous. Leaves palmately (3)5–7-foliolate, the leaflet blade linear, 2–7 cm long, 1–4 mm wide, the apex obtuse, the base cuneate, the margin entire, the upper and lower surfaces glabrous or glabrate, the petiole 1–4 cm long. Flowers in a spike 1–6 cm long, the peduncle 3–10 cm long; calyx 2–3 mm long, glabrous or sparsely pubescent; corolla papilionaceous, 5–7 mm long, purplish blue. Fruit elliptic, 9–11 mm long, 5–6 mm wide, rugose, glabrous, glandular; seeds reniform, 5–7 mm long.

Sandhills. Occasional; northern counties south to Hillsborough and Polk Counties. North Carolina south to Florida, west to Alabama. Spring–summer.

Orbexilum pedunculatum (Mill.) Rydb. [Pedunculate, in reference to the long peduncle.] SAMPSON'S SNAKEROOT.

Hedysarum pedunculatum Miller, Gard. Dict., ed. 8. 1768. *Psoralea pedunculata* (Miller) Vail, Bull. Torrey Bot. Club. 21: 114. 1894; non Ker Gawler, 1817. *Orbexilum pedunculatum* (Miller) Rydberg, in Britton, N. Amer. Fl. 24: 7. 1919.

Trifolium psoralioides Walter, Fl. Carol. 184. 1788. *Psoralea melilotoides* Michaux, Fl. Bor.-Amer. 2: 58. 1803, nom. illegit. *Psoralea melilotus* Persoon, Syn. Pl. 2: 347. 1807, nom. illegit. *Melilotus psoralioides* (Walter) Nuttall, Gen. N. Amer. Pl. 2: 104. 1818. *Lotodes psoralioides* (Walter) Kuntze, Revis. Gen. Pl. 1: 194. 1891. *Psoralea psoralioides* (Walter) Cory, Rhodora 38: 406. 1936. *Psoralea psoralioides* (Walter) Cory var. *typica* F. L. Freeman, Rhodora 39: 425. 1937, nom. inadmiss. *Orbexilum pedunculatum* (Miller) Rydberg var. *psoralioides* (Walter) Isely, Sida 13: 122. 1988. *Orbexilum psoralioides* (Walter) Vincent, Phytoneuron 2014–36: 2. 2014.

Psoralea melilotoides Michaux var. *gracilis* Torrey & A. Gray, Fl. N. Amer. 1: 303. 1838. *Psoralea gracilis* (Torrey & A. Gray) Chapman ex Small, Fl. S.E. U.S. 623. 1903. *Orbexilum gracile* (Torrey & A. Gray) Rydberg, in Britton, N. Amer. Fl. 24: 7. 1919. *Psoralea psoralioides* (Walter) Cory var. *gracilis* (Torrey & A. Gray) F. L. Freeman, Rhodora 39: 427. 1937. *Orbexilum pedunculatum* (Miller) Rydberg var. *gracile* (Torrey & A. Gray) J. W. Grimes, Mem. New York Bot. Gard. 61: 47. 1990. TYPE: FLORIDA: without data, *Chapman s.n.* (holotype: NY).

Erect or ascending perennial herb, to 80 cm; stem sparsely strigose to glabrate. Leaves 3-foliolate, the leaflet blade lanceolate to elliptic, 2–7 cm long, 0.5–2 cm wide, the apex obtuse, mucronate, the base rounded, the upper and lower surfaces glabrous or glabrate, the petiole 0.2–5 cm long. Flowers in a spike 2–13 cm long, the peduncle 4–16 cm long; calyx 4–7 mm long, sparsely pubescent or glabrous; corolla papilionaceous, 5–7 mm long, violet to purple. Fruit obovate, 3–5 mm long, 3–4 mm wide, rugose, glabrous, eglandular or glandular; seeds subrotund-obovate, 3–4 mm long.

Flatwoods and savannas. Rare; central panhandle, Holmes, Jefferson, and Clay Counties. Maryland south to Florida, west to Michigan, Illinois, Kansas, Oklahoma, and Texas. Spring–summer.

Orbexilum virgatum (Nutt.) Rydb. [Long and slender.] PINELAND LEATHERROOT.

Psoralea virgata Nuttall, Gen. N. Amer. Pl. 2: 104. 1818. *Lotodes virgata* (Nuttall) Kuntze, Revis. Gen. Pl. 1: 194. 1891. *Orbexilum virgatum* (Nuttall) Rydberg, in Britton, N. Amer. Fl. 24: 6. 1919.

Erect, perennial herb, to 50 cm; stem appressed-pubescent. Leaves 1-foliolate, the leaflet blade elliptic to narrowly lanceolate, 2–9 cm long, 3–12 mm wide, the apex obtuse, the base rounded, the upper and lower surfaces appressed-pubescent to glabrate, the petiole 1–3.5 cm long. Flowers in a spike 1–4 cm long, the peduncle 2–13 cm long; calyx 3–5 mm long, appressed-pubescent, glandular; corolla papilionaceous, 5–7 mm long, purple. Fruit subrotund-obovate, ca. 4 mm long and wide, rugose, glabrous, glandular; seeds obovate, ca. 2 mm long.

Dry to moist flatwoods and savannas. Rare; northern peninsula. Georgia and Florida. Spring–summer.

EXCLUDED TAXON

Orbexilum simplex (Nuttall ex Torrey & A. Gray) Rydberg—Cited by Correll and Johnston (1970, as *Psoralea simplex* Nuttall ex Torrey & A. Gray) from "Se U.S.," which would presumably include Florida. Excluded from Florida by Isely (1990).

Pachyrhizus Rich. ex DC., nom. cons. 1825.

Perennial herbaceous vines. Leaves alternate, 3-foliolate, stipellate, petiolate, stipulate. Flowers in axillary or terminal pseudoracemes, bracteate, bracteolate; calyx 5-lobed; petals 5, the corolla papilionaceous; stamens 10, diadelphous (9 + 1), the anthers basifixed. Fruit a dehiscent legume.

A genus of 5 species; North America, West Indies, Mexico, Central America, South America, Africa, Asia, and Pacific Islands. [From the Greek *pachy*, thick, and *rhiza*, root.]

Selected reference: Sørensen (1988).

Pachyrhizus erosus (L.) Urb. [Having an irregularly toothed margin, apparently in reference to the leaflets.] YAM BEAN.

Dolichos erosus Linnaeus, Sp. Pl. 2: 726. 1753. *Dolichos bulbosus* Linnaeus, Sp. Pl., ed. 2. 1021. 1763, nom. illegit. *Pachyrhizus angulatus* Richard ex de Candolle, Prodr. 2: 402. 1825, nom. illegit. *Stizolobium bulbosum* Sprengel, Syst. Veg. 3: 252. 1826, nom. illegit. *Pachyrhizus bulbosus* Kunz, J. Asiat. Soc. Bengal 45(2): 246. 1876, nom. illegit. *Cacara erosa* (Linnaeus) Kuntze, Revis. Gen. Pl. 1: 165. 1891. *Pachyrhizus erosus* (Linnaeus) Urban, Symb. Antill. 4: 311. 1905. *Pachyrhizus erosus* (Linnaeus) Urban var. *typicus* R. T. Clausen, Cornell Univ. Agric. Exp. Sta. Mem. 264: 13. 1945, nom. inadmiss.

Trailing, climbing, or somewhat erect herbaceous vine; stem to 5(10) m, strigose or glabrous. Leaves 3-foliolate, the pinna blade ovate to rhombic or ovate-reniform, sometimes palmately 3- to 5-lobed, the terminal blade 5–8(18) cm long, 6–13(20) cm wide, the lateral blades smaller, the apex acute, the base obtuse, the margin entire or coarsely sinuate-dentate on the distal ½, the upper and lower surfaces glabrous or strigose; stipules linear-lanceolate, 5–11 mm long. Flowers in a terminal, erect to spreading pseudoraceme 10–45(70) cm long; bracts and bracteoles setaceous; calyx tubular, 8–12 mm long; corolla papilionaceous, 14–22 mm long, the banner blue-violet to red-purple, white, or bicolored. Fruit 6–15 cm long, 1–2 cm wide, strigose to glabrate, pale brown to olive-green; seeds 4–10, 4-angled with rounded corners or suborbicular, 5–10 mm long, olive-green to brown or reddish brown.

Disturbed pinelands. Rare; Brevard and Miami-Dade Counties. Escaped from cultivation. Florida; West Indies, Mexico, Central America, and South America; Africa, Asia, and Pacific Islands. Native to Mexico and Central America. Fall.

Parkinsonia L. 1753. PALO VERDE

Trees or shrubs. Leaves alternate, 2-pinnate, appearing 1-pinnate due to the obsolete petiole, leaflets alternate, subopposite, or opposite, stipulate. Flowers in axillary racemes, bracteate, bracteolate; calyx actinomorphic, basally connate; petals 5, the corolla caesalpiniaceous; stamens 10, free, the anthers basifixed, apically dehiscent. Fruit an indehiscent legume.

A genus of about 12 species; North America, West Indies, Mexico, Central America, South America, Africa, Asia, Australia, and Pacific Islands. [Commemorates John Parkinson (1567–1650), English herbalist and London apothecary.]

Parkinsonia aculeata L. [With spines.] MEXICAN PALO VERDE; JERUSALEM THORN.

Parkinsonia aculeata Linnaeus, Sp. Pl. 1: 375. 1753. *Parkinsonia spinosa* Kunth, in Humboldt et al., Nov. Gen. Sp. 6: 335. 1824, nom. illegit.

Trees or shrubs, to 10 m; stem with nodal stipular thorns, glabrous or sparsely pubescent. Leaves from spurs, 2-pinnate (appearing 1-pinnate), pinna 2(3), the rachis with glandular areas in the leaf axis, the leaflets 40–76 per pinna, alternate, subopposite, or opposite, the leaflet blade oblanceolate to oblong, 2–8 mm long, 1–3 mm wide, the apex obtuse, apiculate, the base attenuate, the upper surface glabrate, the lower surface sparsely pubescent, the petiole obsolete; stipules obsolescent or spinescent, the spines to 2 cm long. Flowers 2–15 in an axillary raceme, the rachis 8–18 cm long; bracts linear or lanceolate, ca. 2 mm long; bracteoles linear, ca. 2 mm long; calyx lobes oblong, 6–7 mm long, reflexed, sparsely pubescent; corolla caesalpiniaceous, 13–20 mm in diameter, yellow, the upper petal sometimes spotted or tinged with red, the petal margin erose. Fruit subterete, 2–12 cm long, 5–8 mm wide, glabrous; seeds (1)2–5, oblong to suborbicular.

Disturbed sites. Occasional; central and southern peninsula, Escambia County. Escaped from cultivation. South Carolina south to Florida, west to California; West Indies, Mexico, Central America, and South America; Africa, Asia, Australia, and Pacific Islands. Native to the southwestern United States and Mexico. Spring–fall.

Pediomelum Rydb. 1919. INDIAN BREADROOT

Perennial herbs. Leaves alternate, 3-foliolate (rarely 1- or 2-foliolate), estipellate, petiolate, stipulate. Flowers in axillary pseudoracemes, bracteate, ebracteolate; sepals 5, connate, 5-lobed; petals 5, the corolla papilionaceous; stamens diadelphous (9 + 1). Fruit an indehiscent 1-seeded legume.

A genus of 21 species; North America and Mexico. [From the Greek *pedio*, plain, and *melo*, apple, equivalent to the French vernacular name "pomme de prairie," in reference to the edible tuberous roots of *P. esculentum*.]

Pediomelum canescens (Michx.) Rydb. [Grayish, in reference to the foliage.] BUCKROOT.

Psoralea canescens Michaux, Fl. Bor.-Amer. 2: 57. 1803. *Lotodes canescens* (Michaux) Kuntze, Revis. Gen. Pl. 1: 194. 1891. *Pediomelum canescens* (Michaux) Rydberg, in Britton, N. Amer. Fl. 24: 18. 1919.

Erect perennial herb, to 1 m; stem canescent. Leaves 3-foliolate, the upper sometimes 1- or 2-foliolate, the leaflets obovate to elliptic, (2)3–4.5(5) cm long, 2–3 cm wide, the apex rounded, the base cuneate, the upper surface glabrous, the lower surface canescent, the petiole 3–8 mm long, the upper leaves subsessile. Flowers 5–14 in a pedunculate pseudoraceme 3–6 cm long, the pedicel 2–5 mm long; calyx 7–8 mm long, short-campanulate, the lobes triangular, longer than the tube; corolla papilionaceous, 1–1.5 cm long, violet to dark blue, marked with yellowish green. Legume ovate, 8–10 mm long (including the beak), papery, brittle, enclosed in the persistent calyx; seed 1.

Sandhills and flatwoods. Frequent; northern counties, central peninsula. Virginia south to Florida, west to Alabama. Spring–summer.

Peltophorum (Vogel) Benth., nom. cons. 1840.

Trees. Leaves alternate, even-bipinnately compound, estipellate, petiolate, stipulate. Flowers in terminal or axillary panicles, bracteate, ebracteolate; calyx lobes 5; petals 5, the corolla caesalpiniaceous; stamens 10, free. Fruit an indehiscent legume.

A genus of about 6 species; North America, West Indies, South America, Africa, Asia, Australia, and Pacific Islands. [From the Greek *pelte*, peltate, and *phoros*, bearing, in reference to the peltate stigma.]

1. Leaflets 0.5–1.5 cm long; corolla 2–3(3.5) cm in diameter..**P. dubium**
1. Leaflets 1–2 cm long; corolla 3–4 cm in diameter..**P. pterocarpum**

Peltophorum dubium (Spreng.) Taub. [Doubtful.] HORSEBUSH.

Caesalpinia dubia Sprengel, Syst. Veg. 2: 343. 1825. *Peltophorum vogelianum* Bentham, J. Bot. (Hooker) 2: 75. 1840, nom. illegit. *Brasilletia dubia* (Sprengel) Kuntze, Revis. Gen. Pl. 1: 174. 1891. *Peltophorum dubium* (Sprengel) Taubert, in Engler & Prantl, Nat. Pflanzenfam. 3(3): 176. 1892. *Baryxylum dubium* (Sprengel) Pierre, Fl. Forest. Cochinch. 4: t. 390. 1899.

Tree, to 10(25) m; branchlets reddish pubescent when young. Leaves even-bipinnately compound, 15–55 cm long, the pinnae (10)20–44), the leaflets (20)24–60 per pinna, the blade narrowly oblong, 0.5–1.5 cm long, 2–4.5 mm wide, the apex acute, apiculate, the base truncate to cordate, asymmetric, the upper and lower surfaces sparsely pubescent, the rachis reddish pubescent. Flowers in a terminal or axillary panicle 14–40 cm long; calyx lobes deltoid to lanceolate, 6–9 mm long; corolla caesalpiniaceous, 2–3(3.5) cm in diameter, yellow. Fruit narrowly oblong-elliptic, 4–8(10) cm long; seeds 1–2, obovoid.

Disturbed sites. Rare; central peninsula, Miami-Dade County. Escaped from cultivation. Florida; West Indies and South America. Native to South America. Spring–summer.

Peltophorum pterocarpum (DC.) Backer ex K. Heyne [From the Greek *pteron*, wing, and *carpos*, fruit.] YELLOW POINCIANA.

Inga pterocarpa de Candolle, Prodr. 2: 441. 1825. *Peltophorum pterocarpum* (de Candolle) Backer ex K. Heyne, Nutt. Pl. Ned.-Ind., ed. 2. 755. 1927.

Tree, to 10(35) m; branchlets reddish pubescent when young. Leaves even-pinnately compound, 15–50 cm long, the pinnae 14–30 pairs, the leaflets 16–40 per pinna, the blade oblong, 1–2 cm long, 0.5–1 cm wide, the apex rounded to blunt, often retuse, the base rounded to subcordate, the upper surface glabrous, the lower surface sparsely reddish-pubescent, the rachis reddish pubescent. Flowers in a terminal or axillary panicle 20–40 cm long; calyx lobes deltoid to lanceolate, 7–11 mm long; corolla caesalpiniaceous, 3–4 cm in diameter, yellow. Fruit oblong elliptic, 4–12 cm long; seeds 1–4, oblong.

Disturbed sites. Rare; Lee and Miami-Dade Counties, Monroe County keys. Escaped from cultivation. Florida; West Indies; Africa, Asia, Australia, and Pacific Islands. Native to Asia and Australia. Summer–fall.

Phanera Lour. 1790.

Woody vines. Leaves alternate, 2-foliolate, petiolate, estipellate, stipulate. Flowers in axillary racemes, bracteate, bracteolate; sepals 5, connate, 5-lobed, splitting at maturity into 2 segments; petals 5, the corolla caesalpiniaceous; stamens 3, monadelphous, the anthers dorsifixed, dehiscing longitudinally. Fruit a dehiscent legume.

A genus of about 100 species; North America and Asia. [From the Greek *phaner*, manifest, easily seen, in reference to the spreading calyx and corolla.]

Phanera yunnanensis (Franch.) Wunderlin [Of Yunnan, China.] YUNNAN BAUHINIA.

Bauhinia yunnanensis Franchet, Pl. Delavay. 190. 1890. *Lasiobema yunnanensis* (Franchet) A. Schmitz, Bull. Soc. Roy. Bot. Belgique 110: 13. 1977. *Phanera yunnanensis* (Franchet) Wunderlin, Phytoneuron 2011–19: 1. 2011.

Woody vine, to 15 m; branchlets with axillary, paired tendrils, glabrous. Leaves 2-foliolate, the leaflet blade obliquely ovate, 2–4.5(6) cm long, 2–3 cm wide, the apex rounded, the base obliquely rounded, the upper and lower surfaces glabrous, the petiole to 1–3 cm long; stipules lanceolate, ca. 3 mm long, glabrous, caducous. Flowers 10–20 in an axiliary raceme 8–18 cm long; bracts subulate, ca. 2 mm long, caducous; bracteoles similar to the bracts but small, inserted near the middle of the pedicel; hypanthium 6–8 mm long, glabrous; calyx tube 7–8 mm long, splitting into 2 subequal lobes to the hypanthium lip at maturity, glabrous; corolla caesalpiniaceous, the petals narrowly obovate, 15–20 mm long, white to pinkish, the veins usually pink, subequal, clawed; stamens 3, ca. 1.5 cm long, glabrous, short-connate basally with the 7 much-reduced staminodes. Fruit linear, 8–15 cm long, 1.5–2 cm wide; seeds ca. 10, elliptic-oblong, 7–9 mm long, dark brown, lustrous.

Disturbed sites. Rare; Miami-Dade County. Escaped from cultivation. Florida; Asia. Native to Asia. Spring–fall.

Phaseolus L. 1753. BEAN

Annual or perennial herbs. Leaves 3-foliolate, stipellate, petiolate, stipulate. Flowers in axillary pseudoracemes, bracteate, bracteolate; sepals 5, connate; petals 5, the corolla papilionaceous, the keel coiled 2–3 turns; stamens diadelphous (9 + 1). Fruit a dehiscent legume.

A genus of about 65 species; nearly cosmopolitan. [From the Greek *phaselos*, little boat, referring to the shape of the kidney bean, *Phaseolus vulgaris*.]

1. Raceme with a long, slender axis .... **P. polystachios**
1. Raceme with a stout axis.
    2. Fruit linear, subterete; raceme shorter than the subtending leaf .... **P. vulgaris**
    2. Fruit broadly oblong, flattened; the raceme longer than the subtending leaf .... **P. lunatus**

Phaseolus lunatus L. [Crescent-shaped, in reference to the seed.] LIMA BEAN.

Phaseolus lunatus Linnaeus, Sp. Pl. 2: 724. 1753. *Phaseolus lunatus* Linnaeus forma *vulgaris* Hassler, Candollea 1: 440. 1923, nom. inadmiss.

Erect or twining annual herb; stem to 3(5) m, puberulent or glabrate. Leaves with the leaflet blade ovate to deltate, 4–12 cm long, 3–8 cm wide, the apex acute to acuminate, the base rounded to truncate, the margin entire, the upper and lower surface puberulent or glabrate. Flowers numerous in a short or exserted pseudoraceme; calyx 2–4 mm long; corolla papilionaceous, ca. 1 cm long, greenish white to purple. Fruit broadly oblong, laterally compressed, 5–9(12) cm long, 1.5–2 cm wide, glabrate; seeds few.

Disturbed sites. Rare; Santa Rosa County, Monroe County keys. Escaped from cultivation.

Virginia south to Florida, west to Missouri and Mississippi; West Indies, Mexico, Central America, and South America; Africa, Asia, and Pacific Islands. Native to Mexico, Central America, and South America. Summer.

Phaseolus polystachios (L.) Britton et al. [*Poly*, many, and *stachys*, spike, apparently in reference to the many spikelike inflorescences.] THICKET BEAN.

Twining or trailing perennial herb; stem to 6 m, puberulent. Leaves with the leaflet blade deltate to broadly ovate to rhombic, 1.5–8(10) cm long, 1–5 cm wide, the apex acuminate or obtuse or rounded, the base rounded, the margin entire, sinuate, or 3-lobed, the upper and lower surfaces glabrous, reticulate, the petiole 3–7 mm long. Flowers several to many in an exserted, branched raceme 0.5–3 dm long; calyx 3–4 mm long, the lobes shorter than the tube; corolla papilionaceous, 7–11 mm long, lavender-purple, purple and white, or white. Fruit oblong or oblong-falcate, 3–6 cm long, 7–10 mm wide, laterally compressed, glabrous, the stipe 2–3 mm long; seeds few.

1. Leaflet blades entire or nearly so, 4–8(10) cm long .... var. **polystachios**
1. Leaflet blades conspicuously 3-lobed, 1.5–4 cm long .... var. **sinuatus**

Phaseolus polystachios var. polystachios

> *Dolichos polystachios* Linnaeus, Sp. Pl. 2: 726. 1753. *Phaseolus polystachios* (Linnaeus) Britton et al., Prelim. Cat. 15. 1888.
> *Phaseolus perennis* Walter, Fl. Carol. 182. 1788.

Leaflet blades 4–8(10) cm long, 3–5 cm wide, entire or nearly so, the apex acuminate. Flowers in an inflorescence 0.5–2(3) dm long, commonly branched at the base; corolla white and purple, 9–11 mm long.

Moist to mesic hammocks. Occasional; peninsula west to central panhandle. Maine south to Florida, west to Michigan, Iowa, Missouri, Oklahoma, and Texas; West Indies. Native to North America. Summer–fall.

Phaseolus polystachios var. sinuatus (Nutt. ex Torr. & A. Gray) Maréchal et al. [Strongly waved, apparently in reference to the leaflet margin.]

> *Phaseolus sinuatus* Nuttall ex Torrey & A. Gray, Fl. N. Amer. 1: 278. 1838. *Phaseolus polystachios* (Linnaeus) Britton et. al. var. *sinuatus* (Nuttall ex Torrey & A. Gray) Maréchal et al., Taxon 27: 199. 1978. *Phaseolus polystachios* (Linnaeus) Britton et al. subsp. *sinuatus* (Nuttall ex Torrey & A. Gray) Freytag, in Freytag & Debouck, Sida, Bot. Misc. 23: 115. 2002. TYPE: FLORIDA.

Leaflet blades 1.5–4 cm long, 1–3.5 cm wide, conspicuously 3-lobed, the apex obtuse or rounded. Flowers in an inflorescence 1–3 dm long, rarely branched; corolla 7–9 mm long, lavender purple.

Sandhills and dry hammocks. Occasional; nearly throughout. North Carolina south to Florida, west to Mississippi. Summer–fall.

Phaseolus vulgaris L. [Common.] KIDNEY BEAN.

> *Phaseolus vulgaris* Linnaeus, Sp. Pl. 2: 723. 1753. *Phaseolus esculentus* Salisbury, Prodr. Stirp. Chap. Allerton 335. 1796, nom. illegit.

Erect or twining annual herb; stem to 3 m long, inconspicuously pubescent. Leaves with the leaflet blade broadly ovate to rhombic, 6–12 cm long, 4–8 cm wide, the apex acuminate, the base subcordate to rounded, the margin entire, the upper and lower surfaces glabrous, the petiole 6–12 cm long. The flowers in a pseudoraceme shorter than the subtending leaves, mostly clustered at the apex, sometimes few branched at the base; calyx ca. 4 mm long, the lobes shorter than the tube; corolla ca. 1 cm long, pink-purple to nearly white. Fruit linear, 8–20 cm long, ca. 8 mm wide, subterete, glabrous or pubescent; seeds several.

Disturbed sites. Rare; Hardee County. Escaped from cultivation. Maine south to Florida, west to Missouri; also Montana, Wyoming, and Utah; West Indies, Mexico, Central America, and South America; Europe, Africa, Asia, Australia, and Pacific Islands. Native to Mexico, Central America, and South America. Summer–fall.

HYBRID

Phaseolus ×smilacifolius Pollard (*P. polystachios* var. *polystachios* × *P. polystachios* var. *sinuatus*) THICKET BEAN.

> *Phaseolus smilacifolius* Pollard, Bot. Gaz. 21: 233. 1896, pro sp. *Phaseolus polystachios* (Linnaeus) Britton et al. subsp. *smilacifolius* (Pollard) Freytag, in Freytag & DeBouck, Sida, Bot. Misc. 23: 116. 2002. TYPE: FLORIDA: Lake Co.: Lake City, 29–31 Aug 1895, *Nash 2505* (holotype: US; isotypes: G, GH, K, MICH, MSC, MO, NY).

Wet hammocks. Rare; Columbia, Alachua, and Levy Counties. Summer–fall.

Piscidia L., nom. cons. 1759.

Shrubs or trees. Leaves alternate, odd-pinnately compound, petiolate, stipulate. Flowers in terminal and axillary panicles, pedicellate; sepals 5, connate, 5-lobed; petals 5, the corolla papilionaceous, the wings adnate to the keel; stamens 10, monadelphous. Fruit a dehiscent, winged legume.

A genus of about 7 species; North America, West Indies, Mexico, and Pacific Islands. [*Pisces*, fish, and *caedo*, kill; the bark and roots are placed in the water to stun and catch fish.]

Ichthyomethia P. Browne, nom. rej. 1756.

Selected reference: Rudd (1969).

Piscidia piscipula (L.) Sarg. [Little fish.] FLORIDA FISHPOISON TREE; JAMAICAN DOGWOOD.

> *Erythrina piscipula* Linnaeus, Sp. Pl. 2: 707. 1753. *Piscidia erythrina* Linnaeus, Syst. Nat., ed. 10. 1155. 1759, nom. illegit. *Piscidia inebrians* Medikus, Vorles. Churpfälz. Phys.-Okon. Ges. 2: 394. 1787, nom. illegit. *Piscidia toxicaria* Salisbury, Prodr. Stirp. Chap. Allerton 336. 1796, nom. illegit. *Piscidia piscipula* (Linnaeus) Sargent, Gard. & Forest 4: 436. 1891. *Ichthyomethia piscipula* (Linnaeus) Hitchcock ex Sargent, Gard. & Forest 4: 472. 1891.

Ichthyomethia communis S. F. Blake, J. Wash. Acad. Sci. 9: 247. 1919. *Piscidia communis* (S. F. Blake) Harms, Verh. Bot. Vereins Prov. Brandenburg 65: 91. 1923. TYPE: FLORIDA: Monroe Co.: Ramrod Key (fl) and Jewfish Key (lvs, fr), Sep, *Curtiss 685* (holotype: US).

Shrub or tree, to 15 m; branchlets strigillose, glabrescent. Leaves 2–3 dm long, the leaflets 5–9, the blade elliptic-oblong to obovate-oval, 4–10 cm long, 2–3.5 cm long, the apex obtuse or rounded, rarely apiculate, the base obtuse to rounded, the margin entire, the upper surface glabrous or glabrate, the lower surface strigillose, the petiole 3–4 cm long. Flowers in a terminal or axillary panicle 8–20 cm long, the pedicel 2–7 mm long, short-pedunculate; calyx 6–7 mm long, obliquely campanulate, the lobes triangular-ovate, strigillose; corolla papilionaceous, ca. 1.5 cm long, pink or white and red, silky. Fruit 2–9 cm long, 2–4 cm wide, strigillose, the wings 4, much wider than the body, laciniate to undulate or ruffled, glabrate; seeds 1–6, black.

Tropical hammocks, shell middens, and pine rocklands. Occasional; Pinellas, Hillsborough, and Lee Counties, southern peninsula. Florida, West Indies, Mexico, and Central America; Pacific Islands. Spring.

Pisum L. 1753. PEA

Annual herbs. Leaves alternate, even-pinnately compound, the rachis usually terminated in a branched tendril, estipellate, petiolate, stipulate. Flowers in axillary racemes, bracteate, ebracteolate; sepals 5, connate; petals 5, the corolla papilionaceous; stamens 10, diadelphous (9 + 1). Fruit a dehiscent legume.

A genus of about 3 species; North America, South America, Europe, Africa, and Asia. [Ancient Latin name for the common pea.]

Pisum sativum L. [Cultivated.] GARDEN PEA.

Pisum sativum Linnaeus, Sp. Pl. 2: 727. 1753. *Pisum biflorum* Stokes, Bot. Mat. Med. 4: 30. 1812, nom. illegit; non Rafinesque, 1810. *Pisum biflorum* Stokes var. *album* stokes, Bot. Mat. Med. 4: 31. 1812., nom. inadmiss. *Pisum vulgare* Jundzill, Opisan. Rosl. 305. 1830, nom. illegit. *Pisum commune* Clavaud, Actes Soc. Linn. Bordeaux 38: 572. 1884, nom. illegit. *Pisum sativum* Linnaeus var. *typicum* Beck, Fl. Nieder-Oesterreich 887. 1892, nom. inadmiss.

Erect or sprawling annual herb; stem to 2 m, glabrate. Leaves 5–10 cm long, the terminal tendrils well developed, the rachis winged, the leaflets 4–6 pairs, the leaflet blade ovate to lanceolate, 2–5 cm long, 1.5–3 cm wide, the margin entire or dentate, the upper and lower surfaces glabrate, the petiole 2–3 cm long; stipules ovate to lanceolate, 1.5–8 cm long, the base somewhat sagittate. Flowers 1–3 in an axillary raceme 3–10 cm long; bracts caducous; calyx symmetric, the lobes lanceolate, equal, longer than the tube; corolla papilionaceous, 1.5–3 cm long, the banner petal white or pink, the wing and keel petals red-purple, pink, or white. Fruit 5–9 cm long, 1.5–2 cm wide, glabrous; seeds 3–10, spherical, 4–12 mm long and wide, yellow or green, smooth or papillose, sometimes wrinkled.

Disturbed sites. Rare; Gadsden County. Escaped from cultivation. Greenland south to Florida and Louisiana, also Utah, Washington to California; South America; Europe, Africa, and Asia. Native to Europe, Africa, and Asia. Spring.

Pithecellobium Mart., nom. cons. 1837. BLACKBEAD

Trees or shrubs. Leaves alternate, even-bipinnately compound, with a gland on the upper side of the petiole between the lowest pair of pinnae, stipulate. Flowers in axillary, pedunculate heads or in axillary and/or terminal racemes, bisexual or rarely unisexual; sepals 5–6, connate, subequal; petals 5–6, connate, subequal; stamens 18–35, basally connate. Fruit a dehiscent legume, seeds arillate.

A genus of about 18 species; North America, West Indies, Mexico, Central America, Asia, and Pacific Islands. [From the Greek *pithekos*, monkey, and *ellobion*, earring, in reference to contorted fruits of the type species.]

Selected reference: Barneby and Grimes (1997).

1. Plant lacking stipular spines .... *P.* keyense
1. Plant with stipular spines.
    2. Ovary pubescent; peduncles geminate or fasciculate, the longer ones 1.2 cm long or less ... **P. dulce**
    2. Ovary glabrous; peduncles all or almost all solitary, the longer ones 1.5–5(7) cm long.
        3. Leaflets coriaceous, the larger ones usually 2–2.5 cm long, the upper surface glossy; flowers pink to red .... **P. bahamense**
        3. Leaflets chartaceous, the larger ones 3–5 cm long, the upper surface dull; flowers greenish yellow to light pink .... **P. unguis-cati**

Pithecellobium bahamense Northr. [Of the Bahamas.] BAHAMA BLACKBEAD.

Pithecellobium bahamense Northrop, Mem. Torrey Bot. Club 12: 38, pl. 5. 1902.
Shrub, to 3 m; branchlets glabrous. Leaves with 1 pair of pinnae, the leaflets 2 or 4 per pinnae, sessile, the blade obliquely oblong to oblanceolate or obovate, 1–2.5 cm long, 0.5–1 cm wide, the apex obtuse or mucronate, the base unequally rounded, the upper and lower surfaces glabrous, the petiole 1–10 mm long, with a stout-stalked gland; stipular spines 3–7 mm long. Flowers in an axillary head 2–3 cm in diameter, pink or red, the peduncles usually solitary, 1.5–5.5 cm long; calyx ca. 2 mm long, the teeth acute; corolla ca. 4 mm long, pink or red; stamens 2–3 times as long as the corolla, pink or red; ovary glabrous. Fruit 8–12 cm long, ca. 1 cm wide, coiled or much curved; seeds black, arillate.

Pine rocklands. Rare; Monroe County keys. Florida; West Indies. All year.

Pithecellobium bahamense was treated by Barneby and Grimes (1997) as a hybrid of *P. keyensis* and *P. histrix* (A. Richard) Bentham. We retain it here as a distinct species as *P. histrix* does not occur in Florida. Further study is needed.

Pithecellobium dulce (Roxb.) Benth. [Sweet.] MONKEYPOD.

Mimosa dulcis Roxburgh, Pl. Coromandel 1: 67, pl. 99. 1798. *Inga dulcis* (Roxburgh) Willdenow, Sp. Pl. 4: 1005. 1806. *Pithecellobium dulce* (Roxburgh) Bentham, London J. Bot. 3: 199. 1844. *Feuilleea dulcis* (Roxburgh) Kuntze, Revis. Gen. Pl. 1: 184. 1891. *Zygia dulcis* (Roxburgh) Lyons, Pl. Nam., ed. 2. 503. 1907.
Shrub or small tree, to 10 m; branchlets glabrous. Leaves with 1 pinnae pair, the leaflets 2 per pinnae, the blade obliquely obovate to elliptic or suborbicular, 2–6 cm long, 1–2 mm wide,

the apex rounded to emarginate or obtuse, the base unequally rounded, the upper and lower surfaces glabrous, the petiole ca. 2 cm long, with an orbicular gland; stipular spines 3–7 cm long. Flowers in an axillary head 2–3 cm in diameter, white or pinkish, the peduncles geminate or fasciculate, the longer ones 1.2 cm long or less; calyx ca. 1.5 mm long; corolla 3–4 mm long, minutely sericeous; stamens 7–8 mm long; ovary pubescent. Fruit linear, curved or coiled, 8–12 mm wide, usually red, glabrate; seeds black, lustrous, with a pulpy aril.

Disturbed sites. Rare; Lee and Miami-Dade Counties. Escaped from cultivation. Florida and Texas; West Indies, Mexico, Central America, and South America; Asia, and Pacific Islands. Native to tropical America. Spring.

Pithecellobium keyense Britton ex Britton & Rose [Of the Florida keys.] FLORIDA KEYS BLACKBEAD.

Pithecellobium keyense Britton ex Britton & Rose, N. Amer. Fl. 23: 22. 1928.

Shrub or small tree, to 6 m; branchlets glabrous. Leaves with 1 pinnae pair, the leaflets 2 or 4 per pinnae, the blade obliquely obovate to nearly rounded, 3–7 cm long, 1–6 cm wide, sessile, the apex rounded or retuse, the base unevenly rounded, the upper and lower surfaces glabrous, the petiole 2–3 cm long, with a round gland; stipular spines absent. Flowers in an axillary head 2–3 cm in diameter, nearly white to deep pink, the peduncle solitary; calyx ca. 1.5 mm long, 5-toothed; corolla 4–6 mm long; stamens 1–1.5(2) cm long; ovary glabrous. Fruit coiled or curved, 6–15 cm long, 8–10 mm wide; seeds black, lustrous, the aril red.

Coastal hammocks and strands. Occasional; Martin and Palm Beach Counties, southern peninsula. Florida, West Indies, and Mexico. Winter–spring.

Pithecellobium unguis-cati (L.) Benth. [Cat's claw.] CATCLAW BLACKBEAD.

Mimosa unguis-cati Linnaeus, Sp. Pl. 1: 517. 1753. *Inga unguis-cati* (Linnaeus) Willdenow, Sp. Pl. 4: 1006. 1806. *Pithecellobium unguis-cati* (Linnaeus) Bentham, London J. Bot. 3: 200. 1844. *Feuilleea unguis-cati* (Linnaeus) Kuntze, Revis. Gen. Pl. 1: 184. 1891. *Zygia unguis-cati* (Linnaeus) Sudworth, U.S.D.A. Div. Forest. Bull. 14: 248. 1897.

Mimosa guadalupensis Persoon, Syn. Pl. 2: 262. 1806. *Inga guadalupensis* (Persoon) Desvaux, J. Bot. Agric. 3: 70. 1814. *Pithecellobium guadalupensis* (Persoon) Chapman, Fl. South. U.S. 116. 1860. *Zygia guadalupensis* (Persoon) A. Heller, Cat. N. Amer. Pl., ed. 2. 5. 1900.

Shrub or tree, to 8 m; branchlets glabrous. Leaves with 1 pinnae pair, the leaflets 2 per pinnae, the blade obliquely obovate to oblong, 1–5 cm long, 0.5–4 cm wide, the apex obtuse to rounded, the base unevenly rounded, the upper and lower surfaces glabrous, the petiole 5–20 mm long, with a round gland; stipular spines to ca. 2 cm long. Flowers in pedunculate heads in a terminal raceme, greenish white, sessile, the peduncles all or almost all solitary, 1.5–5(7) cm long; calyx ca. 2 mm long; corolla 5–6 mm long; stamens 1–1.5 cm long; ovary glabrous. Fruit coiled or curved, 5–10 cm long, ca. 7 mm wide, compressed, somewhat constricted between the seeds, red; seeds 4–6 mm wide, nearly black, with a white aril.

Shell middens and coastal hammocks. Frequent; Hillsborough and Brevard Counties southward. Florida; West Indies, Mexico, Central America, and South America. All year.

Pueraria DC. 1825. KUDZU

Woody vines. Leaves alternate, 3-foliolate, stipellate, petiolate, stipulate. Flowers in axillary or terminal pseudoracemes or paniculate, bracteate, bracteolate; sepals 5, connate; petals 5, the corolla papilionaceous; stamens 10, diadelphous (9 + 1). Fruit a legume, tardily dehiscent.

A genus of 8 species; nearly cosmopolitan. [Commemorates Marc Nicolas Puerari (1766–1845), Swiss botanist.]

Selected reference: van der Maesen (1985).

Pueraria montana (Lour.) Merr. var. **lobata** (Willd.) Maesen & S. M. Almeida ex Sanjappa & Predeep. [Growing on mountains, the original material obtained from the mountains of Cochinchina, now Vietnam; lobed, in reference to the leaflets.] KUDZU.

> *Dolichos lobatus* Willdenow, Sp. Pl. 3: 1047. 1802. *Vigna lobata* (Willdenow) Endlicher ex Miquel, Fl. Ned. Ind. 1(1): 188. 1855. *Pueraria lobata* (Willdenow) Ohwi, Bull. Tokyo Sci. Mus. 18: 16. 1947. *Pueraria montana* (Loureiro) Merrill var. *lobata* (Willdenow) Maesen & S. M. Almeida ex Sanjappa & Predeep, Legumes India 288. 1992.
>
> *Dolichos hirsutus* Thunberg, Trans. Linn. Soc. London 2: 339. 1794. *Pachyrhizus thunbergianus* Siebold & Zuccarini, Abh. Math.-Phys. Cl. Königl. Bayer. Akad. Wiss. 4(3): 237. 1846, nom. illegit. *Pueraria thunbergiana* Bentham, J. Linn. Soc., Bot. 9: 122. 1867, nom. illegit. *Pueraria hirsuta* (Thunberg) Matsumura, Bot. Mag. (Tokyo) 16: 33. 1902; non Kurz, 1873.

Woody vines, to 30 m; stem producing tubers 18–45 cm long. Leaves with the leaflet blade ovate to orbicular, 8–20(26) cm long, 5–19(22) cm wide, the apex acute to acuminate, the base cuneate, the margin 3-lobed, the upper and lower surfaces appressed pubescent, the stipels lanceolate, 5–8 mm long, the margin ciliate, the petiole 8–13 cm long; stipules peltate, 8–16(25) mm long, entire, 2-fid, or fringed below the point of insertion. Flowers 15–40 in a pseudoraceme or panicle 10–25(35) cm long; bracts ovate to lanceolate, 4–10 mm long; bracteoles lanceolate, 2–4 mm long; calyx tube 3–5 mm long, the lobes 3–13 mm long, unequal; corolla papilionaceous, 10–25 mm long, purplish, blue, or white, with a green or yellow spot. Fruit straight or falcate, 4–13 cm long, 6–13 mm wide, golden brown pubescent, tardily dehiscent; seeds 10–15, ovoid, flattened, 4–5 mm long, reddish brown with black mottling or nearly black, minutely pitted.

Disturbed sites, often invading hammocks along the edge. Frequent; nearly throughout. Escaped from cultivation. New York and Massachusetts south to Florida, west to Nebraska, Kansas, Oklahoma, and Texas, also Washington and Oregon; West Indies, Mexico, Central America, and South America; Europe, Africa, Asia, Australia, and Pacific Islands. Native to Asia and Pacific Islands. Summer–fall.

Pueraria montana var. *lobata* is listed as a Category I invasive species in Florida by the Florida Exotic Pest Plant Council (FLEPPC, 2015).

Rhynchosia Lour., nom. cons. 1790. SNOUTBEAN

Erect, trailing, or twining perennial herbs. Leaves 1- or 3-foliolate, estipellate, petiolate, stipulate. Flowers in axillary pseudoracemes, bracteate, ebracteolate; sepals 5, connate; petals 5, the corolla papilionaceous; stamens diadelphous (9 + 1). Fruit a dehiscent legume, 1- to 2-seeded.

A genus of about 230 species; nearly cosmopolitan. [From the Greek *rhynchos*, beak or snout, in reference to the keel petals of the type species.]

Dolicholus Medik., 1787, nom. rej.; *Leucopterum* Small, 1733; *Pitcheria* Nutt., 1834.

Selected reference: Grear (1978).

1. Leaves 1-foliolate (rarely a few upper ones 3-foliolate).
    2. Plant erect; fruit much exserted from the calyx...**R. reniformis**
    2. Plant trailing; fruit only slightly exserted from the calyx..**R. michauxii**
1. Leaves 3-foliolate.
    3. Corolla shorter than the calyx.
        4. Plant erect..**R. tomentosa**
        4. Plant trailing or prostrate.
            5. Petiole finely pubescent with incurved trichomes; fruit puberulent.........................**R. cinerea**
            5. Petiole hirsute; fruit hirsute and puberulent...**R. difformis**
    3. Corolla clearly exceeding the calyx.
        6. Plant erect; flowers solitary (rarely 2–3) in the leaf axils...**R. cytisoides**
        6. Plant trailing or prostrate; flowers several in a raceme.
            7. Seeds bicolored red and black...**R. precatoria**
            7. Seeds brown or solid red.
                8. Racemes 4–8(12) cm long, usually exceeding the leaves; calyx 2.5–3(4) mm long..............
                ....**R. minima**
                8. Racemes 1–4(5) cm long, distinctly shorter or only slightly exceeding the leaves; calyx 3.5–7 mm long.
                    9. Lower leaflet surface cinereous-villosulous; corolla 8–10 mm long; fruit 1.5–1.8 cm long; seeds brown ....**R. parvifolia**
                    9. Lower leaflet surface puberulent; corolla 6–8 mm lng; fruit 2–4 cmm long; seeds red
                    ....**R. swartzii**

Rhynchosia cinerea Nash [Ash-gray.] BROWNHAIR SNOUTBEAN.

Rhynchosia cinerea Nash, Bull. Torrey Bot. Club 22: 149. 1895. *Dolicholus cinereus* (Nash) Vail, Bull. Torrey Bot. Club 26: 112. 1899. TYPE: FLORIDA: Lake Co.: vicinity of Eustis, 16–31 Jul 1894, *Nash 1336* (holotype: NY; isotypes: F, G, GH, LE, MICH, MO, P, UC, US).

Trailing or twining perennial herb; stem to 1 m, with incurved or subappressed trichomes. Leaves 3-foliolate (sometimes 1-foliolate at the lower nodes), the leaflet blade reniform or oblate to broadly obovate, 1–3(3.5)cm long, 1–3(3.5) cm wide, the apex rounded, retuse, or emarginate, the base rounded, the margin entire, the upper surface glabrous, the lower surface villosulus, resinous-glandular-dotted, the petiole 1–3 cm long. Flowers 3–6(10) in a raceme 1–2 cm long, shorter than the leaves, sometimes elongating in fruit to 4 cm; calyx 8–10 mm long, the tube 1–2 mm, the lobes much exceeding the tube; corolla papilionaceous, subequaling the calyx, yellow. Fruit asymmetrically ovate, 1–2 cm long, 5–8 mm wide, loosely strigose; seeds brown.

Sandhills and scrub. Frequent; peninsula. Endemic. Spring–fall.

Rhynchosia cytisoides (Bertol.) Wilbur [Resembling the genus *Cytisus* (Fabaceae).] ROYAL SNOUTBEAN.

Lespedeza cytisoides Bertoloni, Mem. Reale Accad. Sci. Ist. Bologna 2: 278. 1850. *Rhynchosia cytisoides* (Bertoloni) Wilbur, Rhodora 64: 60. 1962.

Pitcheria galactoides Nuttall, J. Acad. Nat. Sci. Philadelphia 7: 93. 1834. *Rhynchosia galactoides* (Nuttall) Endlicher & Walpers, in Walpers, Repert. Bot. Syst. 1: 790. 1842; non (Kunth) de Candolle, 1825. *Rhynchosia pitcheria* Burkart, Darwinia 11(2): 268. 1957.

Erect perennial herb, to 1 m; stem minutely puberulent. Leaves 3-foliolate, the leaflet blade ovate to elliptic, 1–2 cm long, the petiole 1–4 mm long, 5–10 mm wide, the apex rounded or retuse, the base broadly cuneate, the upper surface glabrous, the lower surface minutely puberulent, resinous-glandular-dotted. Flowers 1(2–3), the peduncle ca. 3 mm long; calyx 5–7 mm long, the lobes longer than the tube; corolla papilionaceous, 7–10 mm long, yellow, purple-striate or reddish suffused. Fruit oblong, straight, (1)1.5–2 cm long, 5–7 mm wide, puberulous; seeds brown.

Sandhills, scrub, and dry, open bluff forests. Frequent; panhandle. Florida, Alabama, and Mississippi. Spring–summer.

Rhynchosia difformis (Elliott) DC. [Irregularly or unevenly formed, in reference to the dimorphic leaves.] DOUBLEFORM SNOUTBEAN.

Glycine tomentosa Linnaeus var. *volubilis* Michaux, Fl. Bor.-Amer. 2: 63. 1803. *Arcyphyllum difforme* Elliott, J. Acad. Nat. Sci. Philadelphia 1: 372. 1818. *Rhynchosia difformis* (Elliott) de Candolle, Prodr. 2: 384. 1825. *Rhynchosia tomentosa* (Linnaeus) Hooker & Arnott var. *volubilis* (Michaux) Torrey & A. Gray, Fl. N. Amer. 1: 284. 1838. *Rhynchosia volubilis* (Michaux) A. W. Wood, Class-Book Bot., ed. 1861. 321. 1861, nom. illegit.; non Loureiro, 1790.

Dolicholus lewtonii Vail, Bull. Torrey Bot. Club 26: 113. 1899. *Rhynchosia lewtonii* (Vail) Small, Man. S.E. Fl. 714. 1933. TYPE: FLORIDA: Orange Co.: without locality, 7 Jul 1894, *Lewton s.n.* (holotype: NY; isotype: NY).

Dolicholus tomentosus (Linnaeus) Vail var. *undulatus* Vail, Bull. Torrey Bot. Club 26: 113. 1899. TYPE: FLORIDA: without data, 1846, *Chapman s.n.* (lectotype: MO; isolectotypes: F, GA, GH, LE, MISSA, MO, NY, PH, S). Lectotypified by Grear (1978: 123).

Twining or trailing (rarely suberect) perennial herb; stem to 1.5 m, with spreading trichomes. Leaves 3-foliolate (1-foliolate at the lower nodes), the leaflet blade suborbicular to elliptic, (2)2.5–5 cm long, (2)2.5–5 cm wide, the apex rounded, the base rounded to broadly cuneate, the margin entire, the upper surface glabrous, the lower surface villosulous, resinous-glandular-dotted, the petiole 1–2 cm long. Flowers (2)4–10(15) in an apically subumbellate raceme usually shorter than the leaves, the peduncle 1–2 cm long, sometimes elongating in fruit to 8 cm; calyx 8–10(12) mm long, the tube 1–2 mm long; corolla papilionaceous, subequaling the calyx, yellow. Fruit ovate to broadly asymmetrically oblong, 1–2 cm long, 6–9 mm wide, villous or puberulous; seeds brown.

Sandhills. Occasional; northern counties south to Hillsborough County. Virginia south to Florida, west to Missouri, Oklahoma, and Texas. Spring–summer.

Rhynchosia michauxii Vail [Commemorates André Michaux (1746–1802), French explorer in North America.] MICHAUX'S SNOUTBEAN.

Rhynchosia michauxii Vail, Bull. Torrey Bot. Club 22: 458. 1895. *Dolicholus michauxii* (Vail) Vail, Bull. Torrey Bot. Club 26: 112. 1899. TYPE: FLORIDA: Seminole Co.: Sanford, 1 Aug 1895, *Nash 2314* (lectotype: NY; isolectotypes: F, G, GH, LE, MICH, MO, P, PH, US). Lectotypified by Grear (1978: 132).

Trailing perennial herb; stem to 1 m, villosulous or short puberulous. Leaves 1-foliolate, the leaflet blade reniform to oblate, 2–4.5 cm long, 3–5 cm wide, the apex rounded, the base cordate, the margin entire, the upper surface glabrous, the lower surface villosulous or puberulous, resinous-glandular-dotted, the petiole 1.5–2(3) cm long. Flowers 2–8 in a raceme shorter than the leaves, the peduncle ca. 2.5 cm long; calyx 9–15 mm long; corolla papilionaceous, shorter than the calyx, yellow. Legume ovate-oblong, 1–1.5 cm long, 5–8 mm wide, villosulous or puberulent; seeds brown.

Sandhills and dry, open hammocks. Frequent; peninsula, eastern panhandle, Okaloosa Couunty. North Carolina, Georgia, and Florida. All year.

Rhynchosia minima (L.) DC. [The smallest.] LEAST SNOUTBEAN.

Dolichos minimus Linnaeus, Sp. Pl. 2: 726. 1753. *Dolicholus minimus* (Linnaeus) Medikus, Vorles Churpfälz. Phys.-Okon. Ges. 2: 354. 1787. *Glycine minor* Lagasca y Segura, Elench. Pl. 8. 1816, nom. illegit. *Rhynchosia minima* (Linnaeus) de Candolle, Prodr. 2: 385. 1825.
Glycine reflexa Nuttall, Gen. N. Amer. Pl. 2: 115. 1818.
Rhynchosia minima (Linnaeus) de Candolle var. *diminifolia* Walraven, Brittonia 22: 85. 1970. TYPE: FLORIDA: Monroe Co.: Big Pine Key, s.d., *Walraven 163* (holotype: GA).

Trailing or twining perennial herb; stem to 2 m, subcinereous. Leaves 3-foliolate, the leaflet blade broadly ovate or rhombic, (0.7)1–3.5 cm long, 1–3.5 cm wide, the apex rounded, the base rounded to broadly cuneate, the margin entire, the upper surface glabrous, the lower surface glabrous or sparsely pubescent, resinous-glandular-dotted, the petiole 2–3 cm long. Flowers 5–15 in a raceme 4–8(12) cm long; calyx 3(4) mm long, the lobes subequaling the tube; corolla papilionaceous, 4–8 mm long, yellow, tinged or striate with purple or brown. Fruit oblong-falcate, 1–2 cm long, 3–5 mm wide, villosulous; seeds brown.

Hammocks and flatwoods. Frequent; nearly throughout. Georgia and Florida west to Texas; West Indies, Mexico, Central America, and South America; Africa, Asia, Australia, and Pacific Islands. All year.

Rhynchosia parvifolia DC. [Small-leaved.] SMALL-LEAF SNOUTBEAN.

Rhynchosia parvifolia de Candolle, Prodr. 2: 385. 1825. *Dolicholus parvifolius* (de Candolle) Vail, Bull. Torrey Bot. Club 26: 108. 1899. *Leucopterum parvifolium* (de Candolle) Small, Man. S.E. Fl. 713. 1933.

Twining perennial herb; stem to 3 m, cinereous, finely villosulous. Leaves 3-foliolate, the leaflet blade elliptic or obovate, 1–2.5(4) cm long, 0.5–1 cm wide, the apex acute to obtuse, mucronate, the base cuneate, the margin entire, the upper surface glabrous, the lower surface villosulous, resinous-glandular-dotted, the petiole 5–10 mm long. Flowers 2–8, loosely disposed in the upper half of a raceme 1–4(5) cm long, this subequaling or shorter than the leaves; calyx 5–7 mm long, the upper and lower lobes longer than the tube; corolla papilionaceous, 8–10 mm long, exceeding the calyx, yellow. Fruit elliptic, 1.5–1.8 cm long, 4–6 mm wide, villosulous; seeds brown.

Pinelands and beaches. Occasional; Miami-Dade County, Monroe County keys. Florida; West Indies. All year.

Rhynchosia parvifolia is listed as threatened in Florida (Florida Administrative Code, Chapter 5B-40).

Rhynchosia precatoria (Humb. & Bonpl. ex Wild.) DC. [Graceful, in reference to the habit.] ROSARY SNOUTBEAN.

Indigofera volubilis J. C. Wendland, Bot. Beob. 55. 1798; non Milne, 1773. *Glycine precatoria* Humboldt & Bonpland ex Willdenow, Enum. Pl. 755. 1809. *Rhynchosia precatoria* (Humboldt & Bonpland ex Willdenow) de Candolle, Prodr. 3: 385. 1825. *Dolicholus precatorius* (Humboldt & Bonpland ex Willdenow) Rose, Contr. U.S. Natl. Herb. 10: 101. 1906.

Trailing or twining, woody perennial vine; stem to 8 m, villous, resinous-glandular-dotted. Leaves 3-foliolate, the leaflet blade ovate, ovate-lanceolate, or rhombic, 2–12 cm long, 1.5–9 cm wide, the apex acute to acuminate, the base rounded to obtuse or cuneate, the margin entire, the upper surface villous or villosulous, the lower surface villous or villosulous, resinous-glandular-dotted, the petiole 1–6 mm long. Flowers 10–30 in a raceme 5–30 cm long, equaling or longer than the leaves; calyx 4–5 mm long, the tube ca. 2 mm long, the lobes slightly exceeding the tube; corolla papilionaceous, 8–9 mm long, yellowish green, streaked or flushed with brown or purple. Fruit ovate-oblong, 2–3 cm long, 7–12 mm wide, villous; seeds bicolored red and black.

Disturbed sites. Rare; Miami-Dade County. Florida; Mexico, Central America, and South America. Summer–fall.

Rhynchosia reniformis DC. [Kidney-shaped.] DOLLARLEAF.

Trifolium simplicifolium Walter, Fl. Carol. 184. 1788. *Glycine tomentosa* (Linnaeus) Hooker & Arnott var. *monophylla* Michaux, Fl. Bor.-Amer. 2: 63. 1803. *Glycine reniformis* Pursh, Fl. Amer. Sept. 486. 1814, nom. illegit. *Arcyphyllum simplicifolium* (Walter) Elliott, J. Acad. Nat. Sci. Philadelphia 2: 115. 1818. *Glycine monophylla* (Michaux) Nuttall, Gen. N. Amer. Pl. 2: 115. 1818, nom. illegit.; non Linnaeus, 1767. *Glycine simplicifolia* (Walter) Elliott, Sketch Bot. S. Carolina 2: 234. 1823. *Rhynchosia reniformis* de Candolle, Prodr. 2: 384. 1825. *Rhynchosia tomentosa* (Linnaeus) Hooker & Arnott, var. *monophylla* (Michaux) Torrey & A. Gray, Fl. N. Amer. 1: 284. 1838. *Phaseolus reniformis* (de Candolle) Eaton & J. Wright, Man. Bot., ed. 8. 353. 1840, nom. illegit. *Rhynchosia simplicifolia* (Walter) A. W. Wood, Class-Book Bot., ed. 1861. 321. 1861; non de Candolle 1825. *Dolicholus simplicifolius* (Walter) Vail, Bull. Torrey Bot. Club 26: 114. 1899.

Rhynchosia tomentosa (Linnaeus) Hooker & Arnott var. *intermedia* Torrey & A. Gray, Fl. N. Amer. 1: 285. 1838. *Dolicholus intermedius* (Torrey & A. Gray) Vail, Bull. Torrey Bot. Club 26: 115. 1899. *Rhynchosia intermedia* (Torrey & A. Gray) Small, Man. S.E. Fl. 715. 1933. *Rhynchosia simplicifolia* (Walter) A. W. Wood, var. *intermedia* (Torrey & A. Gray) F. J. Hermann, J. Wash. Acad. Sci. 38: 238. 1948. TYPE: FLORIDA: Hillsborough Co.: Tampa Bay, s.d., *Burrows s.n.* (holotype: NY).

Erect perennial herb, to 1.5(2); stem villosulous. Leaves 1-foliolate or rarely the upper 3-foliolate, usually 4–6 in number, the leaflet blade reniform to ovate or subcordate, (2)2.5–4(5) cm long, 2–4(5) cm wide, the apex rounded, mucronate or apiculate, the base shallowly cordate, the margin entire, the upper surface glabrous, the lower surface villosulous, resinous-glandular-dotted, the petiole 1.5–3 cm long. Flowers several to numerous in a congested raceme, the peduncle 0.5–2 cm long; calyx 7–10 mm long, the lobes exceeding the tube; corolla papilionaceous, subqualing the calyx, yellow. Fruit short-oblong or elliptic-oblong, 1.2–1.8 cm long, 6–7 mm wide, villosulous; seeds brown.

Sandhills and dry hammocks. Frequent; northern counties, central peninsula, Miami-Dade County. North Carolina south to Florida, west to Texas. Spring–summer.

Rhynchosia swartzii (Vail) Urb. [Commemorates Olof Peter Swartz (1760–1818), Swedish botanist who collected in the West Indies.] SWARTZ'S SNOUTBEAN.

> *Dolicholus swartzii* Vail, Bull. Torrey Bot. Club. 26: 108. 1899. *Rhynchosia swartzii* (Vail) Urban, Repert. Spec. Nov. Regni Veg. 15: 320. 1918. TYPE: FLORIDA: Key West, s.d., *Blodgett s.n.* (lectotype: NY; isolectotypes: GH, NY). Lectotypified by Grear (1978: 62).

Trailing or twining perennial herb; stem to 2 m, puberulous. Leaves 3-foliolate, the leaflet blade ovate, 2–4(6) cm long, 2–4(6) cm wide, the apex acuminate, the base rounded, the margin entire, the upper and lower surfaces glabrous, resinous-glandular-dotted, the petiole 3–4(6) cm long. Flowers 3–10 in the upper ½ of the raceme, the raceme (1)2–4 cm long; calyx 3–5 mm long, the lobes shorter than the tube; corolla papilionaceous, 6–8 mm long, yellow. Fruit oblong-lanceolate or falcate, 2–4 cm long, 5–8 cm wide, puberulous; seeds red.

Hammocks. Rare; Miami-Dade County, Monroe County keys. Florida; West Indies and Mexico. All year.

Rhynchosia swartzii is listed as endangered in Florida (Florida Administrative Code, Chapter 5B-40).

Rhynchosia tomentosa (L.) Hook. & Arn. [Short-pubescent.] TWINING SNOUTBEAN.

Erect perennial herb, to 7(9) dm; stem densely villous. Leaves 3-foliolate, the lowermost 1-foliolate, the leaflet blade broadly ovate or elliptic, 3.5–7 cm long, 2–4 cm wide, the apex acute to obtuse, the base cordate, the margin entire, the upper surface glabrous, the lower surface densely puberulous or tomentose, resinous-glandular-dotted, the petiole 4–5 cm long. Flowers either several and axillary or sometimes in a short terminal raceme 1–3 cm long, or a solitary, strongly exserted, terminal raceme (sometimes also axillary) 5–20 cm long; calyx 6–9 mm long, the lobes much longer than the tube; corolla papilionaceous, subequaling the calyx, yellow. Fruit ovate-oblong or broadly oblong, 1.2–2.5 cm long, 5–8 mm wide, puberulous or villosulous.

1. Inflorescences several, axillary or sometimes with a short terminal raceme, 1–3 cm long; stipules persistent ....var. **tomentosa**
1. Inflorescence a solitary, strongly exserted, terminal raceme (sometimes also axillary), 5–20 cm long; stipules caducous..var. **mollissima**

Rhynchosia tomentosa var. **tomentosa**

> *Glycine tomentosa* Linnaeus, Sp. Pl. 2: 754. 1753. *Rhynchosia tomentosa* (Linnaeus) Hooker & Arnott, Companion Bot. Mag. 1: 23. 1835. *Dolicholus tomentosus* (Linnaeus) Vail, Bull. Torrey Bot. Club 26: 112. 1899.

Trifolium erectum Walter, Fl. Carol. 184. 1788. *Glycine tomentosa* Linnaeus var. *erecta* (Walter) Michaux, Fl. Bor.-Amer. 2: 63. 1803. *Arcyphyllum erectum* (Walter) Elliott, J. Acad. Nat. Sci. Philadelphia 1: 372. 1918. *Glycine erecta* (Walter) Nuttall, Gen. N. Amer. Pl. 2: 114. 1818; non Thunberg, 1800. *Rhynchosia erecta* (Walter) de Candolle, Prodr. 2: 384. 1825. *Glycine caroliniana* Sprengel, Syst. Veg. 3: 197. 1826. *Rhynchosia tomentosa* (Linnaeus) Hooker & Arnott var. *erecta* (Walter) Torrey & A. Gray, Fl. N. Amer. 1: 285. 1838. *Dolicholus erectus* (Walter) Vail, Bull. Torrey Bot. Club 26: 115. 1899.

Inflorescences several, axillary or sometimes with a short terminal raceme, 1–3 cm long; stipules persistent.

Sandhills and flatwoods. Occasional; northern counties south to Lake County. Maryland south to Florida, west to Texas. Spring–fall.

Rhynchosia tomentosa var. **mollissima** (Elliott) Torr. & A. Gray [The softest.]

Glycine mollissima Elliott, Sketch Bot. S. Carolina 2: 235. 1823. *Rhynchosia tomentosa* (Linnaeus) Hooker & Arnott var. *mollissima* (Elliott) Torrey & A. Gray, Fl. N. Amer. 1: 285. 1838. *Rhynchosia mollissima* (Elliott) Shuttleworth ex S. Watson, Smithsonian Misc. Collect. 15: 256. 1878; non G. Don (1832); nec Dalzell (1873). *Dolicholus mollisimus* (Elliott) Vail, Bull. Torrey Bot. Club 26: 116. 1899. TYPE: FLORIDA: without data, *Baldwin s.n.* (holotype: CHARL).

Inflorescence a solitary, strongly exserted, terminal raceme (sometimes also axillary), 5–20 cm long; stipules caducous.

Sandhills and flatwoods. Occasional; northern peninsula south to Hillsborough County, west to central panhandle. South Carolina, Georgia, and Florida. Spring–fall.

EXCLUDED TAXA

Rhynchosia americana (Miller) Metz—This Mexican species was reported for Florida by Chapman (1860, 1883, 1897, all as *Rhynchosia menispermoidea* de Candolle), who misapplied the name to material of *R. michauxii*.

Rhynchosia caribaea (Jacquin) de Candolle—Reported for Florida by Chapman (1860, 1883, 1897), the name of this African species probably misapplied to material of *R. minima*.

Rhynchosia reniformis de Candolle var. *intermedia* "(Torrey & A. Gray) Walraven" nom. nud.—Reported for Florida by Wilhelm (1984). This combination has not been made. Wilhelm's material is *R. reniformis*.

Rhynchosia reticulata (Swartz) de Candolle—This tropical American species was reported for Florida by Chapman (1897), who apparently misapplied the name to *R. difformis*.

Robinia L. 1753. LOCUST

Shrubs or trees. Leaves odd-pinnately compound, stipellate, petiolate, stipulate. Flowers in axillary racemes, bracteate, ebracteolate; sepals 5; corolla papilionaceous; stamens diadelphous (9 + 1). Fruit a dehiscent legume.

A genus of 4 species; North America, Mexico, South America, Europe, Africa, Asia, Australia, and Pacific Islands. [Commemorates Jean Robin (1550–1629), French botanist and his son Vespasian (1579–1662), royal gardeners to Henry IV of France.]

Selected references: Isely and Peabody (1984); Lavin and Sousa (1995).

1. Branchlets glabrous to puberulent; flowers white; fruit glabrous, narrowly winged ....**R. pseudoacacia**
1. Branchlets prickly-hispid; flowers pink to purple; fruit hispid, not winged ....**R. hispida**

Robinia hispida L. [Bristly.] BRISTLY LOCUST.

Robinia hispida Linnaeus, Mant. Pl. 1: 101. 1767. *Pseudacacia hispida* (Linnaeus) Moench, Methodus 145. 1794. *Robinia hispida* Linnaeus var. *typica* C. K. Schneider, Ill. Handb. Laubholzk. 2: 81. 1907, nom. inadmiss.

Shrubs or trees, to 3(8) m; branchlets prickly-hispid, the trichomes 1–5 mm long. Leaves 10–30 cm long, the leaflets (7)9–13(19), the blade ovate to ovate-lanceolate, (1)2–6 cm long, 1.5–2 cm wide, the apex rounded, mucronate, the base rounded, the upper surface glabrous, the lower surface appressed silky to glabrate, the stipels subulate, the petiole and rachis prickly-hispid; stipules spinose, to 1 cm long. Flowers (1)4–11(15) in a raceme; calyx 8–12 mm long, glandular-hispid, the lobes subequaling the tube or longer, the lower one 3–6 mm long; corolla papilionaceous, 1.5–2.5 cm long, various shades of pink to rose-purple or rarely white. Fruit oblong, 3–7 cm long, 1–2 cm wide, densely hispid.

Sandhills. Rare; Alachua County, central panhandle, Escambia County. Escaped from cultivation. Nova Scotia and Maine south to Florida, west to Washington and California. Native to Virginia and Kentucky south to Alabama and Georgia. Spring.

Robinia pseudoacacia L. [False Acacia, a pre-Linnaean name used by Catesby.] BLACK LOCUST.

Robinia pseudoacacia Linnaeus, Sp. Pl. 2: 722. 1753. *Robinia acacia* Linnaeus, Syst. Nat., ed. 10. 1161. 1759, nom. illegit. *Pseudo-Acacia vulgaris* Medikus, Vorles Churpfälz. Phys.-Okon. Ges. 2: 364. 1787. *Pseudacacia odorata* Moench, Methodus 145. 1794, nom. illegit. *Robinia fragilis* Salisbury, Prodr. Stirp. Chap. Allerton 336. 1796, nom. illegit. *Robinia pseudoacacia* Linnaeus forma *normalis* Voss, Vilm. Blumengaertn., ed. 3. 1: 218. 1894, nom. inadmiss. *Robinia pseudoacacia* Linnaeus var. *typica* Ascherson & Graebner, Syn. Mitteleur. Fl. 6(2): 715. 1908, nom. inadmiss. *Robinia pseudo-acacia* Linnaeus forma *vulgaris* Ascherson & Graebner, Syn. Mitteleur. Fl. 6(2): 715. 1908, nom. inadmiss.

Tree, to 25 m; bark rough, dark brown, ridged and furrowed, the branchlets glabrous to puberulent. Leaves 10–30 cm long, the leaflet blade elliptic to oblong, 2–4(6) cm long, ca. 1.5 cm wide, the apex rounded or retuse, mucronate, the base broadly cuneate, the upper and lower surfaces glabrate, the petiole 1.5–2 cm long; stipules spinose, to 1.5 cm long. Flowers 3–20+ in a pendent raceme, the peduncle and pedicel velvety puberulent; calyx 6–9 mm long, velvety puberulent, the lobes shorter than the tube, the lower 2–3 mm long; corolla papilionaceous, 1.5–2 cm long, white. Fruit oblong, laterally compressed, somewhat segmented, the upper margin narrowly winged, the valves thick-chartaceous, glabrous.

Disturbed hammock margins. Occasional; northern counties south to Lake County. Escaped from cultivation. Nearly throughout North America; South America; Europe, Africa, Asia, Australia, and Pacific Islands. Native to eastern North America north of Florida. Spring.

Senna Mill. 1754.

Annual or perennial herbs, shrubs, or small trees. Leaves even-pinnately compound, petiolate, stipulate. Flowers in axillary racemes commonly aggregated into compound, corymbose, or thyrsiform inflorescences or reduced to 1–2 in the leaf axils, bracteate, ebracteolate; sepals 5, free; corolla caesalpiniaceous; stamens 10, all fertile or variously heteromorphic and staminodial, the anthers basifixed, dehiscent by pores. Fruit an indehiscent or dehiscent legume.

A genus of about 300 species; North America, West Indies, Central America, South America, Africa, Asia, Australia, and Pacific Islands. [From the Arabic *sana* or *sanna*, for species with cathartic and laxative properties.]

Adipera Raf., 1838; *Chamaesenna* Raf. ex Pittier, 1928; *Ditremexa* Raf., 1838; *Peiranisia* Raf., 1838; *Psilorhegma* (Vogel) Britton & Rose, 1930.

Selected reference: Irwin and Barneby (1982).

1. Leaves eglandular.
  2. Herb or subshrub; fruit 4-winged...**S. alata**
  2. Shrub or tree; fruit flat...**S. atomaria**
1. Leaves with 2 or more petiolar glands.
  3. Leaf gland(s) at the base of the petiole.
    4. Petiolar gland cylindric to conic ....**S. ligustrina**
    4. Petiolar gland spherical.
      5. Leaflets 4–5(6) pairs; fruit with a conspicuous light margin............................**S. occidentalis**
      5. Leaflets 6–8 pairs; fruit without a conspicuous light margin............................**S. marilandica**
  3. Leaf glands(s) on the rachis between the leaflet pairs.
    6. Herb; fruit quadrangular...**S. obtusifolia**
    6. Tree or shrub; fruit flat or cylindric.
      7. Leaf glands between the lowermost 2 pairs or all pairs of leaflets......................**S. surattensis**
      7. Leaf gland usually solitary between the lowermost pair of leaflets.
        8. Leaflets oblong-lanceolate; leaf gland short-cylindric, acute.........................**S. corymbosa**
        8. Leaflets broadly cuneate-obovate to elliptic-lanceolate; leaf gland spheroidal (rarely substipitate and acute).
          9. Fruit flat, papery, the valves dehiscent; leaflets elliptic-lanceolate ....**S. mexicana**
          9. Fruit cylindric, somewhat woody, the valves indehiscent or breaking irregularly; leaflets cuneate-obovate...**S. pendula**

Senna alata (L.) Roxb. [Winged, in reference to the winged fruits.] CANDLESTICK PLANT.

Cassia alata Linnaeus, Sp. Pl. 1: 378. 1753. *Senna alata* (Linnaeus) Roxburgh, Fl. Ind., ed. 1832. 2: 349. 1832. *Herpetica alata* (Linnaeus) Rafinesque, Sylva Tellur. 123. 1838.

Erect, perennial, suffrutescent herb, to 2(4) m; stem glabrate or villosulous. Leaves 1.5–6 dm long, eglandular, the leaflets 7–12 pairs, the blade obovate to broadly oblong, 5–12 cm long, 2–5 cm wide, the apex rounded or emarginate, the base unequally rounded, the upper and lower surfaces glabrous, short petiolulate, the petiole 1–3 cm long, eglandular; stipules deltate, 1–2 cm long, basally subauriculate. Flowers numerous in a spikelike raceme 3–6(10) cm long;

bracts initially conspicuous and enclosing the flowers in bud, subsequently caducous; sepals oblong-ovate, 10–15 mm long, yellow, petaloid; corolla caesalpiniaceous, the petals 1.5–2 cm long, yellow, with an erose margin; fertile stamens 5. Fruit oblong, 10–15 cm long, laterally compressed, the valves conspicuously 4-winged, dark brown or black, chartaceous, dehiscent; seeds numerous, rhombic, 5–7 mm long, brown.

Disturbed sites. Rare; central and southern peninsula. Escaped from cultivation. Florida west to Texas, also California; West Indies, Mexico, Central America, and South America; Australia and Pacific Islands. Native to South America. Spring–fall.

Senna atomaria (L.) H. S. Irwin & Barneby [Sprinkled with minute particles.] FLOR DE SAN JOSE.

> *Cassia atomaria* Linnaeus, Mant. Pl. 68. 1767. *Senna atomaria* (Linnaeus) H. S. Irwin & Barneby, Mem. New York Bot. Gard. 35: 588. 1982.

Shrub or tree, to 10 m; branchlets pilosulous or tomentulose, soon glabrescent. Leaves (8.5)10–24 cm long, the leaflets (1)2–5 pairs, the blade obovate to elliptic or ovate, (2)3–11(13) cm long, (1)1.5–5.5 cm wide, the apex obtuse, rounded, or emarginate, the base rounded or cuneate, the margin entire, revolute, the upper and lower surfaces sparsely pilose, the petiole 4–6 mm long, eglandular; stipules subulate or triangular-subulate, 2–4 mm long. Flowers (3)5–15(18) in a pendulous raceme, the pedicel 15–20 mm long; bracts ovate or lanceolate, 2–3 mm long, caducous; sepals ovate, 3–4 mm long, greenish or yellow and petaloid, reflexed in anthesis; corolla caesalpiniaceous, the petals 8–16 mm long, orange or yellow, dark-veined; fertile stamens 7. Fruit linear, 20–35 cm long, 8–12 mm wide, straight, laterally compressed, the sutures much thickened, the valves woody, dark brown or black; seeds obovoid or oblong-obovoid, 4–5 mm long, reddish brown, smooth.

Disturbed sites. Rare; Collier County. Escaped from cultivation. Florida; West Indies, Mexico, Central America, and South America. Native to tropical America. Spring–fall.

Senna corymbosa (Lam.) H. S. Irwin & Barneby [Flowers in corymbs.] ARGENTINE SENSITIVE PLANT.

> *Cassia corymbosa* Lamarck, Encycl. 1: 644. 1785. *Cassia falcata* Dumont de Courset, Bot. Cult., ed. 2. 6: 35. 1811, nom. illegit; non Linnaeus 1753. *Chamaefistula corymbosa* (Lamarck) G. Don, Gen. Hist. 2: 451. 1832. *Adipera corymbosa* (Lamarck) Britton & Rose, in Britton, N. Amer. Fl. 23: 242. 1930. *Senna corymbosa* (Lamarck) H. S. Irwin & Barneby, Mem. New York Bot. Gard. 35: 397. 1982.

Shrub or small tree, to 3.5 m; branchlets glabrous. Leaves 3–5 cm long, the leaflets 2–3 pairs, the blade oblong-lanceolate, 2.5–3.5 cm long, ca. 1 cm wide, the apex acute, the base rounded to broadly cuneate, the upper and lower surfaces glabrous, the petiole ca. 2 cm long, with a cylindric, acute gland between the lowermost leaflet pair; stipules linear-lanceolate, 2–5 mm long, caducous. Flowers 4–12 in a subterminal, axillary raceme, forming a corymbiform inflorescence, the pedicel 1–2 cm long; bracts lance-subulate, 1–4 mm long; sepals ovate or elliptic-obovate, 4–8 mm long, brownish; corolla caesalpiniaceous, the petals (8)9–14 mm long, yellow; fertile stamens 7. Fruit oblong-cylindric, 4–10 cm long, 7–10 mm wide, straight, chartaceous-subcoriaceous, glabrous, brownish stramineous; seeds oblong-ellipsoid, 4–5 mm long, brown.

Disturbed sites. Rare; Miami-Dade County. Escaped from cultivation. South Carolina south to Florida, west to Texas; South America; Africa and Asia. Native to South America. Summer–fall.

Senna ligustrina (L.) H. S. Irwin & Barneby [Resembling *Ligustrum* (Oleaceae), in reference to the leaflets resembling the leaves of *Ligustrum*.] PRIVET WILD SENSITIVE PLANT.

Cassia ligustrina Linnaeus, Sp. Pl. 1: 378. 1753. *Ditremexa ligustrina* (Linnaeus) Britton & Rose ex Britton & P. Wilson, Sci. Surv. Porto Rico & Virgin Islands 5: 372. 1924. *Senna ligustrina* (Linnaeus) H. S. Irwin & Barneby, Mem. New York Bot. Gard. 35: 409. 1982, nom. cons.

Cassia bahamensis Miller, Gard. Dict., ed. 8. 1768. *Peiranisia bahamensis* (Miller) Britton & Rose, in Britton, N. Amer. Fl. 23: 266. 1930.

Perennial herb, sometimes suffrutescent, to 3 m; stem glabrate, rarely puberulent. Leaves 10–13 cm long, the leaflets 6–8 pairs, the blade lanceolate, 2.5 6 cm long, 8–10 mm long, the apex acute, the base asymmetrically cuneate, the upper and lower surfaces glabrous, the petiole 1–3 cm long, with a conspicuous pointed gland near the base; stipules lanceolate, 3–6 mm long, caducous. Flowers 5–10(20+) in an axillary, subterminal raceme forming a subcorymiform inflorescence, the pedicel 1–2 cm long; bract lanceolate, elliptic, or ovate, 3–6 mm long, yellowish; sepals obovate, 5–7 mm long, greenish, yellow, or brown; corolla caesalpiniaceous, the petals yellow, subequal, 12–15 mm long; fertile stamens 6. Fruit oblong-falcate, 10–13 cm long, 5–7 mm wide, laterally compressed, brown, glabrate, papery; seeds obovate to oblong-elliptic, 3–4 mm long, olivaceous or gray.

Hammocks and disturbed sites. Frequent; Gilchrist County, central and southern peninsula. Florida; West Indies and Central America; Africa. Native to North America and West Indies. All year.

Senna marilandica (L.) Link [Of Maryland.] MARYLAND WILD SENSITIVE PLANT.

Cassia marilandica Linnaeus, Sp. Pl. 1: 378. 1753. *Cassia acuminata* Moench, Methodus 273. 1794, nom. illegit. *Cassia reflexa* Salisbury, Prodr. Stirp. Chap. Allerton 326. 1796, nom. illegit. *Senna marilandica* (Linnaeus) Link, Handbuch 2: 140. 1831. *Ditremexa marilandica* (Linnaeus) Britton & Rose, in Britton, N. Amer. Fl. 23: 257. 1930.

Cassia marilandica Linnaeus var. *floridana* Chapman, Fl. South. U.S., ed. 3. 124. 1897. TYPE: FLORIDA: St. Johns Co.

Cassia medsgeri Shafer, Torreya 4: 179. 1904. *Ditremexa medsgeri* (Shafer) Britton & Rose, in Britton, N. Amer. Fl. 23: 258. 1930.

Ditremexia nashii Britton & Rose, in Britton, N. Amer. Fl. 23: 258. 1930. TYPE: FLORIDA: Lake Co.: near Eustis, 16–25 Aug 1894, *Nash 1720* (holotype: NY).

Ascending perennial herb, to 1.5 m; stem glabrate. Leaves 10–20 cm long, the leaflets 6–8 pairs, the blade elliptic to elliptic-oblong, 3–5 cm long, 0.5–1.5(2) cm wide, the apex acute, the base cuneate, the upper and lower surfaces glabrate, the petiole (2)3–7 cm long, with a spheroid or ovoid gland near the base; stipules narrowly lanceolate, 5–9 mm long, caducous. Flowers (3)5–10(19) in an axillary raceme 0.7–4 cm long and/or a terminal corymbose-paniculate inflorescence, the pedicel 9–15 mm long; bracts lanceolate, 2–4 mm long, caducous; sepals ovate or oblong-ovate, 4–8 mm long; corolla caesalpiniaceous, the petals 8–13 mm long, yellow; fertile stamens 6. Fruit oblong, 6–10 cm long, ca. 8 mm wide, laterally compressed, straight or downwardly falcate, the valves thinly coriaceous, brown, glabrate, dehiscent; seeds obovoid or oblong, 4–5 mm long, light brown to castaneous, glabrous.

Moist, open hammocks. Occasional; northern and central peninsula west to central panhandle. Massachusetts and New York south to Florida, west to Wisconsin, Nebraska, Kansas, Oklahoma, and Texas. Summer–fall.

Senna mexicana (Jacq.) H. S. Irwin & Barneby var. **chapmanii** (Isely) H. S. Irwin & Barneby. [Of Mexico; commemorates Alvan Wentworth Chapman (1809–1899), American botanist.] CHAPMAN'S WILD SENSITIVE PLANT.

Cassia chapmanii Isely, Mem. New York Bot. Gard. 25: 199. 1975. *Senna mexicana* (Jacquin) H. S. Irwin & Barneby var. *chapmanii* (Isely) H. S. Irwin & Barneby, Mem. New York Bot. Gard. 35: 417. 1982. TYPE: FLORIDA: Monroe Co.: Big Pine Key, 17 Nov 1912, *Small 3983* (holotype: NY).

Spreading or erect shrub, to 2.5(3) m; branchlets puberulent. Leaves 4–7 cm long, the leaflets (2)4–5(6) pairs, the blade elliptic to elliptic lanceolate, 1.5–4.5 cm long, 1–1.5 cm wide, the apex obtuse, mucronate, the base asymmetrically rounded, the upper and lower surfaces glabrous, the petiole 0.8–2 cm long, with a spheroid gland between the lower leaflet pair; stipules subulate, 2–6 mm long, caducous. Flowers 3–15 in an axillary or terminally aggregated raceme 0.5–7 cm long, the pedicel 7–20 mm long; bracts lanceolate, 1–3 mm long, caducous; sepals obovate, oblong-obovate, or oblanceolate, 6–8 mm long, greenish brown, the margin membranous; corolla caesalpiniaceous, the petals 9–14 mm long, yellow; fertile stamens 6. Fruit oblong-falcate, 6–10 cm long, 5–7 mm wide, laterally compressed, thick-chartaceous, brown, dehiscent; seeds obovoid or oblong-ellipsoid, 3–4 mm long, brownish olivaceous.

Pinelands, hammocks, and dunes. Occasional; Miami-Dade County, Monroe County keys. Florida; West Indies. All year.

Senna mexicana var. *chapmanii* is listed as threatened in Florida (Florida Administrative Code, Chapter 5B-40).

Senna obtusifolia (L.) H. S. Irwin & Barneby [With obtuse leaves; in reference to the leaflets.] COFFEEWEED; SICKLEPOD.

Cassia obtusifolia Linnaeus, Sp. Pl. 1: 377. 1753. *Cassia tora* Linnaeus var. *obtusifolia* (Linnaeus) Haines, Bot. Bihar Orissa 3: 304. 1922. *Senna obtusifolia* (Linnaeus) H. S. Irwin & Barneby, Mem. New York Bot. Gard. 35: 252. 1982.

Annual herb, to 1 m; stem glabrate or puberulent. Leaves 5–15 cm long, the leaflets 3 pairs, the blade obovate, 1.5–5(7) cm long, 1–2 cm wide, the apex rounded, mucronate, the base cuneate, the upper surface glabrate, the lower surface glabrate or puberulent, the petiole 2–3 cm long, usually with a fusiform gland between the lower pair of leaflets; stipules linear-subulate falcate, 5–12 mm long, semipersistent. Flowers solitary or paired in the leaf axil or clustered at the stem tip, the pedicel 1–2.5 cm long; bracts ovate or lanceolate, 2–5 mm long, caducous; sepals obovate or oblong-obovate, 6–9 mm long, pale green, glabrous or the margin ciliate; corolla caesalpiniaceous, the petals 9–15 mm long, pale yellow; fertile stamens 10. Fruit linear-falcate,

(5)10–20 cm long, 3–5 mm wide, decurved, quadrangular, chartaceous, dehiscent; seeds rhomboid or subcylindric-oblong, 3–5 mm long, castaneous, lustrous.

Disturbed sites. Frequent; nearly throughout. Massachusetts and New York south to Florida, west to Wisconsin, Nebraska, Kansas, Oklahoma, and Texas, also California; West Indies, Mexico, Central America, and South America; Africa, Asia, and Pacific Islands. Native to tropical America. All year.

Senna occidentalis (L.) Link [Western.] SEPTICWEED.

> *Cassia occidentalis* Linnaeus, Sp. Pl. 1: 377. 1753. *Cassia foetida* Persoon, Syn. Pl. 1: 457. 1805, nom. alt. illegit. *Senna occidentalis* (Linnaeus) Link, Handbuch 2: 140. 1829. *Ditremexa occidentalis* (Linnaeus) Britton & Rose ex Britton & P. Wilson, Sci. Surv. Porto Rico 5: 372. 1924.

Annual herb, to 1(2) m; stems glabrous. Leaves 10–20 cm long, the leaflets 4–5 pairs, the blade ovate to ovate-lanceolate, 3–8 cm long, 2–3 cm wide, the apex acuminate, the base rounded, the upper and lower surfaces glabrous, the petiole with 1(2) spheroid gland(s) near the base; stipules linear-falcate, 1–2 cm long, caducous. Flowers 1–5 in a short axillary or a weakly terminally congested raceme, the pedicel 8–17 mm long; bracts 8–15 mm long, caducous; sepals obovate to oblong-obovate, 5–10 mm long, pink or fuscous-tinged; corolla caesalpiniaceous, the petals 12–16 mm long, yellow; fertile stamens 6. Fruit oblong-falcate, 8–14 cm long, 6–8 mm wide, laterally compressed, the valves thick-chartaceous, brown, the margins light-colored. glabrous, tardily dehiscent; seeds obovoid to suborbicular, 3–5 mm long, olivaceous or tan, rarely castaneous, smooth.

Disturbed sites. Frequent; nearly throughout. New York and Massachusetts south to Florida, west to Iowa, Kansas, Oklahoma, and Texas; West Indies, Mexico, Central America, and South America; Africa, Asia, Australia, and Pacific Islands. Generally considered native to tropical America, but perhaps an Old World introduction. Summer–fall.

Senna pendula (Humb. & Bonpl. ex Willd.) H. S. Irwin & Barneby var. **glabrata** (Vogel) H. S. Irwin & Barneby [Hanging, in reference to the inflorescences; glabrate.] VALAMUERTO.

> *Cassia indecora* Kunth var. *glabrata* Vogel, Gen. Cass. Syn. 19: 1837. *Senna pendula* (Humboldt & Bonpland ex Willdenow) H. S. Irwin & Barneby var. *glabrata* (Vogel) H. S. Irwin & Barneby, Mem. New York Bot. Gard. 35: 382. 1982.
> *Cassia coluteoides* Colladon, Hist. Nat. Med. Casses 102, t. 12. 1816.

Sprawling shrub, to 3 m; branchlets glabrous or rarely puberulent. Leaves 4–8 cm long, the leaflets (2)4–5 pairs, the blade obovate to elliptic, 1.5–3.5 cm long, 1–1.5 cm wide, the apex rounded, the base asymmetrically broadly cuneate to rounded, the upper and lower surfaces glabrous or pilosulous, the petiole 2–2.5 cm long, with a sessile, spheroid or substipitate gland between the lowermost leaf pair; stipules linear-lanceolate, 5–9 mm long, caducous. Flowers 3–12 in a subterminally congested raceme, the pedicel 1–1.5 cm long; bracts lanceolate-subulate, 1–4 mm, caducous; sepals elliptic-lanceolate to elliptic-suborbicular, 3–9 mm long, yellowish to fuscous or reddish castaneous; corolla caesalpiniaceous, the petals 1–2.5 cm long, yellow; fertile stamens 7. Fruit cylindric-moniliform, 7–12(15) cm long, 1–1.5 cm wide, subwoody, brown, glabrous, indehiscent, breaking irregularly into segments; seeds obovoid, 4–6 mm long, brown.

Disturbed sites. Occasional; central and southern peninsula. Escaped from cultivation. Florida, Texas, Arizona, and California; South America; Africa and Australia. Native to South America. Fall–winter.

Senna pendula var. *glabrata* is listed as a category I invasive species in Florida by the Florida Exotic Pest Plant Council (FLEPPC, 2015).

Senna surattensis (Burm. f.) H. S. Irwin & Barneby [Of Surat, India.] GLOSSY SHOWER.

Cassia surattensis Burman f., Fl. Indica 97. 1768. *Senna surattensis* (Burman f.) H. S. Irwin & Barneby, Mem. New York Bot. Gard. 35: 81. 1982.

Cassia suffruticosa J. König ex Roth, Nov. Pl. Sp. 213. 1821. *Cassia glauca* Lamarck var. *koenigii* Kurz, J. Asiat. Soc. Bengal, Pt. 2, Nat. Hist. 45. 284. 1876. *Cassia glauca* Lamarck var. *suffruticosa* (J. König ex Roth) Baker, in Hooker f., Fl. Ind. 2: 265. 1878, nom. illegit. *Psilorhegma suffruticosa* (J. König ex Roth) Britton ex Britton & Rose, in Britton, N. Amer. Fl. 23: 255. 1930. *Cassia surattensis* Burman f. subsp. *suffruticosa* (J. König ex Roth) K. Larsen & S. S. Larsen, Nat. Hist. Bull. Siam. Soc. 23: 205. 1974. *Cassia surattensis* Burman f. var. *suffruticosa* (J. König ex Roth) Sealy ex Isely, Mem. New York Bot. Gard. 25: 209. 1975, nom. illegit.

Shrub or small tree; branchlets puberulent. Leaves 6–13 cm long, the leaflets 6–10 pairs, the blade elliptic or obovate-elliptic, 3–5 cm long, 1–1.5 cm wide, the upper and lower surfaces puberulent or glabrate, the petiole 2–3 cm long, the rachis with a clavate gland between 2 or more of the proximal leaflet pairs; stipules linear-falcate, ca. 1 cm long, caducous. Flowers 5–15 in an axillary or terminally congested raceme, the pedicel 1.5–2.5 cm long; bracts ovate or lanceolate, 3–7 mm long; sepals elliptic-lanceolate, 2–7 mm long, yellowish; corolla caesalpiniaceous, the petals 1.5–2.5 cm long, yellow; fertile stamens 10. Fruit oblong, laterally compressed, 7–10 cm long, 12–14 mm wide, straight of slightly decurved, the valves chartaceous, brown, tardily dehiscent; seeds oblong-elliptic, 4–6 mm long, castaneous, smooth, lustrous.

Disturbed sites. Rare; Miami-Dade County, Monroe County keys. Escaped from cultivation. Florida and Texas; West Indies; Asia, Australia, and Pacific Islands. Native to Asia and Australia. Spring–fall.

EXCLUDED TAXA

Senna angustisiliqua (Lamarck) H. S. Irwin & Barneby—Reported for Florida by Chapman (1860, as *Cassia angustisiliqua* Lamarck), who misapplied the name to our material of *S. ligustrina*.

Senna bicapsularis (Linnaeus) Roxburgh—Reported for Florida by Long and Lakela (1971, as *Cassia bicapsularis* Linnaeus) and Correll and Correll (1982, as *Cassia bicapsularis* Linnaeus), who misapplied the name to material of *S. pendula* var. *glabrata*. Material reported for Florida under this name by Irwin and Barneby (1982) was also placed under *S. pendula* var. *glabrata* by Isely (1990).

Senna didymobotrya (Fresenius) H. S. Irwin & Barneby—Reported for "vacant lots" in Miami-Dade County by Small (1933, as *Chamaesenna didymobotrya* (Fresenius) Small), this followed by Wunderlin (1998) and Wunderlin and Hansen (2003, 2011). There is no real evidence this cultivated species escapes in Florida.

Senna pendula (Willdenow) H. S. Irwin & Barneby var. *ovalifolia* H. S. Irwin & Barneby—Reported as "sparingly escaped from cultivation in s. peninsular Florida" by Irwin and Barneby (1982). No specimens are known, and with the difficulty of keying out any of their 17+ varieties of this species, a misidentification seems likely.

Senna siamea (Lamarck) H. S. Irwin & Barneby—Somewhat ambiguously reported for Florida by Irwin & Barneby (1982), but it is believed they were reporting cultivated material.

Senna tora (Linnaeus) Roxburgh—This Asian species was reported for Flora by Small (1903, 1913a, 1913b, 1913d, all as *Cassia tora* Linnaeus; 1933, as *Emelista tora* (Linnaeus) Britton & Rose) who misapplied the name to material of *S. obtusifolia*.

Sesbania Adans nom. cons. 1763. RIVERHEMP

Annual or perennial herbs, shrubs, or trees. Leaves even-pinnately compound, stipellate or estipellate, petiolate, stipulate. Flowers in axillary racemes, bracteate, bracteolate; sepals 5, connate; petals 5, the corolla papilionaceous; stamens 10, diadelphous (9 + 1). Fruit a dehiscent or indehiscent legume.

A genus of about 60 species; North America, West Indies, Mexico, Central America, South America, Africa, Asia, Australia, and Pacific Islands. [From the Arabic name for the plant, *sisabun* or *sesaban*.]

Agati Adans., nom. rej. 1763; *Daubentonia* DC., 1825; *Glottidium* Desv., 1813; *Sesban* Adans., nom. rej. 1763.

1. Corolla 6–8(10) cm long..**S. grandiflora**
1. Corolla less than 3 cm long.
   2. Fruit conspicuously 4-winged.
      3. Corolla red or orange-red; fruit acuminate or tapering to the beak............................ **S. punicea**
      3. Corolla yellow; fruit blunt or with a short-acuminate beak .... **S. drummondii**
   2. Fruit not winged.
      4. Fruit flattened, (1)2-seeded ....**S. vesicaria**
      4. Fruit quadrangular or subterete, more than 2-seeded.
         5. Fruit 4–5(6) cm long, 6–8 mm wide..**S. virgata**
         5. Fruit (10)15–20 cm long, 3–4(5) mm wide.
            6. Leaves and stem sericeous.. **S. sericea**
            6. Leaves and stem glabrous or glabrate .... **S. herbacea**

Sesbania drummondii (Rydb.) Cory [Commemorates Thomas Drummond (1793–1835), Scottish botanist and collector in North America.] POISONBEAN.

Daubentonia drummondii Rydberg, Amer. J. Bot. 10: 498. 1923. *Sesbania drummondii* (Rydberg) Cory, Rhodora 38: 406. 1936.

Suffrutescent perennial herb or shrub, to 3 m; branchlets glabrate or obscurely pubescent. Leaves 5–20 cm long, the leaflets 10–20 pairs, the blade elliptic-oblong, 1–3 cm long, 4–6 mm wide, the apex obtuse to rounded, mucronate, the base cuneate, the margin entire, the upper and lower surfaces glabrous or obscurely pubescent, the petiole ca. 1 cm long; stipules small, caducous. Flowers 5–20 in an axillary raceme 7–10 cm long, the pedicel 5–10 mm long; calyx campanulate, 4–5 mm long, sinuately 5-lobed; corolla papilionaceous, 13–17 mm long, yellow. Fruit oblong, 3–7 cm long, 1–1.5 cm wide, longitudinally 4-winged, impressed between the seeds, chartaceous-coriaceous, glabrous, indehiscent, the stipe 1–2 cm long; seeds few to several.

Fresh to brackish marshes and disturbed sites. Rare; Escambia and Santa Rosa Counties. South Carolina south to Florida, west to Texas; Mexico. Summer.

Sesbania grandiflora (L.) Pers. [Large-flowered.] VEGETABLE HUMMINGBIRD.

Robinia grandiflora Linnaeus, Sp. Pl. 2: 722. 1753. *Aeschynomene grandiflora* (Linnaeus) Linnaeus, Sp. Pl., ed. 2. 1060. 1763. *Coronilla grandiflora* (Linnaeus) Willdenow, Sp. Pl. 3: 1145. 1802. *Sesbania grandiflora* (Linnaeus) Poiret, in Lamarck, Encycl. 7: 127. 1806. *Sesbania grandiflora* (Linnaeus) Persoon, Syn. Pl. 2: 316. 1807. *Agati grandiflora* (Linnaeus) Desvaux, J. Bot. Agric. 1: 120. 1813. *Emerus grandiflorus* (Linnaeus) Hornemann, Hort. Bot. Hafn. 695. 1815. *Resupinaria grandiflora* (Linnaeus) Rafinesque, Sylva Tellur. 116. 1838. *Agati grandiflora* (Linnaeus) Deauvaux var. *albiflora* Wight & Arnott, Prodr. Fl. Ind. Orient. 1: 215. 1834, nom. inadmiss.

Shrub or small tree, to 6 m; branchlets glabrate. Leaves 10–30 cm long, the leaflets 10–20 pairs, the blade elliptic-oblong, 2–5 cm long, 8–12 mm wide, the apex rounded, the base cuneate, the margin entire, the upper and lower surfaces glabrate, the petiole ca. 1 cm long; stipules small, caducous. Flowers 2–4 in an axillary raceme, the pedicel 1–2 cm long; calyx campanulate, 1.8–2.2 cm long, sinuately 5-lobed; corolla papilionaceous, 6–8(10) cm long, scarlet, pink, or white. Fruit linear, 3–4 dm long, 6–8 mm wide, tortulose, tetragonal or compressed, coriaceous, glabrous, dehiscent, the stipe 2–3 cm long; seeds numerous.

Disturbed sites. Rare; Monroe County keys. Escaped from cultivation. Florida; West Indies, Mexico, Central America, and South America; Africa, Asia, Australia, and Pacific Islands. Native to Asia and Pacific Islands. Winter.

Sesbania herbacea (Mill.) McVaugh [Herbaceous.] DANGLEPOD.

Emerus herbaceus Miller, Gard. Dict., ed. 8. 1768. *Sesbania herbacea* (Miller) McVaugh, Fl. Novo-Galic. 5: 695. 1987.

Aeschynomene emerus Aublet, Hist. Pl. Guiane 775; tabl. noms. 1. 1775. *Sesbania emerus* (Aublet) Urban, Repert. Spec. Nov. Regni Veg. 16: 149. 1919. *Sesban emerus* (Aublet) Britton & P. Wilson, Sci. Surv. Porto Rico & Virgin Islands 5: 395. 1924.

Darwinia exaltata Rafinesque, Fl. Ludov. 106. 1817. *Sesban exaltatus* (Rafinesque) Rydberg, in Britton, N. Amer. Fl. 24: 204. 1924. *Sesbania exaltata* (Rafinesque) A. W. Hill, in B. D. Jackson, Index Kew., Suppl. 7. 223. 1929.

Sesbania macrocarpa Muhlenberg ex Nuttall, Gen. N. Amer. Pl. 2: 112. 1818. *Sesban macrocarpus* (Muhlenberg ex Nuttall) Muhlenberg ex Small, Fl. S.E. U.S. 614. 1903.

Annual herb, suffruticose perennial herb, or shrub, to 2(4) m; branchlets glabrous. Leaves 10–20(30) cm long, the leaflets 15–25 pairs, the blade oblong or narrowly oblong, 1–2.5 cm long, 2–4 mm wide, the apex rounded, mucronate, the base broadly cuneate, the margin entire, the upper and lower surfaces glabrous, the petiole ca. 1 cm long; stipules small, caducous. Flowers 2–5 in an axillary raceme 3–6 cm long; calyx 5–7 mm long, campanulate, the lobes 1–2 mm long; corolla papilionaceous, 1–2.2 cm long, yellowish mottled, rarely red. Fruit linear, 15–20 cm long, 3–5 mm wide, subcoriaceous or pulpy-chartaceous, glabrous, tardily dehiscent; seeds numerous.

Disturbed sites. Frequent; nearly throughout. New York and Massachusetts south to Florida, west to Kansas, Oklahoma, and Texas, also Arizona and California; West Indies, Mexico, and Central America. Summer–fall.

Sesbania punicea (Cav.) Benth. [Crimson, in reference to the flower color.] RATTLEBOX.

Psicidia punicea Cavanilles, Icon. 4: 8, t. 316. 1797. *Daubentonia punicea* (Cavanilles) de Candolle, Prodr. 2: 267. 1825. *Sesbania punicea* (Cavanilles) Bentham, in Martius, Fl. Bras. 15(1): 43. 1859. *Emerus puniceus* (Cavanilles) Kuntze, Revis. Gen. Pl. 1: 181. 1891.

Shrub or small tree, to 2 m; branchlets glabrate. Leaves 7–15 cm long, the leaflets 10–17 pairs, the blade elliptic to elliptic-oblong, 0.8–2.5 cm long, 3–5 mm wide, the apex rounded or truncate, apiculate, the base broadly cuneate to rounded, the margin entire, the upper and lower surfaces glabrate, the petiole 1–1.5 cm long; stipules minute, caducous. Flowers 5–15 in an axillary raceme, the pedicel 0.5–1.2(1.5) cm long; calyx 5–6 mm long, sinuately 5-lobed; corolla papilionaceous, 1.5–2.5 cm long, scarlet or orange-red. Fruit broadly oblong, 4–8 cm long, 1–1.5 cm wide, laterally compressed but thick, impressed between the seeds, longitudinally 4-winged, chartaceous-coriaceous, glabrous, indehiscent, the stipe 0.5–1(1.5) cm long; seeds few to several.

Tidal marshes and moist, disturbed sites. Frequent; northern counties, central peninsula. Escaped from cultivation. Virginia south to Florida, west to Texas, also California; South America; Africa. Native to South America. Spring–fall.

Sesbania punicea is listed as a Category II invasive species in Florida by the Florida Exotic Pest Plant Council (FLEPPC, 2015).

Sesbania sericea (Willd.) Link [Silky.] SILKY SESBAN.

Coronilla sericea Willdenow, Enum. Pl. 773. 1809. *Sesbania sericea* (Willdenow) Link, Enum. Hort. Berol. Alt. 2: 244. 1822. *Agati sericea* (Willdenow) Hitchcock, Rep. (Annual) Missouri Bot. Gard. 4: 75. 1893. *Sesban sericeus* (Willdenow) Britton & P. Wilson, Sci. Surv. Porto Rico & Virgin Islands 6: 395. 1924.

Suffruticose shrub, to 2 m; stem pubescent. Leaves 5–20 cm long, the leaflets 10–20 pairs, the blade oblong to narrowly oblong, 1–4 cm long, 4–6 mm wide, the apex rounded, mucronate, the base broadly cuneate to rounded, the upper surface glabrous, the lower surface subappressed-villous, the petiole 1–1.5 cm long; stipules small, caducous. Flowers 2–5 in an axillary raceme 2–6 cm long, the pedicel 0.5–1 cm long; calyx campanulate, ca. 4 mm long, the lobes triangular, ca. 1 mm long; corolla papilionaceous, 8–11 mm long, orange-yellow or greenish yellow. Fruit linear, 10–18 cm long, 4–5 mm wide, usually falcate, subtetragonal, thick-chartaceous, glabrous; seeds numerous.

Disturbed sites. Rare; Pinellas and Broward Counties, Monroe County keys. Florida; West Indies, Mexico, Central America, and South America; Africa and Asia. Native to Africa and Asia. Spring–fall.

Sesbania vesicaria (Jacq.) Elliott [Bladderlike, inflated.] BLADDERPOD; BAGPOD.

Robinia vesicaria Jacquin, Collectanea 1: 105. 1787. *Phaca floridana* Willdenow, Sp. Pl. 3: 1252. 1802. *Colutea floribunda* Poiret, in Lamarck, Encycl., Suppl. 1: 562. 1810. *Sesbania disperma* Pursh, Fl. Amer. Sept. 485. 1814, nom. illegit. *Sesbania vesicaria* (Jacquin) Elliott, Sketch Bot. S. Carolina 2: 222. 1823. *Glottidium floridanum* (Willdenow) de Candolle, Prodr. 2: 266. 1825, nom. illegit.

Emerus vesicarius (Jacquin) Kuntze, Revis. Gen. Pl. 1: 181. 1891. *Glottidium vesicarium* (Jacquin) C. Mohr, Contr. U.S. Natl. Herb. 6: 568. 1901. TYPE: FLORIDA: cult. in Vienna.

Aeschynomene platycarpa Michaux, Fl. Bor.-Amer. 2: 75. 1803. *Sesbania platycarpa* (Michaux) Persoon, Syn. Pl. 2: 316. 1807. TYPE: "Carolinae, usque ad Floridam."

Glottidium floridanum (Willdenow) de Candolle var. *atrorubrum* Nash, Bull. Torrey Bot. Club 23: 101. 1895. *Glottidium vesicarium* (Jacquin) C. Mohr forma *atrorubrum* (Nash) Small, Fl. S.E. U.S. 615. 1903. *Sesbania vesicaria* (Jacquin) Elliott var. *atrorubra* (Nash) S. C. Brooks, Proc. Amer. Acad. Arts 49: 503. 1913. TYPE: FLORIDA: Hillsborough Co.: Tampa, s.d. *Nash 2415* (holotype: NY?).

Annual herb, to 3(5) m; stem glabrate. Leaves 8–20 cm long, the leaflets 16–36, the blade elliptic to elliptic-oblong, 13.5(4) cm long, 2–5 mm wide, the apex rounded, mucronate, the base cuneate, the margin entire, the upper and lower surfaces glabrate, the petiole 0.5–1 cm long; stipules subulate-filiform, caducous. Flowers 3–10(12) in an axillary raceme 8–10(12) cm long, the pedicel 0.5–1 cm long; bracts subulate-filiform; caducous; bracteoles calycine, caducous; calyx campanulate, ca. 4 mm long, sinuate-truncate, obscurely 5-lobed; corolla papilionaceous, 8–9 mm long, yellowish white, often maroon tipped or tinged, rarely solid red or maroon. Fruit elliptic-oblong or oblong, 4–6 cm long, ca. 1.5 cm wide, laterally compressed, tapering to a sharp beak 5–8 mm long, chartaceous, glabrous, tardily dehiscent, the stipe 1–2 cm long; seeds 1–2.

Moist, disturbed sites. Frequent; nearly throughout. North Carolina south to Florida, west to Oklahoma and Texas. Spring–fall.

Sesbania virgata (Cav.) Poir. [Long and slender, in reference to the habit.] WAND RIVERHEMP.

Aeschynomene virgata Cavanilles, Icon. 3: 47, t. 293. 1796. *Coronilla virgata* (Cavanilles) Willdenow, Sp. Pl. 3: 1148. 1802. *Sesbania virgata* (Cavanilles) Poiret, in Lamarck, Encycl. 7: 129. 1806. *Sesbania virgata* (Cavanilles) Persoon, Syn. Pl. 2: 316. 1807. *Agati virgata* (Cavanilles) Desvaux, J. Bot. Agric. 1: 120. 1813. *Coursetia virgata* (Cavanilles) de Candolle, Ann. Sci. Nat. (Paris) 4: 92. 1824. *Daubentonia virgata* (Cavanilles) Rydberg, in Britton, N. Amer. Fl. 24: 208. 1924.

Sesbania marginata Bentham, in Martius, Fl. Bras. 15(1): 43, t. 7. 1859. *Emerus marginatus* (Bentham) Lindman, Bih. Kongl. Svenska Vetensk.-Akad. Handl. 24(Afd. 3, no. 7): 7. 1898.

Shrub, to 2(3) m; stem strigulose or glabrate. Leaves 10–20 cm long, the leaflets 13–16 pairs, the blade elliptic-oblong, 1.5–3 cm long, 4–7 mm wide, the apex round, mucronate, the base broadly cuneate, the upper surface glabrous, the lower surface strigulose, the petiole ca. 1 cm long; stipules small, caducous. Flowers 10–15 in an axillary raceme ca. 10 cm long, the pedicel 5–7 mm long; calyx campanulate, 3–4 mm long, sinuately 5-lobed; corolla papilionaceous, 9–12 mm long, yellowish, rarely red-purple. Fruit oblong, 4–6 cm long, 6–8 mm wide, quadrangular, angular, or flanged (but not winged), somewhat curved, woody, indehiscent; seeds few.

Coastal marshes. Rare; Escambia, Santa Rosa, Gulf, Pinellas, and Hillsborough Counties. Florida and Mississippi; South America. Native to South America. Spring–summer.

EXCLUDED TAXON

Sesbania longifolia de Candolle—Reported for Florida by Small (1903, 1913a, 1913e, all as *Daubentonia longifolia* (Cavanilles) de Candolle), who misapplied the name to material of *S. drummondii*.

Sigmoidotropis (Piper) A. Delgado 2011.

Perennial herbs. Leaves alternate, 3-foliolate, stipellate, petiolate, stipulate. Flowers in pedunculate, axillary pseudoracemes, bracteate; sepals 5, connate into a tube, unequal; petals 5, the corolla papilionaceous, the keel sigmoid-curved; stamens 10, diadelphous (9 + 1). Fruit a dehiscent legume.

A genus of 9 species; North America, West Indies, Mexico, Central America, South America, and Pacific Islands. [From the Greek *sigma*, the letter S, and *tropis*, the keel of a ship, in reference to the keel petals curved like the letter S.]

1. Calyx 5–6 mm long; corolla ca. 2 cm long... **S. antillana**
1. Calyx 8–10 mm long; corolla 3–4 cm long...**S. speciosa**

Sigmoidotropis antillana (Urb.) A. Delgado [Of the Antilles.] ANTILLES BEAN.

Phaseolus antillanus Urban, Symb. Antill. 4: 309. 1905. *Vigna antillana* (Urban) Fawcett & Rendle, Fl. Jamaica 4: 69. 1920. *Sigmoidotropis antillana* (Urban) A. Delgado, Amer. J. Bot. 98: 1711. 2011.

Climbing perennial herb. Leaves 3-foliolate, the leaflet blade triangular to ovate-oblong, 5–7 cm long, 3–5 cm wide, the apex long-acuminate, the base subtruncate, the lateral ones oblique, the margin entire, the upper surface glabrous, the lower surface short-pilose or glabrous, the petiole 3–4 cm; stipules triangular-oblong, 1–4 mm long, entire or with inconspicuous retrorse lobes. Flowers paired at a swollen node on a pseudoraceme 7–30 cm long; calyx 5–6 mm long, the tube campanulate, the lobes triangular to ovate, shorter than the tube; corolla papilionaceous, ca. 2 cm long, blue, the keel petals curved ca. ⅓. Fruit linear-oblong, 8–13 cm long, 4–5 mm wide, laterally compressed, glabrous.

Disturbed sites. Rare; Miami-Dade County. Florida; West Indies. Native to West Indies. Spring–fall.

Sigmoidotropis speciosa (Kunth) A. Delgado [Showy.] WANDERING COWPEA.

Phaseolus speciosus Kunth, in Humboldt et al., Nov. Gen. Sp. 6: 452. 1824. *Vigna speciosa* (Kunth) Verdcourt, Kew Bull. 24: 552. 1970. *Sigmoidotropis speciosa* (Kunth) A. Delgado, Amer. J. Bot. 98: 1711. 2011.

Climbing perennial herb; stem to 3 m, pubescent. Leaves 3-foliolate, the leaflet blade ovate, 4–10 cm long, 2–6 cm wide, the apex acuminate, the base rounded to truncate, that of the lateral lobes oblique, the margin entire, the upper surface glabrous, the lower surface pubescent, the petiole 2–4 cm long; stipules ca. 3 mm long, entire or with inconspicuous retrorse lobes. Flowers 1–2 at the apex of a 3–20 cm long peduncle; bracteoles ca. 2 mm long; calyx 8–10 mm long, the tube campanulate, the lobes deltate, shorter than the tube; corolla papilionaceous, 3.5–4 cm long, blue or purple, marked with yellow and white, the keel petals incurved ca. 240°. Fruit linear, 10–15 cm long, 4–5 mm wide, glabrous.

Disturbed sites. Occasional; central peninsula. Escaped from cultivation. Florida; Mexico, Central America, and South America; Pacific Islands. Native to Mexico, Central America, and South America. All year.

Sophora L., nom. cons. 1753. NECKLACEPOD

Trees or shrubs. Leaves odd-pinnately compound, estipellate, petiolate, stipulate. Flowers in terminal racemes, bracteate, ebracteolate; sepals 5, connate; corolla papilionaceous; stamens 10, free. Fruit an indehiscent legume.

A genus of about 50 species; nearly cosmopolitan. [From the Arabic *sufayra*, yellow, apparently in reference to the flower color of some species.]

Selected references: Isely (1981); Rudd (1972).

Sophora tomentosa L. [Thickly and evenly covered with short matted trichomes.] YELLOW NECKLACEPOD.

Shrub or small tree, to 3 m. Leaves 10–25 cm, the leaflets 11–21, the blade elliptic or obovate, 2–4(5) cm long, 1–2 cm wide, the apex rounded or truncate, the base broadly cuneate, the upper surface pubescent or glabrate at maturity, the petiole 2–3 cm long; stipules obsolete. Flowers numerous in a terminal pedunculate raceme; calyx campanulate, 8–10 mm long, sinuately lobed; corolla papilionaceous, 2–2.5 cm long, yellow. Fruit oblong, 6–15(20) cm long, 5–10 mm wide, moniliform, with 5–8(12) fertile segments, coriaceous, black; seeds brown, ca. 5 mm long.

1. Leaflets densely pubescent at maturity..var. **occidentalis**
1. Leaflets pubescent to glabrate at maturity ....var. **truncata**

Sophora tomentosa var. occidentalis (L.) Isely [Western.]

Sophora occidentalis Linnaeus, Syst. Veg., ed. 10. 1015. 1759. *Sophora tomentosa* Linnaeus subsp. *occidentalis* (Linnaeus) Brummitt, Kirkia 5: 265. 1966. *Sophora tomentosa* Linnaeus var. *occidentalis* (Linnaeus) Isely, Brittonia 30: 471. 1978.

Leaflets densely pubescent at maturity.

Dry, disturbed sites. Rare; Pinellas, Sarasota, Martin, and Miami-Dade Counties. Escaped from cultivation. Florida and Texas; West Indies, Mexico, Central America, and South America; Africa and Asia. Native to Texas, tropical America, Africa, and Asia. All year.

Sophora tomentosa var. truncata Torr. & A. Gray [Ending abruptly as if cut across, in reference to the leaflets.]

Sophora tomentosa Linnaeus var. *truncata* Torrey & A. Gray, Fl. N. Amer. 1: 389. 1840. TYPE: FLORIDA: Hillsborough Co.: Tampa Bay, 1834, *Leavenworth s.n.* (lectotype: NY). Lectotypified by Isely (1981: 253).

Sophora tomentosa Linnaeus forma *aurea* Yakovlev, Trudy Lenningradsk. Khim.-Farm. Inst. 21: 54. 1967. TYPE: FLORIDA: Monroe Co.: Key Largo, 1940, *Meebold s.n.* (holotype: M).

Sophora tomentosa Linnaeus forma *longifolia* Yakovlev, Trudy Lenningradsk. Khim.-Farm. Inst. 21: 54. 1967. TYPE: FLORIDA: Lee Co.: Bokeelia, 10 Apr 1930, *Moldenke 936a* (holotype: ST; isotype: NY).

Sophora tomentosa Linnaeus subsp. *bahamensis* Yakovlev, Trudy Lenningradsk. Khim.-Farm. Inst. 26: 98. 1968. TYPE: FLORIDA: Brevard Co.: Indian River, s.d., *Curtiss 704* (holotype: LE; isotype: NY).

Leaflets pubescent to glabrate at maturity.

Coastal strands and hammocks. Frequent; central and southern peninsula. Florida; West Indies and South America. All year.

EXCLUDED TAXA

Sophora tomentosa Linnaeus—Because infraspecific categories were not recognized, the typical variety was reported for Florida by implication by Chapman (1860, 1883, 1897), by Small (1903, 1913a, 1913b, 1913d, 1913e, 1933) and by Long and Lakela (1971). All Florida material cited by them is probably var. *truncata*.

Strophostyles Elliott, nom. cons. 1823. FUZZYBEAN

Trailing or climbing annual or perennial vines. Leaves 3-folioliate, stipellate, petiolate, stipulate. Flowers in axillary, pedunculate, subumbellate pseudoracemes, bracteate, bracteolate; sepals 5, connate, the calyx 4-lobed; corolla papilionaceous; stamens 10, diadelphous (9 + 1); ovary sessile, surrounded by a nectariferous sheath. Fruit a dehiscent legume.

A genus of 3 species; North America and Mexico. [From the Greek *strophe*, turning, and *stylos*, pillar, apparently in reference to the spirally twisted style.]

Selected reference: Riley-Hulting et al. (2004).

1. Fruit pubescent at maturity, 2–4 cm long; leaves pubescent on the upper surface; seeds glabrous.........
   ....**S. leiosperma**
1. Fruit glabrate at maturity, 3–8 cm long; leaves usually glabrate on the upper surface; seeds pubescent.
   2. Bracteoles just below the flower subequaling to slightly exceeding the calyx tube; leaflets usually
      lobed...**S. helvola**
   2. Bracteoles just below the flower shorter than the calyx tube; leaflets unlobed .... **S. umbellata**

Strophostyles helvola (L.) Elliott [Pale red, in reference to the flower color.] TRAILING FUZZYBEAN.

Phaseolus helvolus Linnaeus, Sp. Pl. 2: 724. 1753, nom. cons. *Dolichos helvolus* (Linnaeus) Nuttall, Gen. A. Amer. Pl. 2: 112. 1818. *Glycine helvola* (Linnaeus) Elliott, J. Acad. Nat. Sci. Philadelphia 1: 326. 1818. *Strophostyles helvola* (Linnaeus) Elliott, Sketch Bot. S. Carolina 2: 230. 1823.

Phaseolus trilobus Michaux, Fl. Bor.-Amer. 2: 60. 1803; non Aiton, 1789. *Phaseolus diversifolius* Persoon, Syn. Pl. 2: 296. 1807.

Glycine peduncularis Muhlenberg, Cat. Pl. Amer. Sept. 64. 1813. *Phaseolus peduncularis* (Muhlenberg) W.P.C. Barton, Comp. Fl. Philadelph. 2: 81. 1818. *Strophostyles peduncularis* (Muhlenberg) Elliott, Sketch Bot. S. Carolina 2: 230. 1823.

Strophostyles angulosa (Ortega) Elliott var. *missouriensis* S. Watson, in A. Gray, Manual, ed. 6. 145. 1890. *Phaseolus helvolus* Linnaeus var. *missouriensis* (S. Watson) Britton, Mem. Torrey Bot. Club 5: 208. 1894. *Strophostyles helvola* (Linnaeus) Elliott var. *missouriensis* (S. Watson) Britton, in Britton & A. Brown, Ill. Fl. N. U.S. 2: 339. 1897. *Strophostyles missouriensis* (S. Watson) Small, Fl. S.E. U.S. 654, 1332. 1903.

Trailing or climbing perennial vine; stem to 3 m, strigose. Leaves 3-foliolate, the leaflet blade ovate, orbicular, or lanceolate, panduriform, 1.3–7 cm long, 0.6–4.5 cm wide, the apex acute to acuminate, the base broadly cuneate, the margin deeply lobed to entire, the upper surface glabrous, the lower surface strigose, the stipels linear, ca. 1 mm long, the petiole 5–7 cm long; stipules lanceolate, 3–4 mm long. Flowers subsessile in a subterminal axil, the peduncle (2.5)5–20 cm long; calyx tube 1–4 mm long, the lobes 1–3 mm long; bracteoles 2–4 mm long, subequaling or exceeding the calyx tube; corolla papilionaceous, 7–13 mm long, pinkish, the keel petals with a dark purple tip. Fruit cylindric, 3–8 cm long, (3)4–8 mm wide, glabrous; seeds 5–7(10), pubescent.

Dry, open hammocks, fields, and dunes. Occasional; northern counties, Volusia County. Quebec south to Florida, west to Ontario, Minnesota, South Dakota, Nebraska, Kansas, Oklahoma, and Texas. Spring–fall.

Strophostyles leiosperma (Torr. & A. Gray) Piper [Smooth seed.] SLICKSEED FUZZYBEAN.

Phaseolus leiospermus Torrey & A. Gray, Fl. N. Amer. 1: 280. 1838. Strophostyles leiosperma (Torrey & A. Gray) Piper, Contr. U.S. Natl. Herb. 668. 1926.

Trailing or climbing annual or short-lived perennial vine; stem to 1 m, sericeous. Leaves 3-foliolate, the leaflet blade lanceolate to linear-lanceolate, 2–5.5 cm long, 2–22 mm wide, the apex acute, the base broadly cuneate, the margin entire or rarely shallowly lobed, the upper and lower surfaces sericeous, the stipels linear, ca. 1 mm long, the petiole ca. 2 cm long; stipules lanceolate, ca. 2 mm long. Flowers subsessile in a subterminal axil, the peduncle 1–12 cm long; calyx tube 1–2 mm long, the lobes 1–2 mm long; bracteoles 1–2 mm long, subequaling the calyx tube; corolla papilionaceous, 4–7(8) mm long, pink, the keel petals with a dark purple tip. Fruit subcylindrical, somewhat laterally compressed, 2–4 cm long, 2–5 mm wide, sericeous; seeds 4–9, glabrous.

Dry, open hammocks. Rare; Wakulla, Franklin, and Gulf Counties. Connecticut south to Florida, west to North Dakota, South Dakota, Nebraska, Colorado, and Arizona; Mexico. Summer–fall.

Strophostyles umbellata (Muhl. ex Willd.) Britton [Umbrella-like, in reference to the inflorescence.] PINK FUZZYBEAN.

Glycine umbellata Muhlenberg ex Willdenow, Sp. Pl. 3: 1058, nom. cons. Phaseolus umbellatus (Muhlenberg ex Willdenow) Britton, Trans. New York Acad. Sci. 9: 10. 1889. Strophostyles umbellata (Muhlenberg ex Willdenow) Britton, in Britton & A. Brown, Ill. Fl. N. US. 2: 339. 1897.

Trailing or climbing perennial herb; stem to 1.5 m, glabrate. Leaves 3-foliolate, the leaflet blade lanceolate, elliptic-oblong, or linear-lanceolate, 1.5–4(7) cm long, 0.5–2(3) cm wide, the apex obtuse, the base broadly cuneate to rounded, the margin entire or shallowly lobed, the upper surface glabrate, the lower surface strigose, the stipels lanceolate, 1–2 mm long, the petiole 1.5–2.5 cm long; stipules lanceolate, 1–2 mm long. Flowers in a subterminal axil, the peduncle (4.5)6–30 cm long; calyx tube 2–4 mm long, the lobes 1–2.5(3) mm long; bracteoles 1–3 mm long, shorter than the calyx tube; corolla papilionaceous, 7–15 mm long, pinkish, the keel petals with a dark purple tip. Fruit subcylindric, laterally somewhat compressed, (2.5)3–6(7) cm long, 2–5 mm wide, the valves glabrate to strigose; seeds 6–12, pubescent.

Open hammocks. Occasional; northern counties, central peninsula. New York and Rhode Island south to Florida, west to Illinois, Missouri, Oklahoma, and Texas. Spring–fall.

Stylosanthes Sw. 1788. PENCILFLOWER

Perennial herbs. Leaves alternate, 3-foliolate, estipellate, petiolate, stipulate. Flowers in terminal or axillary spikes, bracteate, bracteolate; sepals 5, the calyx bilabiate, 5-lobed; corolla papilionaceous; stamens monadelphous, the anthers versatile or sub-basifixed. Fruit a loment, the segments 2, the proximal segment fertile or sterile, the distal segment fertile, the style persistent as a beak.

A genus of 40 species; North America, West Indies, Mexico, Central America, South America, Africa, Asia, Australia, and Pacific Islands. [From the Greek *stylos*, writing implement or pillar, and *anthos*, flower, in reference to the column-like style.]

Selected reference: Mohlenbrock (1957).

1. Loment beak 0.5–1 mm long ....**S. biflora**
1. Loment beak 1–2.5 mm long.
    2. Loment beak incurved or hooked; loment usually with 2 fertile segments ....**S. hamata**
    2. Loment beak straight or only moderately curved; loment with a single fertile segment ....
    ....**S. calcicola**

Stylosanthes biflora (L.) Britton et al. [Two-flowered.] SIDEBEAK PENCILFLOWER.

Trifolium biflorum Linnaeus, Sp. Pl. 2: 773. 1753. *Stylosanthes hispida* Michaux, Fl. Bor.-Amer. 2: 75. 1803, nom. illegit; non Richard, 1792. *Stylosanthes biflora* (Linnaeus) Britton et al., Prelim. Cat. 118. 1888.

Stylosanthes elatior Swartz, Kongl. Vetensk. Acad. Nya Handl. 11: 296, t. 11(2). 1789. *Stylosanthes hispida* Michaux var. *erecta* Pursh, Fl. Amer. Sept. 480. 1814. *Stylosanthes biflora* (Linnaeus) Britton et al. var. *elatior* (Swartz) Kuntze, Revis. Gen. Pl. 1: 209. 1891, nom. illegit.

Stylosanthes hispida Michaux var. *procumbens* Pursh, Fl. Amer. Sept. 480. 1814. *Stylosanthes elatior* Swartz var. *procumbens* (Pursh) Chapman, Fl. South. U.S., ed. 3. 109. 1897.

Stylosanthes riparia Kearney, Bull. Torrey Bot. Club 24: 565. 1897.

Stylosanthes floridana S. F. Blake, Proc. Biol. Soc. Wash. 33: 51. 1920. TYPE: FLORIDA: Walton Co.: DeFuniak Springs, 3 Jul 1891, *Sudworth s.n.* (holotype: US).

Erect or spreading perennial herb, to 6 dm; stem glabrous, puberulent, or densely hispid. Leaves 3-foliolate, the leaflet blade ovate to elliptic or lanceolate, to 2–4 cm long, to 1–2 cm wide, the apex acute to acuminate, the base cuneate, the margin entire or sometimes spinulose, the upper surface glabrous, the lower surface glabrous, sometimes punctate, the petiole 1–3 mm long, the nerves 3–6 pairs; stipular sheath 8- to 15-nerved, the lobes of the upper stipules usually shorter than the sheath. Flowers solitary or 2–8 in a spike, lacking an axis rudiment; bracts 1- or 3-foliolate, 3–6 mm long, 5–9-nerved, puberulent or densely hispid; outer bracteole 1, 2–3 mm long, the apex glabrous or ciliate, the inner bracteole 1, 2–3 mm long, the apex glabrous or ciliate; calyx 3–5 mm long; corolla papilionaceous, 5–7 mm long, yellow or orange-yellow. Fruit strongly reticulate with vertical nerves, the proximal segment sterile, 2–3 mm long, the distal segment fertile, 3–5 mm long, the beak hooked, 0.5–1 mm long; seed 1–2 mm long.

Sandhills and hammocks. Occasional; northern counties, central peninsula. New York south to Florida, west to Wisconsin, Kansas, Oklahoma, and Texas, also Arizona; Mexico. Summer.

Stylosanthes calcicola Small [Calcium dweller.] EVERGLADES KEY PENCILFLOWER.

Stylosanthes calcicola Small, Man. S.E. Fl. 730, 1505. 1933. TYPE: FLORIDA: Miami-Dade Co.: pinelands about Ross-Castello Hammock, 24 Jun 1915, *Small et al. 6537* (holotype: NY).

Erect perennial herb, to 5 dm; stem minutely pubescent along one side or glabrous, rarely pubescent throughout. Leaves 3-foliolate, the leaflet blade lanceolate to ovate, 1–1.5 mm long, 6–8 mm wide, the apex acute to acuminate, the base cuneate, the margin entire or rarely ciliate, the upper and lower surfaces glabrous, the petiole 2–4 mm long, the nerves 3–5 pairs; stipular sheath 7-nerved, the teeth of the upper stipules usually longer than the sheath. Flowers 2-several in a spike, the axis rudiment to 5 mm long; bracts 1-foliolate, ca. 4 mm wide, 5- to 7-nerved, the margin ciliate, also sometimes bristly on the dorsal surface; outer bracteole 1, 1–3 mm long, the apex ciliate, the inner bracteoles 2, 2–3 mm long, the apex ciliate; calyx 34 mm long, the lobes subequal; corolla papilionaceous, 5–6 mm long, yellow. Fruit conspicuously nerved, the proximal segment sterile, 2–3 mm long, the distal segment fertile, 2–3 mm long, the beak straight or only slightly curved, 2–3 mm long; seed 1–2 mm long.

Pinelands. Rare; Miami-Dade County, Monroe County keys. Florida; West Indies, Mexico, and Central America. All year.

Stylosanthes hamata (L.) Taub. [Barbed, hooked at the tip, in reference to the hooked style.] CHEESYTOES.

Hedysarum hamatum Linnaeus, Syst. Nat., ed. 10. 1170. 1759. *Stylosanthes procumbens* Swartz, Prodr. 108. 1788, nom. illegit. *Stylosanthes hamata* (Linnaeus) Taubert, Verh. Bot. Vereins Prov. Brandenburg 32: 22. 1890.

Ascending, spreading, or prostrate perennial herb; stem to 1 m, with a line of pubescence on one side or occasionally sericeous. Leaves 3-foliolate, the leaflet blade lanceolate to elliptic, 1.5–2 cm long, 4–6 mm wide, the apex obtuse to subacute, the base cuneate, the margin entire, the upper and lower surfaces glabrous or short-pilose proximally, the nerves 3–6 pairs, the petiole 2–6 mm long; stipular sheath 3- to 11-nerved, the lobes of the upper stipules usually shorter than the sheath. Flowering spikes few- to 15-flowered, the axis rudiment to 7 mm long; outer bracts 3-foliolate, the inner bract 1-foliolate, 5- to 7-nerved, the sheath pubescent; outer bracteole 1, 1–5 mm long, the apex ciliate, the inner bracteoles 2, 2–4 mm long, the apex ciliate; calyx 4–8 mm long; corolla 4–5 mm long, yellow. Fruit reticulate, ca. 2 mm wide, the fertile proximal segment fertile, 2–4 mm long, the distal segment fertile, glabrous or puberulent, the beak hooked, 2–5 mm long; seeds 1–2 mm long.

Pinelands. Occasional; Indian River, Martin, Palm Beach, and Lee Counties, southern peninsula. Florida; West Indies, Central America, and South America. Native to tropical America. Spring–summer.

Tamarindus L. 1753. TAMARIND

Trees. Leaves alternate, even-pinnately compound, the leaflets opposite, estipellate, petiolate, stipulate. Flowers in terminal racemes, bracteate, bracteolate; sepals 4, connate; petals 5, the corolla caesalpiniaceous; stamens 3, monadelphous, the anthers dorsifixed; staminodes 7, connate with the stamens. Fruit an indehiscent legume.

A monotypic genus; North America, West Indies, Mexico, Central America, South America, Africa, Asia, Australia, and Pacific Islands. [From the Arabic *tamar*, date, *indis*, India.]

Tamarindus indica L. [Of India.] TAMARIND.

Tamarindus indica Linnaeus, Sp. Pl. 1: 34. 1753. *Tamarindus umbrosa* Salisbury, Prodr. Stirp. Chap. Allerton 323. 1796, nom. illegit. *Tamarindus officinalis* Hooker, Bot. Mag. 77: t. 4563. 1851, nom. illegit.

Tree, to 10(25) m; bark shaggy, brownish gray, the branchlets densely strigose or villous, glabrescent in age. Leaves 12- to 22-foliolate, 4–15 cm long, the leaflet blade oblong, (8)12–18(21) mm long, (3)4–7(9) mm wide, the apex truncate, emarginate, the base rounded, the upper and lower surfaces glabrous, the petiole (3)5–8(11) mm long, glabrous; stipules linear, ca. 4 mm long, caducous. Flowers in a lax raceme; bracts scalelike, ca. 2 mm long; bracteoles elliptic, ca. 2 mm long, caducous; calyx turbinate, 10–12 mm long, the lobes lanceolate, the lower one slightly longer than the others, the upper 2 lobes connate nearly to the apex; corolla caesalpiniaceous, creamy white to yellowish, streaked with red, 3 petals 1–1.5 cm long, the 2 other ones scalelike. Fruit elliptic to oblong, (5)7.5–12(20) cm long, 2–2.5 cm wide, rough, glabrous, the mesocarp pulpy; seeds (1)3–6(14), 1–1.5 cm long, dark brown to black, smooth, lustrous.

Disturbed coastal hammocks. Rare; Manatee and Lee Counties, southern peninsula. Escaped from cultivation. Florida; West Indies, Mexico, Central America, and South America; Africa, Asia, Australia, and Pacific Islands. Native to Africa. Spring–summer.

Tephrosia Pers., nom. cons. 1807. HOARYPEA

Perennial herbs. Leaves odd-pinnately compound, estipellate, petiolate, stipulate. Flowers in terminal and/or axillary pseudoracemes, bracteate, ebracteolate; sepals 5, connate, the upper 2 lobes partly fused, the 3 lower ones free; petals 5, the corolla papilionaceous; stamens diadelphous (9 + 1) or monadelphous. Fruit a dehiscent legume.

A genus of about 350 species; North America, West Indies, Central America, South America, Africa, Asia, Australia, and Pacific Islands. [From the Greek *tephros*, ashen-gray, in reference to the gray pubescence of some species.]

Cracca L., nom. rej. 1753.

Selected references: DeLaney (2010); Wood (1949).

1. Style glabrous; corolla less than 12 mm long ....**T. angustissima**
1. Style barbellate; corolla more than 12 mm long.
   2. Corolla 2-colored, the standard yellow, the wings pink.....................................**T. virginiana**
   2. Corolla not 2-colored, usually whitish pink, turning red in age.

3. Petiole 2–4 times the length of the lowest leaflet; peduncle conspicuously flattened ....
.... **T. florida**

3. Petiole equaling or shorter than the length of the lowest leaflet; peduncle terete.

    4. Raceme foliose, with several reduced or undeveloped leaves that enlarge as the fruits mature.. **T. rugelii**

    4. Raceme not foliose (rarely with a single leaflet).

        5. Fruit finely pubescent (trichomes less than 0.5 mm long).

            6. Leaflets 5–7(9), dorsal surface glabrous, semi-lustrous, bright green.. **T. chrysophylla**

            6. Leaflets (7)9–11(13), dorsal surface microscopically tawny-hirsute, glabrate in age, dull (rarely glabrous or lustrous), olive-green or somewhat brownish olive-green ....
....**T. mysteriosa**

        5. Fruit short villous (trichomes 1–1.5 mm long).

            7. Upper stem and leaf rachis with appressed trichomes; calyx 3–4 mm long ....
.... **T. hispidula**

            7. Upper stem and leaf rachis with spreading trichomes; calyx 6–7 mm long ....
....**T. spicata**

Tephrosia angustissima Shuttlew. ex Chapm. [Most narrow, in reference to the leaflets.] NARROWLEAF HOARYPEA.

Prostrate or ascending perennial herb; stem to 5(8) dm, glabrate or variously pubescent. Leaves 3–6 cm long, with the leaflets 11–15(17), the blade oblong-lanceolate to linear, 1.5–4 cm long, 2–5 mm wide, the apex obtuse to rounded or truncate or retuse, mucronate, the margin entire, the surfaces glabrate or strigulose to villosulous, the petiole 1–3 cm long, the base cuneate; stipules subulate to setiform. Flowers in an axillary pseudoraceme 3–5(8) cm long, the axis terete-tetragonal; calyx 4–4.5 mm long; corolla papilionaceous, 7–11 mm long, rose-purple to pinkish or white; stamens monadelphous; style glabrous. Fruit oblong, 3–4 cm long, 3–4 mm wide, laterally compressed, strigulose or finely canescent; seeds 3–8.

1. Plant finely villous or canescent..var. **corallicola**

1. Plant minutely strigose.

    2. Leaflets 2–8 times longer than wide, the broader leaves with reticulate venation between the ascending lateral nerves .... var. **curtissii**

    2. Leaflets 10–20 times longer than wide, lacking reticulate venation .... var. **angustissima**

Tephrosia angustissima var. angustissima NARROWLEAF HOARYPEA.

Tephrosia angustissima Shuttleworth ex Chapman, Fl. South U.S. 96. 1860. *Cracca angustissima* (Shuttleworth ex Chapman) Kuntze, Revis. Gen. Pl. 1: 174. 1891. *Tephrosia purpurea* (Linnaeus) Persoon var. *angustissima* (Shuttleworth ex Chapman) B. L. Robinson, Bot. Gaz. 28: 201. 1899. TYPE: FLORIDA: Miami-Dade Co.

Plant minutely strigose. Leaflets 10–20 times longer than wide, lacking reticulate venation. Pine rocklands. Rare; Miami-Dade County. Endemic. Last collected in 1912. Spring–fall.

Tephrosia angustissima var. *angustissima* is listed as endangered in Florida (Florida Administrative Code, Chapter 5B-40).

Tephrosia angustissima var. **corallicola** (Small) Isely [Coral dweller.] CORAL HOARYPEA.

Cracca corallicola Small, Bull. Torrey Bot. Club. 36: 160. 1909. *Tephrosia corallicola* (Small) León, Contr. Ocas. Mus. Hist. Nat. "De La Salle" 10: 304. 1951. *Tephrosia angustissima* Shuttleworth ex Chapman var. *corallicola* (Small) Isely, Brittonia 34: 341. 1982. TYPE: FLORIDA: Miami-Dade Co.: between Coconut Grove and Cutler, Nov 1904, *Small 2112* (holotype: NY; isotypes: FLAS, NY).

Plant finely villous or canescent. Leaflets 4–7 times longer than wide.

Pine rocklands. Rare; Collier and Miami-Dade Counties. Florida; West Indies. Spring–fall.

Tephrosia angustissima var. *corallicola* is listed as endangered in Florida (Florida Administrative Code, Chapter 5B-40).

Tephrosia angustissima var. **curtissii** (Small ex Rydb.) Isely [Commemorates Allen Hiram Curtiss (1845–1907), botanist for the U.S. Department of Agriculture who collected in Florida.] CURTISS' HOARYPEA.

Cracca curtissii Small ex Rydberg, in Britton, N. Amer. Fl. 24: 179. 1923. *Tephrosia curtissii* (Small ex Rydberg) Shinners, Sida 1: 60. 1962. *Tephrosia angustissima* Shuttleworth ex Chapman var. *curtissii* (Small ex Rydberg) Isely, Brittonia 34: 341. 1982. TYPE: FLORIDA: Brevard Co.: Cape Malabar, Aug–Sep, *Curtiss 584* (holotype: NY; isotype: NY).

Tephrosia seminole Shinners, Sida 1: 60. 1962. TYPE: FLORIDA: Hendry Co.(?): Godden's Mission, 12 Mar 1919, *Sheehan s.n.* (holotype: NY).

Plant minutely strigose. Leaflets 2–8 times longer than wide, the broader leaves with reticulate venation between the ascending lateral nerves.

Coastal strands, rarely inland. Occasional; central peninsula, Broward County. Endemic. Spring–fall.

Tephrosia angustissima var. *curtissii* is listed as endangered in Florida (Florida Administrative Code, Chapter 5B-40).

Tephrosia chrysophylla Pursh [Golden-leaved.] SCURF HOARYPEA.

Tephrosia chrysophylla Pursh, Fl. Amer. Sept. 489. 1814. *Tephrosia prostrata* Nuttall, Gen. N. Amer. Pl. 2: 120. 1818, nom. illegit. *Galega chrysophylla* (Pursh) Steudel, Nomencl. Bot. 350. 1821. *Cracca chrysophylla* (Pursh) Kuntze, Revis. Gen. Pl. 1: 174. 1891.

Cracca carpenteri Rydberg, in Britton, N. Amer. Fl. 24: 172. 1923. *Tephrosia carpenteri* (Rydberg) Killip, J. Wash. Acad. Sci. 26: 360. 1936. TYPE: FLORIDA: Escambia Co.: near Pensacola, Jun 1838, *Carpenter 44* (holotype: NY).

Cracca chrysophylla (Pursh) Kuntze var. *chapmannii* Vail, Bull. Torrey Bot. Club 22: 34. 1895. *Tephrosia chrysophylla* Pursh var. *chapmannii* (Vail) B. L. Robinson, Bot. Gaz. 28: 198. 1899. *Cracca chapmannii* (Vail) Small, Fl. S.E. U.S. 612, 1331. 1903. TYPE: FLORIDA: Gulf Co.: St. Josephs, s.d., *Chapman s.n.* (holotype: NY; isotype: NY).

Prostrate perennial herb; stem to 5 dm long, strigose or hirtellous. Leaves 1–2.5 cm long, the leaflets (3)5–7, the blade obovate, (0.5)1–2.5 cm long, 5–10 mm wide, the apex round to retuse, the base cuneate, the upper surface glabrate or rarely strigulose, the lower surface strigulose or villosulous, the petiole 1–5 mm long; stipules subulate or setiform. Flowers in an axillary pseudoraceme 2–10(15) cm long, the axis angled to somewhat flattened; calyx 3–5 mm long; corolla papilionaceous, (8)10–14 mm long, white, turning pink, then reddish; stamens diadelphous (9 + 1); style barbellate. Fruit oblong, (2)3–4(5) cm long, 4–5 mm wide, laterally compressed, sparsely strigulose.

Sandhills. Frequent; nearly throughout. Georgia, Florida, Alabama, and Mississippi. Spring–fall.

Tephrosia florida (F. Dietr.) C. E. Wood [Of Florida.] FLORIDA HOARYPEA.

Galega villosa Michaux, Fl. Bor.-Amer. 2: 67. 1803; non Linnaeus, 1759. *Tephrosia villosa* Persoon, Syn. Pl. 2: 329. 1807; non (Linnaeus) Persoon, 1807. *Galega florida* F. Dietrich, Nachtr. Vollst. Lex. Gaertn. 3: 422. 1817. *Tephrosia florida* (F. Dietrich) C. E. Wood, Rhodora 51: 305. 1949. TYPE: "Carolina ad Floridam."

Galega ambigua M. A. Curtis, Boston J. Nat. Hist. 1: 121. 1835. *Tephrosia ambigua* (M. A. Curtis) Chapman, Fl. South. U.S. 96. 1860. *Cracca ambigua* (M. A. Curtis) Kuntze, Revis. Gen. Pl. 1: 174. 1891.

Tephrosia gracilis A. W. Wood, Amer. Bot. Fl. 95. 1870; non Nuttall, 1818. TYPE: "Fla. to La."

Tephrosia ambigua (M. A. Curtis) Chapman var. *gracillima* B. L. Robinson, Bot. Gaz. 28: 201. 1899. *Cracca gracillima* (B. L. Robinson) A. Heller, Cat. N. Amer. Pl., ed. 2. 7. 1900. *Tephrosia gracillima* (B. L. Robinson) Killip, J. Wash. Acad. Sci. 26: 360. 1936. *Tephrosia florida* (F. Dietrich) C. E. Wood var. *gracillima* (B. L. Robinson) Shinners, Sida 1: 61. 1962. TYPE: FLORIDA: Brevard Co.: near Eau Gallie, Indian River.

Prostrate or sprawling-ascending, perennial herb; stem to 8 dm, glabrate or strigulose, rarely villosulous. Leaves 3–15 cm long, the leaflets (5)7–13(15), the blade oblanceolate or obovate-elliptic, (1)1.5–3.5(4.5) cm long, 4–8 mm wide, the apex rounded to truncate, mucronate, the base cuneate, the upper surface glabrate, the lower surface strigulose or puberulent, the petiole (1)2–8 cm long, exceeding the length of the lowermost leaflets; stipules subulate, 5–7 mm long. Flowers in an axillary pseudoraceme 5–15(25) cm long, the axis conspicuously flattened; calyx 3–5 mm long; corolla papilionaceous, 10–14 mm long, yellowish white, becoming streaked, then red-maroon; stamens diadelphous (9 + 1); style barbellate. Fruit oblong, 2.5–4 cm long, ca. 4 mm wide, laterally compressed, sparsely strigulose.

Sandhills and dry, open hammocks. Frequent; nearly throughout. North Carolina south to Florida, west to Louisiana. Spring–fall.

Tephrosia hispidula (Michx.) Pers. [Bristly.] SPRAWLING HOARYPEA.

Galega hispidula Michaux, Fl. Bor.-Amer. 2: 68. 1803. *Tephrosia hispidula* (Michaux) Persoon, Syn. Pl. 2: 329. 1807. *Cracca hispidula* (Michaux) Kuntze, Revis. Gen. Pl. 1: 175. 1891.

Decumbent to erect, perennial herb; stem strigulose or finely pilose (ultimately glabrate). Leaves with the leaflets (9)11–17(19), the blade oblong-lanceolate to elliptic, 1–2(2.5) cm long, 3–7 mm wide, the apex obtuse to truncate or retuse, mucronate, the base cuneate, the upper surface glabrate, the lower surface pubescent, the petiole 2–8(15); stipules subulate or setaceous. Flowers in an axillary or terminal pseudoraceme 1.5–15 cm long; calyx 3–4 mm long; corolla papilionaceous, 10–15 mm long, yellowish white, becoming reddish; stamens diadelphous (9 + 1); style barbellate. Fruit oblong, 3–5 cm long, 5–6 mm wide, laterally compressed, pilose or rarely substrigulose.

Sandhills and flatwoods. Frequent; northern counties, central peninsula. Virginia south to Florida, west to Louisiana. Spring–fall.

Tephrosia mysteriosa DeLaney [Most mysterious.] SANDHILL TIPPITOES.

Tephrosia mysteriosa DeLaney, Bot. Explor. 4: 101, f. 1, 2(A), 3, 6. 2010. TYPE: FLORIDA: Highlands
Co.: W of Avon Park at the Carter Creek Preserve, 3 Aug 2006, *DeLaney 5353* (holotype: USF;
isotypes: FLAS, FSU, FTG, MO, NY, US, USF).

Prostrate, perennial herb; stem to 6 dm long, tawny hirsute. Leaves (2.5)4–5(6) cm long, the
leaflets 7–11(13), the blade elliptic or slightly oblong or obovate, (0.8)1.4–1.8(2.5) cm long,
(0.6)0.8–1.2(1.6) cm wide, the apex retuse or emarginate, mucronate, the base broadly cuneate,
the upper surface microscopically tawny-hirtellous or sometimes glabrate, glabrate or glabrous
in age, the lower surface strigose-sericous, the petiole 1–2(3) mm long; stipules subulate, ca.
2 mm long. Flowers in an axillary pseudoraceme (5)10–16(20) cm long; calyx 6–7 mm long;
corolla papilionaceous, 12–16 mm long, white, turning pink streaked with maroon, finally ma-
roon-red; stamens diadelphous (9 + 1); style dorsally barbellate. Fruit oblong, 4–4.5 cm long,
3–5 mm wide, laterally compressed, minutely strigose; seeds 8–10

Sandhills. Rare; Marion, Lake, Polk, and Highlands Counties. Endemic. Summer–fall.

Tephrosia rugelii Schuttlew. ex B. L. Rob. [Commemorates Ferdinand Xavier Rugel
(1806–1878), German-born American botanist.] RUGEL'S HOARYPEA.

Tephrosia rugelii Shuttleworth ex B. L. Robinson, Bot. Gaz. 28: 197. 1899. *Cracca rugelii* (Shuttleworth
ex B. L. Robinson) A. Heller, Cat. N. Amer. Pl., ed. 2. 7. 1900. TYPE: FLORIDA: Manatee Co:
Manatee River, Jun 1845, *Rugel 156* (holotype: GH; isotype: NY).

Decumbent to ascending, perennial herb; stem to 5 dm, strigulose or hirsutulous. Leaves 2–7
cm long, the leaflets (5)9–15, the blade elliptic to obovate, 1–2 cm long, 5–8 mm wide, the apex
rounded or retuse, mucronate, the base cuneate, the upper surface sericeous-velutinous or
rarely glabrate, the lower surface sericeous-velutinous, the petiole 0.3–1(2) cm long; stipules
subulate, 5–6 mm long. Flowers in a terminal, foliose pseudoraceme 3–6(10) cm long, with
2–5 congested flowering nodes, each subtended by an undeveloped leaf that enlarges as the
pseudoraceme matures and elongates, the axis subterete or angular; calyx 5–7 mm long; corolla
papilionaceous, 14–18 mm long, greenish or yellowish white with red veins, turning crimson;
stamens diadelphous (9 + 1); style barbellate. Fruit oblong, 2.5–4 cm long, 4–5 mm wide, laterally compressed, subappressed-villosulous.

Sandhills. Frequent; peninsula, Jefferson County. Endemic. Spring–fall.

Tephrosia spicata (Walter) Torr. & A. Gray [In spikes, in reference to the
inflorescence.] SPIKED HOARYPEA.

Galega spicata Walter, Fl. Carol. 188. 1788. *Tephrosia spicata* (Walter) Torrey & A. Gray, Fl. N. Amer.
1: 296. 1838. *Cracca spicata* (Walter) Kuntze, Revis. Gen. Pl. 1: 175. 1891.

Tephrosia paucifolia Nuttall, Gen. N. Amer. Pl. 2: 119. 1818. *Galega paucifolia* (Nuttall) M. A. Curtis,
Boston J. Nat. Hist. 1: 122. 1835. TYPE: FLORIDA/GEORGIA: without data, *Baldwin s.n.* (lectotype: GH). Lectotypified by Wood (1949: 292).

Cracca spicata (Walter) Kuntze var. *flexuosa* Vail, Bull. Torrey Bot. Club 22: 30. 1895. *Tephrosia vil-
losa* Persoon var. *flexuosa* (Vail) B. L. Robinson, Bot. Gaz. 28: 200. 1899. *Cracca flexuosa* (Vail) A.
Heller, Cat. N. Amer. Pl., ed. 2. 7. 1900. TYPE: FLORIDA: without data, *Chapman s.n.* (holotype:
NY).

Decumbent or ascending perennial herb; stem to 1 m, strigose or pilose. Leaves (2)3–10 cm long, the leaflets (7)9–13(17), the blade elliptic-obovate to oblong, 1.5–2(4) cm long, 6–12 mm wide, the apex rounded or subretuse, mucronate, the base cuneate, the upper surface subappressed-pilose or glabrate, the lower surface subappressed-pilose, the petiole 0.3–1.5 cm long; stipules setose, 4–5 mm long. Flowers in an axillary or terminal pseudoraceme 0.5–2(4) cm long, the axis terete to somewhat flattened; calyx 6–7 mm long; corolla papilionaceous, 13–18 mm long, yellowish white with red stripes, turning pink and maroon; stamens diadelphous (9 + 1); style barbellate. Fruit oblong, 2–3.5 cm long, 3.5–4.5 mm wide, laterally compressed, villous, usually becoming glabrate.

Sandhills and dry hammocks. Frequent; nearly throughout. Maryland south to Florida, west to Kentucky, Tennessee, Mississippi, and Louisiana. Spring–fall.

Tephrosia virginiana (L.) Pers. [Of Virginia.] GOAT'S RUE.

> *Cracca virginiana* Linnaeus, Sp. Pl. 2: 752. 1753. *Galega virginiana* (Linnaeus) Linnaeus, Syst. Nat., ed. 10. 1172. 1759. *Tephrosia virginiana* (Linnaeus) Persoon, Syn. Pl. 2: 329. 1807.
>
> *Cracca latidens* Small, Fl. S.E. U.S. 609, 1331. 1903. *Tephrosia latidens* (Small) Standley, Publ. Field Mus. Nat. Hist., Bot. Ser. 11: 161. 1936. TYPE: FLORIDA: Lake Co.: vicinty of Eustis, 16–30 Jun 1894, *Nash 1072* (holotype: NY; isotypes: GH, MO, NY, PH, UC, US).
>
> *Cracca mohrii* Rydberg, in Britton, N. Amer. Fl. 24: 163. 1923. *Tephrosia mohrii* (Rydberg) R. K. Godfrey, Brittonia 10: 169. 1958. *Tephrosia virginiana* (Linnaeus) Persoon var. *mohrii* (Rydberg) D. B. Ward, Novon 14: 369. 2004. TYPE: FLORIDA: Walton Co.: near Eucheeanna, Jun 1880, *Mohr s.n.* (holotype: US).

Erect perennial herb; stem to 6 dm, strigulose or villous. Leaves 4–12 cm long, the leaflets (9)13–23(37), the blade elliptic to oblong, 1–2.5(3) cm long, 3–6 mm long, the apex obtuse to acute, mucronate, the base cuneate, the upper surface spreading-pubescent or glabrate, the lower surface spreading pubescent, the petiole 1–5(10) mm long; stipules subulate, 3–4 mm long. The flowers in a terminal or axillary raceme 3–6 cm long, the axis tetragonal or terete; calyx (5)7–8 mm long; corolla papilionaceous, 15–20 mm long, bicolored, the standard yellow, the wings pink, the keel yellow- and pink-striped; stamens monadelphous; style barbellate. Fruit oblong, 3–5 cm long, ca. 4 mm wide, laterally compressed, villous or strigulose to glabrate.

Dry hammocks. Occasional; northern counties south to Hernando, Lake, and Orange Counties. New Hampshire south to Florida, west to Ontario, Minnesota, Nebraska, Kansas, Oklahoma, and Texas. Spring–fall.

HYBRIDS

Tephrosia ×intermedia (Small) G. L. Neson & Zarucchi (*T. chrysophylla × T. florida*) [Intermediate.]

> *Cracca intermedia* Small, Bull. Torrey Bot. Club 21: 303. 1894, pro sp. *Cracca smallii* Vail, Bull. Torrey Bot. Club 22: 33. 1895, nom illegit. *Tephrosia ambigua* (M. A. Curtis) Chapman var. *intermedia* (Small) F. J. Hermann, J. Wash. Acad. Sci. 38: 237. 1948. *Tephrosia ×intermedia* (Small) G. L. Nesom

& Zarucchi, J. Bot. Res. Inst. Texas 3: 157. 2009. TYPE: FLORIDA: Duval Co.: Jacksonville, 31 May & 11 Jul 1893, *Curtiss 4231* (holotype: NY; isotype: NY).

Cracca floridana Vail, Bull. Torrey Bot. Club 22: 35. 1895, pro sp. *Tephrosia floridana* (Vail) Isely, Brittonia 34: 340. 1982, pro hybr. TYPE: FLORIDA: Lake Co.: vicinity of Eustis, Jul 1894, *Nash 1198* (lectotype: NY). Lectotypified by Isley (1982: 340).

Sandhills. Occasional; northern and central peninsula, central and western panhandle. Spring–fall.

Tephrosia ×varioforma DeLaney (*T. florida* × *T. mysteriosa*) [Variable forms.]

Tephrosia ×varioforma DeLaney, Bot. Explor. 4(Suppl. A): 1. 2011. TYPE: FLORIDA: Highlands Co.: SE corner of Wester Avenue and Rich Street, ca. 0.5 mi. E of Lake Verona, 12 Sep 2009, *DeLaney 5735* (holotype: USF; isotypes: FLAS, FSU, FTG, MO, NY, US, USF).

Sandhills. Rare; Highlands County. Spring–fall.

EXCLUDED TAXA

Tephrosia leptostachya de Candolle—Reported for Florida by Chapman (1897), based on a Curtiss collection from Brevard Co., which has now been identified as *T. angustissima* var. *curtissii*.

Tephrosia purpurea (Linnaeus) Persoon—Reported for Florida by Small (1903, 1913a, 1933), the name misapplied to material of *T. angustissima* var. *angustissima*.

Trifolium L. 1753. CLOVER

Annual or perennial herbs. Leaves 3-foliolate, estipellate, petioleate, stipulate. Flowers in axillary or terminal subumbellate, headlike, or spikelike racemes, bracteate; sepals 5, basally connate and tubular, lobed; corolla papilionaceous; stamens diadelphous (9 + 1). Fruit a legume, indehiscent or circumscissile dehiscent.

A genus of about 250 species; nearly cosmopolitan. [*Tri*, three, and *folium*, leaf, in reference to the tri-foliolate leaves.]

Selected reference: Hermann (1953); Zohary and Heller (1984).

1. Corolla bright yellow.
    2. Standard conspicuously striate, flat or nearly so; inflorescence 8–13 mm wide, usually with 20–40 flowers ....**T. campestre**
    2. Standard not striate or inconspicuously so, longitudinally folded; inflorescence 5–8 mm wide, usually with 3–15 flowers ....**T. dubium**
1. Corolla cream-colored, white, pink, or red.
    3. Flowers distinctly pedicellate.
        4. Plant stoloniferous; peduncles arising from the stolon .... **T. repens**
        4. Plant not stoloniferous; peduncles arising from the stem above the ground level.
            5. Calyx glabrous.
                6. Calyx lobes scarious-margined..**T. nigrescens**
                6. Calyx lobes not scarious-margined..**T. hybridum**
            5. Calyx pilose (sometimes sparsely so).
                7. Inflorescence 2–4 cm wide; corolla 9–17 mm long; calyx lobes more than 2 times the length of the tube ....**T. reflexum**

7. Inflorescence 1–1.5 cm wide; corolla 5–7 mm long; calyx lobes shorter than the tube ....
....**T. carolinianum**
3. Flowers sessile or subsessile (pedicels less than 0.5 mm long).
   8. Calyx inflated in fruit.
      9. Calyx equally enlarged in fruit, terminated by 5-setaceous teeth; flowers not resupinate.......
      ....**T. spumosum**
      9. Calyx unequally enlarged in fruit, the upper lip terminated with 2 setaceous teeth; flowers resupinate.
         10. Fruiting heads pedunculate; calyx pubescent to tomentose, finally glabrate, its 2 upper teeth prolonged, evident..**T. resupinatum**
         10. Fruiting heads subsessile; calyx lanate, its 2 upper teeth short, usually concealed ....
         ....**T. tomentosum**
   8. Calyx not inflated in fruit.
      11. Corolla 3–8 mm long, nearly equaling or shorter than the calyx.
         12. Calyx tube glabrous, 20-nerved; calyx lobes several-nerved at the base .....**T. lappaceum**
         12. Calyx tube pubescent, 10-nerved; calyx lobes 1-nerved ....**T. arvense**
      11. Corolla 9–17 mm long, or if shorter, then distinctly longer than the calyx.
         13. Inflorescence involucrate.
            14. Calyx densely subappressed pubescent, the tube 20-nerved ....**T. hirtum**
            14. Calyx sparsely pilose, the tube 10-nerved...**T. pratense**
         13. Inflorescence not involucrate.
            15. Corolla white or cream-colored; inflorescence 2.5–3 cm wide, conspicuously bracteate; calyx inflated in fruit, 20-nerved, cross-reticulate ....**T. vesiculosum**
            15. Corolla red (rarely white); inflorescence 1–1.5(2) cm wide, lacking bracts; calyx not inflated in fruit, conspicuously pilose, 10-nerved, not cross-reticulate.........................
            ....**T. incarnatum**

Trifolium arvense L. [Of fields or cultivated land.] RABBITFOOT CLOVER.

Trifolium arvense Linnaeus, Sp. Pl. 2: 769. 1753. *Trifolium arvense* Linnaeus var. *typicum* Beck, Fl. Nieder-Oesterreich 848. 1892, nom. inadmiss.

Erect annual herb, to 2.5(3.5) dm; stem villous or glabrate. Leaves 3-foliolate, the leaflet blade obovate or oblanceolate to narrowly oblong, 5–20 mm long, 2–4 mm wide, the apex obtuse to rounded or slightly retuse, the base cuneate, the margin slightly denticulate or entire, the upper and lower surfaces appressed-villous to glabrate, the petiole 0.3–2.5 cm long; stipules lanceolate, setose-tipped. Flowers numerous in an obovoid to cylindric, short-pedunculate spike 1.5–3 cm long, 8–12 mm wide; calyx tube campanulate, 2–3 mm long, 10-nerved, villous, the lobes subequal, longer than the tube, with a single nerve, setacous, plumose; corolla papilionaceous, 3–4 mm long, pale pink to white, shorter than the calyx. Fruit ovoid, ca. 2 mm long, membranous, indehiscent.

Disturbed sites. Occasional; northern counties, central peninsula. Escaped from cultivation. Nearly throughout North America; Europe, Africa, Asia, Australia, and Pacific Islands. Native to Europe, Africa, and Asia. Spring.

Trifolium campestre Schreb. [Of plains or flat areas.] FIELD CLOVER; HOP CLOVER.

Trifolium campestre Schreber, in Sturm, Deutschl. Fl. 4(16): t. 253. 1804. *Chrysaspis campestris* (Schreber) Desvaux, Observ. Pl. Angers 164. 1818. *Trifolium procumbens* Linnaeus var. *campestre* (Schreber) Seringe, in de Candolle, Prodr. 2: 205. 1825. *Amarenus agrarius* (Linnaeus) C. Presl var. *campestris* (Schreber) C. Presl, Symb. Bot. 1: 46. 1831. *Trifolium procumbens* Linnaeus var. *maius* W.D.J. Koch, Syn. Fl. Germ. Helv. 175. 1837, nom. illegit. *Trifolium agraricum* Linnaeus var. *campestre* (Schreber) Beck, Fl. Nieder-Oesterreich 845. 1892. *Trifolium campestre* Schreber var. *schreberi* Rouy, Fl. France 5: 73. 1899, nom. inadmiss. *Trifolium campestre* Schreber var. *maius* Hayek, Fl. Steiermark 1: 1049. 1910, nom. inadmiss.

Erect or procumbent, annual herb; stem to 3 dm, subappressed-pilose. Leaves 3-foliolate, the leaflet blade obovate, 8–10(15) mm long, 3–8 mm wide, the apex rounded to retuse, the base cuneate, the margin distally denticulate or sinuate, the upper and lower surface glabrate, the petiole 0.3–2 cm long; stipules ovate, 4–5 mm long. Flowers numerous, ascending, then reflexed, in an ovoid to subspherical, pedunculate head 1–1.5 cm long, (7)8–13 mm wide, the pedicel ca. 1 mm long; calyx tube ca. 1 mm long, 5-nerved, glabrous, the lobes linear-subulate, unequal, the lateral ones longer than the tube; corolla papilionaceous, 4–5(6) mm long, bright yellow, the petals conspicuously striate, the standard somewhat inflated, slightly erose. Fruit ellipsoid, 2–3 mm long, indehiscent; seed 1.

Disturbed sites. Frequent; northern counties south to Manatee County. Escaped from cultivation. Nearly throughout North America; Europe, Africa, and Asia. Native to Europe, Africa, and Asia. Spring.

Trifolium carolinianum Michx. [Of Carolina.] CAROLINA CLOVER.

Trifolium carolinianum Michaux, Fl. Bor.-Amer. 2: 58. 1803. *Amoria caroliniana* (Michaux) C. Presl, Symb. Bot. 1: 47. 1831.

Spreading or procumbent annual herb; stem to 3 dm, slightly villosulous or glabrate. Leaves 3-foliolate, the leaflet blade obovate, 8–15 mm long, 5–12 mm wide, the apex rounded to retuse, the base cuneate, the margin undulate-dentate, the upper and lower surfaces villosulous or glabrate, the petiole 1–4 cm long; stipules ovate, 4–6 mm long, entire or irregularly toothed, bristle-tipped. Flowers in an umbellate, spheroid, pedunculate head 1–1.5 cm long and wide, the pedicel 2–6 mm long; bracts minute; calyx tube 1–2 mm long, 5- to 10-nerved, villosulous, the lobes lanceolate, unequal, shorter than the tube, subfoliaceous, 1- to 3-nerved; corolla papilionaceous, 4–6 mm long, yellowish white or lavender, the standard and wings often denticulate or erose. Fruit obovoid or oblong, 2–3 mm long, pubescent at the apex, indehiscent.

Open hammocks and disturbed sites. Occasional; northern counties, Manatee County. Vermont south to Florida, west to Kansas, Oklahoma, and Texas. Spring.

Trifolium dubium Sibth. [Doubtful.] LOW HOP CLOVER; SUCKLING CLOVER.

Trifolium dubium Sibthorp, Fl. Oxon. 231. 1794. *Trifolium minus* Smith, in Relhan, Fl. Cantab., ed. 2. 290. 1802, nom. illegit. *Chrysaspis dubia* (Sibthorp) Desvaux, Observ. Pl. Angers 165. 1818.

Decumbent to erect annual herb; stem 1.5(2) dm, sparsely villosulous. Leaves 3-foliolate, the leaflet blade 5–15 mm long, 3–10 mm wide, the apex rounded to retuse, the base cuneate, the margin apically denticulate, the upper and lower surfaces glabrate, the petiole 2–10 mm long, usually shorter than the leaflets; stipules ovate. Flowers 5–15(20) in a subcapitate, pedunculate head ca. 1 cm long, 4–8 mm wide, the pedicel ca. 1 mm long; bracts minute; calyx tube ca. 1 mm long, 5-nerved, glabrous, the lobes unequal, the longer linear-subulate, equaling or exceeding the tube; corolla papilionaceous, 3–4 mm long, yellow, weakly striate. Fruit obovoid, ca. 2 mm long, indehiscent; seed 1.

Disturbed sites. Occasional; northern counties. Escaped from cultivation. Nearly throughout North America; Europe, Africa, Asia, Australia, and Pacific Islands. Native to Europe, Africa, and Asia. Spring.

Trifolium hirtum All. [Hairy.] ROSE CLOVER.

Trifolium hirtum Allioni, Auct. Fl. Pedem. 20. 1789.

Erect or ascending annual herb, to 3 dm; stem pilose. Leaves 3-foliolate, with the leaflet blade obovate, 1–2.5 cm long, 8–12 mm wide, the apex rounded, the base cuneate, the margin apically dentate or sinuate, the upper and lower surfaces pilose to glabrate, the petiole 0.8–2 cm long; stipules setaceous-tipped. Flowers sessile in a globose or obovoid, sessile or rarely short-pedunculate head, 1–2(2.5) cm long and wide; calyx tube 3–4 mm long, 20-nerved, subappressed pilose, the lobes setaceous, subequal or the lower longer than the tube, pilose; corolla papilionaceous, (1)1.2–1.5 cm long, pinkish or rarely red. Fruit ovoid, 2–3 mm long, indehiscent; seed 1.

Disturbed sites. Rare; Gadsden County. Escaped from cultivation. Virgina south to Florida, west to Kentucky, Tennessee, and Louisiana, also Oregon and California; Europe, Africa, and Asia. Native to Europe, Africa, and Asia. Spring.

Trifolium hybridum L. [Intermediate between two species.] ALSIKE CLOVER.

Trifolium hybridum Linnaeus, Sp. Pl. 2: 766. 1753. *Trifolium bicolor* Moench, Methodus 111. 1794, nom. illegit. *Amoria hybrida* (Linnaeus) C. Presl, Symb. Bot. 1: 47. 1831.

Sprawling or erect perennial herb; stem to 6 dm, glabrate. Leaves 3-foliolate, the leaflet blade elliptic to obovate, 1–3(4)cm long, 6–15 mm wide, the apex rounded, the base broadly cuneate, the margin serrulate or entire, the upper and lower surfaces glabrate, the petiole 4–8 cm long; stipules ovate-lanceolate, membranous. Flowers in a pedunculate, subumbellate, globose head 1.5–2.5(3) cm long and wide, the pedicel 2–4 mm long; bracts small; calyx tube 1–2 mm long, 10-nerved, glabrous except for a few trichomes at the orifice, lobes lanceolate or subulate, subequaling the tube or longer; corolla papilionaceous, 6–10(11) mm long, white turning pink or pale red. Fruit broadly oblong, 3–4 mm long, indehiscent; seeds 2–4.

Disturbed sites. Rare; northern counties, Miami-Dade County. Escaped from cultivation. Nearly throughout North America; Europe, Africa, Asia, Australia, and Pacific Islands. Native to Europe, Africa, and Asia. Spring–summer.

Trifolium incarnatum L. [Flesh-colored.] CRIMSON CLOVER.

Trifolium incarnatum Linnaeus, Sp. Pl. 2: 769. 1753. *Trifolium incarnatum* Linnaeus var. *sativum* Ducommun, Taschenb. 169. 1869, nom. inadmiss. *Trifolium stellatum* Linnaeus subsp. *incarnatum* (Linnaeus) Gibelli & Belli, Mem. Reale Accad. Sci. Torino, ser. 2. 39: 296. 1889.

Erect, annual herb, to 8 dm; stem pubescent. Leaves 3-foliolate, the leaflet blade obovate or ob-cordate, 1–2.5(3) cm long, 1–2.5(3) cm wide, the apex rounded to retuse, the base cuneate, the margin sinuate-denticulate distally, the upper and lower surfaces glabrate, the petiole 2–6 cm long, stipules ovate-oblong, 1–1.5 cm long, conspicuously nerved, usually dark-tipped. Flowers sessile or subsessile in a lanceolate to cylindric, pedunculate spike or raceme (1)2–6 cm long, 1–1.5(2) cm wide; bracts obsolete; calyx tube 3–5 mm long, 10-nerved, pilose, the lobes seta-ceous, equal, usually slightly longer than the tube; corolla papilionaceous, 1–1.3(1.5) cm long, crimson. Fruit ovoid, 3–4 mm long, indehiscent; seed 1.

Disturbed sites. Frequent; northern counties, Hillsborough County. Escaped from cultiva-tion. Nearly throughout North America; Europe, Africa, Asia, Australia and Pacific Islands. Native to Europe, Africa, and Asia. Spring.

Trifolium lappaceum L. [Burr-like, in reference to the fruit cluster.] BURDOCK CLOVER.

Trifolium lappaceum Linnaeus, Sp. Pl. 2: 768. 1753.

Decumbent to erect annual herb; stem to 2.5(4) dm, thinly pubescent. Leaves 3-foliolate, the leaflet blade obovate to oblanceolate, 5–20 mm long, 3–8 mm wide, the apex rounded, the base cuneate, the margin denticulate, the upper and lower surfaces thinly pubescent, the petiole 1–4 cm long; stipules lanceolate, usually bristle-tipped, usually purple veined, pilose. Flowers subsessile in an initially subsessile, subsequently pedunculate, burr-like globose head 1–1.3 cm long and wide; calyx 7–8 mm long, the tube 2–2.5 mm long, ca. 20-nerved, glabrous except at the orifice where villous, the lobes subequal, much longer than the tube, several-nerved at the base, bristlelike, plumose, the fruiting calyx enlarged, the lobes stellate-spreading; corolla papilionaceous, 7–8 mm long, subequaling the calyx, white, turning pink. Fruit ellipsoid, ca. 2 mm long, circumscissile dehiscent; seed 1.

Disturbed sites. Rare; Escambia and Manatee Counties. Escaped from cultivation. New Jer-sey and Pennsylvania, North Carolina south to Florida, west to Texas; Europe, Africa, Asia, and Australia. Native to Europe, Africa, and Asia. Spring–summer.

Trifolium nigrescens Viv. [Becoming black.] SMALL WHITE CLOVER; BALL CLOVER.

Trifolium nigrescens Viviani, Fl. Ital. Fragm. 12, t. 13. 1808.

Sprawling or ascending annual herb; stem to 4 dm, glabrate. Leaves 3-foliolate, the leaflet blade obovate or obcordate, 0.6–2(2.5) cm long, 5–1.5(2) cm wide, the apex rounded, the base cune-ate, the margin serrulate or subentire, the upper and lower surfaces glabrate, the petiole 2–4 cm long; stipules lanceolate, membranous, scarious bordered at the base. Flowers in a pedunculate, subumbellate, globose head 1–1.5(2) cm long, the pedicel 1–2 mm long; bracts evident; calyx tube 2–2.5 mm long, (5)10-nerved, glabrous, the lobes deltate to lanceolate, unequal, about as long as the tube, proximally scarious-margined, bristle-tipped; corolla papilionaceous, 6–8 mm long, white or pink. Fruit ovoid to oblong, 2–4 mm long, indehiscent; seeds 1–4.

Disturbed sites. Occasional; central and western panhandle, Hernando County. Escaped from cultivation. Tennessee south to Florida and Louisiana; Europe, Africa, Asia, and Australia. Native to Europe, Africa, and Asia. Spring–summer.

Trifolium pratense L. [Growing in meadows.] RED CLOVER.

> *Trifolium pratense* Linnaeus, Sp. Pl. 2: 768. 1753. *Lagopus pratensis* (Linnaeus) Bernhardi, Syst. Verz. 239. 1800. *Trifolium pratense* Linnaeus var. *genuinum* Rouy, Fl. France 5: 119. 1889, nom. inadmiss. *Trifolium pratense* Linnaeus subsp. *eupratense* Ascherson & Graebner, Syn. Mitteleur. Fl. 6(2): 548. 1908, nom. inadmiss.

Erect or ascending perennial herb, to 5 dm; stem pilose or glabrate. Leaves 3-foliolate, the leaflet blade ovate, elliptic, or obovate, (1)2–3.5(4) cm long, 1–2 cm wide, the apex obtuse to rounded or slightly retuse, the base cuneate, the margin subentire or sinuate-denticulate, the upper and lower surfaces pilose or glabrate, often with a central blotch, the petiole 1–5 cm long; stipules ovate to ovate-lanceolate, ca. 1 cm long, bristle-tipped, membranous. Flowers subsessile in a sessile or short pedunculate, globose to ovoid head, 2–3 cm long, subtended by an involucre of 2 leaves; calyx tube 2.5–3.5 mm long, 10-nerved, sparsely pubescent, the lobes subulate-acicular, subequal except for the lower one that exceeds the tube, thinly pilose; corolla papilionaceous, 1.1–1.5 cm long, red-purple. Fruit obconic or ovoid, 2–3 mm long, usually circumscissile dehiscent; seeds 1(2).

Disturbed sites. Occasional; peninsula, central panhandle. Escaped from cultivation. Nearly throughout North America; West Indies; Europe, Africa, Asia, and Pacific Islands. Native to Europe, Africa, and Asia. Spring–fall.

Trifolium reflexum L. [Bent abruptly backward, in reference to the fruit.] BUFFALO CLOVER.

> *Trifolium reflexum* Linnaeus, Sp. Pl. 2: 766. 1753. *Amoria reflexa* (Linnaeus) C. Presl, Symb. Bot. 1: 47. 1831.

Spreading to ascending annual or short-lived perennial herb; stem to 5 dm, villous. Leaves 3-foliolate, the leaflet blade obovate to oblong-oblanceolate, 1.5–2.5 cm long, 1–1.5 cm wide, the apex rounded to retuse, the base cuneate, the margin denticulate, the upper and lower surfaces glabrate, the petiole 3–6 cm long; stipules ovate, 1–1.5 cm long, the apex acuminate or attenuate, foliaceous, palmately nerved. Flowers in a pedunculate, umbellate, spheroid head 2–3.5(4) cm long, the pedicel 4–8 mm long; calyx tube 1–1.5 mm long, 5-nerved, usually pubescent, the lobes subequal, lanceolate to subulate, 2 or more times longer than the tube; corolla papilionaceous, 8–12 mm long, white or bicolored, becoming pink in age, the standard slightly erose. Fruit ovate-curved or broadly oblong, 3–4 mm long, indehiscent; seeds (1)2–4.

Open hammocks and disturbed sites. Occasional; central panhandle, Alachua and Polk Counties. Ontario south to Florida, west to Nebraska, Kansas, Oklahoma, and Texas. Spring–summer.

Trifolium repens L. [Creeping, prostrate.] WHITE CLOVER; DUTCH CLOVER.

Trifolium repens Linnaeus, Sp. Pl. 2: 767. 1753. *Amoria repens* (Linnaeus) C. Presl, Symb. Bot. 1: 47. 1831. *Trifolium repens* Linnaeus subsp. *typicum* Ascherson & Graebner, Syn. Mitteleur. Fl. 6(2): 498. 1908, nom. inadmiss. *Trifolium repens* Linnaeus var. *genuinum* Ascherson & Graebner, Syn. Mitteleur. Fl. 6(2): 498. 1908, nom. inadmiss.

Ascending or erect, caespitose or stoloniferous, perennial herb, to 2(3) dm; stem glabrate. Leaves 3-foliolate, the leaflet blade obovate, 5–25 mm long, 5–25 mm wide, the apex retuse, the base cuneate, the margin denticulate, the upper and lower surfaces glabrate, the petiole 5–10 cm long; stipules lanceolate, white-membranous. Flowers in a long pedunculate, subumbellate, globose head 1–2(3) cm long, the pedicel 2–3 mm long; bracts small but evident; calyx tube 2–3 mm long, 5- to 10-nerved, glabrous, the lobes lanceolate, glabrous, equal to the tube or shorter; corolla papilionaceous, 7–10(12) mm long, white, turning pink. Fruit oblong, 3–5 mm long, indehiscent; seeds 3–5.

Disturbed sites. Frequent; nearly throughout. Escaped from cultivation. Nearly throughout North America; West Indies, Mexico, Central America, and South America; Europe, Africa, Asia, Australia, and Pacific Islands. Native to Europe, Africa, and Asia. All year.

Trifolium resupinatum L. [Reversed, turned upside down, in reference to the flower.] PERSIAN CLOVER; REVERSED CLOVER.

Trifolium resupinatum Linnaeus, Sp. Pl. 2: 771. 1753. *Trifolium folliculatum* Lamarck, Fl. Franc. 2: 599. 1779 ("1778"). *Galearia resupinata* (Linnaeus) C. Presl, Symb. Bot. 1: 50. 1831. *Trifolium resupinatum* Linnaeus var. *genuinum* Rouy, Fl. France 5: 92. 1899, nom. inadmiss.

Erect or ascending to spreading annual herb; stem 4(6) dm, glabrate. Leaves 3-foliolate, the leaflet blade obovate, 7–14(20) mm long, 4–8 mm wide, the apex rounded, the base cuneate, the margin denticulate, the upper and lower surfaces glabrate, the petiole 0.5–2 cm long; stipules lanceolate, 4–5 mm long, the apex attenuate. Flowers resupinate, in a capitate head ca. 1 cm long, the peduncle 2–3 cm long, the pedicel 1 mm long; calyx tube 1–2 mm long, striate-nerved, villous on the upper side, the lobes unequal, shorter than the tube; corolla papilionaceous, 4–6 mm long, pinkish lavender. Fruit ovoid or lenticular, 2 mm long, indehiscent, surrounded by an enlarged, obovoid, inflated calyx, the ventral lobes elongated as divergent bristles; seed 1.

Disturbed sites. Rare; Escambia and Duval Counties, central peninsula. Escaped from cultivation. Nearly throughout North America; Europe, Africa, Asia, and Pacific Islands. Native to Europe, Africa, and Asia. Spring–summer.

Trifolium spumosum L. [Frothy, in reference to the appearance of the inflorescence.] MEDITERRANEAN CLOVER.

Trifolium spumosum Linnaeus, Sp. Pl. 2: 771. 1753. *Mistyllus spumosus* (Linnaeus) C. Presl, Symb. Bot. 1: 49. 1831.

Spreading or procumbent, annual herb; stem 30(50) cm, glabrate. Leaves 3-foliolate, the leaflet blade obovate, 1–2(3) cm long, 8–15 mm wide, the apex shallowly retuse, the base cuneate, the margin denticulate, the upper and lower surfaces glabrate, the petiole 1–3 cm long; stipules lanceolate, 5–6 mm long, membranous. Flowers subsessile, in a globose to ovate head, the peduncle 1–4(10) cm long; bracts conspicuous, shorter than the calyx tube; calyx tube pyriform, 4–6 mm long, much inflated, with transverse and longitudinal striations, the lobes setaceous; corolla papilionaceous, 5–7, slightly exceeding the calyx, pink. Fruit oblong, indehiscent; seeds 3–4.

Disturbed sites. Rare; Santa Rosa County. Escaped from cultivation. Florida; Europe, Africa, and Asia. Native to Europe, Africa, and Asia. Spring–summer.

Trifolium tomentosum L. [Woolly.] WOOLLY CLOVER.

> *Trifolium tomentosum* Linnaeus, Sp. Pl. 2: 771. 1753. *Galearia tomentosa* (Linnaeus) C. Presl, Symb. Bot. 1: 50. 1831. *Trifolium resupinatum* Linnaeus subsp. *tomentosum* (Linnaeus) Gibelli & Belli, Mem. Reale Accad. Sci. Torino, ser. 2. 41: 17. 1890.

Procumbent annual herb; stem 15 cm, glabrate. Leaves 3-foliolate, the leaflet blade obovate, 7–14(20) mm long, 4–8 mm wide, the apex rounded, the base cuneate, the margin denticulate, the upper and lower surfaces glabrate, the petiole 0.5–2 cm long; stipules lanceolate, 4–5 mm long, the apex attenuate. Flowers resupinate, subsessile in a capitate head 7–11(14) mm long, the peduncle 2–3 cm long, the pedicel 1 mm long; calyx tube 1–2 mm long, striate-nerved, villous on the upper side, the lobes unequal, shorter than the tube; corolla papilionaceous, 4–6 mm long, pinkish lavender. Fruit ovoid or lenticular, 2 mm long, indehiscent, surrounded by an enlarged, obovoid, inflated, lanate calyx, the ventral lobes short and usually concealed; seeds 1.

Disturbed sites. Rare; Escambia County. Massachusetts, North Carolina, South Carolina, Florida, and California; Europe, Africa, and Asia. Native to Europe, Africa, and Asia. Spring–summer.

Trifolium vesiculosum Savi [With little bladders, in reference to the inflated calyx.] ARROWLEAF CLOVER.

> *Trifolium vesiculosum* Savi, Fl. Pis. 2: 176. 1798.

Decumbent or erect annual herb; stem to 6 dm, glabrate. Leaves 3-foliolate, the leaflet blade obovate, elliptic, or lanceolate, 1.5–3(5) cm long, 0.5–2 cm wide, the apex obtuse to rounded, apiculate, the base broadly cuneate, the margin denticulate, the upper and lower surfaces glabrate, the petiole 4–6 cm long; the stipules lanceolate, ca. 1.5 cm long, the apex apiculate. Flowers sessile, in a short-pedunculate, spheroid to elongate head 4–6 cm long, 2.5–3 cm wide; bracts evident, multistriate; calyx tube ca. 5 mm long, with ca. 20 nerves, glabrous or apically sparsely villous, the lobes subequal, about as long as the tube, setaceous, erect or irregularly divergent; corolla papilionaceous, 1.5–1.8 mm long, white to ochroleucous, stiff with strongly striate petals. Fruit oblong, indehiscent, subtended by an inflated-saccate calyx; seeds 2–3.

Disturbed sites. Occasional; panhandle. Escaped from cultivation. Virginia south to Florida, west to Oklahoma and Texas, also Washington, Oregon, and California; Europe and Asia. Native to Europe and Asia. Spring–summer.

EXCLUDED TAXA

> *Trifolium glomeratum* Linnaeus—This European species was reported by Isely (1990) as "slightly established in ruderal sites"; his range statement includes Florida. No voucher specimens seen.

Trifolium procumbens Linnaeus—Reported for Florida by Small (1903, 1913a, 1933), the name misapplied to material of *T. campestre*.

Trifolium subterraneum Linnaeus—This Old World species was reported for Florida by Isely (1990). Specimens seen from Florida are of cultivated material.

Trigonella L. 1753. FENUGREEK

Annual herbs. Leaves alternate, 3-foliolate, petiolate, stipulate. Flowers in axillary racemes, pedunculate; calyx 5-lobed; corolla papilionaceous; stamens 10, diadelphous (9 + 1). Fruit an indehiscent legume.

A genus of about 50 species; North America, Europe, Asia, Africa, and Australia. [*Trigonum*, triangular, and *ellus*, diminutive, in reference to the shape of the corolla.]

Trigonella caerulea (L.) Ser. [Blue, in reference to the flower color.] BLUE FENUGREEK.

> *Trifolium caeruleum* Linnaeus, Sp. Pl. 2: 764. 1753. *Trifoliastrum caeruleum* (Linnaeus) Moench, Methodus 123. 1794. *Melilotus caeruleus* (Linnaeus) Desrousseaux, in Lamarck, Encycl. 4: 62. 1797. *Trigonella caerulea* (Linnaeus) Seringe, in de Candolle, Prodr. 2: 181. 1825. *Grammocarpus caeruleus* (Linnaeus) Gasparrini, Rendiconto Adunanza Lav. Accad. Sci. 1: 183. 1853. *Teliosma caerulea* (Linnaeus) Alefeld, Landw. fl. 72. 1866. *Teliosma caerulea* (Linnaeus) Alefeld var. *sativa* Alefeld, Landw. Fl. 73. 1866, nom. inadmiss. *Folliculigera caerulea* (Linnaeus) Pasquale, Cat. Ort. Bot. Napoli 46. 1867. *Telis caerulea* (Linnaeus) Kuntze, Revis. Gen. Pl. 1: 209. 1891. *Trigonella melilotus* Link var. *caerulea* (Linnaeus) Aescherson & Graebner, Fl. Nordostdeut. Flachl. 434. 1898. *Trigonella caerulea* (Linnaeus) Seringe subsp. *sativa* Thellung, Fl. Adv. Montpellier 302. 1912, nom. inadmiss.

Erect annual herb, to 6(10) dm; stem sparsely pubescent. Leaves 3-foliolate, the leaflet blade ovate to oblong, (1)2–4(5) cm long, (0.5)1–2(3.5) cm wide, the apex rounded-emarginate, the base cuneate, the margin denticulate distally, the upper surface glabrous, the lower surface sparsely pubescent, the petiole 4–10 mm long; stipules triangular-lanceolate, toothed. Flowers 20–30 in a raceme, the peduncle 2–5 cm long; calyx ca. 3 mm long, the lobes subequaling the tube; corolla papilionaceous, 5–7 mm long, pale blue or white. Fruit rhomboid-obovate, 4–5 mm long, ca. 3 mm wide, yellow-brown to light brown, with a short, abrupt beak 2–3 mm long; seeds 1–2(3), ovoid, ca. 2 mm long, brownish.

Disturbed sites. Rare; Miami-Dade County. Ontario west to Alberta, also New York, Maryland, and Florida; Europe and Asia. Native to Europe. Summer.

Trigonella caerulea is probably the domesticated cultivar of *T. procumbens* (Besser) Reichenbach, native to Europe and Asia, although the two are usually treated as distinct species.

Vachellia Wight & Arn. 1834.

Shrubs or trees. Leaves alternate, even-bipinnately compound, often with glands on the rachis or petiole, petiolate, stipulate, these sometimes spinescent. Flowers sessile, in pedunculate heads or spikes; sepals 5, connate; petals 5, connate; stamens 10–40, free. Fruit a dehiscent or indehiscent legume.

A genus of about 160 species; nearly cosmopolitan. [Commemorates the Rev. G. H. Vachell (nineteenth century), English missionary who collected in China.]

1. Flowers in a spike surrounded by a spathe-like involucre .... **V. cornigera**
1. Flowers in a globose head.
   2. Pinnae 10–15(25) pairs.. **V. macrantha**
   2. Pinnae 1–8(10) pairs.
      3. Petiole gland elliptic; spines usually fused at the base.
         4. Leaflets with reticulate secondary venation .... **V. tortuosa**
         4. Leaflets lacking secondary venation.. **V. sphaerocephala**
      3. Petiole gland circular; spines (if present) free at the base.
         5. Leaflets 1–2 cm long; fruit laterally compressed .... **V. choriophylla**
         5. Leaflets less than 1 cm long; fruit terete or subterete... **V. farnesiana**

Vachellia choriophylla (Benth.) Seigler & Ebinger [With separate leaves.] CINNECORD; TAMARINDILLO.

Acacia choriophylla Bentham, Hooker's J. Bot. Kew Gard. Misc. 1: 495. 1842. *Lucaya choriophylla* (Bentham) Britton & Rose, in Britton, N. Amer. Fl. 23: 87. 1928. *Vachellia choriophylla* (Bentham) Seigler & Ebinger, Phytologia 87: 150. 2005.

Tree, to ca. 9 m; branches glabrous, unarmed. Leaves even-bipinnately compound, 10–20 cm long, the pinnae in 1–(2)3 pairs, the leaflets in 4–8 pairs, the leaflet blade elliptic to obovate, 1–2 cm long, 8–15 mm wide, the apex rounded to slightly emarginate, the base obtuse, sessile, the margin entire, the upper and lower surfaces glabrous, the petiole 8–15 mm long, glandular. Flowers in an axillary, globose head 6–8 mm long and wide, yellow, the peduncle 2–3.5 cm long; corolla puberulent; stamens about twice as long as the corolla. Fruit oblong, 4–8 cm long, 1.5–2.5 cm wide, straight or somewhat curved, woody, turgid, laterally compressed, the apex acute, black, glabrous, woody, tardily dehiscent, the pulp white; seeds ovoid, ca. 8 mm long, brown.

Pine rocklands. Rare; Miami-Dade County, Monroe County keys. Florida; West Indies. Spring.

Vachellia choriophylla is listed (as *Acacia choriophylla*) as endangered in Florida (Florida administrative Code, Chapter 5B-40).

Vachellia cornigera (L.) Seigler & Ebinger [*Corneus*, horn, and *ger*, bearing, in reference to the hornlike stipular spines.] BULLHORN ACACIA; BULLHORN WATTLE.

Mimosa cornigera Linnaeus, Sp. Pl. 1: 520. 1753. *Acacia cornigera* (Linnaeus) Willdenow, Sp. Pl. 4: 1080. 1806. *Acacia cornigera* (Linnaeus) Willdenow var. *americana* de Candolle, Prodr. 2: 460. 1825, nom. inadmiss. *Tauroceras cornigerum* (Linnaeus) Britton & Rose, in Britton et al., N. Amer. Fl. 23: 86. 1928. *Vachellia cornigera* (Linnaeus) Seigler & Ebinger, Phytologia 87: 150. 2005.

Shrub or small tree, to 5 m; branches glabrous, armed with large, hollow, dark- or light-colored basally fused stipular spines to 3 cm long. Leaves even-bipinnately compound, the pinnae 2–8 pairs, the leaflets 15–20 pairs, the leaflet blade oblong, 5–7 mm long, 1–2 mm wide, the apex rounded, usually with a gland (Beltian body), the base rounded or truncate, asymmetric, the upper and lower surfaces glabrous, the petiole ca. 1 cm long, with an elliptic gland at the distal end, the rachis often with small glands between some of the pinnae. Flowers in a dense spike, yellow or cream, this subtended by a cuplike involucre with the flowers initially concealed, the involucre spathe-like at anthesis, the thick peduncle ca. 1 cm long. Fruit oblanceolate, 2–5 cm long, 1–2 cm wide, indehiscent, pulpy, the apex with a long spinelike beak.

Disturbed sites. Rare; Pinellas and Lee Counties. Escaped from cultivation. Florida; West Indies, Mexico, and Central America. Native to Mexico and Central America. Summer.

In its native habitat, the horns of *Vachellia cornigera* are occupied by acacia-ants that utilize the glands and Beltian bodies on the leaves.

Vachellia farnesiana (L.) Wight & Arn. [Farnese Gardens, Italy, where it was first cultivated.]

Shrubs or trees, to 5 m; branches pubescent or glabrous, unarmed or with paired, straight, pinlike stipular spines to 2.5 cm long. Leaves even-bipinnately compound, 3–8 cm long, the pinnae in 2–6 pairs, the leaflets 10–20 pairs, the leaflet blade linear-oblong, 3–6 m long, 1–2 mm wide, the apex rounded to acute, the base rounded, asymmetric, the upper and lower surfaces glabrous, the petiole 5 mm long or longer, glabrous or pubescent, with a small circular, depressed gland near the middle, the rachis glabrous or pubescent, eglandular or sometimes with a gland between the uppermost pinnae pair. Flowers in a globose head 7–13 cm wide, yellow, the peduncle 1–4 cm long, pubescent. Fruit linear, 3–8 cm long, ca. 1 cm wide, irregularly subterete, falcate to nearly straight, the apex and base tapered, woody, blackish, indehiscent or tardily dehiscent, glabrous.

1. Mature leaflets with evident secondary venation ....var. **farnesiana**
1. Mature leaflets lacking evident secondary venation.. var. **pinetorum**

Vachellia farnesiana (L.) Wight & Arn. var. farnesiana SWEET ACACIA.

> *Mimosa farnesiana* Linnaeus, Sp. Pl. 1: 521. 1753. *Acacia farnesiana* (Linnaeus) Willdenow, Sp. Pl. 4: 1083. 1806. *Vachellia farnesiana* (Linnaeus) Wight & Arnott, Prodr. Fl. Ind. Orient. 272. 1834. *Farnesia odora* Gasparrini, Descr. Nouv. Gen. Leg. Vii. 1836, nom. illegit. *Poponax farnesiana* (Linnaeus) Rafinesque, Sylva Tellur. 118. 1838. *Vachellia farnesiana* (Linnaeus) Wight & Arnott var. *typica* Spegazzini, Bol. Acad. Nac. Ci. 26: 298. 1923, nom. inadmiss.
>
> *Vachellia densiflora* Alexander ex Small, Man. S.E. Fl. 655, 1505. 1933. *Acacia densiflora* (Alexander ex Small) Cory, Rhodora 38: 406. 1936; non Morrison, 1912. *Acacia smallii* Isely, Sida 3: 384. 1969. *Acacia minuta* (M. Jones) Beauchamp subsp. *densiflora* (Alexander ex Small) Beauchamp, Phytologia 46: 7. 1980.

Mature leaflets with evident secondary venation.

Shell middens, coastal hammocks, pinelands, and disturbed sites. Occasional; central and southern peninsula, Franklin, Bay, and Escambia Counties. Georgia south to Florida, west to California; West Indies, Mexico, Central America, and South America; Europe, Africa, Asia, Australia, and Pacific Islands. Native to the New World.

Vachellia farnesiana (L.) Wight & Arn. var. **pinetorum** (F. J. Herm.) Seigler & Ebinger [Of pinelands.] PINELAND ACACIA.

Vachellia peninsularis Small, Man. S.E. Fl. 654, 1505. 1933. *Acacia pinetorum* F. J. Hermann, J. Wash. Acad. Sci. 38: 37. 1948. *Vachellia farnesiana* Linnaeus subsp. *pinetorum* (F. J. Hermann) Ebinger & Seigler, Southw. Naturalist 47: 90. 2002. *Vachellia farnesiana* (Linnaeus) Wight & Arnott, var. *pinetorum* (F. J. Hermann) Seigler & Ebinger, Phytologia 87: 157. 2005. TYPE: FLORIDA: Miami-Dade Co.

Vachellia insularis Small, Man. S.E. Fl. 655, 1505. 1933. TYPE: FLORIDA: Monroe Co.

Mature leaflets lacking secondary venation.

Shell middens, coastal hammocks, and pinelands. Occasional; central and southern peninsula. Florida; West Indies. Spring.

Vachellia macracantha (Humb. & Bonpl. ex Willd.) Seigler & Ebinger [Large-spined.] PORKNUT.

Acacia macracantha Humboldt & Bonpland ex Willdenow, Sp. Pl. 4: 1080. 1806, nom. cons. *Mimosa macracantha* (Humboldt & Bonpland ex Willdenow) Poiret, in Lamarck, Encycl., Suppl. 1: 78. 1810. *Acacia flexuosa* Humboldt & Bonpland ex Willdenow var. *lasiocarpa* Grisebach, Abh. Königl. Ges. Wiss. Göttingen 7: 63. 1857. *Acacia macracantha* Humboldt & Bonpland ex Willdenow var. *glabrescens* Grisebach, Fl. Brit. W.I. 222. 1860, nom. inadmiss. *Poponax macracantha* (Humboldt & Bonpland ex Willdenow) Killip, Caribbean Forest. 9: 248. 1948. *Vachellia macracantha* (Humboldt & Bonpland ex Willdenow) Seigler & Ebinger, Phytologia 87: 160. 2006.

Tree, to ca. 10 m; branchlets puberulent or glabrate, with stipular spines to 3 cm long. Leaves even-bipinnately compound, the pinnae 10–15(25) pairs, the leaflets 20–30 pairs, the leaflet blade linear-oblong, 2–4 mm long, 1–2 mm wide, the apex rounded, the base rounded, asymmetric, the upper and lower surfaces glabrate, the petiole usually with a raised, cupuliform gland, the rachis with a gland between the uppermost pinnae. Flowers in a head 7–12 mm wide, yellow, the peduncle 2–3 cm long. Fruit oblong, 5–8 cm long, 6–10 mm wide, somewhat compressed, tardily dehiscent, glabrous.

Mangroves and coastal hammocks. Rare; Manatee (escaped from cultivation) and Miami-Dade Counties, Monroe County keys. Florida; West Indies, Central America, and South America. Spring–Summer.

Vachellia sphaerocephala (Schltdl. & Cham.) Seigler & Ebinger [With spherical heads.] BEE WATTLE.

Acacia sphaerocephala Schlechtendal & Chamisso, Linnaea 5: 594. 1830. *Vachellia sphaerocephala* (Schlechtendal & Chamisso) Seigler & Ebinger, Phytologia 87: 167. 2005.

Shrub, to 3 m; branches glabrous to slightly pubescent, armed with large, hollow, light-colored, basally fused stipular spines to 3 cm long. Leaves even-bipinnately compound, the pinnae 5–8 pairs, the leaflets 8–24 pairs, the leaflet blade oblong, 4–10 mm long, 1–2 mm wide, the apex rounded, usually with a gland (Beltian body), the base rounded or truncate, asymmetric, the upper and lower surfaces glabrous, the petiole ca. 1 cm long, with an elliptic gland at the distal end, the rachis sometimes with a similar gland between some of the pinnae pairs. Flowers in a subglobose head 5–7 mm wide, slightly longer than wide, yellow, this subtended by a 4-lobed involucre, the peduncle 1–1.5 cm long. Fruit oblong, 3–8 cm long, 1–1.5 cm wide, indehiscent.

Disturbed tropical hammocks. Rare; Collier and Miami-Dade Counties. Escaped from cultivation. Florida; Mexico. Native to Mexico. Spring.

In its native habitat, the horns of *Vachellia sphaerocephala* are occupied by acacia-ants that utilize the glands and Beltian bodies on the leaves.

Vachellia tortuosa (L.) Seigler & Ebinger [Much twisted, in reference to the fruits.] POPONAX.

> *Mimosa tortuosa* Linnaeus, Syst. Nat., ed. 10. 1312. 1759. *Acacia tortuosa* (Linnaeus) Willdenow, Sp. Pl. 4: 1083. 1806. *Poponax tortuosa* (Linnaeus) Rafinesque, Sylva Tellur. 118. 1838. *Vachellia tortuosa* (Linnaeus) Seigler & Ebinger, Phytologia 87: 160. 2006.

Shrub, to 4 m; branchlets pubescent to glabrate, with paired, gray-brown, basally fused, pin-like stipular spines to 4 cm long. Leaves even-bipinnately compound, the pinnae in 4–8(10) pairs, ca. 3 cm long, the leaflets in 10–20 pairs, the leaflet blade linear-oblong, 4–7 mm long, the apex obtuse, the base rounded, asymmetric, the upper and lower surfaces pubescent to glabrate, the petiole ca. 1 cm long, with a gland near the middle. Flowers in a globose head ca. 1 cm wide, yellow, the peduncle 1–4 cm long, with a pair of bracts at the top. Fruit linear, 4–13 cm long, 5–7 mm wide, subterete, pubescent or densely strigose, blackish, tardily dehiscent, glabrous.

Shell middens and disturbed sites. Rare; Collier County. Florida; West Indies. Spring–summer.

Vachellia tortuosa is listed (as *Acacia tortuosa*) as endangered in Florida (Florida Administrative Code, Chapter 5B-40).

Vicia L. 1753. VETCH

Annual or perennial herbs. Leaves alternate, even-pinnately compound, the leaflets alternate or subopposite, the rachis terminating in a tendril, estipellate, petiolate, stipulate. Flowers in axillary racemes, bracteate, ebracteolate; sepals 5, connate into a tube, lobed; corolla papilionaceous; stamens diadelphous (9 + 1). Fruit a dehiscent legume.

A genus of about 160 species; nearly cosmopolitan. [Ancient Latin name for a vetch.] Selected reference: Hermann (1960).

1. Peduncle and raceme axis reduced (flowers solitary or 2–3 clustered in the leaf axils).
    2. Corolla pink-purple..**V. sativa**
    2. Corolla yellow, often with purple streaks.
        3. Calyx pubescent, the lobes unequal; fruit 5–7 mm wide, dark brown, glabrous at maturity........
        ....**V. grandiflora**
        3. Calyx glabrous, the lobes unequal; fruit 8–12 mm wide, tan, hirsute with usually tuberculate-based trichomes..**V. lutea**
1. Peduncle and raceme axis well developed.
    4. Raceme with 1–2(3) flowers at the apex of the peduncle.
        5. Fruit sessile, asymmetrically upcurved at the apex.....................................**V. minutiflora**

5. Fruit short-stipitate, symmetrically rounded at the apex...**V. tetrasperma**
4. Raceme with 3–10 flowers loosely disposed along the axis.
    6. Leaflets 2–6.
        7. Leaflets 1–1.5 cm long, elliptic to suborbicular; fruit 1- to 3-seeded.....................**V. floridana**
        7. Leaflets 1.5–5 cm long, linear to linear-elliptic; fruit (4)8- to 12-seeded.
            8. Leaflets mostly 4, 1.5–2.5 cm long; fruit 2.5–3 cm long; seeds ca. 2 mm long......................
            ....**V. acutifolia**
            8. Leaflets mostly 6, 3–5 cm long; seeds 3–4 mm long...**V. ocalensis**
    6. Leaflets 8–12.
        9. Fruit sessile, hirsute or puberulent...**V. hirsuta**
        9. Fruit stipitate, glabrous.
            10. Calyx distinctly gibbous at the base ....**V. villosa**
            10. Calyx symmetrical at the base or nearly so.
                11. Calyx 2–3 mm long; stipules of the lower leaves entire ....**V. caroliniana**
                11. Calyx 3–4 mm long; stipules of the lower leaves usually lobed ....**V. ludoviciana**

Vicia acutifolia Elliott [With acute leaves.] FOURLEAF VETCH.

Vicia acutifolia Elliott, Sketch Bot. S. Carolina 2: 225. 1823. *Cracca acutifolia* (Elliott) Alefeld, Bonplandia 9: 118. 1861.

Sprawling or climbing perennial herb; stem to 1.5 m, glabrate or slightly puberulous. Leaves
even-pinnately compound, the leaflets 2–4(6), the leaflet blade 1.5–2.5(3) cm long, 2–3 mm
wide, the apex acute, the base cuneate, the margin entire, the upper and lower surfaces glabrate,
the tendril simple, the petiole 1–4 mm long; stipules entire. Flowers (2)5–10 loosely disposed in
an axillary raceme 2(5)–5(7) cm long; calyx 2–3 mm long, lobes subequal or the lower slightly
longer, all much shorter than the tube; corolla papilionaceous, 7–8 mm long, white to lavender.
Fruit oblong, 2–3 cm long, ca. 5 mm wide, laterally compressed, glabrous, the stipe 1–1.5 mm
long.

    Wet hammocks, pond and river margins, and wet, disturbed sites. Common; nearly
throughout. South Carolina, Georgia, Alabama, and Florida; West Indies. Spring.

Vicia caroliniana Walter [Of Carolina.] CAROLINA VETCH.

Vicia caroliniana Walter, Fl. Carol. 182. 1788. *Cracca caroliniana* (Walter) Alefeld, Bonplandia 9: 124.
1861.

Vicia hugeri Small, Bull. Torrey Bot. Club 24: 490. 1897.

Sprawling or climbing perennial herb; stem to 1(1.5) m long, finely puberulous or glabrate.
Leaves even-pinnately compound, the leaflets 10–18, the leaflet blade elliptic to narrowly oblong, 1–2.5 cm long, 2–3 mm wide, the apex acute to obtuse, the base cuneate, the margin
entire, the upper and lower surfaces glabrate, the tendril usually simple, the petiole 2–4 mm
long; stipules entire. Flowers 8–20+, loosely disposed on an axillary raceme 3–10 cm long;
calyx 2–3 mm long, the lobes subequal, ca. 1 mm long; corolla papilionaceous, 8–12 mm, white
or lavender-tinged. Fruit oblong, 1.5–2.5(3) cm long, 4–5 mm wide, laterally compressed, glabrous, the stipe ca. 2 mm long.

    Open hammocks. Rare; Jackson County. New York south to Florida, west to Ontario, Min-
nesota, Missouri, Oklahoma, and Texas. Spring.

Vicia floridana S. Watson [Of Florida.] FLORIDA VETCH.

Vicia floridana S. Watson, Proc. Amer. Acad. Arts 14: 292. 1879. TYPE: FLORIDA: Duval Co.

Decumbent or sometimes climbing perennial herb; stem to 6(8) dm, glabrate. Leaves even-pinnately compound, the leaflets 2–5(6), the leaflet blade elliptic, 1–1.5 cm long, 3–4 mm wide, the apex acute to obtuse, the base cuneate, the margin entire, the upper and lower surfaces glabrous, the tendril simple, the petiole 0.5–1.5 cm long; stipule entire. Flowers 2–8, loosely disposed in an axillary raceme 2–4.5 cm long; calyx ca. 2 mm long, the lobes subequal, shorter than the tube; corolla papilionaceous, 6–8 mm long, pale blue or lavender. Fruit elliptic or oblong, 0.8–1.5 cm long, 3–5 cm wide, laterally compressed, glabrous, the stipe 1–2 mm long.

Wet hammocks, pond and river margins, and wet, disturbed sites. Frequent; northern and central peninsula west to central panhandle. Georgia and Florida. Spring–summer.

Vicia grandiflora scop. [Large-flowered.] LARGE YELLOW VETCH.

Vicia grandiflora Scopoli, Fl. Carniol., ed. 2. 2: 65. 1772. *Vicia grandiflora* Scopoli var. *scopoliana* W.D.J. Koch, Syn. Fl. Germ. Helv. 197. 1835, nom. inadmiss. *Cujunia grandiflora* (Scopoli) Alefeld, Bonplandia 9: 102. 1861.

Vicia sorida Waldstein & Kitaibel, Descr. Icon. Pl. Hung. 2: 143. 1803–1805; non Salisbury, 1796. *Vicia grandiflora* Scopoli var. *kitaibeliana* W.D.J. Koch, Syn. Fl. Germ. Helv. 197. 1805. *Vicia grandiflora* Scopali var. *sorida* Grisebach, Spic. Fl. Rumel. 1: 78. 1843, nom. illegit.

Erect or sprawling annual herb; stem to 6 dm, inconspicuously pubescent. Leaves even-pinnately compound, the leaflets 6–12, the leaflet blade elliptic-oblong or oblong, 1–2.5 cm long, 3–6 mm wide, the apex obtuse or truncate-emarginate, apiculate, the margin entire, the upper and lower surfaces glabrate, the tendril branched, the petiole 2–6 mm long; stipules usually lobed. Flowers (1)2(3) in a reduced raceme in the axis of the upper leaves; calyx 11–13 mm long, sparsely pubescent, the lobes subequal, shorter than the tube; corolla papilionaceous, 2.3–3 cm long, yellow, usually purple-streaked. Fruit oblong, 2.5–4.5 cm long, 5–7 mm wide, laterally compressed, dark brown, puberulent at first, glabrous at maturity, sessile.

Disturbed sites. Rare; Jefferson, Gadsden, and Liberty Counties. Escaped from cultivation. New York and Massachusetts south to Florida, west to Michigan, Missouri, Arkansas, and Louisiana; South America; Europe and Asia. Native to Europe and Asia. Spring.

Vicia hirsuta (L.) Gray [With coarse, erect trichomes.] SPARROW VETCH; TINY VETCH.

Ervum hirsutum Linnaeus, Sp. Pl. 2: 738. 1753. *Vicia hirsuta* (Linnaeus) Gray, Nat. Arr. Brit. Pl. 2: 614. 1821. *Ervilia vulgaris* Godron, Fl. Lorraine 1: 173. 1843, nom. illegit. *Ervilia hirsuta* (Linnaeus) Opiz, Seznam 41. 1852. *Endiusa hirsuta* (Linnaeus) Alefeld, Oesterr. Bot. Z. 9: 360. 1859. *Cracca hirsuta* (Linnaeus) Gennari, Nuovo Giorn. Bot. Ital. 2: 137. 1870.

Sprawling or climbing annual herb; stem to 6(8) dm, glabrate or inconspicuously puberulent. Leaves even-pinnately compound, the leaflets (8)10–16, the leaflet blade linear-elliptic, 0.5–1.5(2) cm long, 1–3 mm wide, the apex obtuse to truncate-emarginate, the base cuneate, the margin entire, sometimes revolute, the upper and lower surfaces glabrate, the tendril usually branched, the petiole 1–2 mm long; the stipules often lacerate. Flowers 2–5(7), loosely disposed in an axillary raceme 1–3 cm long; calyx 2–2.5 mm long, the lobes subequal or the lower slightly longer, subequal to the tube or longer; corolla papilionaceous, 3–5 mm long, white, pale blue, or lavender-tinged. Fruit broadly oblong or oval, 6–10 mm long, 3–5 mm wide, hirsute or puberulent, sessile.

Disturbed sites. Occasional; panhandle, Alachua County. Escaped from cultivation. Nearly throughout eastern and westernmost North America; West Indies, Mexico, and South America; Europe, Africa, Asia, Australia, and Pacific Islands. Native to the Old World. Spring.

Vicia ludoviciana Nutt. ex Torr. & A. Gray [Of Louisiana.] LOUISIANA VETCH; DEERPEA VETCH.

> *Vicia ludoviciana* Nuttall ex Torrey & A. Gray, Fl. N. Amer. 1: 271. 1838. *Cracca ludoviciana* (Nuttall ex Torrey & A. Gray) Alefeld, Bonplandia 9: 119. 1861. *Vicia ludoviciana* Nuttall ex Torrey & A. Gray var. *typica* Shinners, Field & Lab. 16: 23. 1948, nom. inadmiss.

Scandent or sprawling annual herb; stem to 1 m, glabrate or sparsely villous. Leaves even-pinnately compound, the leaflets (4)8–14(18), the leaflet blade elliptic to oblong, 0.8–3.8 cm long, 0.5–1 cm wide, the apex obtuse or emarginate, the base cuneate, the margin entire, the tendrils simple or branched, the upper and lower surfaces glabrous or sparsely villous, the petiole 2–4 cm long; stipules of the lower leaves usually lobed. Flowers (2)4–9(12), loosely disposed in a raceme 1.5–4.5 cm long; calyx 3–4 mm long, the lobes subequaling the tube or the lower exceeding the tube; corolla papilionaceous, 5–8 mm long, blue-purple. Fruit oblong, 1.5–3.5 cm long, 4–7(8) mm wide, glabrous, the stipe 1 mm long.

Disturbed sites. Rare; Escambia and Levy Counties. Georgia and Florida, west to Oregon and California; Mexico. Native to the lower midwestern and western United States and Mexico. Spring.

Vicia lutea L. [Yellow, in reference to the corolla color.] SMOOTH YELLOW VETCH.

> *Vicia lutea* Linnaeus, Sp. Pl. 2: 736. 1753. *Hypechusa lutea* (Linnaeus) Alefeld, Bot. Zeitung (Berlin) 18: 166. 1860.

Erect, ascending, or climbing annual herb; stem to 1 m, subglabrous. Leaves even-pinnately compound, the leaflets 10–18, the leaflet blade linear to narrowly elliptic or oblong, 1–2.5 cm long, 2–6 mm wide, the apex acute to rounded or subretuse, the base cuneate, the margin entire, the upper and lower surfaces glabrous or glabrate, the tendril simple or branched, the petiole 1–2 mm long; stipules 3-lobed. Flowers 1(2) in the leaf axis, subsessile; calyx 10–13 mm long, glabrous, the lobes unequal, the lower longer, slightly longer than the tube, the upper shorter than the tube; corolla papilionaceous, 1.5–2 cm long, yellow. Fruit linear-oblong to rhombic-oblong, 2–3 cm long, 8–12 mm wide, laterally compressed, hirsute with the trichomes usually basally tuberculate, sessile.

Disturbed sites. Rare; Leon County. North Carolina, Florida, Alabama, Lousiana, and Texas, also Oregon and California; Europe, Africa, Asia, Australia, and Pacific Islands. Native to Europe, Africa, and Asia. Spring.

Vicia minutiflora D. Dietr. [Small-flowered.] PIGMYFLOWER VETCH.

> *Vicia micrantha* Nuttall ex Torrey & A. Gray, Fl. N. Amer. 1: 271. 1838; non Lowe, 1833; nec Hooker & Arnott, 1833. *Vicia minutiflora* D. Dietrich, Syn. Pl. 4: 1107. 1847. *Abacosa micrantha* Alefeld, Bonplandia 9: 104. 1861, nom. illegit.

Sprawling or ascending annual herb; stem to 8 dm, glabrate or slightly pubescent. Leaves even-pinnately compound, the leaflets 2–4, the leaflet blade broadly elliptic, 1.5–2.5(3) cm long, 3–5 mm wide, the apex acute to obtuse or rounded, the base cuneate, the margin entire, the upper and lower surfaces glabrate, the tendril simple or branched, the petiole 2–5 mm long, the stipules usually lacerate. Flowers 1(2), loosely disposed in an axillary raceme 0.5–3 cm long; calyx 2–3 mm long, the lobes unequal, shorter than the tube; corolla, papilionaceous, 5–6(7) mm long, pale blue to lavender. Fruit oblong, 1.5–3 cm long, 4–5 mm wide, laterally compressed, asymmetrically upcurved at the apex, glabrous or sparsely pubescent, sessile.

Dry hammocks. Rare; Gadsden, Liberty, Calhoun, and Jackson Counties. Missouri south to Florida and Texas. Spring.

Vicia ocalensis R. K. Godfrey & Kral [Of Ocala.] OCALA VETCH.

> *Vicia ocalensis* R. K. Godfrey and Kral, Rhodora 60: 256. 1958. TYPE: FLORIDA: Marion Co.: along Juniper Springs Creek, NE of Juniper Springs, 3 May 1957, *Godfrey 55537* (holotype: FSU).

Trailing or climbing perennial herb; stem to 1.2 m long, glabrate. Leaves even-pinnately compound, the leaflets 2–6, the leaflet blade narrowly elliptic or oblong, 3–5 cm long, 3–6 mm wide, the apex obtuse or rounded to truncate, apiculate, the base cuneate, the margin entire, the upper and lower surfaces glabrous, the tendril simple or rarely branched, the petiole 5–10 mm long; stipules entire. Flowers (2)5–10(12), loosely disposed in an axillary raceme 3–5 cm long; calyx 3 mm long, the lobes subequal, shorter than the tube; corolla papilionaceous, 8–12 mm long, bicolored, blue or lavender and white. Fruit oblong, 4–4.5 cm long, 7–8 mm wide, laterally compressed, glabrous, the stipe ca. 2 mm long.

Spring runs and stream margins. Rare; Marion and Lake Counties. Endemic. Spring.

Vicia ocalensis is listed as endangered in Florida (Florida Administrative Code, Chapter 5B-40).

Vicia sativa L. [Planted, cultivated.] COMMON VETCH.

> *Vicia sativa* Linnaeus, Sp. Pl. 2: 736. 1753. *Faba sativa* (Linnaeus) Bernhardi, Syst. Verz. 250. 1800. *Vicia sativa* Linnaeus var. *melanosperma* Reichenbach, Fl. Germ. Excurs. 530. 1832, nom. inadmiss. *Vicia sativa* Linnaeus var. *vulgaris* Grenier & Godron, Fl. France 1: 458. 1848, nom. inadmiss. *Vicia sativa* Linnaeus var. *typica* Beck, Fl. Nied.-Oesterr. 876. 1892, nom. inadmiss. *Vicia sativa* Linnaeus var. *genuina* Alefeld, Landw. Fl. 60. 1866, nom. inadmiss. *Vicia communis* Rouy, Fl. France 5: 208. 1899, nom. illegit.
>
> *Vicia sativa* Linnaeus var. *angustifolia* Linnaeus, Fl. Suec., ed. 2. 255. 1755. *Vicia angustifolia* (Linnaeus) Linnaeus, Amoen. Acad. 4: 105. 1759. *Vicia sativa* Linnaeus var. *nigra* Linnaeus, Sp. Pl., ed 2. 1037. 1763, nom. illegit. *Vicia sativa* Linnaeus subsp. *nigra* Ehrhart, Hannover Mag. 15: 229. 1780. *Faba angustifolia* (Linnaeus) Bernhardi, Syst. Verz. 250. 1800. *Vicia sativa* Linnaeus subsp. *angustifolia* (Linnaeus) Ascherson & Graebner, Syn. Mitteleur. Fl. 6(2): 971. 1909, nom. illegit.
>
> *Vicia segetalis* Thuillier, Fl. Env. Paris, ed. 2. 367. 1799. *Vicia sativa* Linnaeus var. *segetalis* (Thuillier) Seringe, in de Candolle, Prodr. 2: 361. 1825. *Vicia angustifolia* (Linnaeus) Linnaeus var. *segetalis*

(Thuillier) W.D.J. Koch, Syn. Fl. Germ. Helv. 197. 1835. *Vicia sativa* Linnaeus subsp. *segetalis* (Thuillier) Gaudin, Fl. Helv. 4: 511. 1849.

Erect or sprawling annual herb; stem to 1 m, glabrate or sparsely villous. Leaves even-pinnately compound, the leaflets (6)8–14, the leaflet blade obovate to oblong-lanceolate to linear, 1.5–3(3.5) cm long, 3–6 mm wide, the apex obtuse to truncate or emarginate, apiculate, the base cuneate, the margin entire, the upper and lower surfaces glabrous, the tendril simple or branched, the petiole 4–6 mm long; stipules lobed to semihastate. Flowers 1–2, subsessile in the upper leaf axils; calyx 7–12(15) mm long, the lobes subequaling the tube; corolla papilionaceous, 1–2.5(3) cm long, pink-purple. Legume oblong, 2.5–6 cm long, 3.5–8 mm wide, laterally compressed, strigulose or puberulent, glabrescent, sessile.

Disturbed sites. Occasional; northern counties south to Hernando and Orange Counties. Escaped from cultivation. Nearly throughout North America; West Indies, Mexico, Central America, and South America; Europe, Africa, Asia, Australia, and Pacific Islands. Native to Europe, Africa, and Asia. Spring.

Vicia tetrasperma (L.) Schreb. [Four-seeded]. LENTIL VETCH.

> *Ervum tetraspermum* Linnaeus, Sp. Pl. 2: 738. 1753. *Vicia tetrasperma* (Linnaeus) Schreber, Spic. Fl. Lips. 26. 1771. *Ervilia tetrasperma* (Linnaeus) Opiz, Seznam 41. 1852.

Ascending or sprawling annual herb; stem to 5 dm, sparsely pubescent. Leaves even-pinnately compound, the leaflets 4–10(12), the leaflet blade elliptic to linear, 0.6–2 cm long, 1–3 mm wide, the apex acute to obtuse, the base cuneate, the margin entire, the upper and lower surfaces glabrous, the tendril simple or branched, the petiole 1–2 mm long; stipules entire or semisagittate. Flowers 1–2(3), distally disposed in an axillary raceme 1–3 cm long; calyx 2–3 mm long, the lobes unequal, the lower subequaling the tube; corolla papilionaceous, 4–6 mm long, light purple or purple-striate. Legume oblong, 1–1.3 cm long, 3–4 mm wide, laterally compressed, symmetrically rounded at the apex, glabrous, the stipe ca. 1 mm long.

Disturbed sites. Occasional; northern counties. Escaped from cultivation. Quebec south to Florida and Texas, British Columbia and Montana south to California; South America; Europe, Africa, Asia, Australia, and Pacific Islands. Native to Europe, Africa, and Asia. Spring.

Vicia villosa Roth [With shaggy trichomes.] HAIRY VETCH; WINTER VETCH.

> *Vicia villosa* Roth, Tent. Fl. Germ. 2(2): 183. 1793. *Cracca villosa* (Roth) Grenier & Godron, Fl. France 1: 470. 1848; non Linnaeus, 1753. *Ervum villosum* (Roth) Trautvetter, Trudy Imp. S.-Peterburgsk. Bot. Sada 3: 47. 1875; non Pomel, 1874. *Vicia varia* Host var. *villosa* (Roth) Arcangeli, Comp. Fl. Ital., ed. 2. 527. 1897. *Vicia villosa* Roth subsp. *euvillosa* Cavillier, Annuaire Conserv. Jard. Bot. Gènève 11/12: 21. 1908, nom. inadmiss.
>
> *Vicia dasycarpa* Tenore, Atti Accad. Pontan. 1: 227. 1830. *Cracca dasycarpa* (Tenore) Alefeld, Bonplandia 9: 121. 1861. *Vicia dasycarpa* Tenore var. *typica* Prospichal, Fl. Oesterr. Küstenl. 2: 423. 1898, nom. inadmiss. *Vicia dasycarpa* Tenore forma *genuina* Pospichai, Fl. Oesterr. Küstenl. 2: 423. 1898, nom. inadmiss. *Vicia villosa* Roth subsp. *dasycarpa* (Tenore) Cavillier, Annuaire Conserv. Jard. Bot. Gènève 11/12: 21. 1908.
>
> *Vicia varia* Host, Fl. Austriac. 2: 332. 1831. *Cracca varia* (Host) Grenier & Godron, Fl. France 1: 469. 1848. *Vicia villosa* Roth subsp. *varia* (Host) Corbiere, Nouv. Fl. Normandie 181. 1893.
>
> *Vicia villosa* Roth var. *glabrescens* W.D.J. Koch, Syn. Fl. Germ. Helv. 194. 1835. *Vicia dasycarpa* Tenore var. *glabrescens* (W.D.J. Koch) Beck, in Reichenbach, Icon. Fl. Germ. Helv. 22: 199. 1903.

Erect, scandent, or sprawling annual herb; stem to 1 m, glabrate or conspicuously villous. Leaves even-pinnately compound, the leaflets (10)14–18, the leaflet blade elliptic, narrowly oblong, or linear, 1–2 cm long, 2–4 mm wide, the apex obtuse, the base cuneate, the margin entire, the upper and lower surfaces glabrate or villous, the petiole 1–3 mm long; stipules entire. Flowers (8)10–20+, loosely disposed in an axillary raceme 3–12(15) cm long, secund; calyx 5–6 mm long, villous to glabrate, the tube conspicuously gibbous and oblique at the base, the lobes unequal, the lower shorter or longer than the tube; corolla papilionaceous, (1)1.2–1.6(1.8) cm long, violet or bicolored violet and white, rarely white. Legume elliptic to oblong, 1.5–4 cm long, 6–10 mm wide, laterally compressed, glabrous, the stipe 1–2 mm long.

Disturbed sites. Occasional; northern counties. Escaped from cultivation. Nearly throughout North America; Europe, Africa, Asia, Australia, and Pacific Islands. Native to Europe, Africa, and Asia. Spring.

Vigna Savi, nom. cons. COWPEA

Perennial or annual herbs. Leaves alternate, 3-foliolate, stipellate, petiolate, stipulate, the stipules peltate. Flowers in pedunculate, nodose pseudoracemes, bracteate, bracteolate; sepals 5, connate into a tube, 4- or 5-lobed; petals 5, the corolla papilionaceous, the keel petals strongly incurved or spirally coiled; stamens 10, diadelphous (9 + 1). Fruit a dehiscent legume; seeds numerous.

A genus of about 100 species; tropical and warm temperate regions of the Old and New World. [Commemorates Dominico Vigni (?1577–1647), Italian botanist.]

Recent phylogenetic studies have shown that the American species of *Vigna* sens. lat. is composed of seven genera, of which three occur in Florida: *Leptospron*, *Sigmoidotropis*, and *Vigna* sens. str. (See Delgado-Salinas et al., 2011).

1. Corolla cream-colored or pink to purple... **V. unguiculata**
1. Corolla yellow (sometimes dark spotted or striate).
    2. Corolla 1.5–1.7 cm long; fruit 4–7 cm long..**V. luteola**
    2. Corolla 8–11 mm long; fruit 1–2 cm long..**V. hosei**

Vigna hosei (Craib) Baker [Commemorates Charles Hose (1863–1929), British colonial administrator and naturalist in Sarawak.] SARAWAK BEAN.

Dolichos hosei Craib, Bull. Misc. Inform. Kew 1914: 76. 1914. *Vigna hosei* (Craib) Backer, in Backer & Slooten, Geill. Handb. Jav. Theeonkr. 153. 1924.

Spreading or twining perennial herb; stem to 2 m, pilose or glabrate. Leaves 3-foliolate, the leaflet blade ovate to ovate-lanceolate, 2.5–4 cm long, 1.5–2.5 cm wide, the apex acute, the base broadly cuneate to rounded, the margin entire, the upper and lower surfaces glabrous, the petiole 3–4 cm long; stipules peltate, ca. 3 mm long, entire. Flowers 2–4 on a peduncle 2–5 cm long; bracteoles minute; calyx 1–2 mm long, the tube campanulate, the lobes deltate, shorter than the tube; corolla papilionaceous, 8–11 mm long, yellow, the keel incurved ca. 90°. Fruit oblong, 1–2 cm long, ca. 4 mm wide, glabrous.

Disturbed sites. Rare; Miami-Dade County. Escaped from cultivation. Florida; West Indies; Africa, Asia, and Australia. Native to tropical Asia. Spring–fall.

Vigna luteola (Jacq.) Benth. [Pale yellow, in reference to the corolla.] HAIRYPOD COWPEA.

Dolichos luteolus Jacquin, Hort. Bot. Vindob. 1: 39, t. 90. 1770. *Vigna glabra* Savi, Bull. Sci. Nat. Geol. 6: 62. 1825, nom. illegit. *Vigna luteola* (Jacquin) Bentham, in Martius, Fl. Bras. 15(1): 194. 1859. *Vigna repens* (Linnaeus) Kuntze var. *luteola* (Jacquin) Kuntze, Revis. Gen. Pl. 1: 212. 1891. *Phaseolus luteolus* (Jacquin) Gagnepain, in Lecomte, Fl. Gen. Indochine 2: 229. 1916.

Dolichos repens Linnaeus, Syst. Nat., ed. 10. 1163. 1759. *Vigna repens* (Linnaeus) Kuntze, Revis. Gen. Pl. 1: 212. 1891; non Baker, 1876.

Spreading or twining perennial herb; stem to 3 m long, glabrate or inconspicuously pubescent. Leaves 3-foliolate, the leaflet blade ovate to narrowly lanceolate, 2.5–4 cm long, 1.5–3 cm wide, the apex acute, the base cuneate, the margin entire, the upper and lower surfaces glabrous, the petiole 3–5 cm long; stipules peltate, 2–4 mm long, entire. Flower 2–6 at the apex of the 10–15 cm long peduncle; bracteoles 1–2 mm long; calyx 3–4 mm long, the tube campanulate, the lobes deltate to lanceolate, subequal to or longer than the tube; corolla papilionaceous, 1.5–1.7 cm long, yellow or dark-blotched or striate, the keel incurved ca. 90°. Fruit oblong, 4–7 cm long, 5–7 mm wide, straight or curved, nearly terete, pubescent.

Beaches, hammocks, and disturbed sites. Frequent; peninsula, central and western panhandle. Pennsylvania south to Florida, west to Texas; West Indies, Mexico, Central America, and South America; Africa and Asia. Summer–fall or all year.

Vigna unguliculata (L.) Walp. [Hoof-shaped, clawed.] BLACKEYED PEA; COWPEA.

Dolichos unguiculatus Linnaeus, Sp. Pl. 2: 725. 1753. *Vigna unguiculata* (Linnaeus) Walpers, Repert. Bot. Syst. 1: 779. 1843. *Phaseolus unguiculatus* (Linnaeus) Piper, Torreya 12: 190. 1912; non Linnaeus, 1754.

Dolichos sinensis Linnaeus, Cent. Pl. 2: 28. 1756. *Vigna sinensis* (Linnaeus) Savi ex Hasskarl, Cat. Hort. Bot. Bogor. 279. 1844.

Erect annual herb; stem to 1 m, glabrate. Leaves 3-foliolate, the leaflets ovate, 6–15 cm long, 3.5–8 cm wide, the apex acute, the base rounded, that of the lateral lobes oblique, the margin entire, the upper and lower surfaces glabrate, the petiole 4–6 cm long; stipules peltate, ovate, 5–15 mm long, with conspicuous retrorse basal lobes. Flowers 2–4 at the apex of a 4–20 cm long peduncle; bracteoles 4–6 mm long; calyx 5–8 mm long, the tube campanulate; corolla papilionaceous, 1.5–2.5 cm long, cream-colored, pink, or purple, the keel incurved ca. 90°. Fruit linear, 7–20 cm long, 7–8 mm wide, cylindric, straight or curved, glabrous.

Disturbed sites. Rare; Leon County, central peninsula. Escaped from cultivation. Pennsylvania south to Florida, west to Michigan, Illinois, Missouri, and Texas; West Indies, Mexico, Central America, and South America; Africa, Asia, Australia, and Pacific Islands. Native to Africa. Summer–fall.

EXCLUDED TAXA

Vigna angularis (Willdenow) Ohwi & Ohashi—Reported for Florida by Ward (1972). According to Isley (1990), this report is based on cultivated specimens. No other Florida specimens seen.

Vigna marina (Burman) Merrill—Reported for Florida by Small (1933), the name apparently misapplied to material of *V. luteola*, and Liogier & Martorell (1982, as *V. retusa* (E. Meyer) Walpers), apparently by mistake. No specimens known.

Wisteria Nutt., nom. cons. 1818.

Woody vines, shrubs, or small trees. Leaves odd-pinnately compound, stipellate, petiolate, stipulate. Flowers in terminal racemes, bracteate, ebracteolate; sepals 5, connate into a tube, lobes 5, the upper two fused most of their length; petals 5, the corolla papilionaceous, the standard petal with 2 basal calli; stamens 10, diadelphous (9 + 1) or monadelphous (vexillary stamen proximally fused with the others). Fruit a dehiscent legume.

A genus of about 6 species; North America, South America, Europe, Asia, and Pacific Islands. [Commemorates Caspar Wistar (1760–1818), American physician and anatomist, philanthropist.]

Kraunhia Raf. ex Britton & A. Br., 1808.

Selected reference: Valder (1995).

1. Ovary or fruit glabrous; pedicels 0.5–1(1.5) cm long..**W. frutescens**
1. Ovary or fruit densely pubescent; pedicel 1.5–2 cm long.
    2. Flowers opening almost simultaneously; corolla 2–2.7 cm long.......................................**W. sinensis**
    2. Flowers opening in a gradual sequence from the base; corolla 1.5–2 cm long ....**W. floribunda**

Wisteria floribunda (Willd. DC. [Profusely flowering.] JAPANESE WISTERIA.

Glycine floribunda Willdenow, Sp. Pl. 3: 1066. 1802, nom. cons. *Wisteria floribunda* (Willdenow) de Candolle, Prodr. 2: 390. 1825. *Dolichos japonicus* Sprengel, Syst. Veg. 3: 252. 1826, nom. illegit. *Phaseoloides floribunda* (Willdenow) Kuntze, Revis. Gen. Pl. 1: 201. 1891. *Kraunhia floribunda* (Willdenow) Taubert, in Engler & Prantl, Nat. Pflanzenfam. 3(3): 271. 1894. *Millettia floribunda* (Willdenow) Matsumura, Bot. Mag. (Tokyo) 16: 64. 1902. *Kraunhia sinensis* (Sims) Makino var. *floribunda* (Willdenow) Makino, Bot. Mag. (Tokyo) 24: 298. 1910. *Kraunhia floribunda* (Willdenow) Taubert var. *typica* Makino, Bot. Mag (Tokyo) 25: 17. 1911, nom. inadmiss. *Rehsonia floribunda* (Willdenow) Stritch, Phytologia 56: 183. 1984.

Woody vine, shrub, or small tree, to 15 m; stem glabrate. Leaves odd-pinnately compound, 1.5–3 dm long, the leaflets 13–19, the leaflet blade ovate to lanceolate, 4–8 dm long, 1–2.5 cm wide, the apex acuminate, the base cuneate, the margin entire, the upper surface glabrate, the lower surface pubescent or glabrate, the petiole 3–8 cm long; stipules small, caducous. Flowers numerous in a terminal raceme 2–5(7) dm long, opening gradually from the base, the pedicel 1.5–2 cm long; calyx tube 3–3.5 mm long, strigose; corolla papilionaceous, 1.5–2 cm long, pink to violet; ovary pubescent. Fruit oblong to oblanceolate, 8–12 cm long, 2–2.5 cm wide, often irregularly torulose, pubescent.

Disturbed sites. Rare; Leon, Putnam, and Marion Counties. Escaped from cultivaiton. Maine south to Florida, west to Illinois, Arkansas, and Texas; South America; Asia. Native to Asia. Spring–summer.

Wisteria frutescens (L.) Poir. [Becoming shrubby.] AMERICAN WISTERIA.

Glycine frutescens Linnaeus, Sp. Pl. 2: 753. 1753. *Apios frutescens* (Linnaeus) Pursh, Fl. Amer. Sept. 474. 1814. *Wisteria speciosa* Nuttall, Gen. N. Amer. Pl. 2: 116. 1818, nom. illegit. *Thyrsanthus frutescens* (Linnaeus) Elliott, J. Acad. Nat. Sci. Philadelphia 1: 371. 1818. *Wisteria frutescens* (Linnaeus) Poiret, in Lamarck, Tabl. Encycl. 3: 674. 1823. *Phaseolus frutescens* (Linnaeus) Eaton & J. Wright, Man. Bot., ed. 8. 354. 1840. *Phaseoloides frutescens* (Linnaeus) Kuntze, Revis. Gen. Pl. 1: 201. 1891. *Kraunhia frutescens* (Linnaeus) Rafinesque ex Greene, Pittonia 2: 175. 1891. *Beadlea frutescens* (Linnaeus) Britton, Man. Fl. N. States 549. 1901.

Thyrsanthus floridanus Croom, Amer. J. Sci. Arts 25: 75. 1834. TYPE: FLORIDA.

Wisteria frutescens (Linnaeus) Poiret var. *macrostachya* Torrey & A. Gray, Fl. N. Amer. 1: 283. 1838. *Beadlea macrostachya* (Torrey & A. Gray) Small ex Britton, Man. Fl. N. States 549. 1901. *Wisteria macrostachya* (Torrey & A. Gray) Nuttall ex B. L. Robinson & Fernald, in A. Gray, Manual, ed. 7. 515. 1908. *Kraunhia macrostachya* (Torrey & A. Gray) Small, Bull. Torrey Bot. Club 25: 134. 1898.

Woody vine; stem to 10(15) m, slightly pubescent. Leaves odd-pinnately compound, 1–2(3) dm long, the leaflets 9–13(15), the leaflet blade elliptic to elliptic-lanceolate, (2)3–6.5(8) cm long, 1.5–2.5 cm wide, the upper and lower surfaces glabrous, the apex acuminate, the base cuneate, the margin entire, the petiole 3–4 cm long; stipules minute, caducous. Flowers numerous in a terminal raceme 4–15(25) cm long, opening nearly simultaneously, the pedicel 5–10(15) mm long; calyx tube 4–6 mm long, glabrate or sometimes with coarsely stalked clavate glands; corolla papilionaceous, 1.5–2 cm long, blue-purple to lilac or bicolored, rarely white; ovary glabrous. Fruit oblong, 4–12 cm long, 1–1.2 cm wide, irregularly tortulose, glabrous.

Stream and river margins and wet hammocks. Occasional; northern counties south to Orange County. New York and Massachusetts south to Florida, west to Michigan, Iowa, Missouri, Oklahoma, and Texas; South America. Native to North America. Spring.

Wisteria sinensis (Sims) Sweet [Of China.] CHINESE WISTERIA.

Glycine sinensis Sims, Bot. Mag. 46: t. 2083. 1819. *Wisteria chinensis* de Candolle, Prodr. 2: 290. 1825, nom. illegit. *Wisteria sinensis* (Sims) Sweet, Hort. Brit. 121. 1826. *Millettia chinensis* Bentham, Pl. Junghuhn. 249. 1852, nom. illegit. *Kraunhia chinensis* Greene, Pittonia 2: 175. 1892, nom. illegit. *Kraunhia sinensis* (Sims) Makino, Bot. Mag. (Tokyo) 25: 18. 1911. *Kraunhia floribunda* (Willdenow) Taubert var. *sinensis* (Sims) Makino, Bot. Mag. (Tokyo) 25: 18. 1911. *Rehsonia sinensis* (Sims) Stritch, Phytologia 56: 183. 1984.

Woody vine; stem to 15 m, glabrate. Leaves odd-pinnately compound, 1–3 dm long, the leaflets 7–13, the leaflet blade ovate or elliptic, 4–10 cm long, 1.5–4 cm wide, the apex acuminate, the base cuneate, the margin entire, the upper and lower surfaces glabrous; stipules minute, caducous. Flowers numerous in a terminal raceme 1–2(3.5) dm long, opening nearly simultaneously, the pedicel 1.5–2 cm long; calyx tube 3–4 mm long, strigose; corolla papilionaceous, 2–2.7 cm long, lavender, pink, or violet; ovary pubescent. Fruit oblong or oblanceolate, 6–10(15) cm long, 2–2.5 cm wide, often irregularly torulose, pubescent.

Disturbed sites. Occasional; northern counties south to Brevard County. Escaped from cultivation. Vermont and New York south to Florida, west to Michigan, Illinois, Missouri, and Texas; South America, Europe, Asia, and Pacific Islands. Native to Asia. Spring.

Wisteria sinensis is listed as a Category II invasive species in Florida by the Florida Exotic Pest Plant Council (FLEPPC, 2015).

HYBRID

Wisteria ×formosa Rehder (*W. floribunda* × *W. sinensis*)

Wisteria ×formosa Rehder, J. Arnold Arbor. 3: 36. 1921. *Rehsonia ×formosa* (Rehder) Stritch, Phytologia 56: 184. 1984.

Disturbed sites. Rare; Wakulla and Hernando Counties. Spring.

Zornia J. F. Gmel 1792.

Perennial herbs. Leaves alternate, palmately compound, 4-foliolate, estipellate, petiolate, stipulate. Flowers in terminal or axillary spikes, the bracts paired, peltate, foliaceous, ebracteolate; calyx zygomorphic, 5-lobed; petals 5, the corolla papilionaceous; stamens 10, monadelphous, the anthers alternately versatile and sub-basifixed. Fruit a loment.

A genus of about 75 species; North America, West Indies, Mexico, Central America, South America, Africa, Asia, and Australia. [Commemorates Johannes Zorn (1739–1799), German apothecary.]

Selected reference: Mohlenbrock (1961).

Zornia bracteata J. F. Gmel [With bracts.] VIPERINA.

Zornia bracteata J. F. Gmel, Syst. Nat. 2: 1096. 1792. *Zornia tetraphylla* Michaux, Fl. Bor.-Amer. 2: 76. 1803, nom. illegit. *Hedysarum heterophyllum* Poiret, in Lamarck, Encycl. 6: 405. 1805, nom. illegit.; non Thunberg (1799).

Erect perennial herb, to 80 cm; stem glabrous to strigillose. Leaves palmately compound, 4-foliolate, the leaflet blade linear to narrowly elliptic, (1)2–2.5 cm long, 2–10 mm wide, the base cuneate, the apex acute, 1-nerved, the upper surface glabrous, the lower surface strigose on the midvein, the petiole 1–2 cm; stipules 10–15 mm long, glabrous to sparsely pubescent. Flowers solitary or 2–10 in a spike 2–17 cm long, 1- to 10-flowered; bracts paired, lance-ovate to ovate, 6–12 mm long, 4–8 mm wide, 5- to 7-nerved, the peltate base auriculate, the margin ciliate, the surfaces glabrous; calyx tube 3–4 mm long, 7- to 10-nerved, strigillose; corolla papilionaceous, 10–14 mm long, yellow. Fruit 2- to 6-segmented, the segments 3–4 mm long, the surface smooth, the margin with bristles; seeds ovoid, ca. 2 mm long, black.

Sandhills and flatwoods. Frequent; nearly throughout. Virginia south to Florida, west to Texas; Mexico. All year.

EXCLUDED GENERA

Andira inermis (W. Wright) Kunth ex de Candolle—Reported as escaped in Florida by Small (1933, as *A. jamaicensis* (W. Wright) Urban), West and Arnold (1946), and Long and Lakela (1971). No specimens seen.

Bituminaria bituminosa (Linnaeus) Stirton—Reported for Florida by Small (1933, as *Aspalthium bituminosum* (Linnaeus) Medikus). No vouchering specimens seen.

Cochliasanthes caracalla (Linnaeus) Trew—Reported for Florida by Wunderlin (1982, as *Vigna caracalla* (Linnaeus) Verdcourt), the name misapplied to material of *Leptospron adenanthum*.

Cojoba graciliflora (S. F. Blake) Britton & Rose—This tropical American plant was reported for Florida by Long and Lakela (1971, as *Pithecellobium graciliflorum* S. F. Blake), but the voucher specimen is from cultivated material.

Cullen americanum (Linnaeus) Rydberg—Reported for Florida by Small (1903, as *Psoralea americana* Linnaeus; 1913a, as *Psoralea americana* Linnaeus; 1933). No vouchering specimens seen. A Chapman specimen at NY (Grimes 1997) is said to be from Florida, but its provenance has not been verified.

Cullen corylifolium (Linnaeus) Medikus—Reported for Florida by Britton (1890, as *Psoralea corylifolia* (Linnaeus) Rydberg), were determined by Rydberg (1919) to be *Cullen americanum*. No vouchering specimens seen.

Glycine max (Linnaeus) Merrill—Reported as escaped in Florida by Radford (1964, 1968), Isely (1990), and Wunderlin (1998). The only specimens seen are ambiguous but almost certainly cultivated, so no good vouchers are known. Anywhere soybeans are grown, they can escape locally, so there is a good possibility this species will eventually be vouchered as a legitimate escape.

SURIANACEAE Arn., nom. cons. 1834. BAY CEDAR FAMILY

Shrubs or trees. Leaves alternate, simple, estipulate. Flowers actinomorphic, bisexual, bracteate, bracteolate; sepals 5, basally connate; petals 5, free; stamens 10, free, the anthers basifixed, opening by a longitudinal slit; ovary superior, carpels 5, free, the style gynobasic, the stigma capitate. Fruit a nutlike mericarp.

A family of 5 genera; North America, West Indies, Mexico, Central America, South America, Africa, Asia, Australia, and Pacific Islands.

Suriana L. 1753.

Shrubs or trees. Leaves alternate, simple, sessile, estipulate. Flowers in axillary cymes or solitary, the pedicel base articulated, bracteate; sepals 5, basally connate; petals 5, free; stamens 10, sometimes 5 reduced; carpels 5, ovules 2 per carpel. Fruit a nutlike mericarp, 3–5 per flower; seed 1.

A monotypic genus; North America, West Indies, Mexico, Central America, South America, Africa, Asia, Australia, and Pacific Islands. [Commemorates Joseph Donat Surian, seventeenth-century French physician and botanist.]

Suriana maritima L. [Growing by the sea.] BAY CEDAR.

Suriana maritima Linnaeus, Sp. Pl. 1: 284. 1753.

Shrubs or small trees, to 3 m; branchlets pubescent. Leaves crowded subterminally on the branches, the blade linear-spatulate to linear oblong, 1–4 cm long, 2–6 mm wide, the apex obtuse, the base cuneate, the upper and lower surfaces pubescent, somewhat succulent, sessile. Flowers (1)2–4 in an axillary cyme, the pedicel ca. 1 cm long; bracts lanceolate, 4–9 mm long, pubescent; sepals ovate to narrowly lanceolate, 6–10 mm long, pubescent; petals obovate, 7–9 mm long, yellow, erose distally, glabrous; stamens 10, 5–10 fertile, the staminodes shorter than the fertile stamens; carpels pubescent, the style glabrous. Fruit of 3–5 dry carpels, these subglobose, 4–5 mm long, finely pubescent, subtended by the persistent calyx.

Coastal beaches and dunes. Occasional; Pinellas and Indian River Counties southward. Florida; West Indies, Mexico, Central America, and South America; Africa, Asia, Australia, and Pacific Islands. All year.

POLYGALACEAE Hoffmanns. & Link 1809. MILKWORT FAMILY

Annual or perennial herbs. Leaves alternate, opposite, or whorled, simple, petiolate, estipulate. Flowers in terminal or axillary racemes, spikes, or compound cymes, zygomorphic, bisexual, bracteate, bracteolate; sepals 5, free or connate; petals 3, free; stamens 6 or 8, monadelphous, the anthers apically transversely dehiscent; carpels 2, inferior, the style simple. Fruit a 2-celled, flattened, dehiscent capsule; seeds 2, arillate.

A family of about 20 genera and about 800 species; nearly cosmopolitan. [From the Greek *Poly*, much, and *gala*, milk, originally applied to some species thought to increase the milk flow of cattle.]

Selected reference: Miller (1971b).

1. Lower petal (keel) lacking a terminal crest..**Asemeia**
1. Lower petal (keel) with a terminal crest ....**Polygala**

Asemeia Raf. 1833.

Annual or perennial herbs. Leaves alternate, simple, entire, the venation pinnate, petiolate, estipulate. Flowers in terminal or axillary racemes; bract 1; bracteoles 2; sepals 5, unequal, the inner lateral ones (wings) enlarged and petaloid; petals 3, adnate to the staminal tube, the lower one (keel) lacking a crest; stamens 8; cleistogamous flowers sometimes also present. Fruit a 2-loculicidal capsule, marginally dehiscent; seeds 2, arillate.

A genus of about 25 species; North America, West Indies, Mexico, Central America, and South America. [An apparent corruption of *assideo*, to resemble, in reference to its similarity to *Polygala*.]

Selected references: Nauman (1981); Pastore and Abbott (2012).

Asemeia violacea (Aubl.) J.F.B. Pastore & J. R. Abbott [Blue-red color, in reference to the flowers.] SHOWY MILKWORT.

Polygala violacea Aublet, Hist. Pl. Guiane 735, t. 294. 1775. *Polygala cinerea* de Candolle, Prodr. 1: 330. 1824, nom. illegit. *Asemeia violacea* (Aublet) J.F.B. Pastore & J. R. Abbott, Kew Bull. 67: 811. 2012.

Polygala grandiflora Walter, Fl. Carol. 179. 1788. *Asemeia grandiflora* (Walter) Small, Man. S.E. Fl. 766, 1505. 1933.

Polygala senega Linnaeus var. *rosea* Michaux, Fl. Bor.-Amer. 2: 53. 1803. *Asemeia rosea* (Michaux) Rafinesque, New Fl. 4: 88. 1838 ("1836").

Polygala grandiflora Walter var. *angustifolia* Torrey & A. Gray, Fl. N. Amer. 1: 671. 1840. TYPE: FLOR-IDA: without data, *Leavenworth s.n.* (lectotype: NY). Lectotypified by Gillis (1975: 39).

Polygala flabellata Shuttleworth ex A. Gray, Smithsonian Contr. Knowl. 3(5): 41. TYPE: FLORIDA: Monroe Co.: Key West, Feb 1846, *Rugel 37* (syntypes: BM, K, P, US).

Polygala krugii Chodat, Mém. Soc. Phys. Genève 31(2, 2): 63, t. 15(37–38). 1893. *Polygala grandiflora* Walter subsp. *krugii* (Chodat) Nauman, Sida 9: 17. 1981.

Polygala corallicola Small, Bull. New York Bot. Gard. 3: 425. 1905. *Asemeia grandiflora* (Walter) Small var. *angustifolia* Small, Man. S.E. Fl. 766. 1933. TYPE: FLORIDA: Miami-Dade Co.: Miami, 27 Oct–13 Nov 1901, *Small & Nash s.n.* (lectotype: NY). Lectotypified by Nauman (1981: 15).

Polygala grandiflora Walter var. *leiodes* S. F. Blake, in Britton, N. Amer. Fl. 25: 339. 1924. *Asemeia leiodes* (S. F. Blake) Small, Man. S.E. Fl. 767, 1505. 1933. TYPE: FLORIDA: Lee Co.: vicinity of Fort Myers, 19 Mar 1916, *Standley 25* (holotype: US; isotypes: MO, NY).

Polygala miamiensis Small ex S. F. Blake, in Britton, N. Amer. Fl. 25: 340. 1924. *Asemeia miamiensis* (Small ex S. F. Blake) Small, Man. S.E. Fl. 767, 1505. 1933. TYPE: FLORIDA: Miami-Dade Co.: Everglades W of Miami, 1–9 Nov 1901, *Small & Nash 289* (holotype: NY).

Polygala cumulicola Small, Bull. Torrey Bot. Club 51: 381. 1924. *Asemeia cumulicola* (Small) Small, Man. S.E. Fl. 766, 1505. 1933. TYPE: FLORIDA: Miami-Dade Co.: sand-dunes opposite Miami, 26 Nov–20 Dec 1913, *Small & Small 4568* (holotype: NY; isotypes: FSU, MO, NY, TEX, US).

Erect or ascending, annual or perennial herb, to 4 dm; stem glabrous or puberulent to pubes-cent. Leaves alternate, the blade linear to lanceolate or elliptic, 1–5 cm long, 2–10 mm wide, the apex acuminate to obtuse or rounded, the base cuneate, the upper and lower surfaces glabrous or puberulent to pubescent, the petiole ca. 1 mm long. Flowers numerous in a loose terminal and axillary raceme to 25 cm long, the pedicel 2–4 mm long, the peduncle 3–9 mm long. Peri-anth purplish and greenish, the upper petals yellow-tipped, the keel purplish, the wings green-ish or purplish tinged; upper sepal elliptic-ovate, 2–3 mm long, ciliate and bearing 3–8 pairs of stipitate glands, the lower ones united for ⅘ their length, the wings strongly inequilateral, quadrate-suborbicular, 5–8 mm long and wide, emarginate, short-clawed; keel petal 5–6 mm long, purplish, plicate, somewhat ciliate; stamens 8. Fruit oval to oblong-oval, 4–5 mm long, 2–3 mm wide, glabrous, short-stipitate; seeds cylindric-ellipsoid, ca. 3 mm long, pilose, the aril 3-lobed, pilose laterally.

Sandhills, flatwoods, and open hammocks. Common; throughout. North Carolina south to Florida, west to Louisiana; West Indies, Mexico, Central America, and South America. Spring–fall.

Polygala L. 1753. MILKWORT.

Annual or perennial herbs. Leaves alternate or whorled, simple, entire, the venation pinnate, petiolate, estipulate. Flowers in terminal or axillary racemes, spikes, or umbellate compound cymes; bract 1; bracteoles 2; sepals 5, unequal, the 2 inner lateral ones (wings) enlarged and petaloid; petals 3, variously connate and/or adnate to the staminal tube, the lower (keel) with a terminal crest; stamens 8 or 6, connate into a tube; cleistogamous flowers sometimes also present. Fruit a 2-loculicidal capsule, marginally dehiscent; seeds 1 in each locule, arillate.

A genus of about 400 species; nearly cosmopolitan. [From the Greek *polys*, much, and *gala*, milk, in reference to the belief that it would increase lactation.]

Galypola Nieuwl., 1914; *Pilostaxis* Raf., 1838.

Selected reference: Smith and Ward (1976).

1. Perianth orange, yellow, or greenish, drying yellow, green, or greenish yellow.
    2. Flowers in compound cymes.
        3. Basal leaves linear to linear-lanceolate, forming a persistent rosette; seeds glabrous ....**P. cymosa**
        3. Basal leaves elliptic to spatulate, not forming a persistent rosette; seeds pubescent .... **P. ramosa**
    2. Flowers in solitary racemes.
        4. Wings long-acuminate; sepals long and narrow.
            5. Wings elliptic, involute at the apex; seed body (excluding the stipe) less than 1 mm long (0.7–0.8 mm)..**P. nana**
            5. Wings oblong-lanceolate, not involute at the apex; seed body (excluding the stipe) more than 1 mm long (1.1–1.2 mm) ....**P. smallii**
        4. Wings cuspidate or mucronate; sepals short and broad.
            6. Perianth orange, drying pale yellow; racemes ca. 1.5 cm wide ....**P. lutea**
            6. Perianth lemon-yellow, drying greenish yellow; racemes ca. 2.5 cm wide................**P. rugelii**
1. Perianth purple, pink, or greenish white.
    7. Flowers in compound cymes... **P. balduinii**
    7. Flowers in solitary racemes.
        8. Leaves whorled (at least the lower ones).
            9. Bracts deciduous from the peduncle after anthesis; flowers sessile or on a pedicel less than 1 mm long.
                10. Seeds glabrous or nearly so, fusiform...**P. leptostachys**
                10. Seeds pubescent, ovoid to narrowly ovoid.
                    11. Wings 2–3 mm long..**P. boykinii**
                    11. Wings 1–1.5 mm long.. **P. verticillata**
            9. Bracts persistent on the peduncle after anthesis; flowers distinctly pedicellate.
                12. Raceme tapering to the apex, sparsely flowered (pedicels easily visible) ....**P. hookeri**
                12. Raceme rounded or abruptly narrowed at the apex, densely flowered (pedicels usually obscured by the crowded flowers).
                    13. Racemes sessile or the peduncle rarely exceeding 1 cm in length; wings deltoid, 3–4 mm long, the apex abruptly narrowed to the cuspidate tip .... **P. cruciata**
                    13. Racemes borne on a peduncle usually 2–8 cm long; wings ovate, 1.5–2.5 mm long, the apex obtuse or minutely apiculate... **P. brevifolia**
        8. Leaves all alternate.
            14. Wings less than ½ as long as the corolla ....**P. incarnata**
            14. Wings equaling or exceeding the corolla.
                15. Flowering portion of the racemes short, the flowers congested, deciduous.

16. Leaves reduced to subulate scales ca. 1 mm long .... **P. setacea**
16. Leaves not reduced to subulate scales, longer than 1 mm.
    17. Bracts deciduous .... **P. mariana**
    17. Bracts persistent on the peduncle.
        18. Plant single-stemmed or few-branched well above mid-stem; raceme 8–10 mm .... **P. chapmanii**
        18. Plant much branched near mid-stem; raceme 5–6 mm wide .... **P. nuttallii**
15. Flowering portion of the racemes elongated, the flowers not congested, persistent.
    19. Plant lacking basal cleistogamous flowers; corolla ca. 1.5 mm long.. **P. appendiculata**
    19. Plant with basal cleistogamous flowers; corolla 3–4 mm long.
        20. Capsule margins crenately winged .... **P. crenata**
        20. Capsule margins entire.
            21. Capsules 2–3 times longer than wide; seeds cylindric, the upper part of the aril a long caruncle .... **P. lewtonii**
            21. Capsules less than 2 times longer than wide; seeds ovoid to cylindric, the aril membranaceous .... **P. polygama**

Polygala appendiculata Vell. [Flowers with an appendix, in reference to the keel petal with lobed crests.] SWAMP MILKWORT.

Polygala appendiculata Vellozo, Fl. Flumin. 292. 1829 ("1925").
Polygala leptocaulis Torrey & A. Gray, Fl. N. Amer. 1: 130. 1838. *Polygala paludosa* A. Saint-Hilaire & Moquin-Tandon var. *exappendiculata* Chodat, Mém. Soc. Phys. Genève 3(2(2)): 226. 1893.

Erect annual herb, to 5.6 dm; stem glabrous. Leaves all alternate, the blade linear, 8–25 mm long, 0.5–1 mm wide, the apex acuminate, the base cuneate, the margin slightly revolute, the upper and lower surfaces glabrous. Flowers in a terminal, loose or dense, cylindric raceme to 13 cm long, ca. 5 mm wide, the pedicel ca. 1 mm long, the peduncle 1–2 cm long; bracts linear-lanceolate, ca. 1 mm long, deciduous; perianth rose or white; sepals ovate, 1 mm long, the wings obovate, 2 mm long; keel petal 2 mm long, the blade with a crest on each side bearing 2–3 lobes. Fruit oblong, 1–2 mm long; seed subcylindric, ca. 1 mm long, pubescent, the aril minute, 2-lobed.

Bogs and pond margins. Rare; northern peninsula, central panhandle. Florida, Mississippi, Louisiana, and Texas; West Indies, Mexico, Central America, and South America. Native to Louisiana, Texas, and tropical America. Summer.

Polygala balduinii Nutt. [Commemorates William Baldwin (1779–1819), American physician and botanist who collected in Georgia and Florida.] BALDWIN'S MILKWORT.

Polygala balduinii Nuttall, Gen. N. Amer. Pl. 2: 90. 1818. *Pilostaxis balduinii* (Nuttall) Small, Man. S.E. Fl. 774, 1505. 1933. TYPE: FLORIDA: Nassau Co.: near St. Mary's, s.d., *Baldwin s.n.* (holotype: PH).
Polygala carteri Small, Bull. New York Bot. Gard. 3: 426. 1905. *Pilostaxis carteri* (Small) Small, Man. S.E. Fl. 774, 1505. 1933. *Polygala balduinii* Nuttall var. *carteri* (Small) R. R. Smith & D. B. Ward, Sida 6: 300. 1976. TYPE: FLORIDA: Miami-Dade Co.: between Cutler and Black Point, 13–16 Nov 1903, *Small & Carter 813* (holotype: NY; isotype: NY).

Erect annual or biennial herb, to 50 cm; stem usually solitary below, branched below the inflorescence. Leaves all alternate, the blade of the basal ones obovate, 1–2 cm long, 4–10 mm wide, the apex obtuse to rounded, the base cuneate, short petiolate, the blade of those of the lower stem obovate to elliptic, slightly smaller than the basal ones, the apex obtuse, the base cuneate, short petiolate, the blade of those of the upper stem lanceolate, 7–25 mm long, 2–9 mm wide, the apex acuminate, mucronate. Flowers numerous in several racemes forming a compound cyme, the racemes densely flowered, subcapitate, ca. 2.5 cm long, 6–9 mm wide, the pedicel winged, ca. 1 mm long; bracts lance-subulate, 2 mm long, persistent; perianth greenish white; sepals subulate-ovate to lance-subulate, ca. 2 mm long, the wings elliptic to oblong-elliptic, 3–5 mm long, ca. 1 mm wide, the apex cuspidate; keel petal 2–3 mm long, with a crest on each side of the blade bearing 2 entire to 3-fid lobes; stamens 8. Fruit depressed-suborbicular, ca. 1 mm long and wide; seed ellipsoid, 0.5 mm long, sparsely pilose, the aril scarious, the two oval lobes appressed.

Bogs, marshes, coastal swales, and wet flatwoods. Frequent; nearly throughout. Georgia south to Florida, west to Mississippi, also Texas. Spring–fall.

Polygala boykinii Nutt. [Samuel Boykin (1786–1846), Georgia plantation owner and botanist.] BOYKIN'S MILKWORT.

Polygala boykinii Nuttall, J. Acad. Nat. Sci. Philadelphia 7: 86. 1834.
Polygala praetervisa Chodat, Mém. Soc. Phys. Genève 31(2(2)): 140. 1891. TYPE: FLORIDA: Monroe Co.: Cudjoe Key, Feb, *Curtiss 503* (holotype: G?; isotypes: GH, NY).
Polygala boykinii Nuttall var. *sparsifolia* Wheelock, Mem. Torrey Bot. Club 2: 121. 1891. *Polygala sparsifolia* (Wheelock) Small, Fl. S.E. U.S. 686, 1333. 1903. TYPE: FLORIDA: Monroe Co.: Cudjoe Key, s.d., *Curtiss 503* (holotype: NY; isotype: GH).
Polygala flagellaris Small, Bull. New York Bot. Gard. 3: 427. 1905. TYPE: FLORIDA: Miami-Dade Co.: pinelands near the Homestead Road between Cutler and Longview Camp, 9–12 Nov 1903, *Small & Carter 1078* (holotype NY).
Polygala boykinii Nuttall var. *suborbicularis* R. W. Long, Rhodora 72: 19. 1970. TYPE: FLORIDA: Lee Co.(?): 10 Mile Camp, near Everglades, 23–26 Mar 1905, *Eaton 1384* (holotype: GH).

Erect annual or biennial herb, to 5 dm; stem simple or branched, glabrous. Leaves in whorls of 4–6 nearly to the top, the blade obovate to elliptic or oblanceolate, the blade of those above alternate, linear or linear-lanceolate, 0.6–3 cm long, 2–7 mm wide, the apex obtuse to acuminate, mucronate, the base cuneate, the upper and lower surfaces glabrous. Flowers in a loose raceme 4–25 cm long, pedicel ca. 1 mm long; bracts subulate, 1–2 mm long, ciliate, deciduous; perianth greenish white; sepals ovate, 1–2 mm long, white, the margin ciliolate, the wings ovate, 2–3 mm long, the apex rounded; keel petal 2–3 mm long, with a crest on each side of the blade bearing 3 linear, simple or bifid lobes; stamens 8. Fruit oval, ca. 2 mm long; seed pilose, the aril with the two oblong lobes appressed.

Flatwoods. Occasional; eastern and central panhandle, central and southern peninsula. Tennessee south to Florida, west to Louisiana. Spring–summer.

Polygala brevifolia Nutt. [Short-leaved.] LITTLELEAF MILKWORT.

Polygala brevifolia Nuttall, Gen. N. Amer. Pl. 2: 89. 1818.

Erect annual herb, to 2.6 dm; stem glandular-puberulent above. Leaves in whorls of 4–5, the uppermost alternate, linear-elliptic, the lowermost spatulate or suborbicular, 6–21 mm long, 1–5 mm wide, the apex obtuse or rounded, usually apiculate, the base cuneate, slightly revolute, the upper and lower surfaces sparsely glandular-pubescent or glabrate, sessile or the petiole to 1 mm long. Flowers numerous in an ovoid or cylindric, dense raceme 1–4 cm long, 7–12 mm wide, the pedicel 2–3 mm long, the peduncle to 8 cm long; bracts triangular-ovate, ca. 1 mm long, the margin ciliate, persistent; perianth rose-purple; sepals ovate, 1–2 mm long, the margin ciliate, the wings triangular-ovate to oblong-ovate, ca. 3 mm long, glabrous; keel petal ca. 3 mm long, with a crest on each side of the deltoid blade bearing 2–3 entire or 2-fid lobes; stamens 8. Fruit suborbicular, ca. 2 mm long, with a short oblique stipe-like base; seed obovoid-ellipsoid, ca. 1 mm long, short-pubescent, the aril ca. 1 mm long, the 2 linear lobes appressed.

Titi swamps and coastal swales. Occasional; panhandle, Gilchrist County. New Jersey south to Florida, west to Mississippi. Spring–summer.

Polygala chapmanii Torr. & A. Gray [Commemorates Alvan Wentworth Chapman (1809–1899), American botanist.] CHAPMAN'S MILKWORT.

Polygala chapmanii Torrey & A. Gray, Fl. N. Amer. 1: 131. 1838. TYPE: FLORIDA: "West Florida," s.d., *Chapman s.n.* (holotype: NY).

Erect annual herb, to 3.5 dm; stem glabrous. Leaves all alternate, linear, 6–17 mm long, ca. 1 mm wide, the apex mucronate, the base cuneate, the margin entire, the upper and lower surfaces glabrous. Flowers in a dense ovoid or ovoid-cylindric raceme 1–3.5 cm long, 8–12 mm wide, the pedicel 1–2 mm long, the peduncle 1–7.5 cm long; bracts ovate-subulate, ca. 1 mm long, the margin ciliate, persistent; perianth pink or purplish; sepals suborbicular-ovate, 1–1.5 mm long, the margin sometimes ciliate, the wings obovate or elliptic-obovate, 3–4 mm long; keel petal 2–3 mm long, with a crest on each side of the blade bearing a single or sometimes 2-fid lobe; stamens 8. Fruit rhombic-suborbicular, ca. 2 mm long; seed pyriform, ca. 1 mm long, short-pilose, the aril with 2 long lobes attached to the base of the seed and loosely appressed to it or slightly spreading.

Wet flatwoods and bogs. Frequent; central and western panhandle. Georgia south to Florida, west to Louisiana. Spring–summer.

Polygala crenata C. W. James [With rounded teeth on the margin, scalloped.] SCALLOPED MILKWORT.

Polygala polygama Walter forma *obovata* S. F. Blake, Rhodora 17: 210. 1915. *Polygala crenata* C. W. James, Rhodora 59: 53: 1957.

Erect perennial herb, to 3(3.5) dm; stem simple, glabrous. Leaves alternate, the blade of the older shoots obovate, 1–2 cm long, 3–8 cm wide, that of the flowering shoots smaller, obovate to elliptic, (5)8–13(17) cm long, (2)3–5(9) mm wide, the apex rounded, obtuse, or those of the upper leaves acute, the base cuneate, the upper and lower surfaces glabrous. Flowers in racemes 7–10(15) cm long, 1–1.5 cm wide, the peduncle 1–2(2.5) cm long, the pedicel 3–4 mm long; bracts ovate, ca. 1 mm long, deciduous; perianth light to deep pink; sepals ovate, 1–2 mm long, the wings orbicular to ovate or obovate; keel petal 3–5 mm long, with a crest on each side of the blade bearing 2–several divided lobes; stamens 8; cleistogamous flowers also present at the base or from leaf axils. Fruit oval, 2–3 mm long, the margin narrowly winged, the wing crenate; seeds 2, ellipsoid-ovate, 2 mm long, pilose, the aril oval, connivent, appressed to the seed.

Wet flatwoods, bogs, and acid swamps. Occasional; panhandle. Florida west to Texas. Spring–summer.

Polygala cruciata L. [Cross-shaped, in reference to four verticillate leaves.] DRUMHEADS.

Polygala cruciata Linnaeus, Sp. Pl. 2: 706. 1753.
Polygala cruciata Linnaeus var. *ramosior* Nash ex B. L. Robinson, in A. Gray, Syn. Fl. N. Amer. 1(1): 458. 1897. *Polygala ramosior* (Nash ex B. L. Robinson) Small, Man. S.E. Fl. 771. 1933. SYNTYPE: FLORIDA.

Erect annual, to 1.4 dm; stem simple or cymosely branched, glabrous. Leaves in whorls of 3–4, sometimes the uppermost alternate, the blade linear to linear elliptic, the upper ones the largest, 8–35 mm long, 1–5 mm wide, the apex obtuse or rounded, apiculate, the base cuneate, the upper and lower surfaces glabrous. Flowers in a thick-cylindric or ovoid-cylindric raceme to 6 cm long, 10–17 mm wide, the peduncle 0.1–5 cm long, the pedicel 2–3 mm long; bracts subulate, 1–3 mm long, the margin ciliolate, persistent; perianth rose-purple; sepals ovate, 1–2 mm long, ciliolate, the wings deltoid-ovate, 4–6 mm long, rose-purple; keel petal 3–4 mm long, with a crest on each side of the blade bearing 2–3 entire or 2-fid lobes; stamens 8. Fruit suborbicular, ca. 1 mm long, oblique and winged on the stipe-like base; seed ellipsoid, pubescent, the aril with the linear lobes appressed.

Bogs and wet flatwoods. Frequent; nearly throughout. Maine south to Florida, west to Ontario, Minnesota, Iowa, Oklahoma, and Texas. Summer–fall.

Polygala cymosa Walter [Flowers in cymes.] TALL PINEBARREN MILKWORT.

Polygala cymosa Walter, Fl. Carol. 179. 1788. *Polygala corymbosa* Michaux, Fl. Bor.-Amer. 2: 54. 1803, nom. illegit. *Pilostaxis cymosa* (Walter) Small, Man. S.E. Fl. 774, 1505. 1933.
Polygala attenuata Nuttall, Gen. N. Amer. Pl. 2: 90. 1818. TYPE: "North Carolina to Florida."
Polygala acutifolia Torrey & A. Gray, Fl. N. Amer. 1: 128. 1838. TYPE: FLORIDA: "Middle Florida," s.d., Chapman s.n. (holotype: NY).

Erect biennial herb, to 10 dm; stem simple below the inflorescence, glabrous. Leaves with the blade linear or linear-lanceolate, the basal and lower ones 3.5–14 cm long, 2–6 mm wide, the apex attenuate, the base cuneate, the upper and lower surfaces glabrous, the blade of the middle and upper ones linear, 0.5–6 cm long, 1–2 mm wide, reduced upward, the apex attenuate, the base rounded, sessile. Flowers in a capitate cymose paniculiform inflorescence, the individual racemes cylindric or ovoid-cylindric, 2–3.5 cm long, 7–13 mm wide, the pedicel 1–2 mm long; bracts lance-subulate to lance-ovate, 1–2 mm long, the margin ciliolate, persistent; perianth yellow, turning green in drying; sepals ovate to lance-ovate, 1–2 mm long, the wings oval, 2–3 mm long; keel petal 2–3 mm long, the blade with a crest on each side bearing 2–3 entire or

2-fid lobes; stamens 8. Fruit depressed hemispheric, 1 mm long; seed subhemispheric, slightly less that 1 mm long, the aril minute, unlobed.

Bogs, acid swamps, and cypress domes. Frequent; northern counties, central peninsula. Maryland, North Carolina south to Florida, west to Louisiana. Spring–fall.

Polygala hookeri Torr. & A. Gray [Commemorates William Jackson Hooker (1785–1865), British botanist.] HOOKER'S MILKWORT.

> *Polygala attenuata* Hooker, J. Bot. (Hooker) 1: 195. 1834; non Nuttall, 1818. *Polygala hookeri* Torrey & A. Gray, Fl. N. Amer. 1: 671. 1840.

Erect annual herb, to 2.5 dm; stem papillose-puberulent above. Leaves in whorls of 3–4, the uppermost alternate, the blade linear, 4–11 mm long, ca. 1 mm wide, the apex acute, the base cuneate, the margin revolute, the upper and lower surfaces glandular-puberulent. Flowers a conic-cylindrical raceme 0.7–4 cm long, 6–9 mm wide, the pedicel ca. 2 mm long, the peduncle 3–7 cm long; bracts triangular-ovate, 1 mm long, the margin ciliolate, persistent; perianth pink; sepals broadly ovate, 1–2 mm long, pink, the margin ciliolate, the wing oblong-ovate, 3 mm long, stipitate-glandular on the upper margin; keel petal ca. 3 mm long, the blade triangular with a crest on each side bearing 2 entire lobes; stamens 8. Fruit globose, 1–2 mm long, strongly winged on the stipe-like base; seed ellipsoid, 1 mm long, pubescent, the aril with two linear appressed lobes.

Wet flatwoods and bogs. Occasional; central and western panhandle. North Carolina south to Florida, west to Texas. Summer.

Polygala incarnata L. [Flesh-colored, in reference to the flower color.] PROCESSION FLOWER.

> *Polygala incarnata* Linnaeus, Sp. Pl. 2: 701. 1753. *Galypola incarnata* (Linnaeus) Nieuwland, Amer. Midl. Naturalist 3: 180. 1914.

Erect annual herb, to 3.5 dm; stem glabrous and glaucous. Leaves all alternate, linear, 4–6 mm long, ca. 0.5 mm wide, the apex acuminate, cuspidate, the base cuneate, the upper and lower surfaces glabrous and glaucous. Flowers in a dense raceme 0.6–3.8 cm long, 1–2 mm wide, the pedicel 0.5 mm long, the peduncle 2–3 cm long; bracts small, deciduous; perianth rose-purple; sepals oblong-ovate to lanceolate, ca. 2 mm long, the margin serrulate, the wings linear-oblong, 7 mm long, the margin somewhat undulate-convolute; upper petals ca. 7 mm long, the keel petal 7 mm long, united with the staminal tube and the upper petals and with a crest on each side bearing usually 3 lobes, these variously lobed or cleft; stamens 8. Fruit ovate-subglobose, 2–3 mm long; seed subglobose, 2 mm long, pilose, the aril erect, slightly lobed.

Sandhills and flatwoods. Frequent; nearly throughout. New York south to Florida, west to Ontario, Wisconsin, Iowa, Kansas, Oklahoma, and Texas. All year.

Polygala leptostachys Shuttlew. ex A. Gray [From the Greek *lepto*, slender, and *stachys*, spike, in reference to the inflorescence.] GEORGIA MILKWORT.

> *Polygala leptostachys* Shuttleworth ex A. Gray, Smithsonian Contr. Knowl. 3(5): 41. 1852. TYPE: FLORIDA: Gadsden Co.: Aspalaga, 1843, *Rugel s.n.* (holotype: GH; isotype: NY).

Erect annual herb, to 4.5 dm; stem glabrous. Leaves in whorls of 4–6, the uppermost sometimes alternate, the blade linear to linear-filiform, 0.8–3 cm long, ca. 1 mm wide, the apex acuminate, cuspidate, the base cuneate, the margin revolute, the upper and lower surfaces glabrous. Flowers in a conic-cylindric raceme 1–8.5 cm long, 3–4 mm wide, the pedicel ca. 0.5 mm long, the peduncle 3–6.5 cm long; bracts lance-subulate, glandular-denticulate, deciduous; perianth whitish or greenish; sepals ovate to ovate-lanceolate, ca. 1 mm long, the margin ciliolate, the wings obovate, 1–2 mm long; keel petal 1–2 mm long, the blade with a crest bearing 1–2 lobes on each side; stamens 8. Fruit oblong-ovate, 2 mm long; seed obconic-cylindric, 1–1.7 mm long, glabrous, the aril with 2 linear-elliptic appressed lobes.

Sandhills. Occasional; northern counties south to Pasco County. Georgia south to Florida, west to Mississippi. Spring–summer.

Polygala lewtonii Small [Commemorates Frederick Lewis Lewton (1874–1959), botanist with the USDA, Smithsonian Institution, and Florida collector.] LEWTON'S MILKWORT; LEWTON'S POLYGALA.

Polygala lewtonii Small, Bull. Torrey Bot. Club 25: 140. 1898. TYPE: FLORIDA: Polk Co.: Frostproof, 19 Mar 1894, Lewton s.n. (holotype: NY).

Erect or ascending biennial herb, to 3 dm; stem glabrous. Leaves alternate, usually with fascicles of smaller ones in their axils, the lowest ones reduced and scalelike, the blade linear-spatulate, 6–22 mm long, 1–3 mm wide, the apex rounded to acute, mucronulate, the base cuneate, the margin slightly revolute, the upper and lower surfaces glabrous. Flowers in a loose cylindric raceme 1–7 cm long, 1–1.7 cm wide, the pedicel 1–2 mm long, the peduncle to 1 cm long; perianth rose-purple; sepals oval, ca. 2 mm long, the wings oval or oval oblong, ca. 5 mm long; keel petal 4–5 mm long, the blade with a crest on each side bearing 2- to several-divided lobes; stamens 8; cleistogamous flowers also present on short leafless branches. Fruit oblong, ca. 4 mm long; seed ellipsoid-cylindric, ca. 3 mm long, pilose, the aril with 2 appressed lobes.

Sandhills. Occasional; Marion County south to Highlands County. Endemic. Spring–summer.

Polygala lewtonii is listed as endangered in Florida (Florida Administrative Code, Chapter 5B-40) and endangered in the United States (U.S. Fish and Wildlife Service, 50 CFR 23).

Polygala lutea L. [Yellow, in reference to the flowers.] ORANGE MILKWORT.

Polygala lutea Linnaeus, Sp. Pl. 2: 705. 1753. Polygala lutea Linnaeus var. elatior Michaux, Fl. Bor.-Amer. 2: 54. 1803, nom. inadmiss. Pilostaxis lutea (Linnaeus) Small, Man. S.E. Fl. 774, 1505. 1933.

Erect or ascending biennial herb, to 3.6 dm; stem glabrous. Leaves all alternate, the basal ones clustered, the blade obovate to obovate-ovate, 2–5.8 cm long, 0.8–1.7 cm wide, the apex rounded or obtuse, the base cuneate, those of the stem elliptic, with the blade obovate or oblanceolate, 2–2.8 cm long, 5–9 mm wide, the apex acute to obtuse, mucronulate, the base cuneate, those of the branches with the blade lanceolate or elliptic-lanceolate, 1–1.6 cm long, 1–3 mm wide, the apex acuminate, cuspidate, the base cuneate, the upper and lower surfaces of the leaves glabrous. Flowers in a capitate or thick cylindric raceme, the pedicel 2–3 mm long, winged, the peduncle 2.5–12 cm long; bracts subulate, 2–3 mm long, deciduous; perianth orange yellow, turning yellow in drying; sepals ovate to ovate-deltoid, ca. 2 mm long, the margin ciliolate, the wings obliquely elliptic, 5–7 mm long; keel petal ca. 4 mm long, the blade with a crest on each side bearing a single 2-fid lobe and an unpaired terminal lobe; stamens 8. Fruit broadly obovate, ca. 2 mm long, narrowly erose-margined; seed oblong-ellipsoid, ca. 1 mm long, pubescent.

Bogs, flatwoods, and cypress pond margins. Common; nearly throughout. New York south to Florida, west to Louisiana. Spring–summer.

Polygala mariana Mill. [Of Maryland.] MARYLAND MILKWORT.

Polygala mariana Miller, Gard. Dict., ed. 8. 1768.
Polygala fastigiata Nuttall, Gen. N. Amer. Pl. 2: 89. 1818.
Polygala harperi Small, Fl. S.E. U.S. 688, 1333. 1903.

Erect annual herb, to 4.8 dm; stem sparsely papillose-puberulent above. Leaves all alternate, the blade linear or the lower spatulate, 6–24 mm long, 1–3 mm wide, the apex acute, mucronate, the base cuneate, the upper and lower surfaces glabrous. Flowers in a capitate or short-cylindric raceme 0.5–3.5 cm long, 6–12 mm wide, the pedicel 2–3 mm long, the peduncle 4–17 mm long; bracts subulate-ovate, ca. 1 mm long, the margin erose, often ciliolate, deciduous; perianth pink or purple; sepals ovate-lanceolate, 1–2 mm long, the margin sometimes ciliolate, the wings obovate to elliptic-obovate, 3–5 mm long; keel petal 2–3 mm long, the blade with a crest on each side bearing a single or 2-fid lobe; stamens 8. Fruit subglobose, 2 mm long, with a stipe-like base; seed subglobose-pyriform ca. 1 mm long, pubescent, the aril with 2 subappressed lobes.

Wet, disturbed sites. Occasional; central and western panhandle, northern peninsula, and Pinellas County. New York south to Florida, west to Texas. Summer.

Polygala nana (Michx.) DC. [From the Greek *nano*, dwarf.] CANDYROOT.

Polygala lutea Linnaeus var. *nana* Michaux, Fl. Bor.-Amer. 2: 54. 1803. *Polygala nana* (Michaux) de Candolle, Prodr. 1: 328, 1824. *Pilostaxis nana* (Michaux) Rafinesque, New Fl. 4: 89. 1838 ("1836").
Polygala hyemalis Rafinesque, New Fl. 4: 89. 1838 ("1836"). *Pilostaxis hyemalis* Rafinesque, New Fl. 4: 89. 1838 ("1836"). TYPE: FLORIDA/GEORGIA.
Polygala nana (Michaux) de Candolle var. *humillima* Chodat, Mém. Soc. Phys. Genève 31(2(2)): 200. 1893. SYNTYPE: FLORIDA.

Erect or ascending annual or biennial herb, to 17.5 cm; stem glabrous. Leaves all alternate, the basal ones tufted, the blade obovate or elliptic obovate, 1.5–4.3 cm long, 3–18 mm wide, the apex rounded, sometimes mucronate, the base cuneate, the upper and lower surfaces glabrous, those of the stem with the blade linear-spatulate, oblanceolate, or obovate, 1.3–4 cm long, 2–8 mm wide, the apex rounded to acuminate, the base cuneate, the upper and lower surfaces glabrous. Flowers subsessile in a dense conic-capitate to short-cylindric raceme 1–3.7 cm long, 1–1.7 cm wide, the peduncle 2.3–7.5 cm long; bracts linear-subulate, 5–6 mm long, the margin ciliolate, deciduous or sometimes persistent; perianth yellow, turning green in drying; sepals elliptic-lanceolate to linear-subulate, 3–5 mm long, the margin ciliate, the wings elliptic, 6–8 mm long, sparsely ciliolate; keel petal 4–6 mm long, with a crest on each side bearing 3 linear entire or 1- to 3-fid lobes; anthers 6. Fruit ovate or subglobose, ca. 2 mm long; seed ellipsoid or subpyriform, 1–2 mm long, pubescent, the aril with 2 linear subappressed lobes.

Flatwoods, bogs, and coastal swales. Common; nearly throughout. North Carolina south to Florida, west to Texas. Spring–summer.

Polygala nuttallii Torr. & A. Gray [Commemorates Thomas Nuttall (1786–1859), British-born American botanist.] NUTTALL'S MILKWORT.

Polygala nuttallii Torrey & A. Gray, Fl. N. Amer. 1: 670. 1840.

Erect annual herb, to 25 cm; stem glabrous. Leaves all alternate, the blade linear or the lower ones spatulate-linear, 6–15 mm long, ca. 1 mm wide, the apex obtuse to acute, usually mucronulate, the base cuneate, the margin revolute, the upper and lower surfaces glabrous. Flowers in a cylindric raceme, 0.4–3 cm long, 4–6 mm wide, the peduncle 0.5–2 cm long, the pedicel 1 mm long; bracts lance-subulate, 1 mm long, the margin ciliate, persistent; perianth rose-purple or greenish white; sepals suborbicular to lance-ovate, 1 mm long, the wings obovate or ellipticobovate, 2–3 mm long; keel petal ca. 2 mm long, the blade with a crest on each side bearing an entire or 2 fid lobe; stamens 8. Fruit suborbicular, 2 mm long, with a short stipe-like base; seed pyriform, 1 mm long, pubescent, the aril 2-lobed, subappressed.

Bogs and wet flatwoods. Rare; Baker, Leon, Wakulla, and Jackson Counties. Massachusetts and New York south to Florida, west to Arkansas. Summer.

Polygala polygama Walter [From the Greek *poly*, many, and *gamo*, united, apparently in reference to the plant having regular and cleistogamous flowers.] RACEMED MILKWORT.

Polygala polygama Walter, Fl. Carol. 179. 1788. *Anthallogea polygama* (Walter) Rafinesque ex Nieuwland, Amer. Midl. Naturalist 3: 180. 1914.

Polygala aborigina Small, Torreya 26: 92. 1926. TYPE: FLORIDA: Volusia Co.: Turtle Mound, 24 May 1926, *Small et al. s.n.* (holotype: NY; isotypes: NY).

Erect or ascending annual or biennial herb, to 3 dm; stem glabrous. Leaves all alternate or the lowermost sometimes opposite, the blade of the lower ones spatulate or obovate, that of the upper linear to linear-spatulate, longer than the lower ones, 12–31 mm long, 2–8 mm wide, the apex of the lower ones obtuse to rounded, mucronulate, that of the upper ones acute, the base cuneate on all. Flowers in a loose cylindric raceme to 13.5 cm long, 9–14 mm wide, the pedicel 1–4 mm long, the peduncle 1–2 cm long; bracts ovate or oblong-ovate, 1 mm long, deciduous; perianth pink or purple-pink; sepals ovate, 1–2 mm long, the wings ovate or ovate-oblong, 3–6 mm long; keel petal 3–5 mm long, the blade with a crest on each side bearing 2- to 3-fid lobes; stamens 8; cleistogamous flowers also present in a loose raceme arising from the base or in late season from the leaf axils. Fruit ovate, 3–4 mm long, ridged; seed ellipsoid, 2–3 mm long, pubescent, the aril with 2 appressed lobes.

Sandhills, flatwoods, and dry hammocks. Frequent; nearly throughout. Quebec south to Florida, west to Ontario, Minnesota, Iowa, Oklahoma, and Texas. Spring–summer.

Polygala ramosa Elliott [Branched.] LOW PINEBARREN MILKWORT.

Polygala ramosa Elliott, Sketch Bot. S. Carolina 2: 186. 1822. *Pilostaxis ramosa* (Elliott) Small, Man. S.E. Fl. 774, 1505. 1933.

Erect annual herb, to 4 dm; stem glabrous. Leaves all alternate, the basal leaves in a small tuft, the blade elliptic or obovate, 0.7–2 cm long, 2–6 mm wide, the apex rounded, the base narrowed and petiole-like, the upper and lower surfaces glabrous, the stem leaves with the blade linear or spatulate to elliptic, 0.7–2.4 cm long, 2–8 mm wide, the apex acute to obtuse, the base cuneate, the upper and lower surfaces glabrous, those on the inflorescence branches linear and reduced. Flowers in terminal and axillary racemes forming a flattened cymose panicle to 14 cm wide, the separate racemes loosely flowered, to 1.8 cm long, 7–11 mm wide, the pedicel 1–2 mm long, narrowly winged; bracts lance-ovate, 1–2 mm long, the margin ciliolate, persistent; perianth yellow, turning dark green in drying; sepals ovate, 1–2 mm long, the margin ciliolate, the wings obovate or elliptic-obovate, ca. 3 mm long, strongly cuspidate; keel petal 2 mm long, the blade with a crest on each side bearing 2- to 3-fid lobes; stamens 8. Capsule suborbicular, ca. 1 mm long; seed ellipsoid, 0.6–0.7 mm long, pubescent, the aril with 2 appressed lobes.

Flatwoods, bogs, and coastal swales. Frequent; peninsula, central and western panhandle. New Jersey south to Florida, west to Texas. Summer–fall.

Polygala rugelii Shuttlew. ex Chapm. [Commemorates Ferdinand Xavier Rugel (1806–1878), German-born American botanist.] YELLOW MILKWORT.

Polygala rugelii Shuttleworth ex Chapman, Bot. Gaz. 3: 4. 1878. *Pilostaxis rugelii* (Shuttleworth ex Chapman) Small, Man. S.E. Fl. 774, 1505. 1933. TYPE: FLORIDA: Hillsborough Co.: Tampa.
Polygala reynoldsiae Chapman, Fl. South. U.S., ed. 2. 613. 1883. TYPE: FLORIDA: Flagler Co.: near St. Augustine, May 1882, *Reynolds s.n.* (holotype: ?; isotype: NY).

Erect or ascending annual or biennial herb, to 7.5 cm; stem glabrous. Leaves all alternate, the basal ones tufted, the blade obovate, 3–3.5 cm long, 6–15 mm wide, the apex obtuse to rounded, the base cuneate, the upper and lower surfaces glabrous, the stem leaves oblanceolate to lanceolate, 2–3.5 cm long, 4–8 cm wide, the apex mucronate-apiculate, the branch leaves narrowly elliptic to linear-lanceolate, 1–2 cm long, 2–4 mm wide. Flowers in an ovoid to thick-cylindric raceme, the pedicel 2–4 mm long, the peduncle 5.5–15 cm long; bracts subulate, 3–5 mm long, the margin ciliolate, deciduous; perianth lemon-yellow, turning dark green in drying; sepals deltoid-ovate or ovate, 2–3 mm long cuspidate, the wings oblong or oblong elliptic, 6–7 mm long, cuspidate; keel petal 5 mm long, the blade with a crest on each side bearing 2 divided lobes and a short terminal segment; stamens 8. Fruit obliquely deltoid, 1–2 mm long, narrowly crenulate-margined; seed ellipsoid, 1–1.5 mm long, pubescent, the aril with 2 appressed lobes.

Wet flatwoods. Common; peninsula, eastern panhandle. Endemic. Spring–fall.

Polygala setacea Michx. [Bristlelike.] COASTALPLAIN MILKWORT.

Polygala setacea Michaux, Fl. Bor.-Amer. 2: 52. 1803.

Erect perennial herb, to 3.3 dm; stem glabrous. Leaves all alternate, the blade obovate-subulate to linear-subulate, 0.5–2 mm long, 0.5 mm wide, the apex mucronate, the upper and lower surfaces glabrous. Flowers in a terminal cylindric, densely flowered raceme 8–24 mm long, 3–6 mm wide, the pedicel ca. 0.5 mm long, the peduncle to 12 mm long; bracts ovate-lanceolate, 1 mm long, deciduous; perianth white; sepals broadly ovate to narrowly lanceolate, 1–2 mm long, the wings obovate or elliptic-obovate, 1–2 mm long; keel petal ca. 2 mm long, the blade with a crest on each side bearing 2–3 entire or 2-fid lobes; stamens 6, the filaments connate in 3s. Fruit ovate, ca. 2 mm long, short-stipitate; seed subglobose-pyriform, ca. 1.5 mm long, pubescent, the aril erect and attached to the seed beak, the 2 lobes touching the seed near the base.

Flatwoods and bogs. Frequent; peninsula west to central panhandle. North Carolina south to Florida. All year.

Polygala smallii R. R. Sm. & D. B. Ward [Commemorates John Kunkel Small (1869–1938), American botanist.] SMALL'S MILKWORT; TINY MILKWORT.

Polygala arenicola Small, Bull. New York Bot. Gard. 3: 426. 1905; non Guerke, 1903. *Pilostaxis arenicola* Small, Man. S.E. Fl. 773, 1505. 1933. *Polygala smallii* R. R. Smith & D. B. Ward, Sida 6: 307. 1976. TYPE: FLORIDA: Miami-Dade Co.: near the unfinished railroad grade between Coconut Grove and Cutler, 31 Oct–4 Nov 1903, *Small & Carter 1276* (holotype: NY).

Erect biennial herb, to 7 cm; stem glabrous. Leaves all alternate, mostly basal, the blade oblong to linear-oblanceolate or subspatulate, 1–4.2 cm long, 2–14 mm wide, the apex acute to obtuse, mucronate, the base cuneate, the upper and lower surfaces glabrous. Flowers in a terminal cylindric raceme 4–30 mm long, 5–18 mm wide, usually surpassed by the basal leaves, the pedicel less than 1 mm, the peduncle to 1 cm long; bracts ovate, 3–4 mm long, deciduous; perianth green or greenish-yellow; sepals lanceolate to lance-ovate or linear-lanceolate, 0.5 mm long, the wings oblong-lanceolate, 5–8 mm long; upper petals 4–5 mm long, the keel petal 4 mm long, the blade with a crest on each side bearing 2-fid lobes, stamens 8. Fruit suborbicular, 2–3 mm long; seed ellipsoid, ca. 2 mm long, pilose, the aril with 2 appressed lobes.

Pinelands. Rare; St. Lucie County south to Miami-Dade County. Endemic. All year.

Polygala smallii is listed as endangered in Florida (Florida Administrative Code, Chapter 5B-40) and endangered in the United States (U.S. Fish and Wildlife Service, 50 CFR 23).

Polygala verticillata L. var. **isocycla** Fernald [Whorled, in reference to the leaves; from the Greek *iso*, equal, and *cyclo*, circle, in reference to the upper leaves equal and in a whorl.] WHORLED MILKWEED.

Polygala verticillata Linnaeus var. *isocycla* Fernald, Rhodora 40: 334, t. 501(2–3). 1838.

Erect annual herb, to 2 dm; stem glabrous. Leaves in whorls of 3–7, the blade lanceolate or linear-lanceolate, 5–25 mm long, 1–4 mm wide, the apex acute to acuminate, cuspidate, the base cuneate, the upper and lower surfaces glabrous. Flowers in a dense or loose conic or cylindric-conic raceme, 0.5–1.3 cm long, 2–4 mm wide, the peduncle 0.5–4 cm long, the pedicel to 0.5 mm long; perianth whitish or greenish; sepals ovate, ca. 1 mm long, the margin ciliolate, the wings obovate-ovate, ca. 2 mm long; keel petal 1–1.5 mm long, the blade with a crest on each side bearing 1–2 lobes. Fruit ovate, 1–1.5 mm long; seed ca. 1 mm long, pubescent, the aril with 2 lobes.

Open hammocks. Occasional; central and southern peninsula, Wakulla County. Maine south to Florida, west to Saskatchewan, Montana, Wyoming, Utah, and Texas. Spring–summer.

EXCLUDED TAXON

Polygala paniculata Linnaeus—Reported for Florida by Wunderlin and Hansen (2011, as *P. tenella* Willdenow), the name misapplied to Florida material of *P. appendiculata*.

Polygala verticillata Linnaeus—Because infraspecific categories were not recognized, the typical variety was reported by implication by Chapman (1860), Small (1903, 1913a, 1933), Radford et al. (1964, 1968), and Correll & Johnston (1970). All Florida material is var. *isocycla*.

ROSACEAE Juss., nom. cons. 1789. ROSE FAMILY

Herbs, shrubs, or trees. Leaves alternate or opposite, simple or odd-pinnately or palmately compound, petiolate or epetiolate, stipulate. Flowers bisexual or unisexual, actinomorphic, the perianth perigynous, often with a glandular disk; sepals 4 or 5, free or connate; petals 4, 5, or sometimes absent; stamens 5–many; carpels 1 to many, free or connate, the styles as many as the carpels, free or connate. Fruit an achene, follicle, drupe, or pome.

A family of about 90 genera and 2,500 species; nearly cosmopolitan.

Amygdalaceae Marquis, nom. cons., 1820; *Malaceae* nom. cons., Small, 1803; *Spiraeaceae* Bertuch, 1801.

Selected Reference: Robertson (1974).

1. Herb.
  2. Leaves simple .... **Aphanes**
  2. Leaves compound.
    3. Leaves 3-foliolate.
      4. Flowers white, bractlet lobes entire ....**Fragaria**
      4. Flowers yellow; bractlet lobes apically 3-toothed ....**Duchesnea**
    3. Leaves pinnately or palmately compound.
      5. Leaves palmately compound; achenes numerous, inserted on an enlarged receptacle ....
        .... **Potentilla**
      5. Leaves pinnately compound; achenes 2, surrounded by a hypanthium armed with hooked
        bristles .... **Agrimonia**
1. Tree or shrub.
  6. Leaves compound; stem with armed or straight prickles.
    7. Leaves with conspicuous, persistent stipules; fruit an aggregation of small drupes .... **Rubus**
    7. Leaves with inconspicuous stipules; fruit an aggregation of achenes surrounded by a fleshy
      hypanthium... **Rosa**
  6. Leaves simple; stem sometimes with true thorns, but never prickles.
    8. Ovary superior; fruit a drupe or of inflated follicles.
      9. Leaves subpalmately veined; fruit consisting of separate, inflated follicles; plant stellate-pubescent (at least on the new growth) ....**Physocarpus**
      9. Leaves pinnately veined; fruit a drupe; plant never stellate-pubescent ....**Prunus**
    8. Ovary inferior; fruit a pome.
      10. Inflorescence racemose; petals narrowly oblong; principal lateral leaf veins appressed to the
        midrib for a short distance (seen with low magnification)................................. **Amelanchier**

10. Inflorescence corymbose or the flower solitary; petals as wide as or wider than long; principal leaf veins not appressed to the midrib.

    11. Endocarp becoming hard and bony, separating from the seed with difficulty; branches usually with evident thorns.

        12. Leaf margin crenate; stipules caducous; thorns leafy and tipping the branches ....**Pyracantha**

        12. Leaf margin lobed or toothed; stipules persistent; thorns naked, axillary.................... ....**Crataegus**

    11. Endocarp cartilaginous or membranaceous, easily separated from the seed; branches rarely with thorns.

        13. Upper leaf surface with purplish red glandular trichomes along the midrib; fruit 6–9 mm in diameter .... **Photinia**

        13. Upper leaf surface without glandular trichomes; fruit 1 cm or more in diameter.

            14. Leaf margin coarsely serrate-toothed; inflorescence densely pubescent.................. ....**Eriobotrya**

            14. Leaf margin finely crenate-serrate, serrate, or entire; inflorescence sparsely pubescent to glabrous.

                15. Leaf apex abruptly cuspidate-acuminate; flowers white; styles free..........**Pyrus**

                15. Leaf apex obtuse; flowers pink; styles united at the base........................... **Malus**

Agrimonia L. 1753. AGRIMONY

Perennial herbs. Leaves alternate, odd-pinnately compound, petiolate, stipulate. Flowers in a terminal and axillary simple or compound raceme, bracteate, bracteolate; sepals 5, free; petals 5, free; stamens 5–15; carpels 2, free, superior. Fruit consisting of an indurate hypanthium enclosing 1(2) achene(s) and topped by the connivent sepals with hooked bristles spreading from the rim.

A genus of 18 species; North America, West Indies, Mexico, Central America, South America, and Asia. [A corruption of the Greek *Argemone*, a plant mentioned by Pliny meaning "spot on eye," alleged to have healing properties of the eye.]

Selected reference: Kline and Sørensen (2008).

1. Hypanthium and the lower leaf surface with sessile, amber glands.. **A. incisa**
1. Hypanthium and the lower leaf surface eglandular ....**A. microcarpa**

Agrimonia incisa Torr. & A. Gray [Deeply cut, in reference to the leaves.] INCISED AGRIMONY; HARVEST LICE.

Agrimonia incisa Torrey & A. Gray, Fl. N. Amer. 1: 431. 1840. *Agrimonia parviflora* Aiton var. *incisa* (Torrey & A. Gray) A. W. Wood, Amer. Bot. Fl. 108. 1871.

Erect perennial herb, to 1 m; stem with sessile, amber glands, pubescent to villous, sometimes lanate or hirsute. Leaves odd-pinnately compound with 3–15 leaflets, the leaflet blade obovate to elliptic, the apex rounded to acute, the base cuneate, the lower surface with sessile amber glands, pubescent to villous and hirsute, the terminal leaflet the largest, 2.5–4 cm long, 1–1.5

cm wide, the minor leaflets 1–3 pairs between each major leaflet pair, the petiole 1.5–2 cm long; stipules half-ovate, deeply incised along the margin, 6–25 mm long. Flowers few–50 in a raceme, the rachis with sessile, glistening, glandular trichomes, pubescent to villous, hirsute at least on the lower part; sepals ovate, 2–3 mm long, the apex acute; petals elliptic, 2–5 mm long, yellow. Fruit enclosed in a broadly turbinate hypanthium, 1.5–3 mm long, 2–3.5 mm wide, the hypanthium shallowly sulcate, with hooked bristles in 3–4 circumferential rows, the lowermost reflexed, the surface with sessile amber glands.

Sandhills. Occasional; northern counties south to Hillsborough and Polk Counties. South Carolina south to Florida, west to Texas. Fall.

Agrimonia incisa is listed as endangered in Florida (Florida Administrative Code, Chapter 5B-40).

Agrimonia microcarpa Wallr. [Small-fruited.] SMALLFRUIT AGRIMONY.

Agrimonia microcarpa Wallroth, Beitr. Bot. 1: 39. 1842. *Agrimonia pubescens* Wallroth var. *microcarpa* (Wallroth) Ahles, J. Elisha Mitchell Sci. Soc. 80: 172. 1964.

Agrimonia pumila Muhlenberg ex E. P. Bicknell, Bull. Torrey Bot. Club 23: 524. 1896. TYPE: FLOR-IDA: without data, *Chapman s.n.* (lectotype: NY). Lectotypified by Kline and Sørensen (2008: 19).

Erect perennial herb, to 1 m; stem pubescent to villous and hirsute. Leaves odd-pinnately compound with 3–9 leaflets, the leaflet blade obovate to elliptic, the apex obtuse to acute, the base cuneate, the lower surface pubescent to pilose and hirsute, the terminal leaflet the largest, 3–7.5 cm long, 1.5–3.5 cm wide, the minor leaflets absent or 1 pair, between the major leaflet pairs, the petiole 1.5–2 cm long; stipules ovate, deeply incised with a few teeth along the margin, 7–30 mm long. Flowers few–50 in a raceme, the rachis pubescent to villous and hirsute; sepals ovate, 1.5–2.5 mm long, the apex acute; petals elliptic, 2.5–3.5 mm long, yellow. Fruit enclosed in an obconic to turbinate hypanthium 2.5–4 mm long, 2.5–4.5 mm wide, the hypanthium usually deeply sulcate, with hooked bristles in 2–3 circumferential rows, the lowest spreading but not reflexed, the surface hirsute, lacking sessile amber glands.

Mesic hammocks. Occasional; northern peninsula, central panhandle. Connecticut, New Jersey, and Pennsylvania south to Florida, west to Illinois and Texas, also Kansas. Summer–fall.

EXCLUDED TAXA

Agrimonia parviflora Aiton—Reported by Small (1933) and Correll and Johnston (1970), the basis unknown. No voucher specimens seen.

Agrimonia rostellata Wallroth—Reported for Florida by Chapman (1860, as *A. eupatoria* Linnaeus), Clewell (1985), Wunderlin (1998), and Wunderlin and Hansen (2003), the name misapplied to material of *A. microcarpa*. Excluded from Florida by Kline and Sørensen (2008).

Amelanchier Medik. 1789. SERVICEBERRY

Trees or shrubs. Leaves alternate, simple, pinnate-veined, petiolate, stipulate. Flowers in terminal clusters or racemes; sepals 5, free; petals 5, free; stamens 20, free; carpels 5, connate, adnate to the hypanthium, the styles 5. Fruit a pome.

A genus of about 25 species; North America, Europe, Africa, and Asia. [Derived from *amel-anchièr*, the vernacular name of the European *A. ovalis*, in reference to the honey-like taste of the fruit.]

Amelanchier arborea (F. Michx.) Fernald [Tree.] COMMON SERVICEBERRY.

Mespilus canadensis Linnaeus var. *cordata* Michaux, Fl. Bor.-Amer. 1: 291. 1803. *Mespilus arborea* F. Michaux, Hist. Arb. Forest. 3: 68, t. 11. 1812. *Amelanchier arborea* (F. Michaux) Fernald, Rhodora 43: 536. 1941.

Tree or shrub, to 20 m; branchlets glabrous at flowering. Leaves with the blade ovate to obovate, 4–10 cm long, 2.2–5 cm wide, the apex acute to acuminate, the base cordate or rounded, the margin toothed nearly to the base with 6–10 teeth/cm, the lateral veins 11–17 pairs, becoming indistinct near the margin and not entering the teeth, the upper surface glabrous, the lower surface sparsely pubescent to glabrous, the petiole 1–2.5 cm long, sparsely pubescent or glabrous. Flowers 4–11 per cluster; hypanthium campanulate, 3–5 mm long, the free sepals tips 2–3 mm long, strongly recurved at flowering, the lower surface pubescent; petals linear to oblong, 10–18 mm long, white or pinkish. Fruit subglobose, 6–10 mm long and wide, maroon-purple.

Dry hammocks. Occasional; central and western panhandle. Quebec south to Florida, west to Ontario, Minnesota, Nebraska, Kansas, Oklahoma, and Texas. Spring.

EXCLUDED TAXA

Amelanchier canadensis (Linnaeus) Medikus—Reported by Chapman (1860, as *A. canadensis* var. *botryapium* (Linnaeus f.) A. Gray) and Small (1903, as *A. botryapium* (Linnaeus f.) Borkhausen, 1913a, 1913c, 1913e), who misapplied the name to material of *A. arborea*.

Amelanchier intermedia Spach—Reported by Small (1913a, 1913c, 1933, all as *A. oblongifolia* (Torrey & A. Gray) M. Roemer) apparently based on material of *A. arborea*.

Aphanes L. 1753. PARSLEY PIERT

Annual herbs. Leaves alternate, simple, pinnate-veined, petiolate, stipulate. Flowers in condensed cymes, bracteate; sepals 4, free; petals absent; stamens 1(2), free; carpel 1, the style basal. Fruit an achene.

A genus of about 15 species; North America, South America, Europe, Africa, Asia, Australia, and Pacific Islands. [From the Greek *aphanes*, unseen, invisible, in reference to the inconspicuous nature of the plants.]

Aphanes australis Rydb. [Southern.] SLENDER PARSLEY PIERT.

Aphanes australis Rydberg, in Britton, N. Amer. Fl. 22: 380. 1908.

Erect or ascending annual herb, to 1(2) dm; stem appressed pubescent. Leaves 5–10 mm long, ca. 5 mm wide, the blade divided into 3 segments, each segment (1)2- to 3(4)-lobed, the lobe apex acute to rounded, aristate, the upper and lower surfaces appressed pubescent, the petiole free in the proximal leaves, adnate to the stipules in the distal ones; stipules (3)4–6(7) mm long, divided about ½ their length into (5)7–10(12) oblong lobes. Flowers in a dense, subsessile cluster, shorter than the opposing leaves and stipules, 1–2 mm long; hypanthium subglobose to ellipsoid, inconspicuously 8-ribbed, pubescent or glabrescent between the ribs; sepals ciliate. Fruit ovate, 1–2 mm long, glabrate.

Disturbed sites. Rare; Escambia County. New York and Maryland south to Florida, west to Missouri, Oklahoma, and Texas, also British Columbia and Washington; Europe. Native to Europe. Spring.

EXCLUDED TAXON

Aphanes microcarpa (Boissier & Reuter) Rothmaler—Reported for Florida by Radford et al. (1964, 1968, as *Achmella microcarpa* Boissier & Reuter), Wilhelm (1984, as *Achmella microcarpa* Boissier & Reuter), Wunderlin (1998), and Wunderlin and Hansen (2003), the name misapplied to material of *A. australis*.

Crataegus L. 1753. HAWTHORN

Shrubs or trees. Leaves alternate, simple, pinnate-veined, petiolate, stipulate. Flowers usually terminal on short shoots, corymbose or solitary, bisexual, bracteate, bracteolate; perigynous; sepals 5, free; petals 5, free; stamens 10 or 20; carpels 5, adnate to the hypanthium, the styles 3–5. Fruit a pome.

A genus of about 200 species; North America, Mexico, Central America, Europe, Africa, Asia, Australia, and Pacific Islands. [From the Greek *krataigos*, a kind of flowering thorn (*kratos*).]

There was a proliferation of *Crataegus* species names, especially by Charles S. Sargent, William W. Ashe, and Chauncey D. Beadle in the early twentieth century and in recent years by James B. Phipps. Many of these are here reduced to synonymy, creating an extensive synonymy for our 11 species.

Selected references: Phipps and Dvorsky (2007, 2008); Phipps et al. (2006).

1. Lateral veins or veinlets of the leaf blades ending in both the marginal teeth or points of the lobes (if present) and the sinuses between them.
    2. Leaf blades (at least those of the flowering or fruiting spur-branches) broadest above the middle .. ..**C. spathulata**
    2. Leaf blades broadest at or just above the base.
        3. Leaf blades of most of the leaves pinnately dissected-lobed .... **C. marshallii**
        3. Leaf blades of most of the leaves with 3 major lobes and appearing maplelike ....**C. phaenopyrum**
1. Lateral veins or veinlets of the blades ending only in the teeth or points of the lobes (if present).
    4. Leaf blades with conspicuous pinhead-like glands on the margin ....**C. michauxii**
    4. Leaf blades without glands or with small, inconspicuous glands on the margin, these often shriveling or falling off in age.
        5. Lower leaf surface with tufts of trichomes in the main vein axils.
            6. Inflorescence 5- to 20-flowered, branched corymbs .... **C. viridis**
            6. Inflorescence 1- to 4-flowered, subumbellate, unbranched corymbs.

7. Leaf blades mostly less than 5 cm long, usually obovate.................................. **C. aestivalis**

7. Leaf blades mostly 5–7 cm long, elliptic to lance-elliptic....................................... **C. opaca**

5. Lower leaf surface glabrous or sparsely pubescent, without tufts of trichomes in the main vein axils.

8. Leaf blades of the spur-branches ovate, ovate-oblong, or broadly elliptic, broadest at or below the middle, long-petiolate.

9. Leaf serrations small, crenate or crenate-serrate; leaves rarely lobed...................... **C. flava**

9. Leaf serrations sharp or coarse; leaves usually lobed .... **C. sargentii**

8. Leaf blades of spur-branches broadest above the middle, tapered to the short petiole.

10. Flowers solitary (sometimes 2–3); leaves of the spur-branches usually 1–3 cm long ....**C. uniflora**

10. Flowers several to many in branched corymbs; leaves of the spur branches usually 3–5 mm long ....**C. crus-galli**

Crataegus aestivalis (Walter) Torr. & A. Gray [Of summer.] MAY HAW; MAY HAWTHORN.

Mespilus aestivalis Walter, Fl. Carol. 148. 1788. *Crataegus aestivalis* (Walter) Torrey & A. Gray, Fl. N. Amer. 1: 468. 1840. *Anthomeles aestivalis* (Walter) M. Roemer, Fam. Nat. Syn. Monogr. 3: 141. 1847.

Crataegus fruticosa Sargent, Trees & Shrubs 1: 13, pl. 7. 1902. TYPE: FLORIDA: Volusia Co.: near Seville, Aug 1900 & Apr 1901, *Curtis 6703* (holotype: A?).

Crataegus luculenta Sargent, Trees & Shrubs 1: 11, pl. 6. 1902. *Crataegus aestivalis* (Walter) Torrey & A. Gray forma *luculenta* (Sargent) Sargent, J. Arnold Arbor. 1: 251. 1920. TYPE: FLORIDA: Volusia Co.: Haw Creek, Jul 1900 & Apr 1901, *Curtiss 6677* (holotype: A?).

Crataegus maloides Sargent, Trees & Shrubs 1: 9, pl. 5. 1902. *Crataegus aestivalis* (Walter) Torrey & A. Gray var. *maloides* (Sargent) Sargent, J. Arnold Arbor. 1: 250. 1920. TYPE: FLORIDA: Volusia Co.: Haw Creek, 6 mi. E of Seville, Jul 1900 & Mar–Jun 1901, *Curtiss 6676* (holotype: A?).

Crataegus cerasoides Sargent, Trees & Shrubs 2: 237. 1913. *Crataegus aestivalis* (Walter) Torrey & A. Gray var. *cerasoides* (Sargent) Sargent, J. Arnold Arbor 1: 250. 1920. TYPE: FLORIDA: Volusia Co.: near Seville, Jun–Jul 1901, *Curtiss 6842* (holotype: A?).

Crataegus monantha Sargent, Trees & Shrubs 2: 237. 1913. TYPE: FLORIDA: Volusia Co.: near Seville, 16 Jul 1900, *Curtiss 6703* (holotype: A?).

Crataegus rufula Sargent, J. Arnold Arbor. 1: 251. 1920.

Shrubs or small trees, to 12 m; branchlets gray turning dark brown, glabrous, with thorns 2–4 cm long, straight or nearly so, often forming short shoots. Leaves with the blade obovate, 3–5 cm long, 1.5–2 cm wide, the apex acute to obtuse, the base cuneate, the margin serrate to crenate distally, the teeth sometimes tipped with small black glands, the lateral veins 4–5 pairs, the upper surface dark green, somewhat shiny, scabrous or glabrous, the lower surface with tufts of pale gray to somewhat rufous trichomes in the axils of the lateral veins and sometimes also along the midvein, otherwise glabrous, the petiole 3–5 mm long, winged distally (sometimes much larger on elongating shoots and then sometimes a few lobed). Flowers (1)2–4 in an umbellate cyme, the axis glabrous or sparsely long-pilose; bracts glandular; perianth 12–30 mm in diameter; hypanthium glabrous; sepals short-triangular, entire or nearly so, glabrous; petals suborbicular, white; stamens 20, the anthers pink; carpels 4–5. Fruit subglobose, 8–20 cm long, red or yellow, the persistent sepals short, blunt; pyrenes 4–5, dorsally grooved, the lateral faces smooth.

Wet hammocks and floodplain forests. Occasional; northern counties, Levy and Volusia Counties. Virginia south to Florida, west to Mississippi. Spring.

Crataegus crus-galli L. [Spur of a cock, apparently in reference to the spines.] COCKSPUR HAWTHORN.

Crataegus crus-galli Linnaeus, Sp. Pl. 1: 476. 1753. *Crataegus crus-galli* Linnaeus var. *splendens* Aiton, Hort. Kew. 2: 170. 1786, nom. inadmiss. *Mespilus crus-galli* (Linnaeus) Castiglioni, Viagg. Stati Uniti 2: 294. 1790. *Mespilus cuneifolia* Moench, Methodus 684. 1794, nom. illegit. *Oxyacantha crus-galli* (Linnaeus) Nieuwland, Amer. Midl. Naturalist 4: 277. 1915.

Crataegus pyracanthoides Beadle, Biltmore Bot. Stud. 1: 136. 1902. TYPE: FLORIDA: Jackson Co.: banks of the Chipola River near Mariana.

Crataegus limnophila Sargent, J. Arnold Arbor. 3: 3. 1922. *Crataegus pyracanthoides* Beadle var. *limnophila* (Sargent) E. J. Palmer, J. Arnold Arbor. 13: 420. 1932. TYPE: FLORIDA: Wakulla Co.: near St. Marks, 13 Apr 1920, *Harbison 5692* (holotype: A?).

Crataegus limnophiloides Murrill, Castanea 7: 22. 1942. TYPE: FLORIDA: Clay Co.: between Middleburg and Doctor's Inlet, 30 Oct 1940, *Murrill & Tisdale 34397* (holotype: FLAS).

Shrub or small tree, to 8 m; branchlets reddish brown, turning gray, the thorns 2–4.5 cm long, straight or a little recurved, brown to blackish, turning gray. Leaves with the blade oblanceolate, narrowly obovate, or narrowly elliptic, 2–4(5) cm long, 1–2 cm wide, the apex obtuse or acute, the base cuneate, the margin crenate to serrate distally, the lateral veins (5)6–8 pairs, the upper surface usually pubescent along the midrib, otherwise glabrous, the lower surface glabrous, the petiole 2–6 mm long. Flowers 5–10 in an umbellate cyme, the axis glabrous or pubescent; bracteoles linear, caducous, the margin sparsely glandular; perianth ca. 10 mm in diameter; hypanthium glabrous; sepals triangular, entire, glabrous; petals suborbicular, white; stamens 10–20, cream or pink; carpels (1)2(3). Fruit globose, 5–10 mm long, red, glabrous; pyrenes (1)2(3), dorsally grooved, the lateral faces smooth.

Creek swamps. Northern peninsula south to Lake and Sumter Counties, west to central panhandle. Quebec south to Florida, west to Ontario, Minnesota, Iowa, Kansas, Oklahoma, and Texas. Spring.

Crataegus flava Aiton [Yellow, apparently in reference to the leaves and fruit.] YELLOWLEAF HAWTHORN.

Crataegus flava Aiton, Hort. Kew. 2: 169. 1789. *Mespilus flava* (Aiton) Dumont de Courset, Bot. Cult. 3: 328. 1802. *Anthomeles flava* (Aiton) M. Roemer, Fam. Nat. Syn. Monogr. 3: 142. 1847.

Shrub or small tree, to 6 m; branchlets slightly zigzag at the nodes, purple-brown, turning gray, glabrous, the thorns straight or slightly curved, 2–3 cm long, purple-brown. Leaves with the blade rhombic-obovate, 5–8 cm long, 3–5 cm wide, the apex subacute to obtuse, the base cuneate, the margin shallowly 1–3-lobed per side, the lobes obtuse to acute, the base cuneate, the lateral veins 3–4 pairs, the upper surface glabrous, the lower surface pubescent, becoming glabrescent, the petioles 0.7–1.5 cm long, glandular, winged distally. Flowers 4–6 in an umbellate cyme, the axis slightly villous or glabrous; bracteoles linear, the margin glandular, caducous; perianth 15–18 mm in diameter; hypanthium glabrous; sepals narrow-triangular, 4–5 mm long, the margin glandular; petals suborbicular, white; stamens 10, the anthers purple; carpels 3–5.

Fruit pyriform-oblong, 8–12 mm long, dull orange, the sepals somewhat reflexed; pyrenes 3–5, dorsally sulcate, the lateral faces smooth.

Open, upland hammocks. Rare; Jackson County. Pennsylvania south to Florida, west to Tennessee and Mississippi. Spring.

Crataegus marshallii Eggl. [Commemorates Humphrey Marshall (1722–1801), American botanist.] PARSLEY HAWTHORN.

> *Mespilus apiifolia* Marshall, Arbust. Amer. 89. 1785. *Mespilus oxyacantha* Crantz var. *americana* Castiglioni, Viegg. Stati Uniti 2: 292. 1790. *Crataegus marshallii* Eggleston, in Britton & Shafer, N. Amer. Trees 473. 1908.
> *Crataegus apiifolia* Michaux, Fl. Bor.-Amer. 1: 287. 1803; non Medikus, 1793. *Crataegus oxyacantha* Linnaeus var. *apiifolia* Regel, Trudy Imp. S.-Peterburgsk. Bot. Sada 1: 119. 1871.

Shrub or small tree, to 8 m; branchlets dark purple-brown, becoming dull gray, the thorns 2–3 cm long, straight or slightly curved, dark purple-brown. Leaves with the blade broadly ovate to deltate, 1.5–3 cm long, 1.5–3 cm wide, the apex acute, the base subtruncate, the margin deeply 3-lobed on each side, the sinus narrow, serrate, the lateral veins 7–8 pairs, the veins or veinlets ending in both the teeth or the apex of the lobes and the sinuses, the upper and lower surfaces pubescent, the petiole 1.5–3 cm long. Flowers 3–8 in an umbellate cyme; bracteoles linear, the margin glandular, caducous; perianth 12–17 mm in diameter; hypanthium pubescent; sepals narrowly triangular; petals elliptic, white; stamens 20, the anthers deep rose to red, rarely yellow; carpels 1–2(3). Fruit ellipsoid, 4–6 mm long, bright red; pyrenes 1–2.

Floodplain forests. Frequent; northern counties, central peninsula. Virginia south to Florida, west to Missouri, Oklahoma, and Texas. Spring.

Crataegus michauxii Pers. [Commemorates André Michaux (1746–1802), French explorer in North America.] MICHAUX'S HAWTHORN.

> *Crataegus glandulosa* Michaux, Fl. Bor.-Amer. 1: 288. 1803; non Georgi, 1775; nec Moench, 1785; nec Aiton, 1789. *Crataegus michauxii* Persoon, Syn. Pl. 2: 38. 1806.
> *Crataegus flava* Aiton var. *integra* Nash, Bull. Torrey Bot. Club 22: 150. 1895. *Crataegus integra* (Nash) Beadle, Biltmore Bot. Stud. 1: 87. 1902. *Crataegus lassa* Beadle var. *integra* (Nash) Lance, Phytoneuron 2011(3): 5. 2011. TYPE: FLORIDA: Lake Co.: Lake Ella, 2 Jul 1894, *Nash 1142* (holotype: NY; isotype A).
> *Crataegus alabamensis* Beadle, Bot. Gaz. 30: 342. 1900.
> *Crataegus aprica* Beadle, Bot. Gaz. 30: 335. 1900.
> *Crataegus senta* Beadle, Bot. Gaz. 30: 341. 1900.
> *Crataegus sororia* Beadle, Bot. Gaz. 30: 336. 1900.
> *Crataegus lacrimata* Small, Torreya 1: 97. 1901. *Crataegus michauxii* Persoon var. *lacrimata* (Small) D. B. Ward, Phytologia 91: 21. 2009. TYPE: FLORIDA: Okaloosa Co.: near Crestview, 8 Apr 1899, *Beadle 17* (lectotype: US). Lectotypified by Phipps and Dvorsky (2008: 1112).
> *Crataegus condigna* Beadle, Biltmore Bot. Stud. 1: 35. 1901. TYPE: FLORIDA: Gadsden Co.: River Junction, 5 Sep 1899, *Beadle 1095* (lectotype: US). Lectotypified by Phipps and Dvorsky (2008: 1124).
> *Crataegus consanguinea* Beadle, Biltmore Bot. Stud. 1: 34. 1901. TYPE: FLORIDA: Leon Co.: W of Tallahassee, 28 Mar 1900, *Beadle 2044* (lectotype: US). Lectotypified by Phipps and Dvorsky (2007: 180).

Crataegus dispar Beadle, Biltmore Bot. Stud. 1: 28. 1901.

Crataegus lassa Beadle, Biltmore Bot. Stud. 1: 29. 1901.

Crataegus lepida Beadle, Biltmore Bot. Stud. 1: 36. 1901.

Crataegus munda Beadle, Biltmore Bot. Stud. 1: 38. 1901.

Crataegus quaesita Beadle, Biltmore Bot. Stud. 1: 33. 1901. TYPE: FLORIDA: Gadsden Co.: River Junction, 3 Apr 1900, *Beadle 2080* (lectotype: US). Lectotypified by Phipps and Dvorsky (2008: 1120).

Crataegus segnis Beadle, Biltmore Bot. Stud. 1: 32. 1901. *Crataegus sororia* Beadle var. *segnis* (Beadle) Lance, Phytoneuron 2011(3): 2. 2011.

Crataegus floridana Sargent, Bot. Gaz. 33: 124. 1902. *Crataegus quaesita* Beadle var. *floridana* (Sargent) Lance, Phytoneuron 2011(3): 5. 2011. TYPE: FLORIDA: Duval Co.: near Jacksonville, 18 Jul [no year], *Curtiss 8* (lectotype: A). Lectotypified by Phipps and Dvorsky (2008: 1120).

Crataegus ravenelii Sargent, Bot. Gaz. 33: 122. 1902. *Crataegus alabamensis* Beadle var. *ravenelii* (Sargent) Lance, Phytoneuron 2011(3): 4. 2011.

Crataegus abdita Beadle, Biltmore Bot. Stud. 1: 75. 1902. TYPE: FLORIDA: Gadsden Co.: River Junction.

Crataegus adunca Beadle, Biltmore Bot. Stud. 1: 87. 1902. TYPE: FLORIDA: Leon Co.: Tallahassee, 28 Aug 1901, *Harbison 4941* (lectotype: US). Lectotypified by Phipps and Dvorsky (2008: 1128).

Crataegus adusta Beadle, Biltmore Bot. Stud. 1: 87. 1902. TYPE: FLORIDA: Alachua Co.: Gainesville, 1901, *Beadle 4065* (lectotype: US). Lectotypified by Phipps and Dvorsky (2008: 1117).

Crataegus amica Beadle, Biltmore Bot. Stud. 1: 97. 1902. TYPE: FLORIDA: Marion Co.: Ocala, 20 Aug 1901, *Beadle 40042* (lectotype: NY). EPITYPE: FLORIDA: Marion Co.: Ocala, 21 Mar 1901, *Beadle 4004* (epitype: US). Lecto- and epitypified by Phipps and Dvorsky (2008: 1138).

Crataegus anisophylla Beadle, Biltmore Bot. Stud. 1: 99. 1902. TYPE: FLORIDA: Duval Co.: banks of the St. Johns River near Jacksonville, 30 Mar 1901, *Beadle 4067* (lectotype: NY; isolectotype: US). Lectotypified by Phipps and Dvorsky (2008: 1120).

Crataegus annosa Beadle, Biltmore Bot. Stud. 1: 83. 1902.

Crataegus attrita Beadle, Biltmore Bot. Stud. 1: 98. 1902.

Crataegus audens Beadle, Biltmore Bot. Stud. 1: 114. 1902. TYPE: FLORIDA: Gadsden Co.: near Chattahoochee, 30 Aug 1902, *Harbison 4963* (lectotype: US). EPITYPE: FLORIDA: Gadsden Co.: Chattahoochee, 8 Apr 1901, *Harbison 4097* (epitype: US). Lecto- and epitypified by Phipps and Dvorsky (2008: 1124).

Crataegus clara Beadle, Biltmore Bot. Stud. 1: 75. 1902. TYPE: FLORIDA: Liberty Co.: Bristol, 4 Apr 1901, *Harbison 6033* (lectotype: NY). Lectotypified by Phipps and Dvorsky (2008: 1124).

Crataegus compitalis Beadle, Biltmore Bot. Stud. 1: 93. 1902. TYPE FLORIDA: Alachua Co.: near Gainesville, 22 Aug 1901, *Beadle 40602* (lectotype: US). Lectotypified by Phipps and Dvorsky (2008: 1124).

Crataegus crocea Beadle, Biltmore Bot. Stud. 1: 113. 1902. TYPE: FLORIDA: Marion Co.: near Citra, 21 Aug 1901, *Beadle 44452* (lectotype: US). Lectotypified by Phipps & Dvorsky (2008: 1117).

Crataegus curva Beadle, Biltmore Bot. Stud. 1: 109. 1902. TYPE: FLORIDA: Duval Co.: Jacksonville, 15 Aug 1901, *Beadle 48022* (lectotype: US). EPITYPE: FLORIDA: Duval Co.: Jacksonville, 30 Mar 1901, *Beadle 4802* (epitype: NY). Lecto- and epitypified by Phipps and Dvorsky (2008: 1117).

Crataegus dapsilis Beadle, Biltmore Bot. Stud. 1: 89. 1902. TYPE: FLORIDA: Lake Co.: Lane Park, near Tavares, 17 Aug 1901, *Beadle 4836* (lectotype: US). Lectotypified by Phipps and Dvorsky (2008: 1128).

Crataegus egans Beadle, Biltmore Bot. Stud. 1: 85. 1902. *Crataegus quaesita* Beadle var. *egans* (Beadle) Lance, Phytoneuron 2011(3): 5. 2011. TYPE: FLORIDA: Liberty Co.: near Bristol, 1 Apr 1901, *Harbison 4037* (lectotype: GH; isotypes: A, US). Lectotypified by Phipps and Dvorsky (2008: 1108).

Crataegus egregia Beadle, Biltmore Bot. Stud. 1: 82. 1902. TYPE: FLORIDA: Liberty Co.: near Bristol, 24 Aug 1901, *Harbison 4942* (lectotype: A). Lectotypified by Phipps and Dvorsky (2007: 178).

Crataegus florens Beadle, Biltmore Bot. Stud. 1: 94. 1902. *Crataegus alabamensis* Beadle var. *florens* (Beadle) Lance, Phytoneuron 2011(3): 4. 2011.

Crataegus furtiva Beadle, Biltmore Bot. Stud. 1: 81. 1902.

Crataegus galbana Beadle, Biltmore Bot. Stud. 1: 74. 1902. TYPE: FLORIDA: Gadsden Co.: SW of River Junction, 3 Apr 1900, *Beadle 2083* (lectotype: NY). Lectotypified by Phipps and Dvorsky (2007: 177).

Crataegus illudens Beadle, Biltmore Bot. Stud. 1: 111. 1902. TYPE: FLORIDA: Marion Co.: near Citra, 28 Mar 1901, *Beadle 4055* (lectotype: NY). Lectotypified by Phipps and Dvorsky (2008: 1138).

Crataegus incana Beadle, Biltmore Bot. Stud. 1: 113. 1902. TYPE: FLORIDA: Liberty Co.: near Bristol, 24 Aug 1901, *Harbison 4918* (lectotype: US). EPITYPE: FLORIDA: Liberty Co.: Bristol, 29 Mar 1901, Harbison 4020 (epitype: NY). Lecto- and epitypified by Phipps and Dvorsky (2008: 1117).

Crataegus inopina Beadle, Biltmore Bot. Stud. 1: 75. 1902. TYPE: FLORIDA: Marion Co.: Ocala, 20 Aug 1901, *Beadle 40012* (lectotype: US). EPITYPE: FLORIDA: Marion Co.: near Ocala, 21 Mar 1901, *Beadle 4001* (epitype: NY). Lecto- and epitypified by Phipps and Dvorsky (2008: 1118).

Crataegus lanata Beadle, Biltmore Bot. Stud. 1: 86. 1902. *Crataegus lassa* Beadle var. *lanata* (Beadle) Lance, Phytoneuron 2011(3): 5. 2011.

Crataegus mira Beadle, Biltmore Bot. Stud. 1: 78. 1902. *Crataegus allegheniensis* Beadle var. *mira* (Beadle) Lance, Phytoneuron 2011(3): 2. 2011.

Crataegus panda Beadle, Biltmore Bot. Stud. 1: 89. 1902. TYPE: FLORIDA: Leon Co.: Tallahassee, Sep 1901, *Harbison 4051* (lectotype: A; isolectotype: NY). Lectotypified by Phipps and Dvorsky (2008: 1138).

Crataegus rava Beadle, Biltmore Bot. Stud. 1: 91. 1902. TYPE: FLORIDA: Leon Co.: near Tallahassee, s.d., *Harbison 4064* (lectotype: US). Lectotypified by Phipps and Dvorsky (2008: 1124).

Crataegus recurva Beadle, Biltmore Bot. Stud. 1: 106. 1902. *Crataegus lassa* Beadle var. *recurva* (Beadle) Lance, Phytoneuron 2011(3): 5. 2011. TYPE: FLORIDA: Marion Co.: Ocala, Mar 1901, *Beadle 40072* (lectotype: A). Lectotypified by Phipps and Dvorsky (2008: 1117).

Crataegus rimosa Beadle, Biltmore Bot. Stud. 1: 107. 1902. TYPE: FLORIDA: Marion Co.: Citra, 21 Aug 1901, *Beadle 40432* (lectotype: US). EPITYPE: FLORIDA: Marion Co.: Citra, 27 Mar 1901, Beadle *4043* (epitype: NY). Lecto- and epitypified by Phipps and Dvorsky (2008: 1124).

Crataegus versuta Beadle, Biltmore Bot. Stud. 1: 112. 1902.

Crataegus viaria Beadle, Biltmore Bot. Stud. 1: 101. 1902. TYPE: FLORIDA: Duval Co.: Jacksonville, 29 Mar 1901, *Beadle 4065* (lectotype: US). Lectotypified by Phipps and Dvorsky (2008: 1120).

Crataegus vicana Beadle, Biltmore Bot. Stud. 1: 104. 1902. TYPE: FLORIDA: Lake Co.: Tavares, 17 Aug 1901, *Beadle 40102* (lectotype: US). EPITYPE: FLORIDA: Lake Co.: Tavares, 21 Mar 1901, *Beadle 4010* (epitype: NY). Lecto- and epitypified by Phipps and Dvorsky (2008: 1118).

Crataegus villaris Beadle, Biltmore Bot. Stud. 1: 108. 1902. TYPE: FLORIDA: Marion Co.: Ocala, 27 Mar 1901, *Beadle 4042* (lectotype: US). Lectotypified by Phipps and Dvorsky (2008: 1117).

Crataegus visenda Beadle, Biltmore Bot. Stud. 1: 79: 1902. *Crataegus sororia* Beadle var. *visenda* (Beadle) Lance, Phytoneuron 2011(3): 3. 2011. TYPE: FLORIDA: Liberty Co.: near Bristol, 29 Mar 1901, *Harbison 4031* (lectotype: A). Lectotypified by Phipps and Dvorsky (2007: 175).

Crataegus leonensis E. J. Palmer, J. Arnold Arbor. 13: 422, f. 1. 1932. TYPE: FLORIDA: Leon Co.: near Tallahassee, 3 Apr 1923, *Harbison 6072* (holotype: A).

Crataegus alachuana Murrill, Castanea 7: 20. 1942. TYPE: FLORIDA: Alachua Co.: a few mi. SE of Gainesville, 4 Aug 1939, *Murrill 34277* (holotype: FLAS).

Crataegus alachuaniformis Murrill, Castanea 7: 21. 1942. TYPE: FLORIDA: Alachua Co.: Sugarfoot, near Gainesville, 27 Aug 1940, *Murrill 34276* (holotype: FLAS; isotype: US).

Crataegus dapsilis Beadle forma *serotina* Murrill, Castanea 7: 21. 1942. TYPE: FLORIDA: Marion Co.: 1 mi. S of Reddick, 30 Aug 1940, *Murrill 34898* (holotype: FLAS).

Crataegus floridana Sargent forma *flava* Murrill, Castanea 7: 22. 1902. TYPE: FLORIDA: Baker Co.: highway E of Macclenny, 22 Sep 1940, *Murrill 34419* (holotype: FLAS).

Crataegus globirimosa Murrill, Castanea 7: 22. 1942. TYPE: FLORIDA: Marion Co.: N of Ocala, 30 Aug 1940, *Murrill 34420* (holotype: FLAS).

Crataegus megapulchra Murrill, Castanea 7: 23. 1942. TYPE: FLORIDA: Alachua Co.: 2 mi. N of Alachua, 4 Aug 1939, *Murrill 34454* (holotype: FLAS).

Crataegus praeformosa Murrill, Castanea 7: 42. 1942. TYPE: FLORIDA; Alachua Co.: ca. 2 mi. S of Warren's Cave, 10 Sep 1940, *Murrill 34618* (holotype: FLAS; isotype: US).

Crataegus pyripulchra Murrill, Castanea 7: 25. 1942. TYPE: FLORIDA: Alachua Co.: Sugarfoot, near Gainesville, 11 Oct 1940, *Murrill 34619* (holotype: FLAS).

Crataegus ravenelii Sargent forma *superba* Murrill, Castanea 7: 26. 1942. TYPE: FLORIDA: Marion Co.: 3 mi. S of Ocala, 4 Aug 1939, *Murrill 34901* (holotype: FLAS).

Crataegus rimosiformis Murrill, Castanea 7: 26. 1942. TYPE: FLORIDA: Alachua Co.: campus of the University of Florida, 30 Mar 1941, *Murrill 34886* (holotype: FLAS).

Crataegus subaudens Murrill, Castanea 7: 27. 1942. TYPE: FLORIDA: Alachua Co.: just W of Gainesville, 25 Jul 1940, *Murrill 34603* (holotype: FLAS).

Crataegus subflavida Murrill, Castanea 7: 28. 1942. TYPE: FLORIDA: Alachua Co.: just W of Gainesville, 25 Aug 1940, *Murrill 34599* (holotype: FLAS).

Crataegus tisdalei Murrill, Castanea 7: 29. 1942. TYPE: FLORIDA: Madison Co.: near the Suwannee River at Ellaville, 22 Aug 1940, *Murrill 34900* (holotype: FLAS).

Crataegus visendiformis Murrill, Castanea 7: 29. 1942. TYPE: FLORIDA: Alachua Co.: Carroll's Field, Gainesville, 31 Jul 1939, *Murrill 34895* (holotype: FLAS).

Crataegus watsonii Murrill, Castanea 7: 30. 1942. TYPE: FLORIDA: Alachua Co.: just W of Gainesville, 19 Sep 1940, *Murrill & Watson 34636* (holotype: FLAS).

Crataegus lancei J. B. Phipps, in J. B. Phipps and Dvorsky, J. Bot. Res. Inst. Texas 2: 1143, f. 30. 2008.

Shrub or small tree, to 5 m; branchlets slightly zigzag, gray or reddish brown, pubescent or glabrous, the thorns 1.5–3 cm long, straight or slightly curved. Leaves with the blade broadly rhombic-obovate or -elliptic to narrowly obovate, 1.5–4 cm long, the apex rounded to obtuse, sometimes mucronate, the base cuneate, the margin obscurely crenate, with conspicuous pinhead-like glands on the margin, sometimes shallowly 1- to 2-lobed per side, the lateral veins 3–4(5) pairs, the upper surface sparsely pilose, becoming glabrescent, the lower surface pilose on the veins, otherwise glabrous, the petiole to 1 cm long, pilose, glandular. Flowers 2–4(7) in an umbellate cyme, the axis glabrous or pilose; bracteoles linear, glandular, caducous; perianth 14–18 mm in diameter; hypanthium glabrous or pilose; sepals triangular, the margin irregularly serrate, glandular; petals suborbicular, white; stamens 20, the anthers purple; carpels (2)3–5. Fruit subglobose to pyriform, 6–12 mm long, orange to red, the sepals recurved; pyrenes (2)3–5, grooved dorsally, the lateral faces smooth.

Sandhills, dry hammocks, and scrub. Common; northern counties, central peninsula. Pennsylvania south to Florida, west to Tennessee and Mississippi. Spring.

Crataegus opaca Hook. & Arn. [Darkened, dull, in reference to the leaf surface.] RIVERFLAT HAWTHORN; WESTERN MAYHAW.

Crataegus opaca Hooker & Arnott, in Hooker, Compan. Bot. Mag. 1: 25. 1835.

Tree, to 8 m; branchlets rufous-tomentose, turning gray, the thorns 2–4 cm long, straight. Leaves with the blade elliptic to lance-elliptic, 5–9 cm long, 3–5 cm wide, the apex acute to obtuse, the base cuneate, the margin unlobed or sinuate-lobed, crenate, gland-dotted, the lateral veins 5–10 pairs, the upper surface scabrate, the lower surface rufous-tomentose to glabrate, the petiole 4–7 mm long, rufous-tomentose. Flowers 3–4(6) in an umbellate cyme, the branches glabrous; bracteoles oblong-linear, glandular, caducous; perianth 12–18 mm in diameter; hypanthium glabrous; sepals triangular, 4 mm long, entire to slightly glandular-serrate; petals suborbicular, pale rose; stamens 20, the anthers reddish or rose; carpels 4–5. Fruit globose, 12–15 mm long, red or yellow, portions of the calyx persistent; pyrenes 4–5, shallowly grooved dorsally, the lateral faces smooth.

River swamps. Rare; Santa Rosa and Escambia Counties. Florida and Alabama west to Texas. Spring.

Crataegus phaenopyrum (L. f.) Medik. [From the Greek *phaeo*, dark, and *pyro*, fiery, in reference to the dark red fruits.] WASHINGTON HAWTHORN.

> *Mespilus phaenopyrum* Linnaeus f., Suppl. Pl. 254. 1782. *Crataegus phaenopyrum* (Linnaeus f.) Medikus, Gesch. Bot. 84. 1793.
>
> *Crataegus youngii* Sargent, J. Arnold arbor. 4: 105, 1923.

Tree, to 10 m; branchlets deep reddish brown, glossy, turning dark gray, glabrous, the thorns 2–5 cm long, straight. Leaves with the blade broadly to narrowly deltate, 3–6(7) cm long, 3–6 cm wide, palmately 3- to 5(7)-lobed, the apex acute, the base truncate or cordate to broadly cuneate, the margin serrulate, the lateral veins 6–7(8) pairs, the veins and the veinlets ending in both the teeth or apex of the lobes and the sinuses, the upper and lower surfaces glabrous, the petiole 1.5–3.5 cm long, usually glandular. Flowers 20–30 in an umbellate cyme; bracteoles linear, the margin glandular, caducous; perianth 10–12 mm in diameter; hypanthium glabrous; sepals broadly triangular, 2 mm long; petals suborbicular, white; stamens 20; carpels 3(4). Fruit globose, 5–8 mm long, the calyx remnants persistent; pyrenes 3(4), dorsally grooved, the lateral faces smooth.

Moist to wet hammocks. Rare; Wakulla, Liberty, Washington, and Walton Counties. Maine south to Florida, west to Ontario, Michigan, Illinois, Missouri, Arkansas, and Louisiana. Spring.

Crataegus phaenopyrum is listed as endangered in Florida (Florida Administrative Code, Chapter 5B-40).

Crataegus sargentii Beadle [Commemorates Charles Sprague Sargent (1841–1927), American botanist and first director of the Arnold Arboretum.] BEAUTIFUL HAWTHORN; SARGENT'S HAWTHORN.

> *Crataegus sargentii* Beadle, Bot. Gaz. 28: 407. 1899.
>
> *Crataegus pulcherrima* Ashe, J. Elisha Mitchell Sci. Soc. 16: 77. 1900. TYPE: FLORIDA: Gadsden Co.: River Junction, 10–11 Aug 1895, *Nash 2377* (lectotype: DOV; isolectotype: US). Lectotypified by Phipps et al. (2006: 983).
>
> *Crataegus venusta* Beadle, Bot. Gaz. 30: 338. 1900.
>
> *Crataegus incilis* Beadle, Biltmore Bot. Stud. 1: 41. 1901. *Crataegus pulcherrima* Ashe var. *incilis* (Beadle) Lance, Phytoneuron 2011(3): 7. 2011.
>
> *Crataegus opima* Beadle, Biltmore Bot. Stud. 1: 40. 1901. *Crataegus pulcherrima* Ashe var. *opima* (Beadle) Lance, Phytoneuron 2011(3): 7. 2011.
>
> *Crataegus tecta* Beadle, Biltmore Bot. Stud. 1: 26. 1901.
>
> *Crataegus abstrusa* Beadle, Biltmore Bot. Stud. 1: 66. 1902. TYPE: FLORIDA: Leon Co.: near

Tallahassee, 29 Aug 1901, *Harbison H4958 [=H4059/2]* (lectotype: US). Lectotypified by Phipps et al. (2006: 990).

Crataegus assimilis Beadle, Biltmore Bot. Stud. 1: 68. 1902. TYPE: FLORIDA: Jackson Co.: near Chattahoochee, 8 Apr 1901, *Harbison H4096* (lectotype: US). Lectotypified by Phipps et al. (2006: 979).

Crataegus concinna Beadle, Biltmore Bot. Stud. 1: 70. 1902. TYPE: FLORIDA: Liberty Co.: near Bristol. 2 Apr 1901, *Harbison H4096* (lectotype: US). Lectotypified by Phipps et al. (2006: 996).

Crataegus contrita Beadle, Biltmore Bot. Stud. 1: 61. 1902. TYPE: FLORIDA: Gadsden Co.: River Junction, 3 Apr 1900, *Beadle 2078* (lectotype: US). Lectotypified by Phipps et al. (2006: 990).

Crataegus gilva Beadle, Biltmore Bot. Stud. 1: 60. 1902. *Crataegus sargentii* Beadle var. *gilva* (Beadle) Lance, Phytoneuron 2011-3: 8. 2011.

Crataegus mendosa Beadle, Biltmore Bot. Stud. 1: 65. 1902.

Crataegus robur Beadle, Biltmore Bot. Stud. 1: 69. 1902. TYPE: FLORIDA: Leon Co.: Tallahassee, 18 Sep 1901, *Beadle 2051* (lectotype: A). Lectotypified by Phipps et al. (2006: 985).

Crataegus robur Beadle forma *megacarpa* Murrill, Castanea 7: 27. 1942. TYPE: FLORIDA: Alachua Co.: Plum Woods, near Gainesville, 4 Aug 1939, *Murrill 34452* (holotype: FLAS).

Crataegus robur Beadle forma *sulphurea* Murrill, Castanea 7: 27. 1942. TYPE: FLORIDA: Alachua Co.: hill W of Buzzard's Roost, 19 Jul 1940, *Murrill 34453* (holotype: FLAS).

Shrub or small tree, to 4 m; branchlets deep reddish brown, turning gray, glabrous, thorns 2–3.5 cm long, straight or slightly curved. Leaves with the blade narrowly ovate, oblong, narrowly obovate or narrowly rhombic, 2.5–5 cm long, 2–4 cm wide, with 2–4 shallow lobes per side, the apex acute, the base cuneate, the margin serrulate, the lateral veins 2–5(6) pairs, the upper surface slightly pubescent along the main veins when young, otherwise glabrous, the lower surface glabrous, the petiole 1–2 cm long, glabrous, with sessile or stipitate glands. Flowers 4–7 in an umbellate cyme, the axis glabrous; bracteoles oblong, the margin stipitate-glandular, caducous; perianth 15–20 mm in diameter; hypanthium glabrous; sepals narrow-triangular, 3–4 mm long, the margins glandular-serrate, glabrous; petals suborbicular, white; stamens 20, the anthers pink; carpels usually 3. Fruit globose, 6–7 mm long, yellow or reddish purple, the sepals reflexed, persistent; pyrenes usually 3, dorsally grooved, the lateral faces smooth.

Upland deciduous woods. Occasional; central panhandle, Jefferson and Alachua Counties. Tennessee, Alabama, Georgia, and Florida. Spring.

Crataegus spathulata Michx. [Spoon shaped, in reference to the leaves.] LITTLEHIP HAWTHORN.

Crataegus spathulata Michaux, Fl. Bor-Amer. 1: 288. 1803. *Mespilus spathulata* (Michaux) Poiret, in Lamarck, Encycl., Suppl. 4: 68. 1816. *Phaenopyrum spathulatum* (Michaux) M. Roemer, Fam. Nat. Syn. Monogr. 3: 155. 1847. *Cotoneaster spathulata* (Michaux) Wenzig, Linnaea 38: 201. 1874.

Shrub or small tree, to 7 m; branchlets purple-brown, turning gray, the thorns 3–4 cm long, straight. Leaves with the blade narrowly or broadly spatulate, 1.5–3 cm long, 0.5–2 cm wide, shallowly to deeply lobed on each side, the apex acute to rounded, the base cuneate, the margin subentire or crenate to serrate, the lateral veins 5–8 pairs, the veins or veinlets ending in both the teeth or the apex of the lobes and the sinuses, the upper surface with long trichomes on the midrib, the lower surface with scattered trichomes, the petiole 0.5–1 cm long. Flowers 20–30 in an umbellate cyme, the axis glabrous; bracteoles stipular or apparently absent, caducous; perianth ca. 10 mm in diameter; hypanthium glabrous; sepals triangular, glabrous; petals suborbicular, white; stamens 20, the anthers usually pale yellow; carpels 3–5. Fruit globose, 3–5 mm long, orange to bright red; pyrenes 3–6, obscurely dorsally grooved, the lateral faces smooth.

Moist to wet hammocks and stream banks. Occasional; central and western panhandle. Virginia south to Florida, west to Missouri, Oklahoma, and Texas. Spring.

Crataegus uniflora Münchh. [One-flowered.] DWARF HAWTHORN.

> *Crataegus uniflora* Münchhausen, Hausvater 5: 147. 1770. *Phaenopyrum uniflorum* (Münchhausen) M. Roemer, Fam. Nat. Syn. Monogr. 3: 153. 1847. *Mespilus uniflora* (Münchhausen) K. Koch, Dendrologie 1: 141. 1869.
>
> *Crataegus parvifolia* Aiton, Hort. Kew. 2: 169. 1789. *Mespilus parvifolia* (Aiton) Willdenow, Enum. Pl. 523. 1809. *Phaenopyrum parvifolium* M. Roemer, Fam. Nat. Syn. Monogr. 3: 152. 1847.
>
> *Crataegus grossiserrata* Ashe, Bull. North Carolina Agric. Exp. Sta. 175: 112. 1900. TYPE: FLORIDA.
>
> *Crataegus choriphylla* Sargent, J. Arnold Arbor. 3: 201. 1923. TYPE: FLORIDA: Columbia Co.: Lake Co·12 Jun 1917, *Harbison 5687* (holotype: A?).
>
> *Crataegus croomiana* Sargent, J. Arnold Arbor. 3: 202. 1923. TYPE: FLORIDA: Leon Co.: near Tallahassee, 15 Apr 1920, *Harbison 5710* (holotype: A?).

Shrub, to 2(5) m; branchlets brown, densely appressed pubescent, turning grayish, the thorns 3–5(8) cm long, straight. Leaves with the blade elliptic, spatulate, oblanceolate, rhombic-elliptic or suborbicular, (1)1.5–3(6) cm long, 1–2(3.5) cm wide, the apex obtuse-rounded, the base cuneate, with variably caduceus, black gland-tipped teeth or eglandular, the margin crenate, the lateral veins 3–4(5) pairs, the upper surface scabrous-pubescent, glossy, the lower surface pubescent, the petiole to 1 cm, glandular or eglandular. Flowers 1 solitary or rarely 3–5 in an umbellate cyme, the pedicel pubescent; bracteoles linear, 1–2, the margin glandular; perianth 10–15 mm in diameter; hypanthium tomentose; sepals narrowly triangular, glandular-serrate, sparsely pubescent; petals suborbicular, white; stamens 20, the anthers white to cream; carpels 5. Fruit suborbicular, 8–10(12) mm long, yellow-green, yellow, or yellow-orange, tomentose, the sepals persistent, spreading; pyrenes 4–5, dorsally furrowed, the lateral faces smooth.

Sandhills and hammocks. Occasional; northern counties south to Polk County. New York south to Florida, west to Missouri, Oklahoma, and Texas. Spring.

Crataegus viridis L. [Green, in reference to the foliage.] GREEN HAWTHORN.

> *Crataegus viridis* Linnaeus, Sp. Pl. 1: 476. 1753. *Mespilus coccinea* (Linnaeus) Marshall var. *viridis* (Linnaeus) Castiglioni, Viagg. Stati Uniti 2: 293. 1790. *Mespilus viridis* (Linnaeus) Sweet, Hort. Brit. 134. 1826. *Crataegus coccinea* Linnaeus var. *viridis* (Linnaeus) Torrey & A. Gray, Fl. N. Amer. 1: 465. 1840. *Phaenopyrum viride* (Linnaeus) M. Roemer, Fam. Nat. Syn. Monogr. 3: 156. 1847.
>
> *Crataegus arborescens* Elliott, Sketch Bot. S. Carolina 1: 550. 1821.
>
> *Crataegus subviridis* Beadle, Biltmore Bot. Stud. 1: 51. 1902. TYPE: FLORIDA: Gadsden Co.: near Chattahoochee.
>
> *Crataegus paludosa* Sargent, Trees & Shrubs 1: 15, pl. 8. 1902. TYPE: FLORIDA: Volusia Co.: Haw Creek, 1 Aug 1900, Mar & Sep 1901, *Curtiss 6678* (holotype (A?).
>
> *Crataegus newelliana* Murrill, Castanea 7: 24. 1942. TYPE: FLORIDA: Levy Co.: vicinity of Gulf Hammock, 7 Apr 1940, *Murrill 34609* (holotype: FLAS).
>
> *Crataegus subpaludosa* Murrill, Castanea 7: 28. 1942. TYPE: FLORIDA: Suwannee Co.: margin of a small limesink near Hildreth, 10 Jul 1940, *Murrill 34892* (holotype: FLAS).

Shrub or tree, to 15 m; branchlets reddish, glabrous, turning gray, the thorns 3–4 cm long, straight, blackish. Leaves with the blade narrowly rhombic, oblong, or suborbicular, 2–6 cm long, 1.5–3(6) cm wide, unlobed or sharply 1- to 3-lobed per side, the apex acute to obtuse, the base cuneate to rounded, the margin serrate or crenate-serrate, the lateral veins 3–5 pairs, the upper surface glabrous, the lower surface glabrous except for tufts of trichomes in the vein axils, the petiole 0.8–2.5 cm long. Flowers 10(30–50) in an umbellate cyme, the axis glabrous; bracteoles linear, filiform, glandular or eglandular, caducous; perianth 10(13–15) mm in diameter; hypanthium glabrous; sepals narrowly triangular; petals suborbicular; stamens 20, the anthers cream or ivory; carpels 3–5. Fruit globose, 5–8 mm in diameter, orange to red; pyrenes 3–5, the lateral faces smooth.

Wet hammocks, floodplain forests, and pond margins. Occasional; northern counties, Levy, Marion, and Volusia Counties. Pennsylvania south to Florida, west to Kansas, Oklahoma, and Texas. Spring.

EXCLUDED TAXA

Crataegus calpodendron (Ehrhart) Medikus—Reported for Florida by Radford et al. (1964, 1968). No specimens known.

Crataegus coccinea Linnaeus—This more northern species was reported for Florida by Chapman (1860), who apparently misapplied the name to our material of *C. phaenopyrum*.

Crataegus flabellata (Bosc ex Desfontaines) K. Koch—Reported for Florida by Radford et al. (1964, 1968). No specimens known.

Duchesnea Sm. 1811.

Perennial herbs. Leaves basal or 1–2 per node on stolons, 3-foliolate, pinnately veined, petiolate, stipulate. Flowers axillary, solitary, ebracteate, ebracteolate, perigynous, epicalyx bractlets 5; sepals 5, free; petals 5, free; stamens 20–30 in 3 whorls, free, the anthers basifixed, the thecae 2, dehiscing by lateral slits; carpels 50–100, the styles lateral or subterminal. Fruit an aggregate, the fleshy torus bearing numerous superficial achenes.

A genus of 2 species; North America, Europe, and Asia. [Commemorates Antoine Nicolas Duchesne (1747–1827), French botanist noted particularly for his work in *Fragaria*.]

Duchesnea indica (Andrews) Teschem. [Of India.] INDIAN STRAWBERRY.

Fragaria indica Andrews, Bot. Repos. 7: pl. 479. 1807. *Duchesnea fragiformis* Smith, Trans. Linn. Soc. London 10: 373. 1811, nom. illegit. *Duchesnea indica* (Andrews) Teschemacher, Hort. Reg. & Gard. Mag. 1: 460. 1835. *Potentilla indica* (Andrews) T. Wolf, in Ascherson & Graebner, Syn. Mitteleur. Fl. 6: 64. 1904.

Perennial herb; stem spreading to weakly erect, the stolons creeping, filiform, (1)3–10(13) dm long, bearing adventitious roots, alternate leaves, and pedicels on a short shoot with each node subtended by opposite, (1)3-lobed bractlets. Leaves 3-foliolate, the leaflet blade elliptic to obovate, (1)2–4(5) cm long, (0.8)1–3 cm wide, the upper surface pilose or glabrous, the lower surface sparsely strigose, short-petiolulate, the petiole 2–20 cm long, pilose; stipules broadly lanceolate to narrowly ovate, 5–8 mm long, adnate to the petiole base, sheathing, deeply 2-lobed, sparsely villous. Flowers solitary, axillary, the epicalyx bractlets obovate, 4–12 mm long and wide, the margin serrate apically, spreading to slightly reflexed; hypanthium cupulate, 3–5 mm in diameter, hirsute; sepals broadly lanceolate, (3)4–10 mm long, the outer surface pilose; petals narrowly obovate, 4–8(11) mm long, the apex rounded to truncate or emarginate, glabrous. Fruit hemispheric or turbinate, 1–1.5 cm long, reddish to purplish brown, fleshy, the achenes numerous, 1–2 mm long, reddish, glabrous.

Disturbed sites. Rare; northern counties south to Hillsborough County. New York and Massachusetts south to Florida, west to Ontario, Nebraska, Oklahoma, and Texas, also British Columbia south to California; Europe and Asia. Native to Asia. Spring.

Eriobotrya Lindl. 1821. LOQUAT

Trees or shrubs. Leaves alternate, simple, pinnate-veined, petiolate, stipulate. Flowers in terminal panicles, bracteate, bracteolate; sepals 5, basally connate; petals 5, free; stamens 20; carpels 5, 5-loculate, connate, ovary inferior and adnate to the hypanthium, the styles 5, free. Fruit a pome.

A genus of about 30 species; North America, West Indies, Central America, South America, Africa, Asia, and Pacific Islands. [From the Greek *erion*, wool, and *botrys*, a bunch of grapes, alluding to the fruit cluster.]

Eriobotrya japonica (Thunb.) Lindl. [Of Japan.] LOQUAT

Mespilus japonica Thunberg, Nova. Acta Regiae Soc. Sci. Upsal. 3: 208. 1780. *Eriobotrya japonica* (Thunberg) Lindley, Trans. Linn. Soc. London 13: 102. 1822. *Photinia japonica* (Thunberg) Hooker f. & Bentham ex Ascherson & Schweinfurth, Ill. Fl. Egypte 73. 1887.

Trees or shrubs, to 10 m; branchlets yellowish brown, grayish or rufous-tomentose. Leaves with the blade oblong-lanceolate to elliptic-oblong, 12–30 cm long, 3–9 cm wide, the apex acute or acuminate, the base cuneate, the lateral veins 15–25 pairs, the margin toothed in the distal ½, the upper surface glabrous, rugose, the lower surface grayish, rufous-tomentose, the petiole 6–10 mm long, grayish brown-tomentose; stipules subulate, 1–1.5 cm long, pubescent. Flowers 20–40 in a terminal panicle, the rachis densely rufous-tomentose; bracts of 1–several reduced leaves; bracteoles narrowly triangular, rufous-tomentose; hypanthium shallowly cupular, 3–4 mm long, rufous-tomentose; sepals triangular-ovate, 2–3 mm long, the outer surface rufous-tomentose; petals oblong or ovate, 8–10 mm long, white; ovary rufous-tomentose apically. Fruit subglobose, 2–3 cm long, apricot-yellow, rufous-tomentose, soon glabrescent, the pedicel 3–8 mm long; seeds 3–5, ovoid, black, glossy.

Disturbed sites. Rare; peninsula, central and western panhandle. Escaped from cultivation. Georgia, Florida, Louisiana, and California; West Indies, Mexico, Central America, and South America; Africa, Asia, Australia, and Pacific Islands. Native to Asia. Spring.

Fragaria L. 1753. STRAWBERRY

Perennial herbs; stems stoloniferous. Leaves alternate, 3-foliolate, petiolate, stipulate. Flowers in terminal cymes, bracteate, unisexual or bisexual (plants dioecious, monoecious, or polygamodioecous); epicalyx bractlets 5; sepals 5, free; petals 5, free; stamens 15 in 3 whorls (sterile staminodes in carpellate flowers); carpels 30–150, free, the style basal. Fruit with a fleshy torus bearing numerous superficial achenes.

A genus of 24 species; nearly cosmopolitan. [*Fraga*, fragrant, in reference to the sweet-smelling fruit.]

Selected reference: Staudt (1999).

Fragaria virginiana Duchesne [Of Virginia.] VIRGINIA STRAWBERRY.

> *Fragaria virginiana* Duchesne, Hist. Nat. Frais. 204. 1766. *Fragaria odora* Salisbury, Prodr. Stirp. Chap. Allerton 363. 1796, nom. illegit.

Perennial herb; stem acaulescent, stoloniferous, the stolons spreading pubescent. Leaves 3-foliolate, the leaflet blade oblong-ovate or slightly rhombic, the apex rounded, the base cuneate, the margin serrate, the upper and lower surfaces spreading-pubescent, the petiole spreading-pubescent. Flowers solitary or 2–10 in a long pedunculate cyme; perianth 12–23 mm in diameter; hypanthium saucer-shaped, 10–20 mm in diameter; sepals broadly lanceolate, 4–8 mm long; petals obovate, white. Fruit a globose to ovoid, 0.5–2 cm long, fleshy torus bearing numerous superficial achenes, the achenes ca. 1 mm long, in deep pits, the bractlets and sepals clasping or the sepals clasping and the bractlets spreading or reflexed.

Open hammocks. Rare; Leon and Jackson Counties. Nearly throughout North America, except Greenland; Europe. Native to North America. Spring.

Malus Mill. 1754. APPLE

Trees. Leaves alternate, simple, pinnate-veined, petiolate, stipulate. Flowers in terminal panicles on short shoots, bracteate, bracteolate; sepals 5, basally connate; petals 5, free, clawed; stamens 20; carpels 5, connate, the ovary inferior. Fruit a pome.

A genus of 55 species; North America, Europe, and Asia. [Greek name for the apple tree.]

Malus angustifolia (Aiton) Michx. [With narrow leaves.] SOUTHERN CRABAPPLE.

> *Pyrus angustifolia* Aiton, Hort. Kew. 2: 176. 1789. *Malus angustifolia* (Aiton) Michaux, Fl. Bor.-Amer. 1: 292. 1803. *Pyrus coronaria* Linnaeus var. *angustifoli*a (Aiton) Wenzig, Linnaea 38: 41. 1874.
> *Malus angustifolia* (Aiton) Michaux forma *pendula* Rehder, J. Arnold. Arbor. 2: 53. 1920. TYPE: FLORIDA: Gadsden Co.: River Junction, 25 Jun 1914, *Harbison 129* (holotype: A).

Trees, to 5(10) m; bark reddish brown to gray, longitudinally fissured with platelike scales, the twigs reddish brown to gray. Leaves on the long shoots with the blade elliptic or ovate, (3.5)4–6(8) cm long, (1.5)3–4(5) cm wide, the apex rounded to broadly acute, the base cuneate, the margin slightly lobed, crenate to serrate or entire, the upper surface glabrous, the lower surface villous only on the veins, the petiole (1)1.5–2.5 cm long, villous, leaves on the short shoots with the blade elliptic or oval, (1)1.5–5 cm long, 1–2(3) cm wide, the margin lacking lobes, the apex rounded, the petiole 3–10(25) mm long; stipules linear-lanceolate, 2–4(5) mm long. Flowers solitary or 3–5(7) in a flat-topped, subsessile panicle, the pedicel (1)2–3(4) cm long, glabrous; bracteoles filiform, (1)3–4(6) mm long; hypanthium urn-shaped, glabrous or rarely slightly villous; calyx lobes triangular (2)3–4(5) mm long; petals oblong to narrowly obovate, (10)12–16(22) mm long, pink, rarely white, the margin entire, sinuate, or fimbriate; stamens with the filaments purple to white, the anthers rose or pink; stigmas and styles rose to pink. Fruit subglobose, 1–2 cm long, green or yellow-green, glaucous; seeds dark brown, smooth.

Dry, often calcareous hammocks. Occasional; eastern and central panhandle. New Jersey and Pennsylvania south to Florida, west to Missouri, Arkansas, and Texas. Spring.

Malus angustifolia is listed as threatened in Florida (Florida Administrative Code, Chapter 5B-40).

EXCLUDED TAXA

Malus coronaria (Linnaeus) Miller—This northern species was reported for Florida by Small (1913a, 1913c, 1933, as *M. bracteata* Rehder), who apparently misapplied the name to material of *M. angustifolia.*

Photinia Lindl. 1820. CHOKEBERRY

Shrubs. Leaves alternate, simple, pinnate-veined, petiolate, stipulate. Flowers in a terminal corymbose inflorescence, bracteate, bracteolate; sepals 5, basally connate; petals 5, free, clawed; stamens 16–22; carpels 5, basally connate, adnate to the hypanthium (inferior), the styles 5, terminal, basally connate, ovules 2 per carpel. Fruit a pome.

A genus of about 50 species; North America, Mexico, and Asia. [From the Greek *photeinos*, shining, in reference to the shiny leaves.]

Photinia pyrifolia (Lam.) K. R. Robertson & J. B. Phipps [With leaves like *Pyrus.*] RED CHOKEBERRY.

Crataegus pyrifolia Lamarck, Encycl. 1: 83. 1783. *Photinia pyrifolia* (Lamarck) K. R. Robertson & J. B. Phipps, in K. R. Robertson et al., Syst. Bot. 16: 391. 1991.

Mespilus arbutifolia Linnaeus, Sp. Pl. 1: 478. 1753. *Pyrus arbutifolia* (Linnaeus) Linnaeus f., Suppl. Pl. 256. 1782. *Aronia arbutifolia* (Linnaeus) Persoon, Syn. Pl. 2: 39. 1906. *Mespilus arbutifolia* Linnaeus var. *erythrocarpa* Michaux, Fl. Bor.-Amer. 1: 292. 1803, nom. inadmiss. *Aronia arbutifolia* (Linnaeus) Persoon var. *tomentosa* Elliott, Sketch. Bot. S. Carolina 1: 557. 1821, nom. inadmiss. *Pyrus arbutifolia* (Linnaeus) Linnaeus f. var. *erythrocarpa* Torrey, Rep. Bot. Dept. Surv. New York Assemb. 50: 135. 1939, nom. inadmiss. *Sorbus arbutifolia* (Linnaeus) Heynhold, Nom. Bot. Hort. 684. 1841. *Adenorachis arbutifolia* (Linnaeus) Nieuwland, Amer. Midl. Naturalist 4: 94. 1915.

Pyrus arbutifolia (Linnaeus) Linnaeus f. var. *oblongifolia* Farwell, Amer. Midl. Naturalist 10: 215. 1927. TYPE: FLORIDA: Lake Co.: Eustis, 12–31 Mar 1894, *Nash 3* (holotype: MICH).

Shrubs, to 2 m; bark gray or brown, smooth, appressed pilose. Leaves with the blade elliptic to obovate, 2.5–7.5(18) cm long, apex acute to acuminate, the margin glandular-serrulate-dentate, the venation pinnate, the upper surface glabrous or glabrescent, the lower surface glandular-pubescent, densely gray-pubescent along the midrib; stipules narrowly triangular, the margin glandular, adnate to the petiole. Flowers 5–12(200) in a corymb; bract 1, as a gland at the base of the pedicel; bracteoles 1–3, as glands on the pedicels; hypanthium campanulate, 1–2 mm long, villous; sepals triangular, erect; petals elliptic to orbicular, 3–6 mm long, white to pale pink. Fruit obovoid to subglobose, red, pilose; seeds 1–8, 2–3 mm long.

Flatwoods, bogs, titi swamps, and along spring runs. Frequent; northern counties, central peninsula. Quebec south to Florida, west to Ontario, Oklahoma, and Texas. Spring.

EXCLUDED TAXON

Photinia melanocarpa (Michaux) K. R. Robinson & J. B. Phipps—Reported for Florida by Chapman (1860, as *Pyrus arbutifolia* (Linnaeus) Linnaeus f. var. *melanocarpa* (Michaux) Hooker) and Small (1903, as *Aronia nigra* (Willdenow) Koehne; 1913e, as *Aronia melanocarpa* (Michaux) Elliott; 1933, as *Aronia melanocarpa* (Michaux) Elliott), apparently based on misidentification of material of *P. pyrifolia*.

Physocarpus (Camb.) Raf., nom. et orth. cons. 1838. NINEBARK

Shrubs. Leaves alternate, simple, palmate-veined, petiolate, stipulate. Flowers in umbel-like racemes on short leafy shoots; bracteate, ebracteolate; hypanthium present; sepals 5, free; petals 5, free; stamens 20–40; carpels 3–5, basally connate, the styles terminal. Fruit an aggregate of dehiscent follicles.

A genus of about 10 species; North America, Europe, and Asia. [From the Greek *physa*, a bladder or bellows, and *karpos*, fruit, in reference to the inflated fruits.]

Opulaster Medik. ex Rydb., 1908.

Physocarpus opulifolius (L.) Maxim. [With leaves like *Opulus* (Caprifoliaceae), apparently in reference to *Viburnum opulus*.] COMMON NINEBARK.

Spiraea opulifolia Linnaeus, Sp. Pl. 1: 489. 1753. *Physocarpus riparius* Rafinesque, New. Fl. 3: 73. 1838 ("1836"), nom. illegit. *Neillia opulifolia* (Linnaeus) Bentham & Hooker f. ex W. H. Brewer & S. Watson, in S. Watson, Bot. Calif. 1: 171. 1876. *Physocarpus opulifolius* (Linnaeus) Maximowicz, Trudy Imp. S.-Peterburgsk. Bot. Sada 6: 220. 1879, nom. cons. *Opulaster opulifolius* (Linnaeus) Kuntze, Revis. Gen. Pl. 2: 949. 1891. *Opulaster opulifolius* (Linnaeus) Kuntze var. *typicus* C. K. Schneider, Handb. Laubholzk. 1: 442. 1904, nom. inadmiss.

Spiraea opulifolia Linnaeus var. *ferruginea* Nuttall ex Torrey & A. Gray, Fl. N. Amer. 1: 414. 1840. *Neillia opulifolia* (Linnaeus) Bentham & Hooker f. ex W. H. Brewer & S. Watson var. *ferruginea* (Nuttall ex Torrey & A. Gray) S. Watson, Bibl. Index N. Amer. Bot. 290. 1878. *Physocarpus opulifolius* (Linnaeus) Maximowicz var. *ferrugineus* (Nuttall ex Torrey & A. Gray) Chapman, Fl. South U.S., ed. 3. 132. 1897. *Opulaster stellatus* Rydberg ex Small, Fl. S.E. U.S. 513. 1903. *Physocarpus ferrugineus* (Nuttall ex Torrey & A. Gray) Daniels, Univ. Missouri Stud., Sci. Ser. 1: 291. 1907, nom. illegit. *Physocarpus stellatus* (Rydberg ex Small) Rehder, J. Arnold Arbor. 1: 256. 1920. *Neillia stellata* (Rydberg ex Small) Bean, Bull. Misc. Inform. Kew 1934: 224. 1934. TYPE: ALABAMA/FLORIDA/GEORGIA.

Shrub, to 30 dm; stem spreading or ascending, producing numerous small upright branches, glabrous. Leaves with the blade ovate to obovate, 6–10 cm long, 4–10 cm wide, 3(5)-lobed, the apex obtuse to rounded, the base broadly cuneate to truncate, the margin irregularly serrate, the upper surface glabrous, the lower surface glabrous or sometimes sparsely pubescent, petiole 1–3 cm long; stipules narrowly ovate, 6–10 mm long, remotely glandular-dentate, sparsely stellate-pubescent. Flowers 30–50 in an umbelliform raceme ca. 5 cm long; bracts elliptic to spatulate or rhombic, ca. 5 mm long, entire, 3-fid, or coarsely erose, gland-tipped; hypanthium cup-shaped, ca. 2 mm long, glabrous or sparsely stellate-pubescent; sepals triangular, 2–3 mm long, pale green to white, gland-tipped, usually stellate-pubescent; petals elliptic to broadly orbiculate, 4–5 mm long, white to pale pink; stamens equaling or exceeding the petals, the anthers purplish. Fruit of inflated, basally connate carpels 5–10 mm long, bright red to brownish red, glabrescent; seeds 2(5), pyriform, ca. 2 mm long, yellow, lustrous.

Mesic hammocks. Rare; Jackson and Calhoun Counties. Quebec south to Florida, west to Manitoba, North Dakota, South Dakota, Nebraska, Colorado, and Oklahoma; Europe. Native to North America. Spring.

Potentilla L. 1753. CINQUEFOIL

Perennial herbs. Leaves basal and cauline, palmately compound, petiolate, stipulate. Flowers cymose, solitary, racemiform, bracteate, ebracteolate, bisexual; epicalyx bractlets 5; hypanthium cupulate; sepals 5, free; petals 5, free; stamens 25–30, carpels numerous, free, the styles subterminal, lateral, or basal. Fruit an aggregate of achenes.

A genus of about 400 species; North America, Mexico, Central America, South America, Europe, Africa, Asia, Australia, and Pacific Islands. [Diminutive of *potens*, powerful, originally applied to *P. anserina* because of its reputed medicinal properties.]

1. Stem erect; flowers numerous, cymose ....**P. recta**
1. Stem prostrate; flowers solitary or rarely 2 from a node.
    2. Leaves of the flowering stem 1–3(5) cm long, the apex rounded; flowers 18–25 mm wide**P. reptans**
    2. Leaves of the flowering stem to 7 cm long, the apex acute; flowers 10–15 mm wide ....**P. simplex**

Potentilla recta L. [Upright.] SULPHUR CINQUEFOIL.

> *Potentilla recta* Linnaeus, Sp. Pl. 1: 497. 1753. *Fragaria recta* (Linnaeus) Crantz, Inst. Rei Herb. 2: 177. 1766. *Potentilla pallens* Moench, Methodus 658. 1794, nom. illegit. *Pentaphyllum rectum* (Linnaeus) Nieuwland, Amer. Midl. Naturalist 4: 62. 1915.

Erect or suberect, perennial herb, to 7 dm; stem simple or few branched, glandular, minutely hispid, and sericeous. Leaves primarily cauline, the leaflets 5–7, the leaflet blade oblong to obovate, 1.5–10 cm long, 0.5–3.5 cm wide, the apex acute, the base narrowly cuneate, the margin remotely serrate, the upper and lower surfaces densely glandular, minutely hispid, and sericeous, the petiole (3)4–8 cm long; stipules of cauline leaves 8–40 mm long, deeply lobed. Flowers in a cyme; epicalyx bractlets narrowly lanceolate to lanceolate, 5–12 mm long; hypanthium cupulate, 5–9 mm in diameter; sepals lanceolate to elliptic, 4–10(12) mm long; petals broadly obcordate, 4–11(13) mm long, pale yellow to yellow, the apex retuse or rarely rounded; stamens 25–30. Fruit 1–2 mm long, yellowish brown, strongly rugose, glabrous.

Disturbed sites. Rare; Leon, Madison, and Alachua Counties. Nearly throughout North America; South America; Europe, Africa, Asia, Australia, and Pacific Islands. Native to Europe, Africa, and Asia. Spring–summer.

Potentilla reptans L. [Creeping.] CREEPING CINQUEFOIL.

Potentilla reptans Linnaeus, Sp. Pl. 1: 499. 1753. *Fragaria reptans* (Linnaeus) Crantz, Inst. Rei. Herb. 2: 179. 1766. *Dynamidium reptans* (Linnaeus) Fourreau, Ann. Soc. Linn. Lyon., ser. 2. 16: 371. 1868.

Prostrate, stoloniferous perennial herb, to 1 m; stem pilose or glabrous. Leaves basal and cauline, the leaflets (3)5–7, the leaflet blade oblanceolate or narrowly obovate to obovate, 2–10(13) cm long, 0.5–2 cm wide, the apex acute, the base cuneate, the margin serrate, the upper and lower surfaces pilose or glabrate, the petiole 2–5(15) cm long on the basal leaves, 0.5–10(15) cm long on the stolon, pilose or glabrate; stipules of the basal leaves narrowly oblong to narrowly ovate, (3)8–15 mm long, entire or 2-lobed, those of the stolon linear to lanceolate, 2–4(5) mm long, entire or 2-lobed. Flowers solitary, axillary; epicalyx bracts narrowly elliptic to oblong, 4–8(10) mm long; hypanthium cupulate, 4–7 mm in diameter; sepals lanceolate, 3–6 mm long; petals broadly obcordate, (6)8–12 mm long, yellow, the apex rounded or retuse; stamens 20. Fruit 1–2 mm long, yellowish brown to dark brown, rugose, glabrous.

Disturbed sites. Rare; Miami-Dade County. Quebec and Ontario south to Virginia and Kentucky, also Florida and California; Europe, Africa, and Aisia. Native to Europe, Africa, and Asia. Spring.

Potentilla simplex Michx. [Unbranched, in reference to the stem.] COMMON CINQUEFOIL.

Potentilla simplex Michaux, Fl. Bor.-Amer. 1: 303. 1803. *Potentilla canadensis* Linnaeus var. *simplex* (Michaux) Torrey & A. Gray, Fl. N. Amer. 1: 443. 1840. *Potentilla simplex* Michaux var. *typica* Fernald, Rhodora 33: 188. 1931, nom. inadmiss.

Prostrate or sprawling, stoloniferous perennial herb; stem to 1 m, villous. Leaves basal and cauline, the leaflets 5, the leaflet blade oblanceolate or elliptic to narrowly obovate, 2–7.5 cm long, 1–3 cm wide, the apex acute, mucronate, the base cuneate, the margin remotely serrate, the upper and lower surfaces villous to nearly glabrous, short petiolulate, the petiole 1–15 cm long on basal leaves, 0–3 cm on the stolon, usually villous; stipules of the basal leaves narrowly lanceolate, 10–18 mm long, entire or 2-lobed, those of the stolon lanceolate to elliptic (5)8–30 mm long, usually deeply lobed. Flowers solitary, axillary on a short shoot of the stolon; epicalyx bracts linear to narrowly lanceolate, 2–5 mm long; hypanthium cupulate, 3–5 mm in diameter; sepals lanceolate, 4–6 mm long; petals obovate, 4–7 mm long, yellow, the apex rounded, truncate, or retuse; stamens 20. Fruit ca. 1 mm long, yellowish brown, obscurely rugose, glabrous.

River floodplains. Rare; Jackson County. Quebec south to Florida, west to Ontario, Minnesota, Nebraska, Kansas, Oklahoma, and Texas. Spring.

Prunus L. 1753. PLUM; CHERRY

Trees or shrubs. Leaves alternate, simple, pinnate-veined, petiolate, stipulate. Flowers on short shoots or axillary in racemes, corymbs, or umbellate fascicles, bracteate, bisexual or unisexual (plants dioecious), perigynous; sepals 5, free; petals 5, free; stamens 10–30; carpel 1, the style terminal, the ovules 2 (1 usually aborted). Fruit a drupe.

A genus of about 400 species; nearly cosmopolitan. [*Prunum*, ancient Latin name of the plum, *Prunus domestica*.]

Amygdalus L., 1753; *Laurocerasus* Duhamel, 1755; *Padus* Mill., 1754.

1. Flowers and fruits borne in racemes.
  2. Leaves chartaceous, deciduous, the margin evenly serrate with incurved teeth .... P. serotina
  2. Leaves coriaceous, evergreen, the margin entire or irregularly toothed.
    3. Leaf margin (at least some) irregularly toothed; fruit ellipsoid or oval .... P. caroliniana
    3. Leaf margin entire; fruit subglobose .... P. myrtifolia
1. Flowers and fruits borne singly, in fascicles, or in umbels.
  4. Flowers sessile or subsessile, the pedicel to 2 mm long, not evident beyond the bud scales.
    5. Leaves more than 4 cm long; stem straight or nearly so; flowers pink; fruit pubescent ....
    .... P. persica
    5. Leaves less than 2.5 cm long; stem conspicuously zig-zag; flowers white; fruit glabrous ....
    .... P. geniculata
  4. Flowers distinctly pedicellate, the pedicel more than 5 mm long, evident beyond the bud scales.
    6. Flowers 2–2.5 cm in diameter.
      7. Pedicels glabrous; fruit yellow or red .... P. americana
      7. Pedicels pubescent; fruit black .... P. subhirtella
    6. Flowers ca. 1 cm in diameter.
      8. Teeth of the leaves tipped with a deciduous gland leaving a callus thickening at maturity.....
      .... P. angustifolia
      8. Teeth of the leaves not gland tipped .... P. umbellata

Prunus americana Marshall [Of America.] AMERICAN PLUM.

Prunus americana Marshall, Arbust. Amer. 111. 1785. *Prunus domestica* Linnaeus var. *americana* (Marshall) Castiglioni, Viagg. Stati Uniti 2: 339. 1790.
Prunus americana Marshall var. *floridana* Sargent, J. Arnold Arbor 2: 113. 1920. TYPE: FLORIDA: Wakulla Co.: near St. Marks, 30 Mar 1914, *Harbison 30* (holotype: A?).

Tree or shrub, to 8 m; branchlets glabrous or pubescent. Leaves with the blade elliptic to ob-ovate, rarely ovate, 5–11 cm long, 2–5.5 cm wide, the apex abruptly long acuminate or short acuminate to acute, the base cuneate to obtuse, occasionally rounded, the margin coarsely and doubly serrate, the upper surface glabrous to appressed pubescent, the lower surface glabrous except along the midrib, sometimes entirely glabrous or pubescent all over, the petiole 4–19 mm long, pubescent, sometimes only on the upper side, rarely glabrous, occasionally with 1–2 dark discoid glands near the blade; stipules linear to lanceolate, the margin toothed to narrowly lobed, usually glandular. Flowers 2–5 in umbellate fascicles, bisexual, the pedicel (4)8–20 mm long, glabrous; hypanthium obconic, 2.5–5 mm long, glabrous or slightly to densely pubescent externally; sepals ovate to lanceolate, 2–4(5) mm long, spreading to reflexed, the margin with several teeth at the apex, eglandular or with a few obscure glands, ciliate, the outer surface pubescent, the inner surface glabrous or pubescent; petals ovate to oblong-obovate, 7–12 mm long, white; ovary glabrous. Fruit subglobose to ellipsoid, 15–30 mm long, red, orange, or yellowish, glabrous, glaucous, the stone ovoid, strongly laterally flattened.

Mesic hammocks and floodplain forests. Occasional; northern peninsula south to Lake County, west to central panhandle. Nearly throughout North America. Spring.

Prunus angustifolia Marshall [With narrow leaves.] CHICKASAW PLUM.

Prunus angustifolia Marshall, Arbust. Amer. 111. 1785.

Tree or shrub, to 5 m; branchlets glabrous. Leaves with the blade lanceolate to narrowly elliptic, 1.5–6 cm long, 0.8–2 cm wide, the apex acute, the base cuneate to obtuse, the margin crenate-serrulate, the teeth glandular, this deciduous and leaving a callus thickening at maturity, the upper surface glabrous, the lower surface with a few trichomes along the midrib and major veins, the petiole 2–14 mm long, usually sparsely pubescent along the upper side, rarely glabrous or pubescent all over, usually eglandular but sometimes with 1–2 glands near the blade; stipules linear to lanceolate, the margin toothed to narrowly lobed, usually glandular. Flowers 2–4 in an umbelliform fascicle, bisexual, the pedicel 3–10 mm long, glabrous; hypanthium campanulate, 2–3 mm long, glabrous or rarely pubescent externally; sepals ovate, 1–2 mm long, erect to spreading, the margin entire, eglandular, ciliate, the outer surface pubescent; petals suborbicular to obovate, 3–6 mm long, white; ovary glabrous. Fruit globose to ellipsoid, 15–20 mm long, red to yellow, glabrous, the stone ovoid, somewhat laterally flattened.

Dry, open hammocks and disturbed sites. Occasional; northern counties south to Pinellas and Hillsborough Counties. Probably introduced in the peninsula. New Jersey and Pennsylvania south to Florida, west to Nebraska, Colorado, and New Mexico, also California. Spring.

Prunus caroliniana (Mill.) Aiton [Of Carolina.] CAROLINA LAURELCHERRY.

Padus caroliniana Miller, Gard. Dict., ed. 8. 1768. *Prunus caroliniana* (Miller) Aiton, Hort. Kew. 2: 163. 1789. *Cerasus caroliniana* (Miller) Michaux, Fl. Bor.-Amer. 1: 285. 1803. *Laurocerasus caroliniana* (Miller) M. Roemer, Fam. Nat. Syn. Monogr. 3: 90. 1847.

Tree or shrub, to 12 m; branchlets glabrous. Leaves with the blade narrowly elliptic to oblanceolate, 5–10 cm long, 2–4 cm wide, the apex acute to short acuminate, the base cuneate to obtuse, the margin entire, undulate, or spinose-serrate, the upper and lower surfaces glabrous, the petiole 5–8 mm long, glabrous, eglandular; stipules linear to lanceolate, the margin toothed to narrowly lobed, usually glandular. Flowers 12–30 in racemes, bisexual or sometimes with staminate at the base, pedicel 1–4 mm long, glabrous; hypanthium cupulate, 2–3 mm long, glabrous externally; sepals half-rounded, ca. 1 mm long, the margin entire or sometimes with a glandular tooth, the surfaces glabrous; petals suborbicular to elliptic, 1–2 mm long, white; ovary glabrous. Fruit ovoid, 9–12 mm long, black, glabrous, the stone ovoid.

Hammocks and disturbed sites. Frequent; nearly throughout. North Carolina south to Florida, west to Texas, also California. Spring.

Prunus geniculata R. M. Harper [Bent sharply like a knee.] SCRUB PLUM.

Prunus geniculata R. H. Harper, Torreya 11: 67. 1911. TYPE: FLORIDA: Lake Co.: ca. 10 mi. S of Tavares, 17 Apr 1911, *Harper 31* (holotype: ?; isotype: NY).

Shrub, to 1(2) m; branchlets often bent slightly at an angle, with thorns, pubescent. Leaves with the blade elliptic, 0.8–2.5 cm long, 0.4–1.3 cm wide, the apex obtuse to rounded, the base obtuse to rounded, the margin crenate-serrulate in the distal ½, the small leaves subentire, the upper and lower surfaces glabrous, the petiole 3–6 mm long, pubescent, sometimes only on the upper side, rarely glabrous, sometimes with 1–2 stalked, discoid glands near the blade; stipules linear to lanceolate, the margin toothed to nearly lobed, usually glandular. Flowers solitary, bisexual, the pedicel 0–3 mm long, glabrous; hypanthium campanulate, 2–3 mm long, glabrous externally; sepals ovate, 1–2 mm long, erect to spreading, the margin ciliate, eglandular, the outer surface pubescent; petals elliptic, ca. 2 mm long, white; ovary glabrous. Fruit ovoid, 12–25 mm long, reddish, glabrous, the stone ovoid, somwhat laterally flattened.

Scrub and turkey oak barrens. Rare; Lake, Orange, Polk, and Highlands Counties. Endemic. Spring.

Prunus geniculata is listed as endangered in Florida (Florida Administrative Code, Chapter 5B-40) and endangered in the United States (U.S. Fish and Wildlife Service, 50 CFR 23).

Prunus myrtifolia (L.) Urb. [With leaves like *Myrtus* (Myrtaceae).] WEST INDIAN CHERRY.

Celastrus myrtifolius Linnaeus, Sp. Pl. 1: 196. 1753. *Prunus sphaerocarpa* Swartz, Prodr. 80. 1788, nom. illegit. *Laurocerasus sphaerocarpa* M. Roemer, Fam. Nat. Syn. Monogr. 3: 89. 1847. *Prunus myrtifolia* (Linnaeus) Urban, Symb. Antill. 5: 93. 1904. *Laurocerasus myrtifolia* (Linnaeus) Britton, in Britton & Shafer, N. Amer. Trees 510. 1908.

Tree, to 12 m; branchlets glabrous. Leaves with the blade elliptic to broadly elliptic, 4–10 cm long, 2–4.5(6.5) cm wide, the apex acute to acuminate, the base cuneate to obtuse or nearly rounded, the margin entire, undulate, the upper and lower surfaces glabrous, with a pair of glands on the lower surface near the blade, the petiole 8–16 mm long, glabrous; stipules linear to lanceolate, the margin toothed to narrowly lobed, usually glandular. Flowers 12–30 in a raceme, bisexual, sometimes with staminate at the base, the pedicel (2)3–6 mm long, glabrous; hypanthium cupulate, 2–3 mm long, glabrous externally; sepals semicircular, ca. 1 mm long, spreading, the margin entire, occasionally with a glandular tooth, the surfaces glabrous; petals obovate to suborbicular, ca. 2 mm long, white; ovary glabrous. Fruit globose to subovoid, 8–12 mm long, black, glabrous, the stone subglobose.

Hammocks and pine rocklands. Rare; Miami-Dade County. Florida; West Indies, Mexico, Central America, and South America. Fall–winter.

Prunus myrtifolia is listed as threatened in Florida (Florida Administrative Code, Chapter 5B-40).

Prunus persica (L.) Batsch [Of Persia.] PEACH.

Amygdalus persica Linnaeus, Sp. Pl. 1: 372. 1753. *Persica vulgaris* Miller, Gard. Dict., ed. 8. 1768. *Prunus persica* (Linnaeus) Batsch, Beytr. Entw. Pragm. Gesg. Naturr. 30. 1801. *Prunus persica* (Linnaeus) Batsch var. *vulgaris* Maximowicz, Bull. Acad. Imp. sci. Saint-Pétersbourg 29: 82. 1883, nom. inadmiss.

Tree, to 6 m; branchlets glabrous. Leaves with the blade oblong to lanceolate, (5)7–15 cm long, 2–4 cm wide, the apex acuminate, the base cuneate to obtuse, the margin crenulate-serrulate, the teeth glandular, the upper and lower surfaces glabrous, the petiole 5–10(15) mm long, glabrous, sometimes with 1–4 sessile discoid glands near the blade; the stipules linear to lanceolate, the margin toothed to narrowly lobed, usually glandular. Flowers solitary or paired, the pedicel 0–3 mm long, glabrous; hypanthium cupulate, 4–5 mm long, glabrous externally; sepals oblong-ovate, 4–5 mm long, spreading, the margin entire, ciliate, the outer surface pubescent, especially along the margin, the inner surface glabrous; petals obovate to suborbicular, 10–17 mm long, dark pink; ovary pubescent. Fruit globose, 4–8 cm long, yellow to orange tinged with red, velutinous, the stone ellipsoid, strongly flattened, deeply pitted and furrowed.

Disturbed hammocks. Rare; central and western panhandle, northern and central peninsula. Escaped from cultivation. Nova Scotia south to Florida, west to Ontario, Wisconsin, Iowa, Kansas, Oklahoma, and Texas, and from Idaho and Oregon south to Arizona and California; West Indies; Europe, Africa, and Asia. Native to Asia. Spring.

Prunus serotina Ehrh. [Late to leaf or flower.]

Tree or shrub, to 25 m; branchlets glabrous or pubescent. Leaves with the blade narrowly elliptic, oblong-elliptic, or obovate, 2–13 cm long, 1–6.5 cm wide, the apex acute to acuminate or obtuse to rounded, the base cuneate to rounded, the margin crenulate-serrulate to serrate with incurved callus-tipped teeth, the lower surface glabrous or with dense trichomes along the midrib toward the blade base, sometimes scattered over the surface, the upper surface glabrous; stipules linear to lanceolate, the margin toothed to narrowly lobed, usually glandular. Flowers 18–25(90) in a raceme, the pedicel 1–10 mm long, glabrous or pubescent; hypanthium cupulate, 2–3 mm long, glabrous externally; sepals semicircular, 1–2 mm long, the margin entire, rarely ciliate and/or with a few glandular teeth, the surfaces glabrous; petals obovate to suborbicular, 2–4 mm long; ovary glabrous. Fruit subglobose, 5–10 mm long, dark purple to nearly black, glabrous, the stone subglobose.

1. Young branchlets, petioles, and inflorescence axes glabrous; leaf blades oval-oblong or elliptic, rarely ovate, the apex acute or acuminate ....var. **serotina**
1. Young branchlets, petioles, and inflorescence axes pubescent; leaf blades usually obovate, the apex obtuse.. var. **alabamensis**

Prunus serotina var. serotina BLACK CHERRY.

Prunus serotina Ehrhart, Gartenkalender 3: 234. 1783–1784, nom. cons. *Padus serotina* (Ehrhart) Borkhausen, Arch. Bot. (Leipzig) 1(2): 38. 1787. *Cerasus serotina* (Ehrhart) Loiseleur-Deslongchamps, in Duhamel du Monceau, Traite Arbr. Arbust., ed. 2. 5: 3. 1812. *Prunus serotina* Ehrhart forma *typica* Schwerin, Mitt. Deutsch. Dendrol. Ges. 15: 3. 1907, nom. inadmiss.

Young branchlets, petioles, and inflorescence axes glabrous; leaf blades oval-oblong or elliptic, rarely ovate, the apex acute or acuminate.

River swamps and disturbed sites. Frequent; northern counties, central peninsula. Quebec south to Florida, west to Ontario, North Dakota, Nebraska, Kansas, and Arizona, also British Columbia and Washington; Mexico, Central America, and South America; Europe. Native to North America, Mexico, and Central America. Spring.

Prunus serotina var. alabamensis (C. Mohr) Little [Of Alabama.] ALABAMA CHERRY.

Prunus alabamensis C. Mohr, Bull. Torrey Bot. Club 26: 118. 1899. *Padus alabamensis* (C. Mohr) Small, Fl. S.E. U.S. 574, 1331. 1903. *Prunus serotina* Ehrhart forma *alabamensis* (C. Mohr) C. K. Schneider ex Schwerin, Mitt. Deutsch. Dendrol. Ges. 15: 3. 1907. *Prunus serotina* Ehrhart var. *alabamensis* (C. Mohr) Little, Phytologia 4: 309. 1953.

238 / FLORA OF FLORIDA

Prunus hirsuta Elliott, Sketch Bot. S. Carolina 1: 541. 1821. *Prunus serotina* Ehrhart subsp. *hirsuta* (Elliott) McVaugh, Britton 7: 299. 1951.

Prunus cuthbertii Small, Bull. Torrey Bot. Club 28: 290. 1901. *Padus cuthbertii* (Small) Small, Fl. S.E. U.S. 574, 1331. 1903.

Young branchlets, petioles, and inflorescence axes pubescent; leaf blades usually obovate, the apex obtuse.

Mesic forests. Occasional; central and western panhandle. North Carolina south to Florida, west to Mississippi. Spring.

Prunus subhirtella Miq. [Slightly hairy.] WINTER-FLOWERING CHERRY.

Prunus subhirtella Miquel, Ann. Mus. Bot. Lugduno-Batavum 2: 91. 1865. *Cerasus subhirtella* (Miquel) S. Y. Sokolov, Trees & Shrubs USSR 3: 734. 1954.

Tree; branchlets pilose. Leaves with the blade narrowly oblong to narrowly obovate, 6–12 cm long, 3–4 cm wide, the apex acuminate, the margin doubly serrate, the upper surface glabrous or sparsely pubescent, the lower surface pubescent, especially on the veins, the petiole pubescent; stipules linear, the margin glandular-incised-serrate. Flowers 2–5 in a sessile umbel, the pedicel 1.5–3 cm long, pubescent; hypanthium ca. 5 mm long, slightly inflated, pubescent externally; petals ca. 12 mm long, pink, the apex retuse. Fruit ellipsoid-globose, 5–10 mm long, black, sparsely pilose to glabrous, the stone subglobose.

Disturbed sites. Rare; Leon County. Escaped from cultivation. Ohio, District of Columbia, and Florida; Asia. Native to Asia. Fall–winter.

Prunus umbellata Elliott [In an umbel, in reference to the inflorescence.] FLATWOODS PLUM; HOG PLUM.

Prunus umbellata Elliott, Sketch Bot. S. Carolina 1: 541. 1821.

Tree or shrub, to 6 m; branchlets glabrous or pubescent. Leaves with the blade elliptic to broadly elliptic or oblanceolate to obovate, 3.5–8 cm long, 1.5–4 cm wide, the apex acute to short acuminate, the base cuneate to obtuse, rarely rounded, the margin serrulate, sometimes doubly serrate, the teeth eglandular, the upper surface glabrous, the lower surface pubescent to subglabrous; stipules linear to lanceolate, the margin toothed to narrowly lobed, usually glandular. Flowers 2–4(6) in an umbellate fascicle, the pedicel 5–22 mm long, glabrous or sometimes pubescent; hypanthium tubular to tubular-urceolate, 2–4 mm long, glabrous or pubescent externally; sepals ovate-oblong, 2–3 mm long, erect-spreading, the margin entire or 2-fid at the apex, ciliate, eglandular, the outer surface pubescent or glabrous, the inner surface pubescent; petals obovate to suborbicular, 3–8 mm long, white, sometimes fading to pink; ovary glabrous. Fruit globose, 10–15 mm long, red or yellow, maturing to dark blue or nearly black, glabrous, the stone ovoid, slightly flattened.

Sandhills and flatwoods. Frequent; northern counties, central peninsula. North Carolina south to Florida, west to Texas. Spring.

EXCLUDED TAXA

Prunus avium (Linnaeus) Linnaeus—Reported by Small (1933), basis unknown. No specimens seen.
Prunus cerasus Linnaeus—Reported by Small (1933), basis unknown. No specimens seen.

Prunus virginiana Linnaeus—Reported by Small (1913a, 1913c, 1933, all as *Padus virginiana* (Linnaeus) Miller, the name misapplied to material of *P. serotina*.

Pyracantha M. Roem. 1847. FIRETHORN

Shrubs. Leaves alternate, fascicled on short shoots, simple, pinnate-veined, petiolate, stipulate. Flowers in terminal panicles, bracteate, bisexual; sepals 5, free; petals 5, free; stamens 15–20, free; carpels 5, inferior. Fruit a pome.

A genus of about 10 species; North America, Europe, Asia, Australia, and Pacific Islands. [From the Greek *Pyr*, fire, and *acantha*, thorn, in reference to the red fruits and thorny branches.]

1. Leaves with the margin crenate-serrate or sometimes with only a few teeth, the apex acute to obtuse . ...**P. fortuneana**
1. Leaves with the margin entire or with a few apical teeth, the apex usually retuse, sometimes rounded .... **P. koidzumii**

Pyracantha fortuneana (Maxim.) H. L. Li [Commemorates Robert Fortune (1812–1880), Scottish botanist who collected in China.] CHINESE FIRETHORN.

Photinia fortuneana Maximowicz, Bull. Acad. Imp. Sci. Saint-Pétersbourg 19: 179. 1873. *Pyracantha fortuneana* (Maximowicz) H. L. Li, J. Arnold Arbor. 25: 420. 1944.

Shrub, to 3 m; branches short, with thorns, the branchlets rusty-pubescent when young, glabrescent. Leaves with the blade obovate or oblong-obovate, 1.5–6 cm long, 0.5–2.5 cm wide, the apex acute or obtuse, emarginate, or short apiculate, the base cuneate, the margin remotely serrulate or crenulate, often entire near the base, the upper and lower surfaces glabrate, the petiole 2–5 mm long, slightly pubescent; stipules lanceolate, 1–8 mm long, remotely toothed. Flowers 6–40 in a flattened panicle 3–4 cm wide, the peduncle and pedicel glabrate, rarely appressed brown pubescent; hypanthium glabrate, rarely slightly pubescent; sepals triangular, 1–2 mm long, the apex obtuse; petals suborbicular, 3–4 mm long, white, the apex rounded, the base slightly clawed. Fruit globose, 3–6 mm long, orange-red to dark red, the pedicel 2–10 mm long; pyrenes 5.

Disturbed sites. Rare; Okaloosa County. Escaped from cultivation. South Carolina, Florida, Alabama, Texas, and California; Europe, Asia, Australia, and Pacific Islands. Native to Asia. Spring.

Pyracantha koidzumii (Hayata) Rehder [Commemorates Gen'ichi Koidzumi (1883–1953), Japanese botanist.] FORMOSA FIRETHORN.

Cotoneaster koidzumii Hayata, J. Coll. Sci. Imp. Univ. Tokyo 30(1): 101. 1911. *Pyracantha koidzumii* (Hayata) Rehder, J. Arnold Arbor. 1: 261. 1920.

Shrub, to 4 m; branches often with spinose short shoots, the branchlets reddish brown pubescent when young, glabrescent. Leaves with the blade oblanceolate or narrowly obovate, 2.5–4.5 cm long, 0.5–1.2 cm wide, the apex truncate or slightly emarginate, sometimes rounded, the base cuneate, the margin entire, rarely with a few minute teeth near the apex, the upper surface glabrous, the lower surface brown pubescent when young, becoming glabrate and slightly glaucescent, the petiole 1–3 mm long, slightly pubescent; stipules lanceolate, 1–8 mm long, remotely toothed. Flowers 6–40 in a flattened panicle 3–4 cm wide, the peduncle and pedicel sparsely brown pubescent; hypanthium densely pubescent; sepals triangular, 1–2 mm long, the apex acute; petals suborbicular, 3–4 mm long, the apex slightly emarginate, the base slightly clawed. Fruit globose, 4–7 mm long, orange-red, the pedicel 5–12 mm long; pyrenes 5.

Disturbed sites. Occasional; central and western panhandle, northern and central peninsula. Escaped from cultivation. South Carolina south to Florida, west to Texas, also Arizona; Europe, Asia, and Pacific Islands. Native to Asia. Spring.

EXCLUDED TAXON

Pyracantha coccinea M. Roemer—Reported by Robertson (1974) as "locally naturalized in the eastern United States from Pennsylvania to Florida and Louisiana; it also persists near old house sites." However, no vouchering specimens have been seen.

Pyrus L. 1753. PEAR

Trees. Leaves alternate, simple, pinnate-veined, petiolate, stipulate. Flowers in umbelliform racemes or simple corymbs, bisexual, epigynous; sepals 5; petals 5; stamens ca. 20; carpels 2–5, connate, adnate to the hypanthium. Fruit a pome.

A genus of about 25 species; North America, Europe, Africa, and Asia. [From the Greek *pyr*, fire, the classical name for the pear (*P. communis*).]

1. Calyx lobes densely compactly woolly-pubescent on the inner surface, not persistent on the fruit; styles (2)3; fruit globular, the surface rough granular..**P. calleryana**
1. Calyx lobes sparsely and loosely woolly-pubescent on the inner surface toward the base, persistent on the fruit; styles 5; fruit pyriform, the surface smooth ....**P. communis**

Pyrus calleryana Decne. [Commemorates Joseph-Marie Callery (1810–1862), Italian-French student of China who sent specimens of the tree to Europe from China.] CALLERY PEAR.

Pyrus calleryana Descaisne, Jard. Fruit. 1: sub. t. 8 1872.

Tree, to 20 m; branches of short shoots, often thorn-tipped, the branchlets reddish brown and tomentose when young, becoming grayish brown and glabrous in age. Leaves with the blade ovate to broadly ovate, the apex acuminate, the base broadly cuneate, the margin obtusely serrate or crenate, the upper and lower surfaces glabrous, the petiole 2–4.5 cm long, glabrous; stipules small, caducous. Flowers with the hypanthium campanulate; sepals lanceolate, ca. 5 mm long, the apex acuminate; petals obovate, 6–7(13) mm long, white; ovary 2- to 3-locular, the styles 2–3. Fruit globose or slightly oblong, 1–1.5 cm long, blackish brown, brown, or yellow-brown with white or tan dots, the pedicel 1–1.5 cm long.

Disturbed sites. Occasional; panhandle, Suwannee County. Escaped from cultivation. Pennsylvania and New Jersey south to Florida, west to Illinois, Oklahoma, and Texas; Europe and Asia. Native to Asia. Spring.

Pyrus communis L. [Common.] COMMON PEAR.

Pyrus communis Linnaeus, Sp. Pl. 1: 479. 1753. *Sorbus pyrus* Crantz, Stirp. Austr. Fasc. 2: 56. 1763, nom. illegit. *Pyrus sylvestris* Moench, Methodus 679. 1794, nom. illegit. *Pyrus communis* Linnaeus var. *sylvestris* de Candolle, in de Candolle & Lamarck, Fl. Franc. 4: 430. 1815, nom. inadmiss.

Tree, to 15(30) m; branches of short shoots, often thorn tipped in young plants, grayish brown or dark reddish brown, glabrous. Leaves with the blade ovate or suborbicular, the apex acute or short-acuminate, the base broadly cuneate to subrotund, the margin obtusely serrate, serrulate, crenate, or entire, densely ciliate, the upper and lower surfaces pubescent when young, glabrescent, the petiole 1.5–5 cm long; stipules small, caducous. Flowers with the hypanthium campanulate; sepals triangular-lanceolate, 5–9 mm long, the apex acuminate; petals obovate, 13–15 mm long, white; ovary 5-locular, the styles (3)5. Fruit globose, subglobose, or pyriform, 3–16 cm long, green, yellowish or reddish green, the pedicel 2–3.5 cm long.

Disturbed hammocks. Rare; central panhandle, Brevard County. Escaped from cultivation. Nearly throughout North America; Europe and Asia. Native to Europe and Asia. Spring.

Rosa L. 1753, nom. cons. ROSE

Shrubs; stems with prickles. Leaves alternate, odd-pinnately compound, petiolate, stipulate. Flowers solitary in leaf axils or in terminal corymbs or panicles, bisexual, bracteate, with an evident floral tube bearing the sepals, petals, and stamens on the rim; sepals 5, free; petals 5 or 10, free; stamens numerous; carpels numerous, free. Fruit of achenes produced within the fleshy floral tube (hip).

A genus of about 120 species; North America, Europe, Africa, Asia, Australia, and Pacific Islands. [Ancient Latin name.]

1. Stipules pectinately fringed or toothed on the margin.
    2. Corolla of numerous ("double") pink petals.. **R. wichuraiana**
    2. Corolla a single whorl of 5 white petals.
        3. Branchlet, pedicel, and hypanthium tomentose... **R. bracteata**
        3. Branchlet, pedicel, and hypanthium glabrous, or nearly so..................................... **R. multiflora**
1. Stipules entire or glandular-ciliate on the margin.
    4. Stipules adnate to the petiole for less than ½ their length; petals white.........................**R. laevigata**
    4. Stipules adnate to the petiole for most of their length; petals pink.
        5. Leaves with 3 leaflets (at least on the flowering and fruiting stem)..............................**R. setigera**
        5. Leaves with 5 or more leaflets.
            6. Stem with the major prickles recurved-hooked, lacking smaller straight prickles; plant of wet areas...**R. palustris**
            6. Stem with the major prickles straight and diverging at a right angle to the stem, smaller straight stem prickles usually present; plant of upland, dry sites.........................**R. carolina**

Rosa bracteata J. C. Wendl. [With evident bracts.] MACARTNEY ROSE.

Rosa bracteata J. C. Wendland, Bot. Beob. 50. 1798.
Rosa lucida Lawrance, Roses pl. 84. 1799; non Ehrhart, 1784.

Arching or clambering shrub, to 2 m; the branchlets with loosely matted grayish and numerous, exserted, stiff, purplish red, stipitate-glandular trichomes, also with a pair of broad-based, straight or curved, prickles at the nodes or sometimes internodally. Leaves with 5–9 leaflets, the blade oblong-elliptic to obovate, those on the flowering branchlets 1–2 cm long, 0.8–1.2 cm wide, those of the vigorous stems larger, the apex rounded, usually mucronate, the base broadly cuneate to rounded, the margin serrate, the teeth tipped with a purplish gland, the upper surface glabrous, lustrous, the lower surface paler and dull green, sparsely pubescent and sometimes with a few stipitate glands along the midrib, sometimes this with a sharply hooked prickle, the rachis, petiolules, and petiole sometimes with a few hooked prickles; stipules adnate to the petiole for a short distance, variously pectinate-toothed or pectinate-branched distally. Flowers solitary or few in a terminal cyme, subtended by an evident involucrate cluster of broad, toothed bracts; floral tube densely matted-pubescent; sepals ovate or oblong-ovate, 1.5–2 cm long, densely matted to sparsely pubescent; petals obovate, 2–3 cm long, white, the apex rounded or emarginate; stigmas pubescent. Fruiting floral tube subglobose, 1.5–2 cm long, dark brown; achenes oblong to obovate, ca. 5 mm long, tan, with straight easily detached trichomes.

Dry hammocks and disturbed sites. Occasional; northern peninsula, Volusia and Citrus Counties, central and western panhandle. Escaped from cultivation. Maryland south to Florida, west to Texas; Europe, Africa, Asia, and Australia. Native to Asia. Spring–summer.

Rosa carolina L. [Of Carolina.] CAROLINA ROSE.

Rosa carolina Linnaeus, Sp. Pl. 1: 492. 1753.
Rosa serrulata Rafinesque, Ann. Gen. Sci. Phys. 5: 218. 1820.

Erect shrub, to 1(1.5) m; stem glabrous, with broad-based, subterete, straight or divergent prickles 5–10 mm long, sometimes also with gland-tipped prickles. Leaves with 3–7(9) leaflets, the blade elliptic to oblanceolate or narrowly obovate, 2–3 cm long, 1–1.5 cm wide, the apex acute to rounded, the base cuneate, the margin serrate from a little above the base, the upper and lower surfaces glabrous or pubescent, the petiole and leaf axes short pubescent and/or stipitate-glandular, the rachis, petiolules, and petiole sometimes with a few hooked prickles; stipules linear, the margin entire or with gland-tipped teeth. Flowers solitary in uppermost leaf axils of branchlets; floral tube glandular; calyx linear-lanceolate, sometimes dilated distally, the margin entire or toothed or lobed, the inner surface densely woolly-pubescent at the base, the outer surface woolly-pubescent and stipitate-glandular; petals broadly obovate, 2–3 cm long, pink, the apex broadly rounded; stigmas pubescent. Fruiting floral tube subglobose, ca. 1 cm long, red; achenes obovoid, ca. 2 mm long, brown, smooth.

Upland hammocks. Rare; central panhandle, Clay and Alachua Counties. Quebec south to Florida, west to Ontario, Minnesota, Nebraska, Kansas, Oklahoma, and Texas. Spring.

Rosa laevigata Michx. [Smooth.] CHEROKEE ROSE.

Rosa laevigata Michaux, Fl. Bor.-Amer. 1: 295. 1803.
Rosa cherokeensis Donn ex Small, Fl. S.E. U.S. 528. 1903. TYPE: FLORIDA/GEORGIA/TEXAS.

Erect or clambering shrub, to 6 m; stem glabrous, with recurved, broad-based, reddish brown prickles. Leaves with 3 leaflets, the blade lance-elliptic, 4–6 cm long, 1.5–2 cm wide, the apex acute, the base broadly cuneate, the margin finely and sharply serrate from a little above the base, the upper surface glabrous, lustrous, the lower surface dull, glabrous, sometimes with a few straight prickles on the midrib, the rachis, petiolules, and petiole usually with a few hooked prickles; stipules adnate to the petiole for only a very short distance, the free tips subulate or foliose, the margin glandular-toothed. Flowers solitary in the uppermost leaf axils of lateral branches; floral tube with divergent, stiff trichomes tipped by minute, deciduous red glands; sepals ovate, gradually narrowed to an acute or expanded to a foliose apex, this entire or apically toothed, the inner surface woolly-pubescent basally, the outer surface woolly-pubescent near the edges to subglabrous, sometimes with a few trichomes like those on the floral tube; petals broadly obovate, 3–4 cm long, white, the margin irregularly wavy and sometimes dentate; stigmas pubescent. Fruiting floral tube pyriform, ca. 3 cm long, red; achenes obovoid, ca. 3 mm long, with 2 flat faces and 1 rounded, 1 stramineous.

Moist hammocks and disturbed sites. Occasional; northern counties, central peninsula. Virginia south to Florida, west to Texas; Europe, Africa, Asia, Australia, and Pacific Islands. Native to North America. Spring.

Rosa multiflora Thunb. [Many-flowered.] MULTIFLORA ROSE.

Rosa multiflora Thunberg, in Murray, Syst. Veg., ed 14. 474. 1784.

Erect or arching shrub, to 3 m; stem glabrous, with subnodal, broad-based, conic-recurved prickles. Leaves with 5–9 leaflets, the blade elliptic or obovate, the apex acute or short-acuminate, the base rounded or broadly cuneate, the margin serrate from near the base, the teeth sometimes gland-tipped, the upper surface glabrous, the lower surface sparsely short-pubescent, primarily along the veins, or glabrous, the rachis, petiolules, and petiole with sessile and stipitate glands, sometimes with a few hooked prickles; stipules adnate to the petiole most of their length, the margin pectinately toothed, the surface variably stipitate-glandular. Flowers numerous in a panicle on a lateral branch, with variable bracts and stipules throughout the inflorescence, the lowermost panicle branches usually subtended by leaves; floral tube glabrous to densely pubescent; sepals ovate or triangular, the margin entire or with a few subulate lobes, the inner surface woolly-pubescent, the outer surface variably glabrous to densely pubescent to stipitate-glandular; petals broadly obovate, 1.5–2 cm long, white or pale pink, the apex emarginate; styles glabrous. Fruiting floral tube subglobose, ca. 7 mm long, red; achenes irregularly angular-obovoid, 3–5 mm long, light brown, the faces slightly papillate.

Disturbed sites. Occasional; central and western panhandle. Escaped from cultivation. Quebec south to Florida, west to Ontario, Minnesota, Nebraska, Kansas, and New Mexico, also British Columbia south to California; Europe, Africa, Asia, Australia, and Pacific Islands. Native to Asia. Spring.

Rosa palustris Marshall [Of swamps.] SWAMP ROSE.

Rosa palustris Marshall, Arbust. Amer. 135. 1785.

Rosa lancifolia Small, Fl. S.E. U.S. 527, 1331. 1903. TYPE: FLORIDA: Lake Co.: vicinity of Eustis, 16–25 Aug 1894, *Nash 1693* (holotype: NY).

Rosa floridana Rydberg ex Small, Shrubs Florida 27, 133. 1913. TYPE: FLORIDA: Duval Co.: near
  Jacksonville, 10 May & 2 Nov 1894, *Curtis 4754* (holotype: NY).

Erect shrub, to 2 m; stem glabrous, with broad-based straight or divergent, somewhat recurved
prickles, most subnodal. Leaves with 5–9 leaflets, the blade elliptic or lanceolate, the apex
acute, the base cuneate to rounded, the margin finely serrate above the base, the upper surface
glabrous, the lower surface densely pubescent, the midrib sometimes with a few small prickles, the rachis, petiolules, and petiole sometimes with a few hooked prickles; stipules narrow,
pubescent (at least at the base and on the margin) or glabrous. Flowers solitary in the axil of
the uppermost leaves of a branchlet or in a few-flowered corymb of the uppermost leaf; floral
tube stipitate-glandular; sepals ovate, triangular, or oblong, often abruptly contracted to a tail-
like apex that is somewhat dilated distally, the inner surface short-pubescent, the outer surface
stipitate-glandular and short-pubescent, at least marginally, the tail-like tip usually sparsely
pubescent and stipitate-glandular; petals broadly ovate, 2–3 cm long, pink, the apex emarginate, the margin somewhat irregularly wavy or crinkled; stigmas pubescent. Fruiting floral tube
broadly ovate, ca. 1 cm long; achenes irregularly angular-obovoid, 7–8 mm long, light brown.

  Cypress swamps, floodplain forests, and river banks. Frequent; northern peninsula south
to Hillsborough and Polk Counties, eastern and central panhandle. Quebec south to Florida,
west to Ontario, Wisconsin, Iowa, Missouri, Arkansas, and Louisiana. Spring.

Rosa setigera Michx. [Bearing bristlelike organs, in reference to the seta on the calyx.] CLIMBING ROSE.

Rosa setigera Michaux, Fl. Bor.-Amer. 1: 295. 1803. *Rosa setigera* Michaux var. *glabra* Torrey & A. Gray,
  Fl. N. Amer. 1: 458. 1840, nom. inadmiss.
Rosa setigera Michaux var. *tomentosa* Torrey & A. Gray, Fl. N. Amer. 1: 458. 1840.

Climbing, leaning, or trailing shrub; stem glabrous, with broad-based recurved prickles. Leaves
with 3–5 leaflets, the blade ovate-oblong, 5–10 cm long, 3–4 cm wide, the apex acute to obtuse,
the base cuneate to rounded, the margin sharply serrate with gland-tipped teeth, the upper surface glabrous, the lower surface glabrous or pilose along the veins, sometimes tomentose, petiolules and petiole glandular hispid, sometimes with a few hooked prickles, the lateral leaflets
subsessile, the terminal one petiolate; stipules linear, the margin entire or glandular-ciliate, the
lobes spreading. Flowers several in a terminal corymb; floral tube stipitate-glandular-hispid;
sepals elliptic-lanceolate, 1–1.5 cm long, stipitate-glandular-hispid, reflexed; petals broadly obovate, 2–3 cm long, rose-pink fading to whitish. Fruiting floral tube subglobose, 8–12 cm long,
bristly.

  Stream banks. Rare; locality unknown. Collected by Chapman from "Florida" (NY) and
not subsequently recollected. New Hampshire south to Florida, west to Ontario, Wisconsin,
Kansas, Oklahoma, and Texas. Spring.

Rosa wichuraiana Crép. [Commemorates Max Ernst Wichura (1817–1866), German botanist.] MEMORIAL ROSE.

Rosa wichuraiana Crépin, Bull. Soc. Roy. Bot. Belgique 15: 204. 1876. *Rosa lucieae* Franchet & Roche-
  brune ex Crépin var. *wichuraiana* (Crépin) Koidzumi, J. Coll. Sci. Imp. Univ. Tokyo 32(2): 234.
  1913.

Trailing or clambering shrub, to 5 m; stem pubescent, soon glabrous, with broad-based, straight or recurved prickles. Leaves with 5–9 leaflets, the blade elliptic to obovate, 2.5–4 cm long, 1.5–2 cm wide, the apex rounded to acute, the base rounded to broadly cuneate, the margin serrate from just above the base, the upper and lower surfaces glabrous; stipules adnate for ½ their length, the margin irregularly lacerate-toothed, the teeth gland-tipped. Flowers solitary or in a paniculiform corymb; floral tube glandular-pubescent; sepals ovate, the inner surface pubescent, the outer surface glabrous; petals numerous, obovate, 3–4 cm long, pink; styles pubescent. Fruiting floral tube globose or subglobose, 6–18 mm long, purple-black glandular-pubescent.

Dry, disturbed sites. Rare; Gulf, Walton, Escambia, and Hardee Counties. Escaped from cultivation. New York and Massachusetts south to Florida, west to Illinois, Arkansas, and Louisiana; Asia. Native to Asia. Spring.

Our plants are the double-flowered "Dorothy Perkins" cultivar.

Rubus L., nom. cons. 1753. BLACKBERRY

Shrubs; stems erect, arching, or trailing, armed with prickles. Leaves alternate, palmately compound, petiolate, stipulate. Flowers axillary or terminal in raceme-like or panicle-like cymes, perigynous, bisexual; sepals 5, free; petals 5, free, stamens numerous; carpels 5–many, free. Fruit an aggregate of few to many drupelets.

A genus of 250–700 species; North America, Europe, and Asia. [*Ruber*, red, in reference to the fruit color.]

Rubus has long presented taxonomic problems, which have resulted in the proliferation of names, notably those of L. H. Bailey. Small (1913) recognized 18 species, 10 of which he attributed to Florida. The treatment here is very conservative with five species recognized.

1. Stem prostrate or trailing.
  2. Stem armed with stout thorns and hispid glandular trichomes; branches of the inflorescence usually armed; flowers solitary ....**R. trivialis**
  2. Stem armed only with stout thorns; branches of the inflorescence usually unarmed; flowers usually 2–5 in a cyme ....**R. flagellaris**
1. Stem erect, arching-erect, or scrambling.
  3. Leaves of the flowering stems with 5–7 leaflets; petals rose-pink.........................**R. niveus**
  3. Leaves of the flowering stems with 3(5) leaflets; petals white or sometimes tinged with pink.
    4. Leaflets densely white or pale grayish tomentose on the lower surface, the apex obtuse or rounded ....**R. cuneifolius**
    4. Leaflets glabrous to densely pubescent on the lower surface, if pubescent, then the trichomes not white or pale gray, the apex acute ....**R. pensilvanicus**

Rubus cuneifolius Pursh [With leaflets tapering.] SAND BLACKBERRY.

Rubus cuneifolius Pursh, Fl. Amer. Sept. 1: 347. 1814.
Rubus probabilis L. H. Bailey, Gentes Herb. 1: 180. 1923.
Rubus cuneifolius Pursh var. *angustior* L. H. Bailey, Gentes Herb. 1: 263. 1925. TYPE: FLORIDA: Marion Co.: near Ocala.

Rubus inferior L. H. Bailey, Gentes Herb. 1: 263. 1925. TYPE: FLORIDA: Marion Co.: near Ocala, s.d., *Bailey s.n.* (holotype: BH).

Rubus audax L. H. Bailey, Gentes Herb. 2: 367. 1932. TYPE: FLORIDA: Hillsborough Co./Manatee Co.

Rubus chapmanii L. H. Bailey, Gentes Herb. 5: 431, f. 195. 1943. TYPE: FLORIDA: Franklin Co.: Apala-chicola, s.d., *Bailey 736* (holotype: BH).

Rubus cuneifolius Pursh var. *austrifer* L. H. Bailey, Gentes Herb. 5: 432, f. 194(c–d). 1943.

Rubus escatilis L. H. Bailey, Gentes Herb. 5: 454, f. 206. 1943.

Rubus humei L. H. Bailey, Gentes Herb. 5: 457, f. 208. 1943. TYPE: Florida: Nassau Co.: E of Yulee, s.d., *Hume 391* (holotype: BH?).

Rubus probativus L. H. Bailey, Gentes Herb. 5: 450, f. 204. 1943.

Rubus randolphiorum L. H. Bailey, Gentes Herb. 5: 447, f. 203. 1943.

Shrub, to 1 m; stem erect or sometimes arching, often densely pubescent when young, the prickles usually dense, hooked, or retrorse, usually broad-based. Leaves with 3–5 leaflets, the terminal obovate to elliptic, 2–6 cm long, 3–4 cm wide, the apex broadly rounded to subtruncate, the base cuneate, the margin serrate or sometimes doubly serrate, entire proximally, often somewhat revolute, the upper surface glabrous, the lower surface strongly gray- to white-pubescent, often with a few small prickles on the midvein; stipules adnate to the petiole base, filiform, linear, or lanceolate, 3–15 mm long. Flowers solitary or 3–5 in a raceme; sepals lanceolate, armed or unarmed, densely pubescent, strongly reflexed; petals elliptic to obovate, 5–15 mm long, white. Fruit globose to cylindric, 0.6–2 cm long, black, the drupelets 15–50, strongly coherent, separating with the receptacle attached.

Sandhills and flatwoods. Common; nearly throughout. Vermont south to Florida, west to Louisiana. Spring.

Rubus flagellaris Willd. [Whiplike, in reference to the stems.] NORTHERN DEWBERRY.

Rubus flagellaris Willdenow, Enum. Pl. 549. 1809.

Rubus enslenii Trattinnick, Rosac. Monogr. 3: 63. 1823. *Rubus villosus* Aiton subsp. *enslenii* (Trattinnick) W. Stone, Pl. New Jersey 480. 1911.

Rubus rhodophyllus Rydberg ex Small, Fl. S.E. U.S. 518, 1331. 1903. *Rubus procumbens* Muhlenberg ex W.P.C. Barton subsp. *rhodophyllus* (Rydberg ex Small) Focke, Biblioth. Bot. 19: 83. 1914.

Shrub, to 1 m; stem creeping or low-arching, the flowering branches erect, glabrous to densely pubescent, usually with sessile to short-stipitate glands, the prickles sparse to dense, hooked to retrorse, broad-based to slender. Leaves with 3–5 leaflets, the terminal ovate or elliptic to suborbicular, 3–11 cm long, 2–7.5 cm wide, the apex acute, acuminate, to short-attenuate, the base cuneate or rounded to shallowly cordate, the margin serrate to doubly serrate or serrate-dentate, the upper surface glabrous, the lower surface sparsely to moderately pubescent, with sessile to short-stipitate glands along the larger veins, with a few small prickles on the midvein or unarmed; stipules usually adnate to the petiole, filiform, linear, or lanceolate, 3–20 mm long. Flowers solitary or 3(5) in a raceme; sepals lanceolate, densely pubescent, sparsely to moderately sessile glandular; petals elliptic, obovate, to oblanceolate, 8–20 mm long, white. Fruit globose to cylindric, 1–2 cm long, black, the drupelets 10–40, strongly coherent, separating with the receptacle attached.

Open hammocks, sandhills, and flatwoods. Occasional; central and western panhandle. Quebec south to Florida, west to Ontario, Minnesota, Nebraska, Kansas, Oklahoma, and Texas. Spring.

Rubus niveus Thunb. [Snowy, in reference to the white lower surface of the leaves.] SNOWPEAKS RASPBERRY.

Rubus niveus Thunberg, Rubo 9. 1813.

Shrub, to 4.5 m; stem arching, sparsely pubescent, glabrescent in age, the prickles dense, falcate to recurved. Leaves with (3)5–7(9) leaflets, the terminal ovate to broadly ovate, 4–6 cm long, (2.5)3.5–5 cm wide, occasionally shallowly lobed, the apex acute to short-acuminate, base shallowly cordate, the margin coarsely serrate to doubly serrate, the upper surface glabrous, the lower surface densely white-tomentose, the midvein with scattered prickles; stipules adnate to the petiole, linear-lanceolate, 4–8 mm long. Flowers solitary or 2–11 in a raceme; sepals ovate to oblong; petals broadly obovate to orbicular, ca. 4 mm long, pink to magenta. Fruit globose, 5–7 mm long, purple-black, the drupelets 60–75, coherent, separating from the receptacle.

Disturbed sites. Rare; Miami-Dade County. Escaped from cultivation. Florida; Asia. Native to Asia. Spring.

Rubus pensilvanicus Poir. [Of Pennsylvania.] SAWTOOTH BLACKBERRY; PENNSYLVANIA BLACKBERRY.

Rubus pensilvanicus Poiret, in Lamarck, Encycl. 6: 246. 1804.
Rubus argutus Link, Enum. Hort. Berol. Alt. 2: 60. 1822.
Rubus floridus Trattinnick, Rosac. Monogr. 3: 73. 1823. *Rubus argutus* Link var. *floridus* (Trattinnick) L. H. Bailey, Sketch Evol. Native Fr. 305. 1898.
Rubus ostryifolius Rydberg, in Britton, Manual Fl. N. States 497. 1901.
Rubus betulifolius Small, Fl. S.E. U.S. 518, 1331. 1903.
Rubus persistens Rydberg ex Small, Fl. S.E. U.S. 519, 1331. 1903.
Rubus ucetanus L. H. Bailey, Gentes Herb. 1: 269. 1925. TYPE: FLORIDA: Hillsborough Co.: Uceta.
Rubus abundifolius L. H. Bailey, Gentes Herb. 3: 141, f. 92–94. 1933.
Rubus penetrans L. H. Bailey, Gentes Herb. 3: 147. 1933.
Rubus floridensis L. H. Bailey, Gentes Herb. 5: 437, f. 198. 1943. TYPE: FLORIDA: Hillsborough Co.: along Memorial Highway 1.5 mi. W of Tampa, s.d., *Hume s.n.* (holotype: BH).
Rubus arrectus L. H. Bailey, Gentes herb. 5: 644, f. 286. 1945.
Rubus harperi L. H. Bailey, Gentes Herb. 5: 648, f. 289. 1945. TYPE: FLORIDA: Leon Co.: near Tallahassee, s.d., *Bailey 6913H* (holotype: BH).
Rubus zoae L. H. Bailey, Gentes Herb. 5: 654, f. 291. 1945. TYPE: FLORIDA: Orange Co.: Winter Park, s.d. *Bailey 999* (holotype: BH).

Shrub, to 3 m; stem erect-arching, glabrous or sparsely to densely pubescent, the prickles sparse to dense, straight or slightly retrorse, usually sparsely to densely stipitate-glandular. Leaves with (3)5(7) leaflets, the terminal ovate to lanceolate, 5–15 cm long, 3–13 cm wide, the apex acuminate to long-attenuate, the base rounded to shallowly cordate, the margin coarsely serrate to doubly serrate, the upper surface glabrous, the lower surface moderately pubescent, usually with a few small prickles along the midvein, moderately glandular along the veins or eglandular. Flowers (2)5–12(16) in a raceme; sepals lanceolate, sparsely to densely pubescent, glandular or eglandular; petals obovate to elliptic or rarely suborbicular, 1–4 cm long. Fruit globose to cylindric, 1–2 cm long, the drupelets 10–100, strongly coherent, separating with the receptacle attached.

Wet hammocks, pond margins, and titi swamps. Common; northern counties, central peninsula. Newfoundland south to Florida, west to Ontario, Minnesota, Iowa, Missouri, Oklahoma, and Texas; Pacific Islands. Native to North America. Spring.

Rubus trivialis Michx. [Commonplace, ordinary.] SOUTHERN DEWBERRY.

> *Rubus trivialis* Michaux, Fl. Bor.-Amer. 1: 296. 1803. *Rubus procumbens* Muhlenberg ex. W.P.C. Barton, Comp. Fl. Philadelph. 1: 233. 1818, nom. illegit.
>
> *Rubus carpinifolius* Rydberg ex Small, Fl. S.E. U.S. 519, 1331. 1903; non Weihe et al., 1824. *Rubus hispidus* Linnaeus subsp. *continentalis* Focke, Biblioth. Bot. 19: 86. 1914. *Rubus continentalis* (Focke) L. H. Bailey, Gentes Herb. 1: 173. 1923.
>
> *Rubus lucidus* Rydberg ex Small, Shrubs Florida 26, 133. 1913. TYPE. FLORIDA: Pinellas Co.: Dunedin, s.d. *Tracy 6855* (holotype: NY; isotype: BH).
>
> *Rubus ictus* L. H. Bailey, Gentes Herb. 1: 231. 1925.
>
> *Rubus mirus* L. H. Bailey, Gentes Herb. 1: 231. 1925. FLORIDA: cult. at Ithaca.
>
> *Rubus okeechobeus* L. H. Bailey, Gentes Herb. 1: 228. 1925. TYPE: FLORIDA: Palm Beach Co.: not far above Canal Point, near shore of Lake Okeechobee.
>
> *Rubus tallasseanus* L. H. Bailey, Gentes Herb. 1: 286. 1925. TYPE: FLORIDA: Leon Co.: Tallahassee.
>
> *Rubus sons* L. H. Bailey, Gentes Herb. 2: 312. 1932.
>
> *Rubus agilis* L. H. Bailey, Gentes Herb. 3: 134, f. 89. 1933. TYPE: FLORIDA: Leon Co.: Tallahassee.
>
> *Rubus magniflorus* L. H. Bailey, Gentes Herb. 3: 137, f. 90–91. 1933. TYPE: FLORIDA: Franklin Co.: near Carabelle.

Shrub, to 4(5) m; stem low arching or creeping, the flowering branches erect, rooting at the tips, glabrous to moderately pubescent, often with glandular bristles, the prickles moderate to dense, hooked to retrorse. Leaves with 3–5 leaflets, the terminal narrowly elliptic or ovate to obovate, 2–8.5 cm long, 0.7–4.5 cm wide, the apex acute to acuminate, the base rounded to narrowly cuneate, the margin moderately to coarsely serrate to doubly serrate, the upper surface glabrous, the lower surface glabrous or rarely moderately pubescent, sparsely short stipate-glandular or eglandular, with a few small hooked prickles on the midvein. Flowers solitary or rarely in a short few-flowered raceme; sepals densely pubescent; petals elliptic to obovate, 10–16(25) mm long, white to pink. Fruit globose to ovoid.

Dry, open hammocks. Common; nearly throughout. Pennsylvania south to Florida, west to Kansas, Oklahoma, and Texas. Spring.

EXCLUDED TAXA

> *Rubus allegheniensis* Porter—Reported for Florida by Small (1903, as *R. nigrobaccus* L. H. Bailey) apparently in error. No specimens known.
>
> *Rubus bifrons* Vest ex Trattinnick—Reported for Florida by Radford (1964, 1968) and Correll and Johnston (1970), but the basis of these reports of the Himalayan blackberry in our flora is unknown.
>
> *Rubus linkianus* Serenge—This taxon is apparently a hybrid between two European species, *R. canescens* de Candolle and *R. discolor* Weihe & Nees von Esenbeck (fide Davis, 1972) and was reported as escaping in Florida by Small (1933), the basis unknown.

ELAEAGNACEAE Juss., nom. cons. 1789. OLEASTER FAMILY

Shrubs or trees. Leaves alternate, simple, pinnate-veined, petiolate, estipulate. Flowers axillary, solitary or in an umbel, bisexual; sepals 4, free, petaloid; petals absent; stamens 4, adnate to the hypanthium, the anthers basifixed or dorsifixed, laterally dehiscent; carpels 1, 1-locular, superior, the style 1, apical, the stigma 1, the ovule solitary. Fruit a drupelike achene, indehiscent; seed 1.

A family of 3 genera and 45 species; North America, Europe, Africa, Asia, Australia, and Pacific Islands.

Elaeagnus L. 1753.

Shrubs or trees. Leaves alternate, simple, pinnate-veined, petiolate, estipulate. Flowers in umbels, solitary or paired, pedicellate; hypanthium elongate; sepals 4, free, petaloid; petals absent; stamens 4, alternate with the sepals; ovary 1, the style stigmatic on 1 side. Fruit a drupelike achene.

A genus of about 45 species; North America, Europe, Africa, Asia, Australia, and Pacific Islands. [From the Greek *elaia*, olive, and *agnos*, willow.]

1. Plant evergreen; lower leaf surface ashy-white, conspicuously flecked with brown scales ....
.... **E. pungens**
1. Plant deciduous; lower leaf surface silvery, with few, if any, brown scales .... **E. umbellata**

Elaeagnus pungens Thunb. [Piercing, terminating in a sharp point, in reference to the thorns.] SILVERTHORN; THORNY OLIVE.

> *Elaeagnus pungens* Thunberg, Fl. Jap. 68. 1784. *Elaeagnus pungens* Thunberg subsp. *eupungens* Servettaz, Bull. Herb. Boisier, ser. 2. 8: 387, nom. inadmiss. *Elaeagnus pungens* Thunberg var. *typica* C. K. Schneider, Ill. Handb. Laubholzk. 2: 413. 1909, nom. inadmiss.

Shrub or small tree, to 5 m; stem with thornlike lateral branches, dark gray lepidote or reddish. Leaves evergreen, the blade elliptic, (3.5)4–8(10) cm long, 1.5–2.5 cm wide, the apex broadly acute to obtuse, the base broadly cuneate to subrotund, the margin entire, undulate, the upper surface dark green or dull silver-green, lustrous, the lower surface densely pubescent with lepidote scales and stellate trichomes, silver-green. Flowers few in an umbel, the hypanthium 2–3 mm long, with lepidote scales and stellate trichomes; calyx ca. 6 mm long, white or cream, with lepidote scales and stellate trichomes on the outer surface. Fruit ellipsoid, 4–8 mm long, reddish brown, with silvery lepidote scales; seeds with 8 striae.

Dry disturbed sites. Occasional; northern counties, Marion and St. Lucie Counties. Escaped from cultivation. Massachusetts, Virginia south to Florida, west to Kentucky, Tennessee, and Louisiana; Europe, Asia, and Australia. Native to Asia. Fall–spring.

Elaeagnus pungens is listed as a Category II invasive species in Florida by the Florida Exotic Pest Plant Council (FLEPPC, 2015).

Elaeagnus umbellata Thunb. [In an umbel, in reference to the inflorescence.] SILVERBERRY; AUTUMN OLIVE.

Elaeagnus umbellata Thunberg, Fl. Jap. 66, t. 14. 1784. *Elaeagnus umbellata* Thunberg subsp. *euumbellata* Servettaz, Bull. Herb. Boissier, ser. 2. 8: 383. 1908, nom. inadmiss. *Elaeagnus umbellata* Thunberg var. *typica* C. K. Schneider, Ill. Handb. Laubholzk. 2: 411. 1909, nom. inadmiss.

Shrub or small tree, to 5 m. stem with thornlike lateral branches when young, the older stems unarmed, the branches silver-green when young, becoming densely lepidote in age. Leaves deciduous, the blade elliptic, (2)3–8(10) cm long, 1–2.5 cm wide, the apex obtuse, the base broadly acute to rounded, the margin entire, sometimes slightly undulate, the upper surface sparsely pubescent, dark green, glossy, the lower surface with silver lepidote scales and stellate trichomes. Flowers in a dense umbel, often several umbels together and flowers appearing to encircle the stem; hypanthium funnelform, 7–8 mm long, silver with lepidote scales and stellate trichomes; calyx 3.5–4 mm long, petaloid, with silver-lepidote scales and stellate trichomes on the outer surface. Fruit ovoid, 6–8 mm long, red or pink, the pedicels ca. 1 cm long; seeds with 8 striae.

Disturbed sites. Rare; Leon and Gadsden Counties. Escaped from cultivation. Maine south to Florida, west to Ontario, Wisconsin, Nebraska, Kansas, Arkansas, and Louisiana, also Montana, Washington, and Oregon; Europe, Asia, Australia, and Pacific Islands. Native to Asia. Spring.

RHAMNACEAE Juss., nom. cons. 1789. BUCKTHORN FAMILY

Trees, shrubs, or woody vines, unarmed or armed with thorns or stipular spines. Leaves simple, alternate, subopposite, or opposite, simple, pinnate or pinnipalmate-veined, petiolate, stipulate. Flowers in axillary fascicles, umbelliform or corymbiform cymes, or axillary and/or terminal racemiform or spiciform thyrses, with a distinct hypanthium, bisexual or unisexual; sepals 4–5, free; petals 4–5, free, or absent; stamens 4–5, adnate to the petals at the base and inserted at or below the disk margin, the anthers dorsifixed, 2-locular, dehiscent by longitudinal slits; staminodes present in carpellate flowers; nectariferous disk intrastaminal, hypogynous or epigynous; carpels 2–3, connate, locules (1)2–3, ovary superior to inferior, the style 1, the stigmas 1–3. Fruit a drupe with 2 or 3 pyrenes, or a schizocarp; seeds arillate.

A family of about 50 genera and about 900 species; nearly cosmopolitan.

Frangulaceae DC., 1805.

Selected reference: Brizicky (1964).

1. Leaves opposite or subopposite.
    2. Leaves with serrate or serrulate margins, the apex acute or acuminate; petals present ..... **Sageretia**
    2. Leaves usually notched at the apex, otherwise entire; petals absent.
        3. Sepals keeled; leaves chartaceous or subcoriaceous..**Krugiodendron**
        3. Sepals not keeled; leaves coriaceous..**Reynosia**
1. Leaves alternate.
    4. Woody vines.

5. Vine climbing by twining; leaf margins entire or undulate-crenate; fruit a drupe..... **Berchemia**

5. Vine climbing by solitary tendrils; leaf margins coarsely and remotely crenate-serrate; fruit a 3-winged schizocarp.. **Gouania**

4. Trees or shrubs.

6. Plant armed with stipular spines or axillary thorns..**Ziziphus**

6. Plant unarmed.

7. Fruit fleshy; ovary superior; inflorescence a corymbiform fascicle...........................**Rhamnus**

7. Fruit leathery; ovary semi-inferior; inflorescence a thyrse.

8. Inflorescence many-flowered, longer than the subtending leaves.....................**Ceanothus**

8. Inflorescence few-flowered, much shorter than the subtending leaves..............**Colubrina**

Berchemia Neck ex DC., nom. cons. 1825. SUPPLEJACK

Woody vines. Leaves deciduous, alternate, chartaceous, pinnate-veined, the margin entire or undulate-crenate. Flowers in racemiform thyrses composed of corymbiform terminal cymes, unisexual (plants dioecious); sepals 5, free; petals 5, free; stamens 5; nectariferous disk fleshy, filling the hypanthium; carpels 2, 2-loculate, the ovary superior, the stigmas 2. Fruit a single-stoned, indehiscent, 2-locular drupe; seeds 1–2.

A genus of about 12 species; North America, Mexico, and Central America; Asia, Malaysia, Africa. [Commemorates Jacob Pierre Berthoud van Berchem, eighteenth-century Dutch mineralogist and naturalist.]

Berchemia scandens (Hill) K. Koch [Climbing.] ALABAMA SUPPLEJACK; RATTAN VINE.

Rhamnus scandens Hill, Hort. Kew. 453, pl. 20. 1768. *Berchemia scandens* (Hill) K. Koch, Dendrologie 1: 602. 1869. *Berchemia undulata* Rafinesque, Sylvia Tellur. 33. 1838; nom. illegit.

Rhamnus volubilis Linnaeus f., Suppl. Pl. 152. 1782. *Oenoplea volubilis* (Linnaeus f.) Schultes, in Roemer & Schultes, Syst. Veg. 5: 332. 1819. *Berchemia volubilis* (Linnaeus f.) de Candolle, Prodr. 2: 23. 1825.

Berchemia repanda Rafinesque, Sylva Tellur. 33. 1838.

Climbing-scandent and twining woody vine; stem to 10 cm wide, the bark smooth, glabrous. Leaves with the blade ovate to elliptic-ovate or elliptic, 3–6(8) cm long, 2–3.5(4) cm wide, the apex acuminate to obtuse or rounded, the base truncate to rounded or obtuse, the margin entire or shallowly undulate-crenate, the lateral veins 8–12 pairs, the upper surface glabrous, the lower surface glabrous or pubescent, sometimes only along the veins, the petiole 8–15 mm long. Flowers in a racemiform thyrse of corymbiform terminal cymes, pedicellate; hypanthium cupulate, 1–2 mm long; sepals triangular to lanceolate or linear, ca. 1 mm long, slightly keeled; petals spatulate to lanceolate, subequaling the sepals, short-clawed, greenish white; nectariferous disk with 10 irregular lobes, free at the margin; stamens subequaling the petals; ovary deeply immersed in the disk, the style undivided, the stigmas emarginate or 2-fid. Fruit ellipsoid to narrowly obovoid, 5–14 mm long, fleshy, bluish black, the stone 1- to 2-seeded; seeds thin, partly adnate to the endocarp.

Wet hammocks, floodplain forests, and wet flatwoods. Frequent; nearly throughout. Maryland south to Florida, west to Missouri, Oklahoma, and Texas; Mexico and Central America. Spring.

Ceanothus L. 1753.

Shrubs. Leaves alternate, 1- to 3-veined from the base, the margin serrulate or crenate-serrulate to subentire, petiolate. Flowers terminal and/or axillary racemiform or corymbiform thyrses of umbelliform cymes or axillary cymes, bisexual; sepals 5, free; petals 5, free; stamens 5; nectariferous disk annular, between insertion of the stamens and upper part of the ovary; carpels 3, 3-locular, the ovary half-inferior, the stigmas 3. Fruit a dehiscent drupe with 3 pyrenes.

A genus of about 55 species; North America, Mexico, and Central America. [From the Greek *keanothus*, name used by Dioscorides for some plant.]

1. Leaves 2–10 cm long, the margin serrate to serrulate ....C. americanus
1. Leaves 0.3–1 cm long, the margin entire...C. microphyllus

Ceanothus americanus L. [Of America.] NEW JERSEY TEA; REDROOT.

Ceanothus americanus Linnaeus, Sp. Pl. 1: 195. 1753. *Ceanothus officinalis* Rafinesque, New Fl. 3: 54. 1838; nom. illegit.

Ceanothus intermedius Pursh, Fl. Amer. Sept. 167. 1814. *Ceanothus americanus* var. *intermedius* (Pursh) Torrey & A. Gray, Fl. N. Amer. 1: 264. 1838. *Ceanothus virgatus* Rafinesque, New Fl. 3: 56. 1838; nom. illegit.

Shrub, to 1.5 m; branches erect to ascending, green, puberulent. Leaves deciduous, the blade ovate to ovate-oblong, 2–10 cm long, 1–6.4 cm wide, the apex acuminate to acute or obtuse, the base rounded, the margin serrate to serrulate, 3-veined from the base, the upper and lower surfaces puberulent, especially on the major veins, the petiole 4–13 mm long. Flowers terminal or axillary in a paniculiform thyrse; sepals lanceolate to deltate, minute, incurved, white; petals 1–2 mm long, hooded, each hood enfolding a developing stamen, slender clawed, white; disk white. Fruit globose, 4–6 mm long, smooth or 3-lobed above the middle, the leathery exocarp sloughing off prior to dehiscence, the valves dorsally crested, somewhat rugulose; seeds ovoid, somewhat compressed, 3–5 mm long, gray to black.

Sandhills and dry hammocks. Frequent; northern counties, central peninsula. Quebec south to Florida, west to Minnesota, Nebraska, Kansas, Oklahoma, and Texas. Spring–summer.

Ceanothus microphyllus Michx. [Small-leaved.] LITTLE BUCKBRUSH.

Ceanothus microphyllus Michaux, Fl. Bor.-Amer. 1: 154. 1803. TYPE: "Georgiae et Floridae."

Ceanothus serpyllifolius Nuttall, Gen. N. Amer. Fl. 1: 154. 1818. *Ceanothus microphyllus* Michaux var. *serpyllifolius* (Nuttall) A. W. Wood, Class-Book Bot., ed. 1861. 291. 1861. TYPE: FLORIDA: Nassau Co.: "around the town of St. Marys" [Georgia], s.d., *Baldwin s.n.* (holotype: PH?).

Shrub, to 1 m; branches erect to ascending, reddish green or yellow, puberulent. Leaves deciduous, the blade elliptic, ovate elliptic, or narrowly obovate, 2–10 mm long, 1–6 mm wide, the apex rounded or obtuse, the base cuneate to rounded, the margin entire, obscurely 1- to

3-veined from the base, the upper surface glabrous, the lower surface puberulent on the veins, the petiole ca. 1 mm long. Flowers in an axillary or terminal, umbelliform or racemiform thyrse or cyme; sepals lanceolate to deltate, minute, incurved, white; petals 1–2 mm long, hooded, each hood enfolding a developing stamen, slender clawed, white; disk white. Fruit globose, 3–5 mm long, smooth or 3-lobed above the middle, the leathery exocarp sloughing off prior to dehiscence, the valves dorsally crested, smooth; seeds ovoid, somewhat compressed, 2–4 mm long, gray to black.

Sandhills and flatwoods. Frequent; northern counties, central peninsula. Georgia, Alabama, and Florida.

EXCLUDED TAXA

Ceanothus atropurpureus Rafinesque—The name a provisional name not accepted by the author and thus not validly published, based on a Florida specimen whose identity is very doubtful.

Ceanothus cubensis (Jacquin) Lamarck—Misapplied to *Colubrina cubensis* var. *floridana.*

Ceanothus herbaceus Rafinesque—Reported for Florida by Small (1903, 1913a, 1933, all as *C. ovatus* Desfontaines), misappled to material of *C. americanus.*

Colubrina Rich. ex Brongn., nom. cons. 1827. NAKEDWOOD

Shrubs or trees. Leaves alternate, pinnate-veined or 3-veined from the base. Flowers in axillary umbelliform or corymbiform cymes in thyrses, bisexual; sepals 5, free; petals 5, free; stamens 5, free; nectariferous disk large, fleshy, pentagonal or shallowly 10-lobed, surrounding the ovary and adnate to its lower half, carpels 3, 3-locular, the ovary semi-inferior, the stigmas 3. Fruit a drupe, the stone dehiscent septicidally into 3 pyrenes.

A genus of about 30 species; North America, West Indies, Mexico, Central America, South America, Africa, Asia, Australia, and Pacific Islands. [*Coluber*, snakelike, in reference to the twisted, deep furrows on the stems of some species.]

Selected reference: Johnston (1971).

1. Leaves with the margin serrate, the blade 3-nerved from the base; scandent shrub .... **C. asiatica**
1. Leaves with the margin entire, the blade pinnately nerved; tree or shrub.
    2. Leaves glabrous or nearly so, chartaceous .... **C. elliptica**
    2. Leaves tomentose or tomentulose, at least on the lower surface, coriaceous.
        3. Leaves pubescent on both surfaces; peduncles longer than the petioles.................... **C. cubensis**
        3. Leaves tomentulose on the lower surface, glabrate on the upper surface; peduncles shorter than the petioles .... **C. arborescens**

Colubrina arborescens (Mill.) Sarg. [Becoming treelike.] GREENHEART.

Rhamnus colubrina Jacquin, Enum. Syst. Pl. 16. 1760. *Ceanothus colubrina* (Jacquin) Lamarck, Tabl. Encycl. 2: 90. 1797. *Colubrina ferruginosa* Brongniart, Ann. Sci. Nat. (Paris) 10: 369. 1827. *Marcorella colubrina* (Jacquin) Rafinesque, Sylva Tellur. 31. 1838. *Colubrina americana* Nuttall, N. Amer. Sylv. 2: 47. 1846; nom. illegit. *Colubrina colubrina* (Jacquin) Millspaugh, Publ. Field Columbian Mus., Bot. Ser. 2: 69. 1900; nom. inadmiss.

Ceanothus arborescens Miller, Gard. Dict., ed. 8. 1768. *Colubrina arborescens* (Miller) Sargent, Trees & Shrubs 2: 167. 1911.

Rhamnus ferruginea Nuttall, J. Acad. Nat. Sci. Philadelphia 7(1): 90. 1834. *Perfonon ferrugineum* (Nuttall) Rafinesque, Sylva Tellur. 29. 1838.

Shrub or tree, to 8 m; branches densely reddish tomentose. Leaves persistent, the blade ovate to ovate-oblong to elliptic or oblong-obovate, 5–15 cm long, 4–8 cm wide, pinnate-veined with (4)5–9 vein pairs, the apex acute to acuminate, the base rounded to subcordate, with 2 basal glands, the margin entire, the upper and lower surfaces red-brown tomentose, the upper surface soon glabrescent or glabrous, the lower surface remaining tomentose along the veins, the petiole 5–20 mm long. Flowers 10–30 in an axillary cyme or thyrse, the peduncle (3)5–10 mm long; sepals ovate, ca. 2 mm long, pubescent; petals spatulate, shorter than the sepals, clawed, greenish yellow. Fruit 6–8 mm long, black; seeds ca. 4 mm long, black, lustrous.

Tropical hammocks. Occasional; Collier and Monroe Counties. Florida; West Indies, Mexico, and Central America. All year.

Colubrina asiatica (L.) Brongn. [Of Asia.] LATHERLEAF; ASIAN NAKEDWOOD.

Ceanothus asiaticus Linnaeus, Sp. Pl. 1: 196. 1753. *Rhamnus asiatica* (Linnaeus) Lamarck ex Poiret, in Lamarck, Encycl. 4: 474. 1796. *Colubrina asiatica* (Linnaeus) Brongniart, Ann. Sci. Nat. (Paris) 10: 369. 1827.

Clambering or scandent shrub, to 5 m; stem glabrous. Leaves persistent, the blade ovate to broadly ovate, 4–8 cm long, 2–5 cm wide, 3-veined from the base, the apex short acuminate, the base obtuse or rounded to subcordate, the margin crenate-serrate, the upper and lower surfaces glabrous or glabrate, the petiole 7–17 mm long. Flowers solitary or 2–7 in a short-pedunculate, axillary thyrse; sepals triangular-ovate, ca. 2 mm long, glabrate; petals spatulate, about as long as sepals, clawed, white. Fruit subglobose, 7–8 mm long; seeds gray, smooth.

Coastal hammocks and beaches. Frequent; Martin and Lee Counties, southern peninsula. Escaped from cultivation. Florida; West Indies; Africa, Asia, Australia, and Pacific Islands. Native to Africa, Asia, Australia, and Pacific Islands. All year.

Colubrina asiatica is listed as a Category I invasive species in Florida by the Florida Exotic Pest Plant Council (FLEPPC, 2015).

Colubrina cubensis (Jacq.) Brongn. var. floridana M. C. Johnst. [Of Cuba; of Florida.] CUBAN NAKEDWOOD.

Colubrina cubensis (Jacquin) Brongniart var. *floridana* M. C. Johnston, Wrightia 3: 96. 1963. *Colubrina cubensis* (Jacquin) Brongniart subsp. *floridana* (M. C. Johnston) Borhidi, Acta Bot. Acad. Sci. Hung. 19: 44. 1973. TYPE: FLORIDA: Miami-Dade Co.: Lewis-Nixon Hammock, Redlands district, 2 Feb 1930, *Moldenke 553* (holotype: US; isotypes: NY, S, W).

Shrub or tree, to 8 m; branches puberulent-tomentose. Leaves deciduous, the blade elliptic-oblong to ovate-oblong or obovate-oblong, 4–10 cm long, 1–5 cm wide, pinnate-veined with 6–12 vein pairs, the apex acute to rounded or emarginate, the base rounded to obtuse, the margin entire or obscurely serrulate to crenate, the upper surface villous-strigose, glabrescent, the lower surface tawny tomentose, the petiole 5–15 mm long. Flowers 20–50(70) in an axillary, corymbiform thyrse, the pedicel 4–9 mm long; sepals triangular-ovate, ca. 2 mm long, densely pubescent; petals spatulate, about as long as sepals, clawed, yellow. Fruit globose, 7–9 mm long.

Tropical hammocks and pine rocklands. Occasional; Miami-Dade County, Monroe County keys. Florida; West Indies. All year.

Colubrina elliptica (Sw.) Brizicky & W. L. Stern [Elliptic, referring to the leaves.] SOLDIERWOOD.

Rhamnus elliptica Swartz, Prodr. 50. 1788. *Diplisca elliptica* (Swartz) Rafinesque, Sylva Tellur. 31. 1838. *Colubrina elliptica* (Swartz) Brizicky & W. L. Stern, Trop. Woods. 109: 95. 1958.

Ceanothus reclinatus L'Héritier de Brutelle, Sert. Angl. 6. 1789. *Colubrina reclinata* (L'Héritier de Brutelle) Brongniart, Ann. Sci. Nat. (Paris) 10: 369. 1827.

Shrub or tree, to 6(10) m; bark orange-brown, furrowed, exfoliating in thin layers, the branches strigose, soon glabrate. Leaves deciduous, the blade broadly elliptic to elliptic-ovate or ovate-lanceolate, 4–8 mm long, 2–5 cm wide, pinnate-veined with 5–8 vein pairs, the apex acute to acuminate, the base cuneate to rounded, with a prominent pair of glands at the base, the margin entire, the upper surface glabrous, the lower surface sparsely strigose to glabrescent, the petiole 5–25 mm long. Flowers 8–20 in an axillary thyrse, the peduncle 6–7 mm long; sepals triangular-ovate, ca. 1 mm long, glabrous; petals spatulate, shorter than the sepals, clawed. Fruit globose, orange-red, 7–10 mm long; seeds ca. 5 mm long, blackish, lustrous.

Tropical hammocks. Occasional; Miami-Dade County, Monroe County keys. Florida; West Indies, Mexico, Central America, and South America. All year.

Gouania Jacq. 1763. CHEWSTICK

Woody vines, climbing by solitary tendrils. Leaves alternate, pinnate-veined, sometimes indistinctly 3-nerved at the base. Flowers in axillary and terminal racemiform thyrses, unisexual (plants polygamomonoecious); sepals 5, free; petals 5, free; stamens 5, free; nectariferous disk epigynous, fleshy, 5-lobed; carpels 3, the ovary inferior, the stigmas 3. Fruit a winged schizocarp, splitting septicidally through each wing into 3, 2-winged, indehiscent mericarps.

A genus of about 50 species; North America, West Indies, Mexico, Central America, South America, Africa, Asia, Australia, and Pacific Islands. [Commemorates Antoine Gouan (1733–1821), French botanist and ichthyologist at Montpellier.]

Gouania lupuloides (L.) Urb. [Resembling *Humulus lupulus* (Cannabaceae).] CHEWSTICK; WHITEROOT.

Banisteria lupuloides Linnaeus, Sp. Pl. 1: 427. 1753. *Gouania domingensis* Linnaeus, Sp. Pl., ed. 2. 1663. 1763; nom. illegit. *Lupulus lupuloides* (Linnaeus) Kuntze, Revis. Gen. Pl. 1: 119. 1891. *Gouania lupuloides* (Linnaeus) Urban, Symb. Antill. 4: 378. 1910.

Rhamnus domingensis Jacquin, Enum. Syst. Pl. 17. 1760. *Gouania glabra* Jacquin, Select. Stirp. Amer. Hist. 264. 1763; nom. illegit. *Lupulus lupuloides* (Linnaeus) Kuntze var. *domingensis* (Jacquin) Kuntze, Revis. Gen. Pl. 1: 119. 1891.

Scrambling and climbing woody vines, to 7 m; stem with coiled tendrils modified from branch tips at the base of the inflorescences; branches loosely pubescent, soon glabrous. Leaves deciduous, the blade ovate to oblong, elliptic, or elliptic-obovate, 3–9 cm long, 1–5 cm wide, the apex acute-acuminate, the base rounded to subcordate, the margin serrate to crenate-serrate, the upper surface glabrous, the lower surface sparsely puberulent along the veins, the petiole 5–15 mm long. Flowers in a terminal or axillary raceme 5–15 cm long, the rachis pubescent; hypanthium cupulate; sepals ovate-triangular, 1–2 mm long, keeled, green, pubescent; petals spatulate, 1–2 mm long, hooded, enfolding the stamens, yellowish; stamens shorter than the petals. Fruit 6–13 mm long, 8–10 mm wide, longitudinally 3-winged, the wings wider than the body, reticulate-veined, glabrous.

Coastal hammocks. Occasional; Clay County, central and southern peninsula. Florida; West Indies, Mexico, Central America, and South America. Spring.

Krugiodendron Urb. 1902.

Trees or shrubs. Leaves persistent, opposite or subopposite, pinnate-veined, the margin entire. Flowers in axillary, umbelliform cymes, bisexual; hypanthium shallow; sepals 5, free; petals absent; stamens 5, free; nectariferous disc fleshy, annular, 5-lobed, surrounding the base of the ovary; carpels 2, 2-locular, the ovary superior, the stigmas 2. Fruit a single-stoned drupe, the stone 2-locular, 1(2)-seeded.

A monotypic genus; North America, West Indies, Mexico, and Central America. [Commemorates Carl Wilhelm Leopold Krug (1833–1898), collaborator with Ignatz Urban on the West Indian flora, and from the Greek *dendron*, tree.]

Krugiodendron ferreum (Vahl) Urb. [*Ferreus*, iron, in reference to the hard wood.] BLACK IRONWOOD; LEADWOOD.

Rhamnus ferrea Vahl, Symb. Bot. 3: 41, t. 58. 1794. *Ceanothus ferreus* (Vahl) de Candolle, Prodr. 2: 30. 1825. *Scutia ferrea* (Vahl) Brongniart, Ann. Sci. Nat. (Paris) 10: 363. 1827. *Condalia ferrea* (Vahl) Grisebach, Fl. Brit. W.I. 100. 1859. *Rhamnidium ferreum* (Vahl) Sargent, Gard. & Forest 4: 16. 1891. *Krugiodendron ferreum* (Vahl) Urban, Symb. Antill. 3: 314. 1902.

Shrub or tree, to 10 m; branches glabrous. Leaves with the blade ovate to elliptic to suboval, 2–6 cm long, 1–5 cm wide, the apex acute, rounded to obtuse, or truncate-emarginate, the base rounded, the margin entire, undulate, the upper and lower surfaces glabrous, the petiole 3–6 mm long. Flowers 3–5 in an axillary, umbelliform cyme, the pedicel 1–6 mm long; sepals triangular-ovate, ca. 2 mm long, greenish yellow; nectariferous disk fleshy, filling the hypanthium, the margins crenate. Fruit globose, 5–12 mm long, purplish red to nearly black.

Coastal hammocks. Frequent; Brevard County southward, southern peninsula. Florida; West Indies, Mexico, and Central America. Spring.

Reynosia Griseb. 1866. DARLINGPLUM

Shrubs or trees. Leaves opposite, pinnate-veined, the margin entire. Flowers in axillary, umbelliform, cymose fascicles or solitary, bisexual; sepals 5, free; petals absent; stamens 5, free; nectariferous disk fleshy, lining the hypanthium; carpels 2, 2-locular, the ovary superior, the stigmas 2. Fruit a single-stoned drupe, 1-locular by reduction, 1-seeded.

A genus of about 15 species; North America, West Indies, and Central America. [Commemorates Alvaro Reynoso (1830–1888), Cuban agriculturist and chemist.]

Reynosia septentrionalis Urb. [Northern.] DARLINGPLUM.

Reynosia septentrionalis Urban, Symb. Antill. 4: 356. 1899. SYNTYPE: Florida.

Shrubs or trees, to 7 m; branches glabrous. Leaves persistent, the blade elliptic-oblong to oval or obovate, 2–4 cm long, 1–3 cm wide, the apex acuminate to rounded or truncate-emarginate, the base cuneate to truncate, the margin entire, the upper and lower surfaces glabrous, the petiole 1.5–3 mm long. Flowers solitary or 2–4 in an axillary, umbelliform cyme; hypanthium cupulate, 2–4 mm long; sepals 1–2 cm long, yellow green. Fruit globose to ovoid or ellipsoidal, dark purple to black, 10–20 mm long.

Hammocks and mangrove margins. Occasional; Miami-Dade County, Monroe County Keys. Florida; West Indies. Spring–summer.

Rhamnus L. 1753. BUCKTHORN

Shrubs or trees. Leaves deciduous, alternate, pinnate-veined, the margin toothed. Flowers in axillary umbelliform cymes or solitary, bisexual and/or unisexual (plants polygamodioecious); sepals 4–5, free; petals 4–5, free; stamens 4–5, free; nectariferous disk adnate to and lining the wall of the hypanthium; carpels 3, 3-locular, the ovary superior, the stigmas 3. Fruit a drupe with 3 pyrenes, each 1-seeded.

A genus of about 150 species; nearly cosmopolitan. [From the Greek *rhamnos*, an ancient name for some species of the genus.]

Rhamnus caroliniana Walter [Of Carolina.] CAROLINA BUCKTHORN.

Rhamnus caroliniana Walter, Fl. Carol. 101. 1788. *Sarcomphalus carolinianus* (Walter) Rafinesque, Sylva Tellur. 29. 1838. *Frangula caroliniana* (Walter) A. Gray, Gen. Amer. Bor. 2: 178. 1849.

Shrub or tree, to 6 m; stem gray, glabrous or pubescent. Leaves with the blade oblong to elliptic or obovate-elliptic, (3)5–13 cm long, (1)3–8 cm wide, the veins 6–9(10) pairs, the apex acute to obtuse, the base rounded to obtuse, the margin serrulate or crenulate, the upper surface glabrous, dark green, lustrous, the lower surface paler and dull, sparsely pubescent on the veins, the petiole 0.8–2 cm long. Flowers 1 or 2–14 in a pedunculate, umbelliform cyme, the pedicel 3–6 mm long; hypanthium campanulate to cupulate, 2–3 mm long; sepals ovate triangular, 2–3 mm long, green or greenish yellow, keeled; petals spatulate, shorter than the sepals, short-clawed, yellow or whitish. Fruit globose, 8–10 mm long, black.

Hammocks. Occasional; northern and central peninsula, central and western panhandle. Maryland south to Florida, west to Illinois, Missouri, Oklahoma, and Texas. Spring–fall.

Sageretia Brongn. 1826. MOCK BUCKTHORN

Shrubs or trees; branches with thorns. Leaves persistent to tardily deciduous, opposite or subopposite, pinnate-veined, the margin serrulate. Flowers in terminal and axillary spiciform thyrses, bisexual; sepals 5, free; petals 5, free; stamens 5, free; nectariferous disk fleshy, collar-like, confluent with the hypanthium at the base; carpels 3, 3-locular, the stigmata 3, the ovary superior. Fruit a drupe with (2)3 pyrenes, each 1-seeded.

A genus of about 30 species; North America, Mexico, Central America, South America, Africa, Asia, Australia, and Pacific Islands. [Commemorates Augustin Sageret (1763–1851) French botanist.]

Sageretia minutiflora (Michx.) C. Mohr. [Small-flowered.] SMALLFLOWER MOCK BUCKTHORN.

> *Rhamnus minutiflora* Michaux, Fl. Bor.-Amer. 1: 154. 1803. *Sageretia michauxii* Brongniart, Ann. Sci. Nat. (Paris) 10: 360. 1827; nom. illegit. *Afarca parviflora* Rafinesque, Sylva Tellur. 30. 1838; nom. illegit. *Sageretia minutiflora* (Michaux) C. Mohr, Contr. U.S. Natl. Herb. 6: 609. 1901. TYPE: "a Carolina septentrionali ad Floridam."

Sprawling or climbing shrubs, to 8 m; branchlets thorn-tipped. Leaves with the blade ovate to elliptic-ovate, 1.5–4(6) cm long, 1–2 cm wide, the apex acute to acuminate, the base rounded, the margins shallowly serrulate, the upper and lower surfaces tomentulose, soon glabrescent. Flowers in an umbelliform cyme, sessile or subsessile; hypanthium shallowly cupulate, 1–2 mm long; sepals triangular, 1–2 mm long, keeled, yellowish green; petals spatulate, emarginate, shorter than the sepals, white to yellow; nectariferous disk cupulate, the outer margin free; stamens subequaling the petals. Fruit subglobose to obovoid, 5–9 mm long, purplish.

Calcareous floodplain forests and hammocks. Frequent; nearly throughout. North Carolina south to Florida, west to Mississippi. Spring.

Ziziphus Mill. 1754. JUJUBE

Shrubs or trees. Leaves alternate, 3-nerved, the margin crenate-serrulate. Flowers axillary, solitary or in corymbiform cymes or in axillary and terminal thyrses, bisexual; sepals 5, free; petals 5, free; stamens 5, free; nectariferous disk fleshy, surrounding the ovary and adherent to it; carpels 2, 2-locular, the ovary superior, the styles 2. Fruit a 1-stoned drupe, 1-seeded.

A genus of about 100 species; nearly cosmopolitan. [Latinized Arabic vernacular name "zizouf" for *Z. jujuba*, the type species.]

1. Branchlets, inflorescences, and lower surface of the leaves densely tomentose ....**Z. mauritiana**
1. Branchlets, inflorescences, and lower surface of the leaves glabrous or glabrate.
    2. Leaves with 3 main veins from the base, the margin crenate..**Z. jujuba**
    2. Leaves pinnate-veined, the margin entire..**Z. celata**

Ziziphus celata Judd & D. W. Hall [Hidden, concealed, in reference to its not having been discovered earlier.] FLORIDA JUJUBE; SCRUB ZIZIPHUS.

> *Ziziphus celata* Judd & D. W. Hall, Rhodora 86: 382. 1984. TYPE: FLORIDA: Highlands Co.: vicinity of Sebring, 18 Mar 1949, *Garrett s.n.* (holotype: FLAS).

Shrubs or trees, to 2 m; branches spinescent, gray, glabrous. Leaves deciduous, the blade oblong-elliptic to elliptic-ovate, or elliptic-obovate, 5–10 mm long, 3–5 mm wide, pinnate-veined, the apex rounded to shallowly emarginate, the base cuneate, the margin entire, the upper and lower surfaces glabrous, the petiole 1–2 cm long; stipules narrowly triangular, minute. Flowers solitary on a short, stout pedicel; hypanthium cupulate; sepals ovate-triangular, ca. 1 mm long, keeled, green; petals obovate or spatulate, ca. 1 mm long, clawed, hooded, each clasping a stamen, white; disk shallow, fleshy, 5- to 10-lobed. Fruit globose to ovoid or oblong, ca. 1 cm long, yellow or orange.

Scrub. Rare; Polk and Highlands Counties. Spring. Endemic.

Ziziphus celata is listed as endangered in Florida (Florida Administrative Code, Chapter 5B-40) and in the United States (U.S. Fish and Wildlife Service, 50 CFR 23).

Ziziphus jujuba Mill. [Arabic vernacular name.] COMMON JUJUBE.

Rhamnus zizyphus Linnaeus, Sp. Pl. 1: 194. 1753. *Ziziphus jujuba* Miller, Gard. Dict., ed. 8. 1768, nom. cons. *Ziziphus officinarum* Medikus, Bot. Beob. 1782. 333. 1784; nom. illegit. *Ziziphus sativa* Gaertner, Fruct. Sem. Pl. 1: 202. 1788; nom. illegit. *Ziziphus vulgaris* Lamarck, Encycl. 3: 316. 1789; nom. illegit. *Ziziphus zizyphus* (Linnaeus) H. Karst, Deut. Fl., ed. 2. 2: 438. 1894; nom. rej.

Shrub or tree, to 12 m; branches glabrous. Leaves deciduous, the blade ovate to ovate-lanceolate or elliptic-oblong, 3–6 cm long, 2–5 cm wide, 3-veined from the base, the apex obtuse to rounded, rarely acute, the base oblique, the margin crenate-serrate, the upper and lower surfaces glabrous, the petiole ca. 5 mm long; stipules spinescent, 1.5–4 cm long, straight or curving. Flowers axillary, solitary or 2–8 in a cyme; hypanthium cupulate; sepals ovate-triangular, ca. 2 mm long, yellow-green; petals ovate to spatulate, ca. 2 mm long, clawed, hooded, each clasping a stamen, yellow or white. Fruit ellipsoid to narrowly ovoid, 1.5–2(3) cm long, dark red or reddish purple.

Disturbed sites. Rare; Gadsden, Bay, and Okaloosa Counties. Escaped from cultivation. Georgia, Alabama, Florida, Louisiana, Texas, Utah, Arizona, and California; Europe and Asia. Native to Europe and Asia. Spring.

Ziziphus mauritiana Lam. [Of Mauritius.] INDIAN JUJUBE.

Ziziphus mauritiana Lamarck, Encycl. 3: 319. 1789.

Shrub or tree, to 10(15) m; branches tomentose. Leaves persistent, the blade oblong to elliptic or elliptic-ovate, 2.5–8 cm long, 2–6 cm wide, 3-veined from the base, the apex rounded, sometimes emarginate, the base obtuse to rounded, often oblique, the margin serrulate, the upper surface glabrous, the lower surface whitish to tawny tomentose, the petiole 1–2 cm long; stipules spinescent, 2–3 mm long, straight to recurved. Flowers 2–8 in an axillary cyme; sepals ovate-triangular, ca. 2 mm long, greenish to greenish white; petals ovate to spatulate, 1–2 mm long, clawed, hooded, each clasping a stamen, white. Fruit globose to ovoid or oblong, 2–3 cm long, yellow to orange or red.

Disturbed sites. Rare; Broward and Miami-Dade Counties. Escaped from cultivation. Florida and California; West Indies, South America; Europe, Africa, Asia, Australia, and Pacific Islands. Native to Asia. Spring.

ULMACEAE Mirb., nom. cons. 1815. ELM FAMILY

Deciduous trees or shrubs. Leaves alternate, simple, 2-ranked, petiolate, stipulate, the blade pinnate-veined, the base symmetrical or asymmetrical, the margin serrate or entire. Flowers solitary, cymose, or in axillary fasciculate aggregations, bisexual or unisexual (plants monoecious or polygamomonoecious), actinomorphic or slightly zygomorphic; sepals connate, 4- to 9-lobed; petals absent; stamens as many as the calyx lobes and opposite them, arising from the base of the calyx, the filaments free, the anthers 2-locular, longitudinally dehiscent; carpels 2, connate, superior, 1-locular, ovule 1, the styles 2, entire or 2-cleft, stigmatic on the upper, inner surface. Fruit a samara or nutlike; seed 1.

A family of ca. 6 genera with ca. 36 species; nearly cosmopolitan.

Celtis and *Trema*, sometimes included here or in the segregate Celtidaceae, are now placed in the Cannabaceae.

Selected reference: Elias (1970).

1. Leaves nearly symmetrical at the base; fruit nutlike ....**Planera**
1. Leaves conspicuously asymmetrical at the base; fruit a samara ....**Ulmus**

Planera J. F. Gmelin 1791. PLANERTREE

Trees or shrubs; branches unarmed. Leaf blades pinnate-veined, the margin irregularly serrate, the base slightly asymmetrical. Flowers unisexual (plants monoecious) or rarely bisexual (plants polygamomonoecious). Staminate flowers in fascicles; calyx 4- or 5-lobed; corolla absent; stamens as many as calyx lobes; pistil rudimentary. Carpellate flowers solitary or 2–3 in a group; calyx 4- to 5-lobed; corolla absent; stamens small and nonfunctional or absent; ovary stalked, the styles 2, elongate, entire. Fruit a short-stipitate nutlike drupe.

A monotypic genus; North America. [Commemorates Johann Jacob Planer (1743–1789), German botanist and physician.]

Selected reference: Barker (1997).

Planera aquatica J. F. Gmelin [Aquatic.] WATERELM; PLANERTREE.

Planera aquatica J. F. Gmelin, Syst. Nat. 2: 150. 1791. *Abelicea aquatica* (J. F. Gmelin) Rafinesque, New Fl. 3: 41. 1838.

Shrub or tree, to 18(20) m; bark scaly, sloughing in long, grayish brown plates exposing the reddish brown inner bark; branchlets unarmed, hirtellous when young, glabrate in age. Leaves with the blade ovate or ovate-oblong, 3–8 cm long, 2–4 cm wide, the apex acute to short acuminate or rounded, the base cuneate to rounded, symmetrical or asymmetrical, the margin irregularly coarse-serrate to somewhat doubly serrate, the upper surface glabrous, the lower surface paler, glabrous or sparsely pubescent mainly on the veins, the petiole 3–6 mm long. Staminate flowers in a fascicle; calyx campanulate, ca. 2 mm long, 4- or 5-lobed from middle to near the base, the lobes rounded apically, greenish yellow, subscarious. Carpellate flower with the calyx as in the staminate; ovary stipitate, tuberculate. Bisexual flowers usually with only 1 stamen. Fruit ovate to oblong, 8–10 mm long, somewhat compressed, with irregular ribs bearing numerous fleshy protuberances, often enclosed basally by the persistent calyx; seeds plano-compressed, ovoid, often oblique.

Floodplain forests and swamps. Occasional; northern counties, Levy County. North Carolina south to Florida, west to Missouri, Oklahoma, and Texas. Spring.

Ulmus L., 1753. ELM

Trees; branchlets unarmed. Leaves pinnate-veined, the bases symmetrical or asymmetrical. Flowers in axillary subsessile or pedicellate cymes or racemes, bisexual; calyx 3- to 7-lobed; stamens the same number as the calyx lobes; ovary sessile or stipitate, the styles entire, flattened. Fruit a samara, 2-beaked with the persistent styles.

A genus of ca. 40 species; nearly cosmopolitan. [The classical Latin name.]

Selected reference: Sherman-Broyles (1997).

1. Leaf margin mostly simply serrate; flowering in late summer and fall.................................U. **parvifolia**
1. Leaf margins doubly serrate; flowering in early spring.
  2. Petiole 1–3 mm long; branches sometimes corky-winged; leaves not cuspidate.
    3. Leaves scabrous on the upper surface, the apex bluntly acute or obtuse; samaras ciliolate, 8–10 mm long .... U. **crassifolia**
    3. Leaves smooth on the upper surface, the apex sharply acute; samaras long-ciliate, 6–8 mm long .... U. **alata**
  2. Petiole 4–10 mm long; branches never corky-winged; leaves cuspidate.
    4. Leaves harshly scabrous on the upper surface; samaras orbicular, 14–17 mm long; bud scales long-pubescent ....U. **rubra**
    4. Leaves smooth or lightly scabrous on the upper surface; samaras ovate, 10–13 mm long; bud scales glabrous or short-pubescent ....U. **americana**

Ulmus alata Michx. [Winged, in reference to the corky outgrowths on the branches.] WINGED ELM.

Ulmus alata Michaux, Fl. Bor.-Amer. 1: 173. 1803. *Ulmus americana* var. *alata* (Michaux) Spach, Ann. Sci. Nat., Bot., Ser. 2. 15: 364. 1841.

Tree to 20 m; bark gray-brown, with shallow furrows and flat-topped ridges; branchlets red-brown, pubescent, commonly with corky-ridged wings, the buds narrowly ovoid, 3.5–4.5 mm long, flattened laterally, the apex acute, the scales red-brown, puberulent, the margin dark and often with a few cilia. Leaves with the blade narrowly ovate to elliptic, 3–7 cm long, 1–3 cm wide, the apex acute to acuminate, the base cuneate to rounded, symmetrical or slightly asymmetrical, the margin doubly or triply serrate, the upper surface glabrous or somewhat scabrous, sometimes with a few trichomes on the midvein, the lower surface slightly paler, pubescent, at least on the main veins, the petiole 2–4 mm long, puberulent. Flowers 6–10 in a raceme, the pedicel 2–3 mm long, glabrous; calyx turbinate, ca. 2 mm long, 5- to 6-lobed, symmetrical, brown; stamens 5–6, the anthers red. Fruit ovate, 7–8 mm long, 3–4 mm wide, flat, the wing narrow, brown, pubescent, densely ciliate, the persistent styles 1–2 mm long, curved inward, touching or overlapping; seed slightly thickened.

Moist to wet hammocks. Occasional; northern counties south to Pasco County. Maryland south to Florida, west to Kansas, Oklahoma, and Texas. Spring.

Ulmus americana L. [Of America.] AMERICAN ELM; FLORIDA ELM.

Ulmus americana Linnaeus, Sp. Pl. 1: 226. 1753.
Ulmus americana Linnaeus var. *aspera* Chapman, Fl. South. U.S. 416. 1860. TYPE: FLORIDA: Franklin Co.: Apalachicola River swamps, *Chapman s.n.* (holotype: NY?).
Ulmus floridana Chapman, Fl. South. U.S. 416. 1860. *Ulmus americana* Linnaeus var. *floridana* (Chapman) Little, Phytologia 4: 306. 1953. TYPE: FLORIDA: *Barratt s.n.* (holotype: NY; isotype: GH).

Tree, to 25 m; bark thick, dark ashy-gray, furrowed, the ridges short, flat-topped or somewhat angular, anastomosed; branchlets red-brown to gray-brown, glabrous or pubescent, the buds narrow, 2–3 mm long, acute. Leaves with the blade ovate to elliptic, 3 15 cm long, 2–8 cm wide, the apex acuminate or acute, the base cuneate or rounded, asymmetrical, the margin coarsely doubly or triply serrate, the upper surface glabrous or scabrous, the lower surface paler, pubescent or glabrate, the petiole 2–6 mm long, glabrous or pubescent. Flowers 2–5 in raceme, the pedicel 10–15 mm long, glabrous; calyx tube campanulate, asymmetrical, 1–2 mm long, greenish or reddish, with a few trichomes at the tips of the lobes, 7- to 9-lobed; stamens 5–9, the anthers red. Fruit ovate, 10–14 mm long, 9–12 mm wide, flattened, winged, the apex a notch with the incurved or straight, sharp points of the styles, the surface reticulate, glabrous except for ciliate margin.

Floodplain forests. Frequent; northern counties south to Collier County. Quebec south to Florida, west to Saskatchewan, Montana, Wyoming, Colorado, and Texas. Spring.

Ulmus crassifolia Nutt. [Thick-leaved.] CEDAR ELM.

Ulmus crassifolia Nuttall, Trans. Amer. Philos. Soc., ser. 2. 5: 169. 1835.

Tree, to 25 m; bark scaly-ridged, light brown; branchlets often with corky ridged wings, the buds brown, the apex acute, the scales dark brown, glabrous. Leaves with the blade elliptic, oval, or ovate, 2.5–5 cm long, 1.3–2 cm wide, the apex obtuse to rounded, the base symmetrical or somewhat asymmetrical, the margins irregularly shallowly serrate or doubly serrate, the upper surface glabrous, the lower surface pubescent, the petiole ca. 1.5 mm long. Flowers 2–5 in a raceme, the pedicel 7–10 mm long; calyx campanulate, 4- to 6-lobed, pubescent; stamens 4–6, the anthers red. Fruit elliptic to ovate, 7–10 mm long, flattened, winged, strongly asymmetrical, the apex notched by the 2 short, incurved styles, the surface sparsely pubescent, rugose-veiny, the margins ciliate; seed somewhat thickened.

Floodplain woods. Rare; Suwannee and Lafayette Counties south to Hernando County. Florida, Tennessee south to Mississippi, west to Missouri, Oklahoma and Texas; Mexico. Fall.

Ulmus parvifolia Jacq. [Small-leaved.] CHINESE ELM.

Ulmus parvifolia Jacquin, Hort. Schoenbr. 3: 6, pl. 262. 1798. *Planera parvifolia* (Jacquin) Sweet, Hort. Brit., ed. 2. 464. 1830.

Small tree, to 15 m; bark exfoliating in platelets giving the trunk and branches a mottled appearance; branchlets glabrous or pubescent, the buds acute to obtuse, the scales brown, pubescent. Leaves with the blade elliptic to ovate or obovate, 3–4 cm long, 1.5–2.5 cm wide, the apex acute or subobtuse, the base cuneate, asymmetrical, the margin serrate, the upper surface glabrous, the lower surface pubescent when young, glabrescent in age, the petiole 1–6 mm long. Flowers 2–8 in a raceme, the pedicel 8–10 mm long; calyx campanulate, reddish brown, 3- to 5-lobed, glabrous; stamens 3–4, the anthers reddish. Fruit elliptic-ovate, green to light brown, winged, ca. 1 cm long, the apex notched; seed thickened.

Disturbed sites. Rare; Escambia and Highlands Counties. Escaped from cultivation. Maine south to Florida, west to Kentucky, also California; Asia. Native to Asia. Summer–fall.

Ulmus rubra Muhl. [Red, in reference to the bark color.] SLIPPERY ELM.

Ulmus rubra Muhlenberg, Trans. Amer. Philos. Soc. 3: 165. 1793.
Ulmus fulva Michaux, Fl. Bor.-Amer. 1: 172. 1803.

Tree, to 20 m; bark thick, gray-brown, with shallow fissures and long, flat, often loose plates; branchlets grayish or brown, pubescent, the buds narrowly ovoid, obtuse. Leaves with the blade ovate to oval, 10–13 cm long, 5–7 cm wide, the apex acute, the base rounded to cuneate, asymmetrical, the margin coarsely doubly serrate, the upper surface often rugose, scabrous, the lower surface paler, pubescent and with axillary tufts, the petiole 3–5 mm long, pubescent. Flowers 8–12 in a raceme, the pedicel 1–2 mm long, pubescent; calyx campanulate, symmetrical, 2.6–4 mm long, pubescent, 5- to 9-lobed, ca. 1 mm long, green, fringed with red-brown trichomes; stamens 5–9, the anthers dark red. Fruit circular to obovate, 8–20 mm long, 13–15 mm wide, broadly winged, obscurely reticulate, rusty tomentose on the seed-body, otherwise glabrous, the apex notched with the tips rounded and overlapping; seed flattened, thickened.

Bluff forests, floodplain forests, and hammocks. Rare; Jefferson County, central panhandle. Maine south to Florida, west to North Dakota and Texas. Spring.

EXCLUDED TAXON

Ulmus pumila Linnaeus—Reported for Florida by Sherman-Broyles (1997), apparently based on misidentification of *U. parviflora*.

CANNABACEAE Endl., nom. cons. 1837. HEMP FAMILY

Shrubs, trees, or annual herbs. Leaves opposite or alternate, simple or palmately compound, petiolate, stipulate. Flowers unisexual (plants dioecious, monoecious, or polygamomonoecious), actinomorphic, bracteate. Staminate flowers in axillary, cymose panicles (thyrses) or fasciculate, pedicellate; sepals 4–5, free or connate; petals absent; stamens 4–5, the anthers 2-locular, longitudinally dehiscent. Carpellate flowers solitary or in axillary, cymose pseudospikes or fascicles; sepals 4–5, free or connate; petals absent; stamens small and sterile or absent; carpels 2, superior, 1-locular, stigmas 1–2, elongate. Fruit an achene or drupe.

A family of 10 genera and about 170 species; nearly cosmopolitan.

Celtis and *Trema*, sometimes put in the Ulmaceae or Celtidaceae, are placed here in the expanded Cannabaceae (see Yang et al. 2013).

Selected references: Elias (1970); Miller (1970).

Celtidaceae Endl. 1841.

1. Annual herbs; leaves palmately compound..**Cannabis**
1. Trees or shrubs; leaves simple.
    2. Leaves glabrous or nearly so on the lower surface, the margin entire or irregularly (rarely regularly) and coarsely serrate; flowers and fruits solitary or few in unbranched cymes..........................**Celtis**
    2. Leaves pubescent on the lower surface, the margin uniformly and finely serrate; flowers and fruits 12–20 in branching cymes ....**Trema**

Cannabis L. 1753. HEMP

Annual herbs. Leaves opposite near the base, alternate above, palmately compound, persistent, the margin serrate, stipulate. Flowers unisexual (plants dioecious or monoecious), bracteate, bracteolate. Staminate flowers in axillary, cymose panicles, pedicellate; sepals 5, free; petals absent; stamens 5, free; ovary rudimentary. Carpellate flowers in axillary pseudospikes; sepals 5, connate; petals absent; styles 2, erect. Fruit an achene.

A genus of 1 species; Asia. [The ancient Greek name for hemp, perhaps from the Persian *Kannab* or the Arabic *kinnab*.]

Selected reference: Small (1997).

Cannabis sativa L. [Cultivated.] HEMP; MARIJUANA.

Cannabis sativa Linnaeus, Sp. Pl. 2: 1027. 1753. *Cannabis sativa* Linnaeus var. *vulgaris* Alefeld, Landw. Fl. 1866, nom. inadmiss.

Annual herb, to 2 m; stem strigose. Leaves 3- to 9-foliolate, the leaflets narrowly lanceolate, 6–14 cm long, 0.3–1.5 cm wide, the margin coarsely serrate, the upper surface scabrid with bulbous-based trichomes, the lower surface strigose, with yellowish brown resinous dots, the petiole 2–7 cm long, stipules linear, 4–6 mm long. Staminate flowers green, the pedicel 1–3 mm long; sepals ovate to lanceolate, 3–4 mm long, puberulent; stamens somewhat shorter than the sepals. Carpellate flowers sessile or nearly so; bracts foliaceous, 4–15 mm long, glandular-pubescent; bracteoles linear, ca. 2 mm long; stamens small and sterile or absent; styles 2–3 mm long, calyx appressed to and surrounding the base of the ovary. Fruit broadly oval, compressed, 2–5 mm long, white or greenish, mottled with purple, enclosed by the enlarged persistent calyx.

Disturbed sites. Occasional; panhandle, central and southern peninsula. Escaped from cultivation. Nearly cosmopolitan. Native to Asia. All year.

Celtis L. 1753. HACKBERRY

Trees or shrubs; branches unarmed or with stipular spines. Leaves simple, pinnipalmate-veined, the margins serrate or entire, the base asymmetrical, petiolate. Flowers unisexual (plants monoecious) or rarely bisexual (plants polygamomonoecious). Staminate flowers cymose or fascicled; calyx connate, 4- or 5-lobed; petals absent; stamens as many as the calyx lobes, inserted on the receptacle; ovary minute and rudimentary. Carpellate flowers solitary or in few-flowered cymes in the upper leaf axils; calyx 4- or 5-lobed; stamens small and sterile or absent; petals absent; ovary sessile, the styles 2, elongate, reflexed, entire or 2-cleft. Fruit a drupe.

A genus of about 80 species; North America, West Indies, Central America, South America, Europe, Africa, and Asia. [Classical Latin name of *Celtis australis*, the "lotus" of the ancient world.]

Momisia F. Dietr., 1819.

Selected reference: Sherman-Broyles et al. (1997).

1. Branches armed with stipular spines.
    2. Leaf margin crenulate-dentate, the blade somewhat succulent, the upper surface decidedly scabrous to the touch...**C. pallida**
    2. Leaf margin serrulate, the blade chartaceous, the upper surface smooth or nearly so ....
    ....**C. iguanaea**
1. Branches unarmed.
    3. Leaves with the bases mostly conspicuously oblique, the surfaces gray-green, often darker above, the blade mostly ovate, the apex abruptly acuminate...**C. occidentalis**
    3. Leaves with the bases slightly oblique, the surfaces yellowish green, the blade mostly elliptic-lanceolate to ovate-lanceolate, the apex acute or gradually acuminate ....**C. laevigata**

Celtis iguanaea (Jacq.) Sarg. [Named for the iguana lizard that is said to eat the fruit.] IGUANA HACKBERRY.

Rhamnus iguanaeus Jacqin, Enum. Syst. Pl. 16. 1760. *Celtis aculeata* Swartz, Prodr. 53. 1788, nom. illegit. *Ziziphus iguanaea* (Jacquin) Lamarck, Encycl. 3: 318. 1789. *Momisia aculeata* Klotzsch, Linnaea 20: 539. 1847, nom. illegit. *Celtis iguanaea* (Jacquin) Sargent, Silva N. Amer. 7: 64. 1895.

Scrambling shrub, to 3 m; bark gray, smooth, the branchlets flexuous, with stout, paired or solitary stipular thornlike spines to 2.5 cm long, puberulent. Leaves with the blade chartaceous, ovate to elliptic, 3.5–7 cm long, 1–2 cm wide, conspicuously 3-veined from the base, the apex obtuse to acute, the base cuneate to rounded, asymmetrical, the margin entire or irregularly serrulate toward the apex, the upper surface sparsely pubescent to glabrous, smooth to the touch, the petiole 3–5 mm long. Flowers in a lax or dense cyme subequaling the petiole; styles 2-cleft. Fruit ovoid, 8–12 mm long, orange to red, the pedicel 2–3 mm long.

Shell middens. Rare; Manatee, Charlotte, Lee, and Collier Counties. Florida, Alabama, and Texas; West Indies, Mexico, Central America, and South America. Spring.

Celtis laevigata Willd. [Smooth.] SUGARBERRY; HACKBERRY.

Celtis laevigata Willdenow, Berlin. Baumz., ed. 2. 81. 1811.
Celtis mississippiensis Bosc ex Spach, Ann. Sci. Nat., Bot., ser. 2. 16: 42. 1841.
Celtis smallii Beadle, in Small, Fl. S.E. U.S. 365, 1329. 1903. *Celtis laevigata* Willdenow var. *smallii* (Beadle) Sargent, Bot. Gaz. 67: 223. 1919.

Tree, to 30 m; bark light gray, smooth or with corky-warty outgrowths, with broad, flat furrows in age, the branchlets unarmed, light red-brown, pubescent at first, then glabrous. Leaves with the blade chartaceous to subcoriaceous, ovate to ovate-lanceolate, 4–12 cm long, 3–4 cm wide, conspicuously 3-veined from the base, the apex long-acuminate, often falcate, the base broadly cuneate to rounded or subtruncate, asymmetrical, the upper surface glabrous or nearly so, the lower surface usually pubescent on the veins and with a tuft of trichomes in the vein axils, the margins entire or with a few teeth on the upper half, the petiole 6–10 mm long. Flowers solitary or in a few-flowered cyme, the styles entire. Drupes oval to subglobose, 5–8 mm long, orange to brownish red, smooth, glaucous, the pedicel 5–15 mm long.

Floodplain forests. Frequent; nearly throughout. Maryland south to Florida, west to Washington, Oregon, and California; Mexico. Spring.

Celtis occidentalis L. [Western.] HACKBERRY.

> Celtis occidentalis Linnaeus, Sp. Pl. 2: 1044. 1753. Celtis obliqua Moench, Methodus 344. 1794, nom. illegit.
>
> Celtis crassifolia Lamarck, Encycl. 4: 138. 1797. Celtis occidentalis Linnaeus var. crassifolia (Lamarck) A. Gray, Manual, ed. 2. 397. 1856.
>
> Celtis pumila Pursh, Fl. Amer. Sept. 1: 200. 1814. Celtis occidentalis Linnaeus var. pumila (Pursh) A. Gray, Manual, ed. 2. 397. 1856. Celtis mississippiensis Bosc ex Spach var. pumila (Pursh) Mackenzie & Bush, Man. Fl. Jackson County 72. 1902. Celtis occidentalis Linnaeus forma pumila (Pursh) Seymour, Fl. New England 228. 1969.
>
> Celtis occidentalis Linnaeus var. integrifolia Nuttall, Gen. N. Amer. Pl. 1: 202. 1818.
>
> Celtis tenuifolia Nuttall, Gen. N. Amer. Pl. 1: 202. 1818. Celtis occidentalis Linnaeus subsp. tenuifolia (Nuttall) E. Murray, Kalmia 12: 19. 1982.
>
> Celtis georgiana Small, Bull. Torrey Bot. Club 24: 439. 1897. Celtis pumila Pursh var. georgiana (Small) Sargent, Bot. Gaz. 67: 227. 1919. Celtis tenuifolia Nuttall var. georgiana (Small) Fernald & B. G. Schubert, Rhodora 50: 160. 1948. Celtis occidentalis Linnaeus var. georgiana (Small) Ahles, J. Elisha Mitchell Sci. Soc. 80: 172. 1964. Celtis occidentalis Linnaeus subsp. georgiana (Small) E. Murray, Kalmia 12: 19. 1982.

Tree, to 25 m; bark thick, gray-brown, warty, deeply furrowed and with shallow furrows and platelike ridges in age, the branchlets unarmed, gray-brown, glabrate or pubescent. Leaves with the blade chartaceous to coriaceous, ovate to ovate-lanceolate, 5–12 cm long, 3–6 cm wide, conspicuously 3-veined at the base, the apex acuminate, the base subcordate, asymmetrical, the margin entire or with a few teeth near the apex or coarsely serrate, upper surface glabrate or scabrous, lower surface pale, pubescent, the petiole 6–12 mm long. Flowers solitary or in a few-flowered cyme; styles entire. Drupe globose, the apex rounded or mucronate, 8–10 mm long, brownish to dark purple, the pedicel 5–15 mm long.

Dry to mesic upland rocky woods. Occasional; panhandle, Suwannee and Alachua Counties. Quebec south to Florida, west to Manitoba, Montana, Wyoming, Utah, and New Mexico. Spring.

Celtis pallida Torrey [Pale.] SPINY HACKBERRY; DESERT HACKBERRY.

> Celtis pallida Torrey, in Emory, Rep. U.S. Mex. Bound. 203, t. 50. 1859. Celtis tala Gillies ex Planchon var. pallida (Torrey) Planchon, in de Candolle, Prodr. 17: 191. 1873. Momisia pallida (Torrey)

Planchon ex Small, Fl. S.E. U.S. 366. 1903. *Celtis spinosa* Sprengel var. *pallida* (Torrey) M. C. Johnston, Southw. Naturalist 2: 172. 1958 ("1957").

Scrambling shrub, to 3 m; bark gray, smooth, the branches flexuous, with stout paired or solitary stipular thornlike spines to 2.5 cm long, spreading white-puberulent. Leaves with the blade coriaceous, ovate to ovate-oblong or elliptic, 2–3 cm long, 1.5–2 cm wide, conspicuously 3-veined from the base, the apex rounded to acute, the base rounded to broadly cuneate, asymmetrical, the margin entire or sparingly crenate-dentate, the surfaces scabrous, the petiole 3–5 mm long. Flowers in a few-flowered cyme longer than the petiole; styles 2-cleft. Fruit ovoid, ca. 6 mm long and wide, glabrous, orange, yellow, or red, the pedicel 2–3 mm long.

Shell middens. Rare; Charlotte and Lee Counties. Florida, Texas to Arizona; Mexico. Spring.

Trema Lour. 1790. NETTLETREE

Trees or shrubs; branches unarmed. Leaves simple, conspicuously 3-veined at the base, pinnipalmate-veined, the margin serrate or denticulate, the base asymmetrical, petiolate. Flowers in axillary, subsessile cymes, unisexual (plants monoecious) or rarely bisexual (plants polygamomonoecious). Staminate flowers subsessile; calyx 5-lobed; petals absent; stamens 4–5; ovary rudimentary. Carpellate flowers pedicellate; calyx as in the staminate; petals absent; ovary sessile, the styles erect, entire. Fruit a drupe.

A genus of about 15 species; North America, West Indies, Mexico, Central America, South America, Africa, and Asia. [From the Greek *trema*, a hole, in reference to the fruit of the type species, whose surface reticulations appear as perforations.]

Selected reference: Schultz (1997).

1. Leaves 1–3 cm long..T. **lamarckiana**
1. Leaves 6 cm long or longer.
   2. Fruit orange-red to yellow at maturity... T. **micrantha**
   2. Fruit black at maturity... T. **orientalis**

Trema lamarckiana (Schultes) Blume [Commemorates Jean Baptiste Antoine Pierre Monnet de Lamarck (1744–1829), French biologist and evolutionist.] PAIN-IN-BACK; WEST INDIAN TREMA.

Celtis lima Lamarck, Encycl. 4: 140. 1797; non Swartz, 1788. *Celtis lamarckiana* Schultes, in Roemer & Schultes, Syst. Veg. 6: 311. 1820. *Sponia lamarckiana* (Schultes) Decaisne ex Planchon, Ann. Sci. Nat., Bot., ser. 3. 10: 322. 1848. *Trema lamarckiana* (Schultes) Blume, Mus. Bot. 2: 58. 1856 "1852."

Shrub or tree, to 6 m; bark gray-brown, smooth, the branchlets glabrous or pubescent. Leaves with the blade ovate to lanceolate, 1–3 cm long, 2–2.5 cm wide, conspicuously 3-veined from the base, the apex subobtuse to acute, the base rounded to truncate, slightly asymmetric, the margin serrulate, the surfaces scabrous with pustulate-based trichomes, prominently reticulate-veined, finely tomentose, the petiole 3–5 mm long. Flowers in an axillary subsessile cyme subequaling the petiole, pink to white; style erect, entire. Fruit ovoid, 2–3 mm long, smooth, pink.

Tropical hammocks and shell middens. Occasional; Collier, Miami-Dade, and Monroe counties. Florida; West Indies. All year.

Trema micrantha (L.) Blume [Small-flowered.] NETTLETREE.

> *Rhamnus micranthus* Linnaeus, Syst. Nat., ed. 10. 937. 1759. *Celtis micrantha* (Linnaeus) Swartz, Prodr. 53. 1788. *Sponia micrantha* (Linnaeus) Decaisne ex Planchon, Ann. Sci. Nat., Bot., ser. 3. 10: 333. 1848. *Trema micrantha* (Linnaeus) Blume, Mus. Bot. 2: 58. 1856 ("1852").
>
> *Celtis mollis* Humboldt & Bonpland ex Willdenow, Sp. Pl. 4: 996. 1806. *Sponia mollis* (Humboldt & Bonpland ex Willdenow) Decaisne ex Planchon, Ann. Sci. Nat., Bot., ser. 3. 10: 331. 1848. *Trema mollis* (Humboldt & Bonpland ex Willdenow) Blume, Mus. Bot. 2: 58: 1856 ("1852").
>
> *Trema floridana* Britton ex Small, Fl. S.E. U.S. 366, 1329. 1903. *Trema micrantha* (Linnaeus) Blume var. *floridana* (Britton ex Small) Standley & Steyermark, Publ. Field Mus. Nat. Hist., Bot. Ser. 23: 40. 1944. TYPE: FLORIDA: Miami-Dade Co.: Miami, 27 Oct-13 Nov 1901, *Small & Nash 32* (holotype: NY).

Tree, to 8 m; bark dark brown, smooth when young, developing warty projections in age, the branchlets gray-brown, densely pubescent. Leaves with the blade oblong-lanceolate to ovate-lanceolate, 6–12 cm long, 2.5–5 cm wide, conspicuously 3-veined from the base, the apex long-attenuate-acuminate, the base rounded to subcordate, slightly asymmetric, the margin serrate, the upper surface scabrous, the lower surface white pubescent, prominently reticulate-veined, the petiole 8–10 mm long. Flowers in an axillary, subsessile cyme subequaling the petiole, greenish white; styles erect, entire. Fruit 2–4 mm long, orange-red.

Hammocks, frequently in disturbed sites. Occasional; Pinellas and Martin Counties southward. Florida; West Indies, Mexico, Central America, and South America. All year.

Trema orientalis (L.) Blume [Of the east.] ORIENTAL TREMA.

> *Celtis orientalis* Linnaeus, Sp. Pl. 2: 1044. 1753. *Sponia orientalis* (Linnaeus) Planchon, Ann. Sci. Nat., Bot., ser. 3. 10: 323. 1848. *Trema orientalis* (Linnaeus) Blume, Mus. Bot. 2: 62. 1856 ("1852").

Tree, to 20 m; bark gray, smooth, irregularly fissured in age, the branchlets gray-brown, pubescent. Leaves with the blade narrowly ovate, 10–18 cm long, 5–9 cm wide, conspicuously 3-nerved from the base, the apex acuminate to acute, the base cordate, the margin denticulate, slightly asymmetric, the upper surface scabrous, the lower surface pubescent, the petiole 1–2 cm long. Flowers in an axillary, subsessile cyme, greenish yellow; styles erect, entire. Fruit ovoid to subglobose, 3–5 mm long, slightly compressed, black.

Disturbed sites. Rare; Palm Beach, Broward, and Miami-Dade Counties. Escaped from cultivation. Florida; Africa, Asia, Australia, and Pacific Islands. Native to Africa, Asia, Australia, and Pacific Islands. Winter–spring.

EXCLUDED GENUS

Humulus lupulus Linnaeus—Reported by Small (1903, 1913a). No specimens seen.

MORACEAE Link, nom. cons. 1831. MULBERRY FAMILY

Deciduous or evergreen trees, shrubs, herbs, or vines, often with milky sap, the branches with axillary thorns or unarmed. Leaves alternate, the blade simple, lobed or unlobed, pinnate- or palmate-veined, petiolate, stipulate. Flowers in axillary heads, racemes, or cymes, unisexual (plants monoecious or dioecious), actinomorphic; sepals 2–6, free or connate, sometimes absent; petals absent. Staminate flowers with as many stamens as sepals or calyx lobes (sometimes fewer) and opposite them, the anthers 1- or 2-locular. Carpellate flowers with the ovary superior or inferior, 2-carpellate or 1-carpellate by abortion, 1-locular, ovule 1, styles 1–2. Fruits indehiscent or dehiscent drupes, multiple (syncarps).

A family of about 40 genera and about 1,100 species; nearly cosmopolitan.

Artocarpaceae Dumort., 1829.

Selected reference: Wunderlin (1997).

1. Herbs.
  2. Plant an acaulescent, rhizomatous perennial; inflorescence long-pedunculate from the base ....
  ....**Dorstenia**
  2. Plant a caulescent annual; inflorescence short-pedunculate from the leaf axils ....**Fatoua**
1. Trees, shrubs, or vines.
  3. Flowers enclosed in an urceolate receptacle with a small opening at the apex (syconium) ..... **Ficus**
  3. Flowers not covered by the receptacle or with only a single female flower immersed in the receptacle.
    4. Leaf margins serrate or dentate, often lobed; venation appearing palmate or weakly 3-veined from the base.
      5. Carpellate inflorescence globose; styles unbranched ....**Broussonetia**
      5. Carpellate inflorescence cylindrical; styles 2-branched..**Morus**
    4. Leaf margins entire, never lobed; venation pinnate.
      6. Branches with axillary spines; inflorescence unisexual (staminate flowers in a loose raceme, the carpellate in a globose head); fruit a globose syncarp 8–12 cm long ....**Maclura**
      6. Branches unarmed; inflorescence bisexual (carpellate flower solitary, immersed in a globose receptacle covered with staminate flowers); fruit a globose syncarp 1.5–2 cm long...............
      ....**Brosimum**

Brosimum Sw. 1788, nom. cons. BREADNUT

Evergreen trees, the sap milky, the branches unarmed. Leaves pinnate-veined, the margin entire. Flowers unisexual (plants monoecious), the flowers in globular heads with a solitary carpellate flower immersed in a fleshy receptacle covered with numerous staminate flowers interspersed with peltate interfloral bracts. Staminate flowers with sepals 3–4; stamen 1, the anther theca fused and peltate, with circumscissile dehiscence. Carpellate flowers lacking sepals; ovary inferior, the stigma 2-lobed. Fruit an indehiscent drupe embedded in a fleshy receptacle (syncarp).

A genus of about 24 species; North America, West Indies, Mexico, Central America, and South America. [From the Greek *brosimos*, edible.]

Brosimum alicastrum Sw. [An old generic name for the breadnut, from the Latin *alica*, a classical name for spelt (wheat), and the suffix, *astrum*, false.] BREADNUT.

Brosimum alicastrum Swartz, Prodr. 12. 1788. *Alicastrum brownei* Kuntze, Revis. Gen. Pl. 2: 623. 1891.

Tree, to 30 m; bark gray-brown, slightly flaking, the branchlets gray-brown, smooth, glabrous. Leaves with the blade subcoriaceous, ovate to oblong-elliptic, 5–15 cm long, 2–6 cm wide, the apex obtuse to short-subcuspidate or acuminate, the base broadly obtuse to rounded, the veins 12–18 pairs, the surfaces glabrous, the petioles 5–7 mm long, the stipules amplexicaul, ca. 4 mm long. Flowers in a subglobose head 3–6 mm long, with numerous staminate flowers and a solitary carpellate flower, the peduncle slender, about as long as the head or somewhat shorter. Staminate flowers with a vestigial calyx, stamen 1, the anther peltate, ca. 1 mm in diameter. Carpellate flower lackling sepals, the style elongate, the stigma 2-lobed. Syncarp ca. 1.5 cm in diameter, yellow to red, glabrous.

Disturbed tropical hammocks. Rare; Miami-Dade County, Monroe County keys. Escaped from cultivation. Florida; West Indies, Mexico, Central America, and South America. Native to tropical America. All year.

Broussonetia L'Hér. ex Vent., nom. cons. 1799. PAPER MULBERRY

Deciduous trees, the sap milky, the branches unarmed. Leaves weakly palmately 3-veined from the base, pinnately veined above, the margins dentate. Staminate and carpellate flowers on different plants. Staminate flowers in axillary, pedunculate spikes with interspersed interfloral bracts; sepals 4, united at the base; stamens 4. Carpellate flowers in axillary, pedunculate, globose heads with interspersed interfloral bracts; sepals 4, tubular, the ovary stipitate, the stigma 2-lobed. Fruit an aggregation of drupes (syncarp), each drupe indehiscent, surrounded by the enlarged perianth.

A genus of 8 species; North America, South America, Africa, Asia, and Pacific Islands. [Commemorates Pierre Marie Auguste Broussonet (1761–1807), French biologist at Montpellier.] *Papyrius* Lam. ex Kuntze, 1891.

Broussonetia papyrifera (L.) Vent. [Paper bearing, referring to the use of the inner bark in paper making.] PAPER MULBERRY.

Morus papyrifera Linnaeus, Sp. Pl. 2: 986. 1753. *Broussonetia papyrifera* (Linnaeus) Ventenat, Tabl. Regne Veg. 3: 547. 1799. *Papyrius polymorphus* Cavanilles, Descr. Pl. 343, 620. 1802, nom. illegit. *Papyrius japonicus* Lamarck ex Poiret, in Lamarck, Encycl. 5: 3. 1804, nom. illegit. *Papyrius papyriferus* (Linnaeus) Kuntze, Revis. Gen. Pl. 2: 629. 1891.

Tree, to 15 m; sap milky; bark tan, smooth or moderately furrowed, the branchlets brown, spreading pubescent, the terminal bud absent, the axillary buds dark brown, short-pubescent. Leaves with the blade ovate, 6–20 cm long, 5–15 cm wide, unlobed or 1- to 5-lobed, the apex acuminate, the base shallowly cordate or truncate to broadly rounded, the margin serrate, the upper surface scabrous, the lower surface densely gray-pubescent, the petiole shorter than or equaling the blade; stipules ovate to ovate-oblong, the apex attenuate, caducous. Staminate flowers in a pedunculate, cylindric spike 6–8 cm long, the peduncle 2–4 cm long; sepals pubescent. Carpellate flowers in a short-pedunculate, globose head ca. 2 cm in diameter, villous; style elongate-filiform. Syncarp globose, 2–3 cm in diameter; fruit oblanceolate, red or orange, exserted from the calyx, the style persistent.

Disturbed thickets. Frequent; nearly throughout. New York and Massachusetts south to Florida, west to Missouri, Oklahoma, and Texas; South America; Africa, Asia, and Pacific Islands. Native to Asia. Escaped from cultivation. Spring.

Broussonetia papyrifera is listed as a Cateory II invasive species in Florida by the Florida Exotic Pest Plant Council (FLEPPC, 2015).

Dorstenia L. 1753.

Subacaulescent perennial herbs, the sap milky. Leaves pinnate-veined. Staminate and carpellate flowers embedded in a long-pedunculate, flat receptacle (plants monoecious), ebracteate. Staminate flowers numerous; calyx minute, 2- to 3-lobed; stamens 2–3. Carpellate flowers usually few, immersed into the center of the receptacle; calyx tubular, 2- to 3-lobed; stigma 2-lobed. Fruit a dehiscent drupe embedded in an enlarged flattened, fleshy receptacle (syncarp).

A genus of about 100 species; North America, West Indies, Mexico, Central America, South America, Africa, and Asia. [Commemorates Theodor Dorsten (1492–1552), German herbalist and professor of medicine at Marburg.]

Dorstenia contrajerva L. [From the Spanish *contra yerba*, the vernacular name of the plant.] TORTUS HERB; TUSILLA.

> *Dorstenia contrajerva* Linnaeus, Sp. Pl. 1: 121. 1753. *Dorstenia quadrangularis* Stokes, Bot. Mat. Med. 4: 338. 1812, nom. illegit. *Dorstenia quadrangularis* Stokes var. *pinnatifida* Stokes, Bot. Mat. Med. 4: 341. 1812, nom. inadmiss.

Subacaulescent perennial herb, to 4.5 dm; stem short, covered with the persistent petiole bases. Leaves with the blade oblong-ovate, deltoid-ovate, or orbicular, 6–20 cm long, 7–22 cm wide, the margin entire or deeply pinnate-lobed, pubescent, the petioles 8–25 cm long. Flowers on one side on a flattened or curved receptacle, on a peduncle 7–25 cm long, the receptacle with an undulate margin, quadrangular or lobed, to 3.5 cm long and wide. Drupes few; seeds explosively expelled, yellowish.

Moist, disturbed sites; often a weed in greenhouses or around nurseries, rarely in the wild. Rare; Clay County. Florida; West Indies, Mexico, Central America, and South America. Native to tropical America. All year.

Fatoua Gaudich. 1830. CRABWEED

Annual herbs, the sap not milky, the branches unarmed. Leaves 3-veined from the base, the margin crenate. Flowers unisexual (plants monoecious), the staminate and carpellate flowers in axillary, short-pedunculate, capitate cymes, bracteate. Staminate flowers with the calyx 4-lobed; stamens 4. Carpellate flowers with the calyx 4-lobed; stigma unbranched. Fruit a dehiscent drupe, surrounded by an enlarged, persistent calyx.

A genus of 2 species; North America, West Indies, Africa, Asia, Australia, and Pacific Islands. [From the Greek *fatou*, foolish, the connotation unknown.]

Fatoua villosa (Thunb.) Nakai [Soft-hairy.] HAIRY CRABWEED; MULBERRY WEED.

Urtica villosa Thunb., Fl. Jap. 70. 1784. *Fatoua villosa* (Thunb.) Nakai, Bot. Mag. (Tokyo) 4: 516. 1927. Annual herb, to ca. 8 dm; stem erect, usually little-branched, pilose. Leaves with the blade chartaceous, broadly ovate, 2.5–10 cm long, 1–7 cm wide, the apex acute to acuminate, the base cordate to truncate, the margin crenate-dentate, the upper and lower surfaces appressed-hirsute, the petiole 1–6 cm long, the stipules linear to lanceolate, 2–3 mm long. Flowers light green, the staminate and carpellate together in a dense, pedunculate, axillary cyme 4–8 mm long, subtended by a narrow bract, the inflorescence pilose. Staminate flowers with the calyx campanulate; stamens exserted. Carpellate flowers with the calyx boat-shaped; style filiform, puberulent. Fruit compressed-trigonous, ca. 1 mm long, white, minutely muricate, enclosed in the persistent perianth.

Disturbed sites. Occasional; nearly throughout. Virginia south to Florida, west to Missouri, Oklahoma, and Texas, also Connecticut, Washington, and California; West Indies; Asia, Australia, and Pacific Islands. Native to Asia. Summer–fall.

Ficus L. 1753. FIG

Evergreen or deciduous trees, shrubs, or woody vines, the sap milky, the branches unarmed, often with adventitious roots, these sometimes descending to the ground and forming pillar-roots. Leaves alternate, the blade pinnate- or palmate-veined; stipules enclosing the naked buds, caducous. Flowers unisexual (plants monoecious or dioecious), the staminate and carpellate borne on the inner wall of an urceolate fruit-like receptacle (syconium); interfloral bracteoles usually present, often bristlelike. Staminate flowers sessile or pedicellate; sepals 2–6, free or connate; petals absent; stamens 1, or if 2(3), then with the filaments connate at the base. Carpellate flowers sessile or pedicellate, of 2 kinds, some with long styles, these developing fruits (seed-flowers), others with short styles that host a larva (gall-flowers) of the pollinating wasp; ovary 1-locular, the style lateral, the stigmas 1–2. Fruit a small indehiscent drupe imbedded in the inner surface of a thick-walled, globose receptacle (syconium) resembling a fruit, subtended at the base by 2–3 bracts, with an apical ostiole closed by 3–4 small bracts that form a disk.

A genus of about 800 species; nearly cosmopolitan. [An old name for the edible fig, *Ficus carica*, derivation unknown.]

The unique wasp-specific pollination biology of the figs has been studied in much detail (e.g., Tomlinson, 1980).

1. Plant climbing, attaching by nodal adventitious roots, or trailing; leaves dimorphic ....**F. pumila**
1. Plant erect or essentially so; leaves monomorphic.
   2. Leaf blades palmately 3- to 5-lobed ....**F. carica**
   2. Leaf blades entire.
      3. Apex of the leaf blade abruptly long caudate or long acuminate, ca. ½ as long as the blade ....
      .... **F. religiosa**

3. Apex of the leaf blade obtuse to acute, if caudate, then much in proportion to the blade.

    4. Basal leaf veins 2–3 pairs.

        5. Syconia pubescent; leaves usually elliptic-oblong, finely and evidently veined .... F. benghalensis

        5. Syconia glabrous; leaves usually ovate, not finely and evidently reticulate veined ....F. altissima

    4. Basal leaf veins 1(2) pairs.

        6. Leaves obdeltoid or obovate, the midrib often branching above ....F. deltoidea

        6. Leaves elliptic to ovate, the midrib never branching above.

            7. Leaves with more than 10 pairs of lateral veins, these uniformly spaced ....F. benjamina

            7. Leaves with up to 10 pairs of lateral veins, or if more than 10, these not uniformly spaced.

                8. Syconia on a peduncle 2–15(25) mm long.

                    9. Petioles (7)15–60 mm long; leaf base usually rounded to cordate; syconia often spotted...F. citrifolia

                    9. Petioles 4–10 mm long; leaf base cuneate to subobtuse; syconia never spotted .. .. F. americana

                8. Syconia sessile or subsessile (rarely with a peduncle to 5 mm long in *F. aurea*).

                    10. Leaf blade 6–12(15) cm long; syconia 15–20 mm long........................... F. aurea

                    10. Leaf blade 4–6(11) cm long; syconia 9–11 mm long .... F. microcarpa

Ficus altissima Blume [The highest, in reference to habit.] COUNCIL TREE.

Ficus altissima Blume, Bijdr. 444. 1826.

Evergreen tree, to 25 m, epiphytic when young; bark gray-brown, the branches with pendent adventitious roots, these not forming pillar-roots; branchlets grayish, glabrous. Leaves with the blade coriaceous, ovate, 13–25 cm long, 4.5–16 cm wide, the basal veins 2–3 pairs, the lateral veins 3–5 pairs, not uniformly spaced, the apex acuminate, the base rounded or subcordate, the margin entire, the upper surface glabrous, the lower surface glabrous or puberulent on the midvein, the petiole 3–8 cm long; stipules lanceolate, 4–5 cm long, pubescent. Plant monoecious. Staminate flowers with 1 stamen. Carpellate flowers with 1 stigma. Syconia paired, sessile, ovoid, 25–30 mm long, 15–20 mm wide, orange-red or yellow, glabrous, the basal bracts broadly ovate, 3–7 mm long, pubescent, persistent, the ostiole disk 2–3 mm wide, slightly umbonate.

Disturbed sites. Occasional; Lee and Palm Beach Counties, southern peninsula. Escaped from cultivation. Florida; South America; Asia. Native to Asia. All year.

Ficus altissima is listed as a category II invasive species by the Florida Exotic Pest Plant Council (FLEPPC, 2015).

Ficus americana Aubl. [Of America.] WEST INDIAN LAUREL FIG.

Ficus americana Aublet, Hist. Pl. Guiane 952. 1775. *Ficus perforata* Linnaeus, Pl. Surin. 17. 1775, nom. illegit.

Evergreen tree, to 12 m, epiphytic when young; bark grayish brown, the branches with pendent adventitious roots, these not forming pillar-roots; branchlets grayish, glabrous. Leaves with the blade subcoriaceous, elliptic to obovate, 2–8 cm long, 1–4 cm wide, the basal veins 1(2) pairs, the lateral veins 6–14 pairs, not uniformly spaced, the apex acute, obtuse, or short-apiculate, the base cuneate to rounded, the margin entire, the upper and lower surfaces glabrous, the petiole 4–10 mm long; stipules lanceolate, 7–9 mm long, glabrous. Plant monoecious. Staminate flowers with 1 stamen. Carpellate flowers with 1 stigma. Syconia paired, on a peduncle 2–5 mm long, globose, 3–7 mm long and wide, red, glabrous, the basal bracts ovate, 1–1.5 mm long, the ostiole disk ca. 2 mm wide, not umbonate.

Disturbed tropical hammocks. Rare; Miami-Dade County. Escaped from cultivation. Florida; West Indies, Mexico, Central and South America. Native to tropical America. All year.

Ficus aurea Nutt. [Golden, in reference to the synconium.] STRANGLER FIG; GOLDEN FIG.

> *Ficus aurea* Nuttall, N. Amer. Sylv. 2: 4, pl. 43. 1846. TYPE: FLORIDA: Monroe Co.: Key West, *Blodgett s n.* (holotype: PH?).
>
> *Ficus aurea* Nuttall var. *latifolia* Nuttall, N. Amer. Sylv. 2: 4. 1846. TYPE: FLORIDA: Monroe Co.: Key West, *Blodgett s.n.* (holotype: PH?).

Tree, to 20 m; bark gray, smooth, the branches with pendent adventitous roots, these sometimes forming pillar-roots; branchlets yellow, glabrous. Leaves with the blade coriaceous, ovate to oblong or obovate, 6–12(15) cm long, 3.5–6 cm wide, the basal veins 1(2) pairs, the lateral veins 6–10 (rarely more), not uniformly spaced, the apex obtuse or shortly and bluntly acuminate, the base rounded to cuneate, the margin entire, the upper and lower surfaces glabrous, the petiole 1–6 cm long, glabrous; stipules lanceolate, 1–1.5 cm long, glabrous. Plant monoecious. Staminate flowers with 1 stamen. Carpellate flowers with 1 stigma. Syconia usually paired, sessile or rarely on a peduncle to 5 mm long, obovoid, 6–15 mm wide, red or yellow, glabrous, the basal bracts 3–5 mm long, glabrous, the ostiole disk 3–4 mm wide, slightly umbonate.

Hammocks and borders of mangrove swamps. Frequent; central and southern peninsula. Florida; West Indies, Mexico, and Central America. Spring–summer.

Ficus benghalensis L. [Of the Bay of Bengal, India.] BANYAN TREE.

> *Ficus benghalensis* Linnaeus, Sp. Pl. 2: 1059. 1753. *Perula benghalensis* (Linnaeus) Rafinesque, Sylva Tellur. 59. 1838. *Urostigma benghalense* (Linnaeus) Gasparrini, Ric. Caprifico 82. 1845.

Evergreen tree, to 30 m; bark brown, smooth, the branches with pendent adventitious roots, these often forming pillar-roots; branchlets puberulent, glabrescent in age. Leaves with the blade coriaceous, elliptic-oblong to ovate, 10–30 cm long, 7–20 cm wide, the basal veins (2)3–4 pairs, ⅓–½ the blade length, the lateral veins 5–6(7) pairs, the apex obtuse, the base cuneate to rounded, the upper surface glabrous, the lower surface puberulent, glabrescent in age, finely and evidently reticulate, the petiole 1.5–7 cm long; stipules stout, 1.5–2.5 cm long, pubescent. Plants monoecious. Staminate flowers with 1 stamen. Carpellate flowers with 1 stigma. Syconia paired, sessile, depressed globose, 15–20 mm long, 20–25 mm wide, orange to red, pubescent, the basal bracts broadly ovate, 3–7 mm long, pubescent, the ostiole disk 3–4 mm wide, flat or subumbonate.

Disturbed hammocks. Rare; Broward and Miami-Dade Counties. Escaped from cultivation. Florida; Asia. Native to Asia. All year.

Ficus benjamina L. [Probably in reference to the supposed relation of the plant to the source of a resin or benzoin early procured from the East.] WEEPING FIG.

Ficus benjamina Linnaeus, Mant. Pl. 129. 1767. *Ficus pyrifolia* Salisbury, Prodr. Stirp. Chap. Allerton 16. 1796, nom. illegit. *Urostigma benjaminum* (Linnaeus) Miquel, London J. Bot. 6: 583. 1847.

Evergreen tree, to 10 m, epiphytic when young; bark gray, smooth, the branches occasionally with pendent adventitious roots, these not forming pillar roots; branchlets brown, glabrous. Leaves with the blade subcoriaceous, oblong, elliptic, lanceolate, or ovate, 4–6(11) cm long, 1.5–6 cm wide, the basal veins 1(2) pairs, short, the lateral veins (6)12(14) pairs, uniformly spaced, the secondary nerves prominent, the apex acuminate or cuspidate, the base rounded or cuneate, the margin entire, the upper and lower surfaces glabrous, the petiole 10–20 mm long; stipules lanceolate, 8–12 mm long, puberulent or glabrate. Plant monoecious. Staminate flowers with 1 stamen. Carpellate flowers with 1 stigma. Syconium single or paired, sessile or subsessile, subglobose, 8–12 mm long, 7–10 mm wide, orange, yellow, or dark red, glabrous, the basal bracts crescent-shaped, 0.5–1.5 mm long, the ostiole disk 1.5–2 mm wide, slightly umbonate.

Disturbed sites. Rare; St. Lucie, Lee, and Monroe Counties. Escaped from cultivation. Florida; West Indies, Mexico, Central America, and South America; Africa, Asia, Australia, and Pacific Islands. Native to Asia, Australia, and Pacific Islands. All year.

Ficus carica L. [From Caria, an ancient region in Asia Minor on the Aegean Sea.] COMMON FIG.

Ficus carica Linnaeus, Sp. Pl. 2: 1059. 1753. *Ficus latifolia* Salisbury, Prodr. Stirp. Chap. Allerton 16. 1796, nom. illegit.

Deciduous shrub or small tree, to 3 m; bark gray, slightly roughened, the branches lacking adventitious roots; branchlets grayish, pubescent. Leaves with the blade obovate, suborbicular, or ovate, 15–30 cm long and wide, palmately 3- to 5-lobed, the basal veins 3–5 pairs, the lateral veins irregularly spaced, the apex acute to obtuse, the base cordate, the margin undulate or irregularly dentate, the upper and lower surfaces scabrous-pubescent, the petiole 2–10 cm long; stipules broadly lanceolate, 10–12 mm long, pubescent. Plant dioecious or structurally gynodioecious (two tree morphs occur, one called the edible fig producing syconia with long-styled flowers (female tree) and the other the caprifig producing syconia with short-styled flowers and male flowers (male tree)). Staminate flowers with 2 stamens. Carpellate flowers with 2 stigmas. Syconium solitary, on a peduncle ca. 10 mm long, pyriform, 5–8 cm long, green or red-purple, the basal bracts ovate, 1–2 mm long, the ostiole disk 3–4 mm wide, umbonate.

Disturbed sites. Rare; central panhandle, central and southern peninsula. Escaped from cultivation. Florida; Massachusetts south to Florida, west to Michigan, Kentucky, Tennessee, and Texas, also in California; West Indies, Mexico, Central America, and South America; Europe, Africa, Asia, Australia, and Pacific Islands. Native to Asia. Spring–summer.

Ficus carica is sometimes found spreading vegetatively or by parthenocarpic means outside of cultivation in Florida.

Ficus citrifolia Mill. [With leaves like *Citrus* (Rutaceae).] WILD BANYAN TREE.

Ficus citrifolia Miller, Gard. Dict., ed. 8. 1768.
Ficus pedunculata Aiton, Hort. Kew. 3: 450. 1789. *Urostigma pedunculatum* (Aiton) Miquel, London J. Bot. 6: 540. 1847.
Ficus laevigata Vahl, Enum. Pl. 2: 183. 1805.
Ficus populnea Willdenow, Sp. Pl. 4: 1141. 1806.
Ficus brevifolia Nuttall, N. Amer. Sylv. 2: 3, pl. 42. 1846. *Ficus populnea* Willdenow var. *brevifolia* (Nuttall) Warburg, in Urban, Symb. Antill. 3: 473. 1903. *Ficus laevigata* Vahl var. *brevifolia* (Nuttall) Warburg ex Rossberg, Notizbl. Bot. Gart. Berlin-Dahlem 12: 583. 1935. *Ficus citrifolia* Miller var. *brevifolia* (Nuttall) D'Arcy, Phytologia 25: 116. 1973. TYPE: FLORIDA: Monroe Co.: Key West, *Blodgett s.n.* (holotype: PH?).
Ficus populnea Willdenow subvar. *floridana* Warburg, in Urban, Symb. Antill. 3: 473. 1903. TYPE: FLORIDA: Miami-Dade Co.: Meigs Key, *Curtiss 2548* (holotype: ?).

Evergreen shrub or tree, to 15 m; bark brownish, smooth, the branches with pendent adventitious roots, these not forming pillar-roots; branchlets yellowish brown, glabrous or sparingly pubescent. Leaves with the blade subcoriaceous, ovate to elliptic or obovate, 3–14(20) cm long, 1.5–8(12) cm wide, the basal veins 1(2) pairs, the lateral veins 4–8, not uniformly spaced, the apex obtuse to acute or acuminate, the base cordate to truncate or broadly cuneate, the upper and lower surfaces glabrous, the petiole (0.7)1.5–6 cm long, the stipules lanceolate, 5–20 mm long, glabrous. Plant moneocious. Staminate flowers with 1 stamen. Carpellate flowers with 1 stigma. Syconium solitary or paired, on a peduncle 5–15 mm long, globose to ovoid, 8–18 mm long, 6–12 mm wide, red to yellow or yellow with red spots, glabrous, the basal bracts deltate or broadly rounded, 2–3 mm long, glabrous or puberulent, the ostiole disk 2–3 mm wide, slightly umbonate.

Tropical hammocks. Occasional; Hillsborough County, southern peninsula. Florida; West Indies, Mexico, Central America, and South America. Spring–summer.

Ficus deltoidea Jack [Delta-shaped, in reference to the leaf.] MISTLETOE RUBBERPLANT.

Ficus deltoidea Jack, Malayan Misc. 2(7): 71. 1822.

Evergreen shrub to 2 m; bark grayish, smooth; branchlets glabrous. Leaves with the blade coriaceous, broadly obovate-cuneate, 6–15 cm long, 4–10 cm wide, the basal vein 1(2) pairs, the lateral veins fewer than 10, not uniformly spaced, the midvein often forked distally, the apex truncate to rounded, the base cuneate, the upper and lower surfaces glabrous, the petiole 5–10 mm long, the stipules lanceolate, 5–8 mm long, glabrous. Plant dioecious. Staminate flowers with 2 stamens. Carpellate flowers with 2 stigmas. Syconium solitary or paired, on a peduncle 5–25 mm long, globose, pyriform, or fusiform-ellipsoid, yellow, orange, or red, 10–15 mm long, glabrous, the basal bracts ovate, 1–2 mm long, glabrous, the ostiole disk 2–3 mm wide, umbonate.

Disturbed site on shell middens; sometimes epiphytic. Rare; Brevard County. Escape from cultivation. Florida; Africa, Asia, Australia, and Pacific Islands. Native to Asia. All year.

Ficus microcarpa L. f. [Small-fruited.] INDIAN LAUREL.

Ficus microcarpa Linnaeus f., Suppl. Pl. 442. 1782. *Urostigma microcarpum* (Linnaeus f.) Miquel, London J. Bot. Kew. 6: 583. 1847.

Evergreen tree, to 30 m, epiphytic when young; bark gray, the branches with pendent aventitious roots, sometimes developing pillar-roots; branchlets brown, glabrous. Leaves with the blade subcoriaceous, elliptic, obovate to oval, 3–11 cm long, 1.5–6 cm wide, the basal pair of veins 1(2), conspicuous and ascending, the lateral veins 5–9 pairs, not uniformly spaced, the apex subacute to acuminate, the base obtuse to cuneate, the margin entire, the upper and lower surfaces glabrous, the petiole 5–10 mm long, puberulent; stipules lanceolate, 7–9(15) mm long, glabrous. Plant monoecious. Staminate flowers with 1 stamen. Carpellate flowers with 1 stigma. Syconia paired, sessile, obovoid, pyriform, or subglobose, 9–11 mm long, 5–6 mm wide, purple or black, the basal bracts ovate-lanceolate, 1.5–3.5 mm long, the apex obtuse to subacute, the ostiole disk 2–2.5 mm wide, umbonate.

Disturbed sites. Rare; central and southern peninsula. Escaped from cultivation. Florida and California; West Indies, Mexico, Central America, and South America; Africa, Asia, Australia, and Pacific Islands. Native to Africa, Asia, Australia, and Pacific Islands. All year.

Ficus microcarpa is listed as a category I invasive species by the Florida Exotic Pest Plant Council (FLEPPC, 2015).

Ficus pumila L. [Dwarf.] CLIMBING FIG.

Ficus pumila Linnaeus, Sp. Pl. 2: 1060. 1753. *Varinga repens* Rafinesque, Sylva Tellur. 58. 1838, nom. illegit.

Evergreen, woody vine or sprawling shrub; stem and branches closely appressed to the substrate or loosely ascending, pubescent, glabrous in age, with adventitious nodal roots. Leaves dimorphic, those of the loose extended stem with the blade coriaceous, oblong to ovate-elliptic or obovate, 4–10 cm long, 2.5–4.5 cm wide, the basal veins 3–5, ½–⅓ the length of the blade, the lateral veins 3–6 pairs, straight, the secondary veins prominent, reticulate, the apex obtuse to subacute, the base obtuse to rounded, the margin recurved, the upper surface glabrous, the lower surface glabrous or puberulent on the veins, strongly raised-reticulate with the areoles deeply sunken and minutely foveolate, the petiole 8–20 mm long, with brown appressed trichomes; stipules lanceolate, 3–8 mm long, pubescent, the blade of the appressed climbing stem distichous, ovate, 1.2–3.5 cm long, 0.7–2 cm wide, the apex obtuse, the base cordate, asymmetrical, the lower surface with the areoles flat and not foveolate. Plant monoecious. Staminate flowers with 2–3 stamens. Carpellate with 2 stigmas. Syconium solitary, on a peduncle 8–15 mm long, oblong, obovoid, pyriform, or subglobose, 3–4 cm long and wide, the apex subtruncate, purple to black, glaucous, slightly villous, glabrescent in age, the basal bracts ovate-lanceolate, 5–7, deciduous, the ostiole disk 2–3 mm wide, umbonate.

Disturbed sites. Rare; central and western panhandle, central peninsula. Escaped from cultivation. Georgia, Florida, and Louisiana; Mexico, Central America, and South America; Asia, Australia, and Pacific Islands. Native to Asia. All year.

Ficus pumila is commonly planted in urban areas on walls. The plants are often invasive, forming large colonies outside of cultivation, and developing abundant mature syconia but not developing fertile fruits.

Ficus religiosa L. [Of religous significance; under this tree Buddha was said to have arrived at true insight.] BO TREE; SACRED FIG.

> *Ficus religiosa* Linnaeus, Sp. Pl. 2: 1059. 1753. *Ficus caudata* Stokes, Bot. Mar. Med. 4: 358. 1812, nom. illegit. *Urostigma religiosum* (Linnaeus) Gasparrini, Ric. Caprifico 82. 1845.

Evergreen tree, to 20 m, epiphytic when young; bark gray, the branches lacking adventitious roots; branchlets gray, glabrous. Leaves with the blade subcoriaceous, ovate to ovate-cordate, 7–25 cm long, 4–16 cm wide, the basal veins 2–3 pairs, the lateral veins 6–9 pairs, unevenly spaced, with several obscure secondary veins, the apex caudate-acuminate with the tip 2.5–9 cm long, the base truncate, the margin often sinuous, the surfaces glabrous, the petiole 3.5–13 cm long; stipules lanceolate, 10–15 mm long, glabrous. Plant monoecious. Staminate flowers in a single ring at the ostiole; stamen 1. Carpellate flowers scattered throughout; stigma 1. Syconia paired, sessile, subglobose, 8–11 mm long and wide, reddish, purple, or black, glabrous, the basal bracts ovate, 3–5 mm long, silky-puberulous, glabrescent in age, the ostiole disk 2–3 mm wide.

Disturbed thickets. Rare; Miami-Dade County. Escaped from cultivation. Florida; Asia. Native to Asia. All year.

EXCLUDED TAXON

> *Ficus elastica* Roxburgh ex Hornemann—Reported for Florida as an escape from cultivation by Small (1933), Long and Lakela (1971), and Wunderlin (1982, 1997). All specimens seen are of cultivated material.

Maclura Nutt., nom. cons. 1818. OSAGE ORANGE

Deciduous trees, the sap milky, the branches with axillary thorns. Leaves pinnate-veined. Flowers unisexual (plants dioecious). Staminate flowers in loose, short racemes clustered on short spur branches; sepals 4; petals absent; stamens 4, the anthers with a short connective. Carpellate flowers in dense heads; sepals 4, the outer 2 wider than the inner ones; petals absent; ovary 1-locular, 1-ovulate, the stigmas 1. Fruit an indehiscent drupe enclosed in its enlarged, fleshy calyx, aggregated into a globose syncarp.

A genus of 11 species; nearly cosmopolitan. [Commemorates William Maclure (1763–1840), Scottish-born American geologist, cartographer, and philanthropist.]

> *Ioxylon* Raf., nom. rej., 1817.

Maclura pomifera (Raf.) C. K. Schneid. [*Pome*, apple, and *ferre*, to bear, in reference to the apple-like fruit.] OSAGE ORANGE; HEDGE APPLE.

> *Ioxylon pomiferum* Rafinesque, Amer. Monthly Mag. & Crit. Rev. 2: 118. 1817. *Ioxylon maclura* Rafinesque, New Fl. 3: 43. 1838, nom. illegit. *Maclura pomifera* (Rafinesque) C. K. Schneider, Ill. Handb. Laubholzk. 1: 806. 1906.

Maclura aurantiaca Nuttall, Gen. N. Amer. Pl. 2: 234. 1818. *Ioxylon aurantiacum* (Nuttall) Rafinesque, Med. Fl. 2: 260. 1830.

Tree, to 20 m; bark dark orange-brown, shallowly furrowed and with flat ridges, often peeling into long, thin strips; branchlets greenish yellow becoming orange-brown, the spines stout, straight, to 3.5 cm long. Leaves with the blade ovate to elliptic or lanceolate, 7–15 cm long, 3–8 cm wide, the apex acuminate, the base truncate or rounded, the margin entire, the upper surface glabrous, the lower surface glabrous or sparsely pubescent, especially on the veins, the petiole 3–5 cm long, pubescent; stipules lanceolate, 1.5–2 mm long. Plants dioecious. Staminate flowers in a globose or cylindric head 13–23 mm long, the peduncle 10–15 mm long, pubescent, the pedicel 2–10 mm long, glabrate; sepals yellow-green, lanceolate, ca. 1 mm long, pubescent. Carpellate flowers in an axillary globose head ca. 1.5 cm long, the peduncle 2–2.5 mm long, glabrous or pubescent, sessile on an obconic receptacle; calyx green, obovate, ca. 3 mm long, enclosing and closely appressed to the ovary, hoodlike, ciliate near the tip; style filiform, ca. 3 mm long, green, glabrous, the stigma filiform, 4–6 mm long, papillose, yellowish. Syncarp globose, 8–12 cm in diameter, the surface irregular, yellow-green, the peduncle short, glabrous or pubescent; drupe completely covered by the accrescent, thickened calyx and deeply imbedded in the receptacle; seed cream-colored, oval to oblong, 8–12 mm long, 5–6 mm wide, the base truncate or rounded with 1–3 minute points, the apex rounded, mucronate, the surface minutely striated or pitted, the margin with a narrow groove.

Disturbed sites. Rare; central and western panhandle, Marion and Volusia Counties. Escaped from cultivation. Nearly throughout North America; Europe. Native to Arkansas, Oklahoma, and Texas. Spring.

Morus L. 1753. MULBERRY

Deciduous trees or shrubs, the sap milky, the branches unarmed. Leaves pinnipalmate-veined. Flowers unisexual (plants monoecious or dioecious), borne in pedunculate spikes. Staminate flowers with the sepals 4, free or connate; petals absent; stamens 4. Carpellate flowers with the sepals 4, free, dimorphic; petals absent; ovary 2-locular, 1-ovulate, the stigmas 2. Fruit an indehiscent drupe surrounded by the fleshy, enlarged calyx, this aggregating and forming a blackberry-like syncarp.

A genus of 11 species; nearly cosmopolitan. [*Morum*, mulberry, the classical Latin name.]

1. Leaves with the upper surface glabrous, smooth to the touch, the lower surface glabrous or with pubescence only along the major veins or in tufts in the axils of the principal lateral veins and the midvein..**M. alba**
1. Leaves with the upper surface with short, stiff, antrorsely appressed trichomes, scabrid, the lower surface pubescent or puberulent ....**M. rubra**

Morus alba L. [White, in reference to the fruit color.] WHITE MULBERRY.

Morus alba Linnaeus, Sp. Pl. 2: 986. 1753.
Morus tatarica Linnaeus, Sp. Pl. 2: 986. 1753. *Morus alba* Linnaeus var. *tatarica* (Linnaeus) Seringe, Descr. Muriers 202. 1855.

Shrub or tree, to 15 m; bark brown tinged with red or yellow, thin, shallowly furrowed with long, narrow ridges, the branchlets orange-brown or dark green with a reddish cast, pubescent or occasionally glabrous. Leaves with the blade ovate, (6)8–10 cm long, 3–6 cm wide, often deeply and irregularly lobed, the apex acute to short-acuminate, the base cuneate, truncate, or cordate, the margin coarsely serrate to crenate, the upper surface glabrous, the lower surface glabrate or sparsely pubescent along the veins and in the vein axils, the petiole 2.5–5 cm long; stipules ovate to lanceolate, 5–9 mm long, pubescent. Plants monoecious or dioecious. Staminate flowers in a spike 2.5–4 cm long; sepals ovate, ca. 1.5 mm long, free, green with a red tip, pubescent; stamens ca. 2.7 mm long, white. Carpellate flowers in a spike 5–8 mm long; sepals enclosing the ovary, green, ciliate; style branches 0.5–1 mm long, red-brown, divergent. Syncarp cylindric, 1.5–2.5 cm long, ca. 1 cm wide; drupe obconic, ca. 4 mm long, ca. 1.5 mm wide, subtended by the persistent, fleshy sepals, red when immature, becoming black, purple, or nearly white at maturity, the pyrene ovoid, 2–3 mm in diameter, light brown.

Disturbed sites. Occasional; central and western panhandle, northern and central peninsula. Escaped from cultivation. Nearly throughout North America; Europe and Asia. Native to Asia. Spring–summer.

Morus rubra L. [Red, in reference to the fruit color.] RED MULBERRY.

Morus rubra Linnaeus, Sp. Pl. 2: 986. 1753.

Shrub or tree, to 20 m; bark gray-brown with an orange tint, the furrows shallow, the ridges flat, broad, tight or occasionally loose; branchlets red-brown to light greenish brown, glabrous or puberulent. Leaves with the blade broadly ovate, 10–18(36) cm long, 8–12(15) cm wide, entire or irregularly lobed, the apex abruptly acuminate, the base rounded to subcordate, the margin serrate or crenate, the upper surface with short antrorse trichomes, scabrous, the lower surface sparsely to densely pubescent, the petiole 2–2.5 cm long, glabrous or pubescent; stipules linear, 10–13 mm long, pubescent. Plants dioecious. Staminate flowers in a spike 3–5 cm long, ca. 1 cm wide, pendent, the peduncle ca. 1 cm long, pubescent; sepals 2–3 mm long, connate at the base, green tinged with red, pubescent outside, ciliate toward the tip; stamens 3–3.5 mm long, white. Carpellate flowers in a spike 8–12 mm long, 5–7 mm wide, erect or drooping, the peduncle 3–5 m long, pubescent; sepals ovate, 2–3 mm long, free, tightly enclosing the ovary, green, ciliate at the apex; stigmas linear, ca. 1.5 mm long, divergent, whitish. Syncarp cylindric, 1.5–2.5 cm long, ca. 1 cm wide, the peduncle 9–12 mm long, pubescent; drupe black or deep purple, the calyx becoming enlarged and succulent, the pyrenes oval, flattened, smooth, ca. 2 mm long, yellowish.

Hammocks and floodplain forests. Frequent; nearly throughout. Vermont and Massachusetts south to Florida, west to Ontario, Minnesota, South Dakota, Nebraska, Kansas, and New Mexico. Spring–summer.

EXCLUDED TAXON

Morus nigra Linnaeus—Reported for Florida by Small (1903, 1913a, 1913c, 1913e) and Long and Lakela (1971), based on misapplication of the name to our material of *M. alba*.

URTICACEAE Juss., nom. cons. 1789. NETTLE FAMILY

Trees, shrubs, subshrubs, or herbs, sometimes with stinging trichomes. Leaves alternate or opposite, the blade simple, pinnate-, palmate-, or pinnipalmate-veined, often with cystoliths in the epidermis, petiolate, stipulate or estipulate, stipules (if present) either paired at the petiole base or intrapetiolar and partly to entirely connate. Flowers in axillary or rarely terminal, cymose spikes, racemes, or panicles, bisexual or unisexual (plants monoecious, dioecious, or polygamous). Staminate flowers actinomorphic, sessile or short-stalked; sepals 2–5, free or connate; petals absent; stamens 2–5, the anthers 2-locular, longitudinally dehiscent. Carpellate flowers actinomorphic or zygomorphic, sessile or subsessile; sepals 2–4, free or connate; petals absent; staminodes present or absent; ovary superior, 1-carpellate, 1-locular, the style 1. Fruit an achene, sometimes drupe-like, often enclosed by the persistent calyx; seed 1.

A family of about 54 genera and about 2,600 species; nearly cosmopolitan.

Cecropia traditionally has been placed in the Moraceae or more recently in the segregate family Cecropiaceae. However, recent molecular studies (e.g., Hadiah et al., 2008; Monro, 2006; Sytsma et al., 2002) show that it is best placed in the Urticaceae.

Cecropiaceae C. C. Berg 1978.

Selected References: Boufford (1997); Miller (1971).

1. Trees ....**Cecropia**
1. Shrubs, subshrubs, or herbs.
   2. Leaves opposite (sometimes alternate or subopposite above).
      3. Plants with stinging trichomes (trichomes with a distinct bulbous base and stiff, translucent apex) .... **Urtica**
      3. Plants without stinging trichomes.
         4. Inflorescence of many small glomerules borne spikelike; plants with trichomes . **Boehmeria**
         4. Inflorescence a dense, secund panicle; plants glabrous .... **Pilea**
   2. Leaves all alternate.
      5. Leaf margins dentate or serrate.
         6. Leaves densely white pubescent with non-stinging trichomes, especially when young; inflorescence axillary .... **Boehmeria**
         6. Leaves sparsely to densely pubescent with stinging trichomes and stipitate-glandular, non-stinging trichomes; inflorescences terminal or in the upper leaf axils .... **Laportea**
      5. Leaf margins entire.
         7. Plants suffrutescent; sepals connate, appressed to and tightly enclosing the achene ....
         .... **Pouzolzia**
         7. Plants herbaceous; sepals free or only basally connate, loosely enclosing the achene.
            8. Stigma long, filiform .... **Rousselia**
            8. Stigma short, penicillate .... **Parietaria**

Boehmeria Jacq. 1760. FALSE NETTLE

Perennial herbs, subshrub, or shrubs, with nonstinging trichomes. Leaves opposite or alternate, pinnipalmate-veined, the margins serrate or dentate, the cystoliths rounded, restricted to the upper epidermis, petiolate, stipulate. Flowers unisexual (plants monoecious), in an axillary, spiciform inflorescence with the carpellate and the staminate flowers occurring mixed in remote to congested clusters along the main axis or paniculate and composed of either staminate or carpellate flowers, actinomorphic, bracteate. Staminate flowers short-pedicellate; sepals 4, connate; petals absent; stamens 4; rudimentary pistil present. Carpellate flowers sessile; sepals 4, connate; petals absent; staminodes absent; style central, prolonged into a filiform stigma that is papillose on 1 side. Fruit an achene enclosed by the persistent calyx.

A genus of about 60 species; nearly cosmopolitan. [Commemorates Georg Rudolph Boehmer (1723–1803), German botanist.]

Ramium Kuntze, 1891.

1. Leaves opposite (sometimes subopposite or alternate dorsally), sparsely to moderately pubescent when young; inflorescence spiciform..**B. cylindrica**
1. Leaves alternate throughout, densely white pubescent when young; inflorescence paniculate **B. nivea**

Boehmeria cylindrica (L.) Sw. [Cylindric, in reference to the spiciform inflorescence.] FALSE NETTLE; BOG HEMP.

Urtica cylindrica Linnaeus, Sp. Pl. 2: 984. 1753. *Boehmeria cylindrica* (Linnaeus) Swartz, Prodr. 34. 1788. *Boehmeria cylindrica* (Linnaeus) Swartz var. *genuina* Weddell, Arch. Mus. Hist. Nat. 9: 1856, nom. inadmiss. *Boehmeria cylindrica* (Linnaeus) Swartz subvar. *gymnostachya* Weddell, Arch. Mus. Hist. Nat. 9: 362. 1856, nom. inadmiss. *Ramium cylindricum* (Linnaeus) Kuntze, Revis. Gen. Pl. 2: 631. 1891.

Boehmeria drummondiana Weddell, Ann. Sci. Nat., Bot., ser. 4. 1: 201. 1854. *Boehmeria cylindrica* (Linnaeus) Swartz subvar. *tomentosa* Weddell, Arch. Mus. Hist. Nat. 9: 363. 1856. *Boehmeria cylindrica* (Linnaeus) Swartz var. *drummondiana* (Weddell) Weddell, in de Candolle, Prodr. 16(1): 202. 1869.

Boehmeria cylindrica (Linnaeus) Swartz var. *scabra* Porter, Bull. Torrey Bot. Club 16: 21. 1889. *Boehmeria scabra* (Porter) Small, Fl. S.E. U.S. 358, 1329. 1903.

Boehmeria decurrens Small, Fl. S.E. U.S. 358, 1329. 1903. TYPE: FLORIDA: *Chapman s.n.* (holotype: NY).

Boehmeria longifolia Gandoger, Bull. Soc. Bot. France 66: 287. 1920 ("1919"). SYNTYPE: FLORIDA.

Perennial herb, to 1.5 m; stem puberulent or scabrid. Leaves opposite, rarely subopposite or alternate dorsally, the blade chartaceous or thickened rugose, ovate to ovate-lanceolate or elliptic, (5)10–15(18) cm long, (2)4–7(10) cm wide, the apex acuminate, the base broadly cuneate to rounded, the margin dentate or serrate, pinnipalmate-veined, the upper surface glabrate or scabrous, the lower surface glabrate or short-pilose or puberulent, the petiole 5–15 cm long, pubescent; stipules linear-lanceolate, 3–12 mm long, deciduous. Flowers clustered in an axillary, continuous or interrupted spiciform inflorescence, the staminate and carpellate intermingled or the plants wholly staminate proximally and carpellate distally. Staminate flowers brown or greenish; calyx with hooked and straight trichomes on the outer surface. Carpellate flowers red or brown; calyx with the lobes ciliate or toothed, the outer surface pubescent with hooked or straight trichomes. Fruit ovoid to elliptic or suborbicular, 1–1.5 mm long, somewhat laterally compressed, closely enclosed by the persistent calyx; seed surrounded with corky tissue except at the base.

Marshes and swamps. Common; nearly throughout. Quebec south to Florida, west to Ontario, Minnesota, South Dakota, and California; West Indies, Mexico, Central America, and South America. Summer–fall.

Boehmeria nivea (L.) Gaudich. [Snowy, in reference to the white lower leaf surface.] RAMIE.

> *Urtica nivea* Linnaeus, Sp. Pl. 2: 985. 1753. *Boehmeria nivea* (Linnaeus) Gaudichaud-Beaupré, Voy. Uranie 499. 1830. *Ramium niveum* (Linnaeus) Kuntze, Revis. Gen. Pl. 2: 632. 1891.

Shrub or subshrub, to 3 m; stem with appressed or somewhat spreading trichomes. Leaves all alternate, the blade ovate to suborbicular or lanceolate, 5–15(30) cm long, 3–12(20) cm long, the apex acuminate, the base rounded, truncate, or shallowly cordate, the margin serrate or dentate, the upper surface slightly scabrous, the lower surface densely white tomentose, the petiole 1–6 cm long, appressed white-pubescent; stipules linear-lanceolate, 3–12 mm long, deciduous. Flowers in an axillary panicle with numerous sessile glomerules, the staminate in the axils of the proximal leaves, the carpellate in the axils of the distal leaves. Staminate flowers with the calyx ca. 1 mm long, pubescent. Carpellate flowers with the calyx ca. 1 mm long, contracted at the mouth, enclosing the ovary, hirsute; style exserted, pubescent. Fruit ovoid or ellipsoidal, ca. 1 mm long, laterally compressed, enclosed in the hirsute, persistent calyx; seed not surrounded with corky tissue.

Disturbed sites. Rare; Seminole, Orange, and Miami-Dade Counties. Escaped from cultivation. Maryland and Virginia, South Carolina south to Florida, west to Texas; Mexico, Central America, and South America; Europe, Africa, and Asia. Native to Asia. Summer–fall.

Cecropia Loefl., nom. cons. 1758.

Trees. Leaves alternate, palmate-veined and -lobed, peltate, the margin entire, petiolate, with trichilia (dense patches of trichomes forming "Müllerian bodies") at the base, stipulate. Flowers unisexual (plants dioecious), in axillary, pedunculate, digitate spikes, subtended by a spathe-like bract, actinomorphic. Staminate flowers with the sepals connate; petals absent; stamens 2, free. Carpellate flowers with the sepals connate; petals absent; style short, the stigma penicillate. Fruit enclosed within the persistent calyx, the whole forming a syncarp.

A genus of ca. 60 species; North America, West Indies, Central America, South America, Africa, Asia, and Pacific Islands. [In reference to the inflorescence resembling the larva of the cecropid moth.]

Cecropia palmata Willd. [Palmate, in reference to the leaf shape.] TRUMPET TREE.

> *Cecropia palmata* Willdenow, Sp. Pl. 4: 652. 1806. *Ambaiba palmata* Kuntze, Revis. Gen. Pl. 2: 624. 1891.

Tree, to 10 m; stem slender with a terminal umbrella-shaped top; branchlets puberulent to hispidulous, with a thin spot at the internode (prostoma) that can be easily perforated by ants. Leaves with the blade subcoriaceous, 20–60 cm long and wide, 7- to 11-lobed ¾ or more to the petiole, the margin entire, the upper surface scabridulous, the lower surface densely white-arachnoid-pubescent, the petiole ca. 60 cm long, with trichilia at the base; stipules 7–15 cm long, connate, the blade palmate-lobed, puberulous. Staminate flowers in 4–6 pendent, digitate, stipitate spikes 8–15 cm long, 8–18 mm wide, the peduncle 6–18 cm long, the basal spathe-like bract 10–17 cm long, white-arachnoid-pubescent; calyx tubular, arachnoid-pubescent proximally, glabrous distally. Carpellate flowers in 4 pendent, digitate, sessile spikes much as the staminate, 8–20 cm long, 5–8 mm wide, the peduncle 20–25 cm long, white-arachnoid-pubescent, the spathe-like bract 10–22 cm long, white-arachnoid. Syncarp fleshy, the fruit ca. 3 mm long, slightly tuberculate.

Cecropia palmata is listed as a Category II invasive species in Florida by the Florida Exotic Pest Plant Council (FLEPPC, 2015).

Disturbed tropical hammocks. Rare; Broward and Miami-Dade Counties. Escaped from cultivation. Florida; West Indies, Mexico, Central America, and South America; Africa, Asia, and Pacific Islands. Native to South America.

Laportea Gaudich., nom. cons. 1830. WOODNETTLE

Annual or perennial herbs, with stinging and nonstinging trichomes. Leaves alternate, pinnate-veined, cystoliths rounded, petiolate, stipulate. Flowers unisexual (plants monoecious), in axillary and terminal cymose panicles, bracteate, both staminate and carpellate in the same inflorescence or the staminate and carpellate flowers in separate inflorescences with the staminate borne lower on the stem than the carpellate. Staminate flowers actinomorphic; sepals 5, free; petals absent, stamens 5; pistil rudimentary. Carpellate flowers zygomorphic, borne on a thick pedicel that develops lateral wings as the fruit matures; sepals 4, free; petals absent; staminodes absent; ovary on a short disk, the stigma subulate. Fruit an achene.

A genus of about 22 species; North America, West Indies, Mexico, Central America, South America, Africa, Asia, and Pacific Islands. [Commemorates Francois Louis de Laporte (Comte de) Castelnau (1810–1880), French naturalist.]

Urticastrum Heister ex Fabr., 1759, nom. rej.

1. Leaf base slightly auriculate; plant with stinging trichomes and also with stipitate-glandular trichomes; fruiting pedicels with a dorsiventral orientation, lacking lateral wings.................................**L. aestuans**
1. Leaf base not auriculate; plant with stinging trichomes, but lacking stipitate-glandular trichomes; fruiting pedicels lacking a dorsiventral orientation, with distinct lateral wings ....**L. canadensis**

Laportea aestuans (L.) Chew [*Aestuo*, to burn, in reference to the burning sensation caused by the stinging trichomes.] WEST INDIAN WOODNETTLE.

Urtica aestuans Linnaeus, Sp. Pl., ed. 2. 1397. 1763. *Fleurya aestuans* (Linnaeus) Gaudichaud-Beaupré ex Miquel, in Martius, Fl. Bras. 4(1): 196. 1853. *Fleurya aestuans* (Linnaeus) Gaudichaud-Beaupré ex Miquel var. *linnaeana* Weddell, in de Candolle, Prodr. 16(1): 72. 1869, nom. inadmiss. *Laportea aestuans* (Linnaeus) Chew, Gard. Bull. Straits Settlem., ser. 3. 21: 200. 1965.

Annual herb, to 1 m; stem with stinging trichomes and stipitate-glandular, nonstinging trichomes. Leaves with the blade broadly ovate to suborbicular, 9–15(20) cm long, 6–12(16) cm wide, the apex acute to acuminate, the base cuneate, rounded to subcordate, slightly auriculate, the margin coarsely dentate or crenate, the upper and lower surfaces hispid with stinging trichomes, the upper surface also bullate, the petiole 0.5–10(16) cm long; stipules linear-lanceolate, 2–6 mm long. Flowers paniculate, pedunculate, usually shorter than the leaves, divaricately branched, both staminate and carpellate in the same inflorescence or the staminate inflorescences lower. Staminate flowers ca. 1 mm long, green; sepals spreading, equal in length. Carpellate flowers ca. 1 mm long; sepals with the inner pair ca. ½ the length of the ovary, the inner 2 lateral, larger than the outer 2, the lower 1 becoming hoodlike; style hooked and beaklike. Fruit ovate or suborbicular, 1–1.5 mm long, strongly compressed, stipitate, asymmetrical, partially covered by the enlarged lateral sepals.

Disturbed sites. Rare; Alachua, Volusia, Palm Beach, and Miami-Dade Counties. South Carolina, Alabama, Florida, and California; West Indies, Mexico, Central America, and South America; Africa and Asia. Native to tropical America. Fall–winter.

Laportea canadensis (L.) Wedd. [Of Canada.] CANADIAN WOODNETTLE.

 Urtica canadensis L., Sp. Pl. 2: 985. 1753. *Fleurya canadensis* (Linnaeus) Bentham, in Hooker, Niger Fl. 517. 1849. *Laportea canadensis* (Linnaeus) Weddell, Ann. Sci. Nat., Bot., ser. 4. 1: 183. 1854. *Urticastrum divaricatum* (Linnaeus) Kuntze var. *canadense* (Linnaeus) Kuntze, Revis. Gen. Pl. 2: 635. 1891. *Urtica divaricata* Linnaeus, Sp. Pl. 2: 985. 1753. *Urticastrum divaricatum* (Linnaeus) Kuntze, Revis. Gen. Pl. 2: 634. 1891. *Laportea divaricata* (Linnaeus) Lunell, Amer. Midl. Naturalist 4: 301. 1916.

Perennial herb, to 1 m; stem with stinging trichomes and nonglandular nonstinging trichomes. Leaves with the blade ovate, 6–30 cm long, 3–18 cm wide, the apex acuminate, the base truncate, rounded, or broadly cuneate, not auriculate, the margin serrate, the upper and lower surfaces with appressed whitish nonstinging trichomes and stinging trichomes, the petiole 6–30 cm long, with stinging trichomes; stipules connate, bifid, caducous. Flowers in an axillary or terminal panicle, the staminate and carpellate panicles separate, the staminate lower on the plant than the carpellate. Staminate flowers 1–1.5 mm long; sepals saccate, with longitudinal ridges, greenish. Carpellate flowers to 0.5 mm long, borne on a thick pedicel that develops lateral wings as the fruit matures; sepals with the inner lateral pair as long as the ovary, the inner 2 lateral, larger than the outer 2, the lower 1 becoming hoodlike; style straight. Fruit suborbicular, 2–3 mm long, strongly compressed, stipitate, asymmetrical, partially covered by the enlarged lateral sepals.

Calcareous hammocks and floodplain forests. Rare; Liberty and Jackson Counties. Quebec south to Florida, west to Saskatchewan, North Dakota, South Dakota, Nebraska, Kansas, Oklahoma, and Louisiana; Mexico. Summer–fall.

Parietaria L. 1753. PELLITORY

Annual or perennial herbs, with nonstinging trichomes. Leaves alternate, pinnate- or pinni-palmate-veined, the margins entire, cystoliths rounded, petiolate, estipulate. Flowers in axillary, few-flowered cymes, 1 cymule on each side of the petiole base, subtended by 1–3 bracts, the lower inflorescences usually bisexual and staminate, the upper ones carpellate; sepals 4, connate at the base; petals absent; stamens 4; ovary short stipitate or estipitate, the stigma penicillate. Fruit an achene, enclosed by the persistent sepals.

A genus of 30 species; nearly cosmopolitan. [*Paries*, wall, the habitat of the original species.]

1. Leaves ovate, the lowest pair of lateral veins arising at or very near the leaf base.
    2. Achene usually less than 1 mm long, with a flanged stipe .... **P. floridana**
    2. Achene usually more than 1 mm, lacking a flanged stipe .... **P. praetermissa**
1. Leaves ovate-lanceolate to rhombic-lanceolate, the lowest pair of lateral veins arising above the leaf base.
    3. Mature fruit light to dark brown, the apical mucro absent or minute...................... **P. pensylvanica**
    3. Mature fruit black, the apical mucro distinct..**P. judaica**

Parietaria floridana Nutt. [Of Florida.] FLORIDA PELLITORY.

Parietaria floridana Nuttall, Gen. N. Amer. Pl. 2: 208. 1818. *Parietaria debilis* G. Forster var. *floridana* (Nuttall) Weddell, Arch. Mus. Hist. Nat. 9: 516. 1857. *Parietaria pensylvanica* Muhlenberg ex Willdenow var. *floridana* (Nuttall) Weddell, in de Candolle, Prodr. 16(1): 235. 1869. TYPE: FLORIDA [?]: "near St. Mary's [Georgia], West [?] Florida—Dr. Baldwyn," s.d., *Baldwin s.n.* (holotype: PH). *Parietaria nummularia* Small, Man. S.E. Fl. 434, 1504. 1933. TYPE: FLORIDA: Seminole Co.: Sanford, 4 Nov 1929, *Rapp s.n.* (holotype: NY).

Ascending or spreading annual herb, to 40 cm; stem simple or much-branched near the base, finely to densely pubescent with short, hooked trichomes. Leaves with the blade ovate to subrhombic, 1–3 cm long, 0.5–1.7 cm wide, the apex obtuse to short-acuminate, the base broadly cuneate to rounded, the margin entire, short-ciliate, the lowest pair of veins arising at or very near the base, the upper and lower surfaces glabrous or pubescent, the petiole 1–3 cm long; bracts linear to linear-lanceolate, 1–2 mm long, subequaling the calyx, obtuse to acute, covered with long, straight and short, hooked trichomes; calyx lobes ovate to linear-oblong, 1–2 mm long, acute. Fruit ovoid, to 1 mm long, light brown, the apex symmetrical, with a short mucro, the stipe with a flange.

Hammocks and open disturbed sites. Frequent; nearly throughout. Delaware south to Florida, west to Texas; West Indies, Mexico, and South America. Summer.

Parietaria judaica L. [Jewish, in reference to the country of the Jews, Judea or Palestine.] SPREADING PELLITORY.

Parietaria judaica Linnaeus, Fl. Palaest. 32. 1756. *Parietaria ramiflora* Moench, Methodus 327. 1794, nom. illegit. *Parietaria officinalis* Linnaeus var. *ramiflora* Ascherson, Fl. Brandenburg 1: 610. 1864. *Parietaria judaica* Linnaeus var. *typica* Halacsy, Consp. Fl. Graec. 3: 119. 1904, nom. inadmiss. *Parietaria officinalis* Linnaeus subsp. *judaica* (Linnaeus) Beguinot, Nuovo Giorn. Bot. Ital., ser. 2. 15: 342. 1908.

Procumbent or ascending perennial herb, to 80 cm; stem much-branched, pubescent. Leaves with the blade ovate-lanceolate, 1–9 cm long, 1–4.5 cm wide, the apex acuminate, the base cuneate or rounded, the margin entire, the lowest pair of lateral veins arising above the base, the upper and lower surfaces pubescent, the petiole shorter than the blade; bracts 1–3 mm long, pubescent; sepals 2–4 cm long, pubescent. Fruit ovate, ca. 1 mm long, and wide, black or dark brown, symmetrical, the mucro absent or minute, lacking a flanged stipe.

Dry, open, disturbed sites. Rare; Escambia County. Not collected in this century. Scattered localities throughout the United States; Europe, Africa, and Asia. Native to Europe and Asia. Summer.

Parietaria pensylvanica Muhl. ex Willd. [Of Pennsylvania.] PENNSYLVANIA PELLITORY.

Parietaria pensylvanica Muhlenberg ex Willdenow, Sp. Pl. 4: 955. 1805. *Parietaria debilis* Forster f. var. *pensylvanica* (Muhlenberg ex Willdenow) Weddell, Arch. Mus. Hist. Nat. 9: 516. 1857. *Helxine pensylvanica* (Muhlenberg ex Willdenow) Nieuwland, Amer. Midl. Naturalist 3: 235. 1914.

Parietaria obtusa Rydberg ex Small, Fl. S.E. U.S. 359, 1329. 1903. *Parietaria pensylvanica* Muhlenberg ex Willdenow var. *obtusa* (Rydberg ex Small) Shinners, Field & Lab. 18: 42. 1950.

Decumbent, ascending, or erect annual herb, to 60 cm; stem usually simple, sometimes with a few branches from near the base, sparsely to densely pubescent, mostly with short, hooked trichomes. Leaves with the blade lanceolate to rhombic-lanceolate or ovate, 2–9 cm long, 0.5–3 cm wide, the apex acuminate or obtuse to rounded, the base cuneate, the lowest pair of lateral veins arising above the base, the surface minutely scabrous, the petiole shorter than the blade. Bracts linear, obtuse, 4–5 mm long, much exceeding the flowers, pubescent with long, straight trichomes; calyx lobes linear-oblong, 1–2 mm long, pubescent. Achene ovoid, ca. 1 mm long, symmetrical, light brown, slightly apiculate, lacking a flanged stipe.

Wet hammocks and disturbed sites. Rare; central and western panhandle. Nearly throughout North America; Mexico. Spring.

Parietaria praetermissa Hinton [An omission, in reference to the recent description of the taxon.] CLUSTERED PELLITORY.

Parietaria praetermissa Hinton, Sida 3: 192. 1968. TYPE: FLORIDA: Brevard Co.: Merritt Island, Feb 1889, *Canby s.n.* (holotype: GH).

Decumbent or ascending annual herb, to 50 cm; stem simple or branched at or near the base, finely to densely pubescent with short, hooked trichomes. Leaves with the blade ovate, deltoid, or broadly rhombic, 2–6 cm long and wide, the apex obtuse to attenuate, the base broadly cuneate to truncate, the margin short-ciliate, the lowest pair of lateral veins arising at or very near the base, the surfaces glabrous or pubescent, the petiole 0.5–4 cm long. Bracts lanceolate to linear, 1–5 mm long, with long, straight and short, hooked trichomes; calyx lobes oblong, 2–3 mm long. Achenes oval, 1.2–1.4 mm long, light brown, symmetrical, the apex obtuse, the mucro subapical, lacking a flanged stipe.

Moist hammocks and floodplain forests. Occasional; peninsula, central panhandle. North Carolina south to Florida, west to Alabama, also Louisiana. Spring.

EXCLUDED TAXA

Parietaria debilis G. Forster—Reported for Florida by Chapman (1860, 1883, 1897), misapplied to material of *P. praetermissa*.

Parietaria officinalis Linnaeus—Reported by Small (1933), based on a misidentification of material of *Pouzolzia zeylanica* (fide Miller, 1971).

Pilea Lindl., nom. cons. 1821. CLEARWEED

Annual or perennial herbs, without stinging trichomes. Leaves opposite, isophyllous (equal) or anisophyllus (unequal), pinnate- or pinnipalmate-veined, cystoliths linear, the margins entire or toothed, petiolate, stipulate. Flowers in axillary cymes, unisexual (plants monoecious), the staminate and carpellate borne at the same node but usually on separate branches, bracteate. Staminate flowers pedicellate, actinomorphic; sepals 4, basally connate; petals absent; stamens 4, free, opposite the sepals; rudimentary ovary present. Carpellate flowers zygomorphic or actinomorphic; sepals 3, equal or 1 strongly hooded; petals absent; staminodes 3; stigma penicillate. Fruit an achene, sessile, laterally compressed, partly covered by a hoodlike sepal, ejected by the staminodes at maturity.

A genus of about 600 species; nearly cosmopolitan. [*Pileus*, a felt cap of the Romans, in reference to the hooded sepal covering the achene.]

Adicea Raf. ex Britton & A. Br. 1896

1. Leaves less than 2 cm long, the petiole less than 1 cm long, the margins entire.
    2. Leaf pairs equal or only slightly unequal, clustered at the ends of the branches, the blades suborbicular to reniform-orbicular.. **P. herniarioides**
    2. Leaf pairs distinctly unequal, not clustered at the ends of the branches, the blades oblanceolate to elliptic...**P. microphylla**
1. Leaves more than 2 cm long, the petiole more than 1 cm long, the margins crenate, dentate, or serrate.
    3. Leaf blades broadly ovate to suborbicular .... **P. nummulariifolia**
    3. Leaf blades ovate to elliptic.
        4. Achene light brown, often with slightly raised darker lines or mottling, longer than broad........
        ....**P. pumila**
        4. Achene dark brown to black with paler margins, slightly roughened but without lines or mottling, as long as broad ....**P. fontana**

Pilea fontana (Lunell) Rydb. [Of springs.] LESSER CLEARWEED.

Adicea fontana Lunell, Amer. Midl. Naturalist 3: 7. 1913. *Pilea fontana* (Lunell) Rydberg, Brittonia 1: 87. 1913. *Pilea pumila* (Linnaeus) A. Gray forma *fontana* (Lunell) B. Boivin, Naturaliste Canad. 93: 434. 1966.

Erect or ascending annual herb, to 40 cm; stem simple or slightly branched, pellucid-dotted, appressed-pubescent to glabrate. Leaves 3-veined from the base, with the blade of the lower ones suborbicular, 3–5 mm long and wide, the apex obtuse, the base rounded to subtruncate, the margin entire, the upper and lower surfaces short-pilose to glabrate, pellucid-dotted, the petiole subequaling the blade, the upper leaves with the blade ovate or elliptic, 1–6 cm long, the apex short-acuminate, the base cuneate or rounded, the margin dentate, the surfaces as in the lower leaves, the petiole ⅕–½ the length of the blade. Flowers in a crowded or lax cyme 0.5–5 cm long, shorter than the petiole, the staminate flowers usually innermost when mixed with the carpellate; calyx ca. 1 mm long, the lobes 3, linear. Fruit broadly ovate, 1–2 mm long, nearly as wide as long, dark brown to black with paler margins, slightly pebbled or warty, but without raised lines or mottling, the persistent calyx shorter than to slightly exceeding the achene.

Moist hammocks. Rare; Wakulla and Alachua Counties. Vermont and New Hampshire south to Florida, west to North Dakota, South Dakota, and Nebraska. Summer–fall.

Pilea herniarioides (Sw.) Lindl. [Resembling *Herniaria* (Caryophyllaceae).] CARIBBEAN CLEARWEED.

Urtica herniarioides Swartz, Kongl. Vetensk. Acad. Nya Handl. 8: 64. 1787. *Pilea herniarioides* (Swartz) Lindley, Coll. Bot. sub t. 4, index. 1826 ("1821"). *Pilea muscosa* Lindley var. *herniarioides* (Swartz) Weddell, Arch. Mus. Hist. Nat. 9: 174. 1856. *Pilea microphylla* (Linnaeus) Liebmann var. *herniarioides* (Swartz) Weddell, in de Candolle, Prodr. 16(1): 106. 1869. *Adicea microphylla* (Linnaeus) Kuntze var. *herniarioides* (Swartz) Kuntze, Revis. Gen. Pl. 2: 622. 1891. *Adicea herniarioides* (Swartz) Small, Fl. S.E. U.S. 358, 1329. 1903.

Prostrate annual or short-lived perennial herb, to 10 cm; stem few- to many-branched, rooting at the nodes, pellucid-dotted, glabrous or sparsely strigose. Leaves usually clustered at the ends of branches, 1-nerved, the pairs equal or only slightly unequal, the blade suborbicular to reniform-orbicular, 1–8 mm long and wide, the apex obtuse to rounded, the base truncate to somewhat decurrent, the margin entire, the upper surface sparsely strigose to glabrous, the crystoliths prominent, with transverse orientation, the lower surface glabrous, the petiole 0.5–2(5) mm long. Flowers in a crowded cyme. Staminate flowers with the calyx turbinate, ca. 1 mm long, purplish, the pedicel to 2 mm long. Carpellate flowers with the calyx scalelike, ca. 0.5 mm long, green, sessile. Fruit elliptic-ovoid, ca. 0.5 mm long, slightly compressed, light brown, pebbled, the margins hyaline.

Moist, disturbed sites. Rare; Miami-Dade and Monroe Counties. Florida; West Indies, Mexico, Central America, and South America. All year.

Pilea microphylla (L.) Liebm. [Small-leaved.] ARTILLERY PLANT; ROCKWEED.

Parietaria microphylla Linnaeus, Syst. Nat., ed. 10. 1308. 1759. *Urtica microphylla* (Linnaeus) Swartz, Kongl. Vetensk. Acad. Nya Handl. 8:66. 1787. *Pilea muscosa* Lindley, Coll. Bot. t. 4. 1821, nom. illegit. *Pilea microphylla* (Linnaeus) Liebmann, Kongel. Danske Vidensk. Selsk. Skr., Naturvidensk. Math. Afd., ser. 5. 2: 296. 1851. *Pilea muscosa* Lindley var. *microphylla* (Linnaeus) Weddell, Arch. Mus. Hist. Nat. 9: 174. 1856, nom. inadmiss. *Pilea trianthemoides* (Swartz) Lindley var. *microphylla* (Linnaeus) Weddell, in de Candolle, Prodr. 16(1): 107. 1869. *Adicea microphylla* (Linnaeus) Kuntze, Revis. Gen. Pl. 2: 622. 1891.

Erect or sprawling annual or short-lived perennial herb, to 20 cm; stem much-branched, succulent or occasionally somewhat woody at the base, glabrous. Leaves often developing from axillary buds so that 4–6 leaves are often clustered at a node, the pairs distinctly unequal, 1-nerved, the larger leaves with the blade oblanceolate, 2–12 mm long, 1–6 mm wide, the apex obtuse to rounded, the base cuneate, the margin entire, the upper surface with conspicuous transverse cystoliths, the petiole 1–4 mm long, the smaller leaves with the blade elliptic to oblanceolate to suborbicular, 1–3 mm long, 1–2.5 mm wide, the petiole 0.5–1 mm long, otherwise similar to larger ones. Flowers in a crowded cyme, the cymes usually paired in the axils, sessile or the peduncle to 1.5 mm long. Staminate flowers with the calyx ca. 1 mm long, white tinged with red, the pedicel ca. 0.5 mm long. Carpellate flowers with the calyx 0.5–1 mm long, white tinged with red, sessile. Fruit ovoid, ca. 0.5–1 mm long, slightly compressed, light brown, smooth or slightly pebbled.

Moist, shaded, often disturbed sites. Frequent; peninsula, central and western panhandle. South Carolina, Georgia, Florida, and Louisiana; West Indies, Mexico, Central America, and South America; Europe, Africa, Asia, Australia, and Pacific Islands. Native to North America and tropical America. All year.

Pilea nummulariifolia (Sw.) Wedd. CREEPING CHARLIE.

> *Urtica nummulariifolia* Swartz, Kongl. Vetensk. Nya Handl. 8: 63, t. 1(2). 1787. *Pilea nummulariifolia* (Swartz) Weddell, Ann. Sci. Nat., Bot., ser. 3. 18: 225. 1852. *Adicea nummulariifolia* (Swartz) Kuntze, Revis. Gen. Pl. 2: 623. 1891.

Prostrate annual or perennial herb; stem rooting at the nodes, with tangled, clear, unicellular trichomes. Leaves a subequal pair, the blade suborbicular to ovate, 9–24 mm long and wide, the apex rounded, the base rounded to cordate, the margin crenate, dentate, or serrate, the upper and lower surfaces hispid with clear, unicellular trichomes, the cystoliths irregularly arranged, the petiole 4–15 mm long, pilose. Staminate and carpellate flowers usually in separate inflorescences, the staminate cymes dense, sessile, ca. ½ as long as the subtending leaves, the axil tomentose, the carpellate cyme dense, on a peduncle 2–10 mm long and equaling or shorter than the subtending leaves, the axil glabrous. Staminate flowers with the calyx 1.5–3 mm long, pilose, the pedicel ca. 1.5 mm long. Carpellate flowers with the calyx ca. 0.5 mm long, sessile or on a pedicel to 0.5 mm long. Fruit ovate, slightly compressed, ca. 0.5 mm long.

Moist disturbed hammocks. Rare; Palm Beach County. Florida; West Indies and South America; Asia. Native to West Indies and South America. All year.

Pilea pumila (L.) A. Gray [Dwarf.] CANADIAN CLEARWEED.

> *Urtica pumila* Linnaeus, Sp. Pl. 2: 984. 1753. *Adicea pumila* (Linnaeus) Rafinesque, First Cat. Gard. Transylv. Univ. 13. 1824. *Adike pumila* (Linnaeus) Rafinesque, New Fl. 1: 63. 1836. *Pilea pumila* (Linnaeus) A. Gray, Manual 437. 1848.
>
> *Adicea deamii* Lunell, Amer. Midl. Naturalist 3: 10. 1913. *Pilea pumila* (Linnaeus) A. Gray var. *deamii* (Lunell) Fernald, Rhodora 38: 169. 1936.

Erect or decumbent annual herb, to 70 cm; stem simple or rarely few-branched from the base, succulent and nearly translucent. Leaves an equal pair, the blade ovate to broadly elliptic, 2–13 cm long, 1–9 cm wide, 3-veined, the apex acuminate, the base broadly cuneate to rounded, the margin crenate-serrate or dentate, the petiole to 8 cm long, as long as the blade. Flowers in a crowded or lax cyme, the staminate and carpellate mixed. Staminate flowers with the calyx united in bud but separating into 4 sepals at anthesis; stamens 4. Carpellate flowers with 3 somewhat unequal sepals, each subtending a scalelike staminode. Fruit ovate, 1–2 mm long, slightly compressed, light brown or green, with slightly raised purple lines or mottling, subtended by the persistent calyx.

Wet hammocks and floodplain forests. Occasional; central and western panhandle, Putnam, Flagler, Volusia, and Orange Counties. Prince Edward Island south to Florida, west to North Dakota, South Dakota, Kansas, Oklahoma, and Texas. Summer–fall.

EXCLUDED TAXA

Pilea serpyllifolia (Poiret) Weddell—Reported from Florida by Small (1933). Apparently based on a misidentification of material of *P. microphylla*.

Pilea tenerrima Miquel—Reported for Florida by Britton & Wilson (1924) and Liogier (1985). No specimens seen.

Pouzolzia Gaudich. 1830. POUZOLZ'S BUSH

Perennial herbs or subshrubs, with nonstinging trichomes. Leaves opposite or alternate, pinnipalmate-veined, the margin entire, the cystoliths rounded, petiolate, stipulate. Flowers in axillary cymose glomerules, bracteate, unisexual (plants monoecious), the staminate and carpellate in the same inflorescence, actinomorphic. Staminate flowers with the sepals 4, connate; petals absent; stamens 4. Carpellate flowers with the sepals 4, connate; petals absent; staminodes absent; stigma filiform. Fruit an achene, enclosed by the persistent calyx.

A genus of about 35 species; North America, West Indies, Mexico, Central America, South America, Africa, Asia, Australia, and Pacific Islands. [Commemorates Pierre Marie Casimir de Pouzolz (1785–1858), French botanist.]

Selected reference: Wilmot-Dear and Friis (2006).

Pouzolzia zeylanica (L.) Bennett [Of Ceylon (Sri Lanka).] POUZOLZ'S BUSH.

Parietaria zeylanica L., Sp. Pl. 2: 1052. 1753. *Urtica alienata* Linnaeus, Syst. Nat., ed. 12. 2: 622. 1767, nom. illegit. *Boehmeria alienata* Willdenow, Sp. Pl. 4: 341. 1805, nom. illegit. *Pouzolzia zeylanica* (L.) Bennett, in Bennett et al., Pl. Jav. Rar. 67. 1838. *Pouzolzia indica* (Linnaeus) Wight var. *alienata* Weddell, Arch. Mus. Hist. Nat. 9: 399. 1856. *Pouzolzia zeylanica* (Linnaeus) Bennett var. *alienata* (Weddell) Sasaki, List Pl. Form. 163. 1928.

Ascending, suffrutescent perennial herb, to 0.5 m; stems much-branched, pilose and/or strigose. Leaves with the blade chartaceous, ovate, elliptic, or lanceolate, 2.5–4 cm long, 0.5–2 cm wide, the apex acute to subobtuse, the base rounded to subcuneate, 3-veined, with the lateral veins reaching ⅔–¾ the length of the blade, the surfaces pilose and/or strigose, the cystoliths evident, white, punctiform-callose, the petiole 3–15 mm long. Flowers in an axillary, cymose glomerule, subsessile, the staminate and carpellate in same inflorescence; bracteate. Staminate flowers with the calyx ca. 3 mm long, the lobes ovate, coarsely pilose. Carpellate flowers 1–2 mm long; sepals ribbed. Fruit ovoid, ca. 1 mm long, the apex acute, slightly compressed, black, lustrous, smooth, the persistent calyx ca. 1 mm long, several-ribbed, coarsely pilose.

Disturbed sites. Occasional; peninsula. Florida; Africa, Asia, Australia, and Pacific Islands. Native to Asia. Summer.

Rousselia Gaudich. 1830.

Annual or perennial herbs, stem without stinging trichomes. Leaves alternate, the margin entire, petiolate, stipulate. Flowers unisexual (plants monoecious). Staminate flowers in axillary racemes; sepals 4, free; petals absent; stamens 4, free. Carpellate flowers paired, sessile, bracteate; sepals 4, free; petals absent; stigma filiform. Fruit an achene, surrounded by the persistent bracts.

A genus of 2 species; North America, West Indies, Mexico, Central America, and South America. [Commemorates Alexandre Victor Roussel (1795–1874), French physician and botanist.]

Rousselia humilis (Sw.) Urb. [Low-growing.]

Urtica humilis Swartz, Kongl. Vetensk. Acad. Nya Handl. 6: 34. 1785. *Rousselia humilis* (Swartz) Urban, Symb. Antill. 4: 205. 1905.

Prostrate or ascending annual or perennial herb; stem to 4 dm long, appressed-pubescent. Leaves with the blade membranous, broadly ovate to elliptic, 0.5–4 cm long, 3-veined, the apex acute to acuminate or obtuse, the base rounded to broadly cuneate, the margin entire, the upper and lower surfaces sparsely pilose, the petiole 2–7 mm long; stipules minute. Staminate flowers in a short, axillary raceme, short-pedicellate, sepals 4, free; petals absent; stamens 4, free, ovary rudimentary. Carpellate flowers paired, sessile; sepals 4, free; petals absent; stigma filiform, plumose on 1 side; bracts 2, ovate, 3–4 mm long, the margin entire, ciliate, foliose. Fruit ovate, 1–2 mm long, laterally compressed, lustrous, surrounded by the persistent bracts.

Wet, disturbed sites. Rare; Broward County. Florida; West Indies, Mexico, and Central America. Native to West Indies, Mexico, and Central America. Summer–fall.

Urtica L. 1753. NETTLE

Annual or perennial herbs with stinging and nonstinging trichomes. Leaves opposite, the blades 3- to 7-veined from near the base, the margin crenate or serrate, the cystoliths rounded or elongate, petiolate, stipulate. Flowers in axillary, paniculiform or spiciform, bracteate, cymes, unisexual, the staminate and carpellate intermixed in the same inflorescence (plants monoecious) or in different inflorescences on different plants (plants dioecious). Staminate flowers pedicillate; sepals 4, equal; petals absent; stamens 4; rudimentary ovary present. Carpellate flowers subsessile; sepals 4, the inner 2 longer than the outer 2; petals absent; stigma penicillate, sessile. Fruit an achene, compressed, loosely enclosed by the enlarged 2 inner sepals.

A genus of 50 species; North America, Mexico, Central America, South America, Europe, Africa, Asia, Australia, and Pacific Islands. [*Uro*, to burn, in reference to the stinging trichomes.]

Selected reference: Woodland (1982).

1. Inflorescences usually longer than the petiole ....**U. dioica**
1. Inflorescences shorter than the petiole.
    2. Inflorescences subglobose to short-spicate, either sessile or short-pedunculate; leaves reduced upward, the blade crenate-dentate; achene ovate, 1–1.5 mm long ....**U. chamaedryoides**
    2. Inflorescences elongate and lax, the flowers in clusters along the floral axis; leaves little reduced upward, the blade sharply incised; achene 1.5–2.5 mm long ....**U. urens**

Urtica chamaedryoides Pursh [Resembling *Chamaedrys* (=*Teucrium*, Lamiaceae).] HEARTLEAF NETTLE.

Urtica chamaedryoides Pursh, Fl. Amer. Sept. 1: 113. 1814.

Erect or sometimes sprawling, annual herb, to 1 m; stem, simple or branched at base, sparsely armed with stinging trichomes, otherwise glabrous. Leaves with the blade broadly ovate to subrotund or lanceolate, 1–6 cm long, 1–4 cm wide, the apex acute to obtuse, the base cordate to truncate, sometimes cuneate in the distal ones, the margin crenate-serrate, the upper and lower surfaces with stinging trichomes, cystoliths elongate, the petiole subequaling or shorter than the blade; stipules linear-lanceolate, 1–4 mm long. Flowers solitary or 2–3 in an axillary, globose to short-spicate inflorescence 3–6 mm long, shorter than the subtending petiole, the staminate and carpellate in the same inflorescence. Carpellate flowers with outer pair of sepals linear, ca. 0.5 mm long, the inner pair ovate, 1–1.5 mm long. Fruit ovate-elliptic, 1–1.5 mm long, tan to brown.

Mesic hammocks. Occasional; eastern and central panhandle, northern and central peninsula. New York and Ohio south to Florida, west to Kansas, Oklahoma, and Texas; Mexico. Spring–summer.

Urtica dioica L. [Plants dioecious.] STINGING NETTLE.

Urtica dioica Linnaeus, Sp. Pl. 2: 984. 1753. *Urtica major* Kanitz, Bot. Zeitung (Berlin) 21: 54. 1863, nom. illegit. *Urtica major* Kanitz var. *vulgaris* Kanitz, Bot. zeitung (Berlin) 21: 54. 1863, nom. inadmiss. *Urtica dioica* Linnaeus var. *vulgaris* Weddell, in de Candolle, Prodr. 16(1): 50. 1869, nom. inadmiss. *Urtica dioica* Linnaeus subsp. *eudioica* Selander, Svensk Bot. Tidskr. 41: 271. 1947, nom. inadmiss.

Erect or sprawling, perennial herb to 2 m; stem simple or branched, rhizomatous, with stinging trichomes, otherwise glabrous or sparsely puberulent. Leaves with the blade ovate to ovate-lanceolate, 5–15 cm long, 2–8 cm wide, the apex acuminate, the base rounded to subcordate or cuneate, the margin serrate, the upper and lower surfaces with appressed stinging trichomes, the lower surface also hispid with nonstinging trichomes, the petiole 1–6 cm long, shorter than blade; stipules linear-lanceolate, 5–15 mm long. Flowers in a pedunculate cymose panicle usually longer than the subtending petioles, the staminate and carpellate on different plants (plants dioecious), the staminate inflorescence ascending, the carpellate lax or recurved. Carpellate flowers with the outer pair of sepals linear to narrowly spatulate, ca. 1 mm long, the inner pair ovate, 1.5–2 mm long. Fruit ovoid, 1–1.5 mm long, compressed, tan.

Disturbed hammocks. Rare; Alachua County. Nearly throughout North America; Mexico and South America; Europe, Africa, Asia, and Pacific Islands. Native to Europe, Africa, and Asia. Summer.

Urtica urens L. [Burning or stinging.] BURNING NETTLE; DWARF NETTLE.

Urtica urens Linnaeus, Sp. Pl. 2: 984. 1753. *Urtica minor* Lamarck, Fl. Franc. 2: 194. 1779, nom. illegit.

Erect annual herb, to 0.5 m; stem simple or branched at the base, with stinging trichomes. Leaves with the blade elliptic, 2–9 cm long, 2–5 cm wide, the apex acute, the base rounded to broadly cuneate, the margin deeply incised-serrate, sometimes with lateral lobes, the upper and lower surfaces glabrate, the cystoliths rounded, the petiole subequaling the blade; stipules oblong, 1–3 mm long. Flowers in a sessile or short-pedunculate spiciform or paniculiform inflorescence, these usually much shorter than the petioles, the staminate and carpellate in the same inflorescence, white or pale yellow. Carpellate flowers with the outer pair of sepals ovate, ca. 0.5 mm long, the inner sepals broadly ovate, 0.5–1 mm long. Fruit ovoid, 1.5–2 mm long, compressed, tan.

Disturbed sites. Rare; central panhandle, northern and central peninsula. Scattered localities nearly throughout North America; Europe, Africa, Asia, Australia, and Pacific Islands. Native to Europe. Summer.

Literature Cited

Anderson, L. C. 1988. Noteworthy plants from north Florida. III. Sida 13: 93–100.

Barker, W. T. 1997. *Planera*. *In*: Flora of North America Editorial Committee. Flora of North America North of Mexico. 3: 376. New York/Oxford: Oxford University Press.

Barneby, R. C. 1964. Atlas of North American *Astragalus*. Mem. New York Bot. Gard. 13(1–2): 1–1188.

———. 1977. Daleae imagines. Mem. New York Bot. Gard. 27: 1–891.

———. 1991. Sensitivae censitae: A description of the genus *Mimosa* Linnaeus (Mimosaceae) in the New World. Mem. New York Bot. Gard. 65: 1–835.

———. 1998. Silk tree, guanacaste, monkey's earring: A generic system for the synandrous Mimosaceae of the Americas. Part 3. *Calliandra*. Mem. New York Bot. Gard. 74(3): 1–223.

Barneby, R. C., and J. W. Grimes. 1996. Silk tree, guanacaste, monkey's earring: A generic system for the synandrous Mimosaceae of the Americas. Part 1. *Abarema*, *Albizia*, and allies. Mem. New York Bot. Gard. 74(1): 1–292.

———. 1997. 1996. Silk tree, guanacaste, monkey's earring: A generic system for the synandrous Mimosaceae of the Americas. Part 2. *Pithecellobium*, *Cojoba*, and *Zygia*. Mem. New York Bot. Gard. 74(2): 1–161.

Boufford, D. E. 1997. Urticaceae. *In*: Flora of North America Editorial Committee. Flora of North America North of Mexico. 3: 400–413. New York/Oxford: Oxford University Press.

Brenan, J.P.M., and R. K. Brummitt. 1965. The variation of *Dichrostachys cinerea* (L.) Wight & Arn. Bol. Soc. Brot., ser. 2. 39: 61–115.

Breteler, F. J. 1960. Revision of *Abrus* Adanson (Pap.) with special reference to Africa. Blumea 10: 607–24.

Britton, N. L. 1890. New or noteworthy North American phanerogams—III. Bull. Torrey Bot. Club 17: 310–16.

Britton N. L., and J. N. Rose. 1928. Mimosaceae. *In*: N. L. Britton. North American Flora. 23: 1–194. New York: New York Botanical Garden.

Britton, N. L., and P. Wilson. 1924. Urticaceae. Scientific Survey of Porto Rico and the Virgin Islands: Botany of Porto Rico and the Virgin Islands. 5: 241–52. New York: New York Academy of Sciences.

Brizicky, G. K. 1964. The genera of Rhamnaceae in the southeastern United states. J. Arnold Arbor. 45: 439–63.

———. 1965. The genera of Vitaceae in the southeastern United States. J. Arnold Arbor. 46: 48–67.

Chapman, A. W. 1860. Flora of the Southern United States. New York: Ivison, Phinney & Co.

———. 1883. Flora of the Southern United States. 2nd ed. New York: Ivison, Blakeman, Taylor & Co.

———. 1897. Flora of the Southern United States. 3rd ed. New York: American Book Co.

Clewell, A. F. 1966. Native North American species of *Lespedeza* (Leguminosae). Rhodora 68: 339–405.

Clewell, A. F. 1985. Guide to the Vascular Plants of the Florida Panhandle. Tallahassee: University Presses of Florida/Florida State University Press.

Correll, D. S., and M. C. Johnston. 1970. Manual of the Vascular Plants of Texas. Renner: Texas Research Foundation.

Davis, P. H. 1972. Flora of Turkey and the East Aegean Islands. Volume 4. Edinburgh: Edinburgh University Press.

DeLaney, K. R. 2010. *Tephrosia mysteriosa* (Fabaceae: Millettieae), a new species from central Florida. Bot. Explor. 4: 99–140.

Delgado-Salinas, A., M. Thulin, R. Pasquet, N. Weeden, and M. Lavin. 2011. *Vigna* (Leguminosae) sensu lato: The names and identities of the American segregate genera. Amer. J. Bot. 98: 1694–1715.

Duncan, W. H. 1977. A new species of *Galactia* (Fabaceae) in the southeastern United States. Phytologia 37: 59–61.

Dunn, D. B. 1971. A case of long range dispersal and "rapid speciation" in *Lupinus*. Trans. Missouri Acad. Sci. 5: 26–38.

Elias, T. S. 1970. The genera of Ulmaceae in the southeastern United States. J. Arnold Arbor. 51: 18–40.

———. 1974. The genera of Mimosoideae (Leguminosae) in the southeastern United States. J. Arnold Arbor. 55: 67–118.

Fantz, P. R. 1977. A monograph of the Genus *Clitoria* (Leguminosae-Glycineae). PhD dissertation, University of Florida.

Fantz, P. R., and S. V. Predeep. 1992. Comments on four legumes (*Clitoria, Centrosema*) reported as occurring in India. Sida 15: 1–7.

Fernald, M. L. 1941. Another century of additions to the flora of Virginia (continued). Rhodora 43: 559–630.

Florida Exotic Pest Plant Council (FLEPPC). 2015. Florida Exotic Pest Plant Council's 2013 List of Invasive Plant Species. www.fleppc.org.

Freytag, G. F., and D. G. Debouck. 2002. Taxonomy, distribution and ecology of the genus *Phaseolus* (Leguminosae-papilionoideae) in North America, Mexico, and Central America. Sida 23: 1–300.

Gillis, W. T. 1975. Bahama Polygalaceae and their Greater Antillean affinities—a preliminary treatment. Phytologia 32: 35–44.

Godfrey, R. K. 1988. Trees, Shrubs, and Woody Vines of Northern Florida and Adjacent Georgia and Alabama. Athens: University of Georgia Press.

Godfrey R. K., and J. W. Wooten. 1981. Aquatic and Wetland Plants of Southeastern United States: Dicotyledons. Athens: University of Georgia Press.

Grear, J. W. 1978. A revision of the New World species of *Rhynchosia* (Leguminosae-Faboideae). Mem. New York Bot. Gard. 31: 1–168.

Grimes, J. W. 1997. A revision of *Cullen* (Leguminosae: Papilionoideae). Austral. Syst. Bot. 10: 565–648.

Gunn, C. R., E. M. Norman, and J. S. Lassiter. 1980. *Chapmannia floridana* Torrey & Gray (Fabaceae). Brittonia 32: 178–85.

Hadiah, J. T., B. J. Conn, C. J. Quinn. 2008. Infra-familial phylogeny of Urticaceae, using chloroplast sequence data. Austral. Syst. Bot. 21: 375–85.

Hermann, F. J. 1953. A Botanical Synopsis of the Cultivated Clovers (*Trifolium*), Agriculture Monograph No. 22. Washington, D.C.: United States Department of Agriculture.

———. 1960. *Vicia*: Vetches of the United States—Native, Naturalized, and Cultivated, Agriculture Handbook No. 168. Washington, D.C.: United States Department of Agriculture.

Hitchcock, C. L. 1952. A revision of the North American species of *Lathyrus*. Univ. Washington Publ. Bot. 15: 1–104.

Hughes, C. E. 1998. A monograph of *Leucaena* (Leguminosae; Mimosoideae). Syst. Bot. Monogr. 55: 1–224.

Irwin, H. S., and R. C. Barneby. 1982. The American Cassiinae: A synoptic revision of Leguminosae tribe Cassieae subtribe Cassiinae in the New World. Mem. New York Bot. Gard. 35(1–2): 1–918.

Isely, D. 1955. The Leguminosae of the north-central United States. II. Hedysareae. Iowa State Coll. J. Sci. 30: 33–118.

———. 1981. Leguminosae of the United States. III. Subfamily Papalionoideae: Tribes Sophoreae, Podalyrieae, Loteae. Mem. New York Bot. Gard. 25(3)1–264.

———. 1982. New combinations and one new variety among the genera *Indigofera*, *Robinia*, and *Tephrosia* (Leguminosae. Brittonia 34: 339–41.

———. 1990. *In*: Massey et al. Vascular Flora of the Southeastern United states. 3(2): 1–258. Chapel Hill/London: University of North Carolina Press.

———. 1998. Native and Naturalized Leguminosae (Fabaceae) of the United States (Exclusive of Alaska and Hawaii). Monte L. Bean Life Science Museum, Provo: Brigham Young University.

Isely, D., and F. J. Peabody. 1984. *Robinia* (Leguminosae: Papilionoideae). Castanea 49: 187–202.

Johnston, M. C. 1971. Revision of *Colubrina* (Rhamnaceae). Brittonia 23: 2–53.

Kline, G. J., and P. D. Sørensen. 2008. A revision of *Agrimonia* (Rosaceae) in North and Central America. Brittonia 60: 11–33.

Kort, de, I., and G. Thijsse. 1984. A revision of the genus *Indigofera* (Leguminosae-Papilionoideae) in southeastern Asia. Blumea 30: 89–151.

Krapovickas, A., and W. C. Gregory. 1994. Taxonomia de genero *Arachis* (Leguminosae). Bonplandia 8(1–4): 1–186.

Lakela, O., and R. P. Wunderlin. 1980. Trees of Central Florida. Miami: Banyan Books.

Lassen, P. 1989. A new definition of the genera *Coronilla*, *Hippocrepis*, and *Securigera* (Fabaceae). Willdenowia 19: 49–62.

Lavin, M., and M. Sousa. 1995. Phylogenetic systematic and biogeography of the tribe Robinieae. Syst. Bot. Monogr. 45: 1–165.

Lavin, M., M. F. Wojciechowski, P. Gasson, C. Hughes, and E. Wheeler. 2003. Phylogeny of Robinioid legumes (Fabaceae) revisited: *Coursetia* and *Gliricidia* recircumscribed, and a biogeographical appraisal of the Caribbean endemics. Syst. Bot. 28: 387–409.

Lewis, G. P. 2005. Caesalpinieae. *In*: G. Lewis, B. Schrire, B. Mackinder, and M. Lock. Legumes of the World. Pp. 127–61. Richmond: Royal Botanical Gardens, Kew.

Lewis, G. P., B. Schrire, B. Mackinder, and M. Lock. 2005. Legumes of the World. Richmond: Royal Botanical Gardens, Kew.

Lievens, A. W. 1992. Taxonomic treatment of *Indigofera* L. (Fabaceae: Faboideae) in the New World. PhD dissertation, Louisiana State University and Agricultural and Mechanical College.

Liogier, H. A. 1985. Urticaceae. Descriptive Flora of Puerto Rico and Adjacent Islands. 1: 71–89. Rio Piedras: Editorial de la Universidad de Puerto Rico.

Liogier, H. A., and L. F. Martorell. 1982. Flora of Puerto Rico and Adjacent Islands: A Systematic Synopsis. Río Piedras: Editorial de la Universidad de Puerto Rico.

Long, R. W., and O. Lakela. 1971. A Flora of Tropical Florida. Miami: University of Miami Press.

Luckow, M. 1993. Monograph of *Desmanthus* (Leguminosae-Mimosoideae). Syst. Bot. Monogr. 38: 1–166.

Miller, N. G. 1970. The genera of the Cannabaceae in the southeastern United States. J. Arnold Arbor. 51: 185–203.

———. 1971a. The genera of the Urticaceae in the southeastern United States. J. Arnold Arbor. 52: 40–68.

———. 1971b. The Polygalaceae in the southeastern United States. J. Arnold Arbor. 52: 267–84.

Mohlenbrock, R. H. 1957. A revision of the genus *Stylosanthes*. Ann. Missouri Bot. Gard. 44: 299–354.

———. 1961. A monograph of the leguminous genus *Zornia*. Webbia 16: 1–141.

Monro, A. K. 2006. The revision of species-rich genera: A phylogenetic framework for the strategic revision of *Pilea* (Urticaceae) based on cpDNA, nrDNA, and morphology. Amer. J. Bot. 93: 426–41.

Moore, M. O. 1991. Classification and systematics of eastern North American *Vitis* L. (Vitaceae) north of Mexico. Sida 14: 339–67.

Moura, T. M., V. F. Mansano, B. M. Torke, G. P. Lewis, and A.G.A. Tozzi. 2013. A taxonomic revision of *Mucuna* (Fabaceae: Papilionoideae: Phaseoleae) in Brazil. Syst. Bot. 38: 631–37.

Nauman, C. E. 1981. *Polygala grandiflora* (Polygalaceae) Walter re-examined. Sida 1–18.

Nicolson, D. H., and C. Jarvis. 1984. *Cissus verticillata*, a new combination for *C. sicyoides* (Vitaceae). Taxon 33: 726–27.

Ohashi, H., and R. R. Mill. 2000. *Hylodesmum*, a new name for *Podocarpum*, Edinburgh J. Bot. 57: 171–68.

Pastore, J. F. B., and I. R. Abbott. 2012. Taxonomic notes and new combinations for *Asemeia* (Polygalaceae). Kew Bull. 67: 801–13.

Pennell, F. W. 1917. Notes on plants of the southern United States—III. Bull. Torrey Bot. Club 44: 337–87.

Phipps, J. B., and K. A. Dvorsky. 2007. Review of *Crataegus* series Apricae, ser. nov., and *C. flava* (Rosaceae). J. Bot. Res. Inst. Texas. 1: 171–202.

———. 2008. A taxonomic revision of *Crataegus* series Lacrimatae (Rosaceae). J. Bot. Res. Inst. Texas. 2: 1101–62.

Phipps, J. B., R. J. O'Kennon, and K. A. Dvorsky. 2006. Review of *Crataegus* series Pulcherrimae (Rosaceae). Sida 22: 973–1007.

Polhill, R. M. 1982. *Crotalaria* in Africa and Madagascar. Rotterdam: A. A. Balkema.

Porter, D. M. 1969. The genus *Kallstroemia* (Zygophyllaceae). Contr. Gray Herb. 198: 41–153.

———. 1972. The genera of Zygophyllaceae in the southeastern United states. J. Arnold Arbor. 53: 531–52.

Radford, A. E., H. E. Ahles, and C. R. Bell. 1964. Guide to the Vascular Plants of the Carolinas. Chapel Hill: University of North Carolina Book Exchange.

———. 1968. Manual of the Vascular Flora of the Carolinas. Chapel Hill: University of North Carolina Press.

Reveal, J. L., and F. R. Barrie. 1991. On the Identity of *Hedysarum violaceum* Linnaeus (Fabaceae). Phytologia 71: 456–61.

Rico Arce, M. L., and S. Bachman. 2006. A taxonomic revision of *Acaciella* (Leguminosae, Mimosoideae). Anales Jard. Bot. Madrid 63: 189–244.

Riley-Hulting, E. T., A. Delgado-Salinas, and M. Larvin. 2004. Phylogenetic systematics of *Strophostyles* (Fabaceae): A North American temperate genus within a neotropical diversification. Syst. Bot. 29: 627–53.

Robertson, K. R. 1973. The Krameriaceae in the southeastern United States. J. Arnold Arbor. 54: 322–27.

———. 1974. The genera of the Rosaceae in the southeastern United States. J. Arnold Arbor. 55: 303–32, 344–401, 611–22.

Robertson, K. R., and Y.-T. Lee. 1976. The genera of Caesalpinioideae (Leguminosae) in the southeastern United States. J. Arnold Arbor. 57: 1–53.

Rudd, V. E. 1955. The American species of *Aeschynomene* (Fabaceae). Contr. U.S. Natl. Herb. 32: 1–172.

———. 1969. A synopsis of *Piscidia* (Leguminosae). Phytologia 18: 473–99.

———. 1972. Leguminosae-Faboideae-Sophoreae. N. Amer. Fl., ser. 2. 7: 1–53.

Rydberg, P. A. 1919. Fabaceae: Psoraleae. *In*: N. L. Britton, North American Flora. 24: 1–126. New York: New York Botanical Garden.

Sauer, J. 1964. Revision of *Canavalia*. Brittonia 16: 106–81.

Schot, A. M. 1994. A revision of *Callerya* Endl. (including *Padbruggea* and *Whitfordiodendron*) (Papilionaceae: Millettieae). Blumea 39: 1–40.

Schrire, B. D. 2005. Tribe Phaseoleae. *In*: G. Lewis, B. Schrire, B. Mackinder, and M. Lock. Legumes of the World. Pp. 393–431. Richmond: Royal Botanical Gardens, Kew.

Schubert, B. G. 1950. *Desmodium*: Preliminary studies—III. Rhodora 52: 135–55.

Schultz, L. M. 1997. *Trema*. *In*: Flora of North America Editorial Committee. Flora of North America North of Mexico. 3: 379–80. New York/Oxford: Oxford University Press.

Sherman-Broyles, S. L. 1997. *Ulmus*. *In*: Flora of North America Editorial Committee. Flora of North America North of Mexico. 3: 369–75. New York/Oxford: Oxford University Press.

Sherman-Broyles, S. L., W. T. Barker, and L. M. Schulz. 1997. *Celtis*. *In*: Flora of North America Editorial Committee. Flora of North America North of Mexico. 3: 376–79. New York/Oxford: Oxford University Press.

Simpson, B. B. 1989. Krameriaceae. Fl. Neotrop. 49: 1–108.

Small, E. 1997. Cannabaceae. *In*: Flora of North America Editorial Committee. Flora of North America North of Mexico. 3: 381–87. New York/Oxford: Oxford University Press.

Small, E., and M. Jomphe. 1989. A synopsis of the genus *Medicago* (Leguminosae). Canad. J. Bot. 67: 3260–94.

Small, J. K. 1903. Flora of the Southeastern United States. New York: Published by the author.

———. 1913a. Flora of the Southeastern United States. 2nd ed. New York: Published by the author.

———. 1913b. Flora of Miami. New York: Published by the author.

———. 1913c. Florida Trees. New York: Published by the author.

———. 1913d. Flora of the Florida Keys. New York: Published by the author.

———. 1913e. Shrubs of Florida. New York: Published by the author.

———. 1933. Manual of the Southeastern Flora. New York: Published by the author.

Smith, R. R., and D. B. Ward. 1976. Taxonomy of the genus *Polygala* series Decurrents (Polygalaceae). Sida 6: 284–310.

Sørensen, M. 1988. A taxonomic revision of the genus *Pachyrhizus* (Fabaceae: Phaseoleae). Nordic J. Bot. 8: 167–92.

Staudt, G. 1999. Systematics and Geographic Distribution of the American Strawberry Species: Taxonomic Studies in the Genus *Fragaria* (Rosaceae: Potentilleae). Publications in Botany 81. Berkeley: University of California Press.

Stevenson, G. A. 1969. An agronomic and taxonomic review of the genus *Melilotus* Mill. Canad. J. Bot. 49: 1–20.

Sytsma, K. J., J. Morawetz, J. C. Pires, M. Nepokroeff, E. Conti, M. Zjhra, J. C. Hall, and M. W. Chase. 2002. Urticalean rosids: Circumscription, rosid ancestry, and phylogenetics based on *rbcl*, *trnL-F*, and *ndhf*. Amer. J. Bot. 89: 1531–46.

Thompson, J. 1980. A revision of *Lysiloma* (Leguminosae). PhD dissertation. Southern Illinois University, Carbondale.

Thulin, M. 1999. *Chapmannia* (Leguminosae-Stylosanthinae) extended. Nord. J. Bot. 19: 597–607.

Tomlinson, P. B. 1980. The Biology of Trees Native to Tropical Florida. Allston, MA: Harvard University Printing Office.

Turner, B. L. 2006. Overview of the genus *Baptisia* (Leguminosae). Phytologia 88: 253–68.

———. 2008. Revision of the genus *Orbexilum* (Fabaceae: Psoraleeae). Lundellia 11: 1–17.

Turner, B. L., and O. S. Fearing. 1964. A taxonomic study of the genus *Amphicarpaea* (Leguminosae). Southw. Naturalist 9: 207–18.

Uttal, L. J. 1984. The type localities of the Flora Boreali-Americana of André Michaux. Rhodora 86: 1–66.

Valder, P. 1995. Wisterias: A Comprehensive Guide. Portland, OR: Timber Press.

van der Maesen, L.J.G. 1985. Revision of *Pueraria* DC. with some notes on *Teyleria* Backer. Wageningen Agric. Univ. Pap. 85(1): 1–132.

van Wyk, B.-E. 1991. A synopsis of the genus *Lotononis* (Fabaceae: Crotalarieae). Contr. Bolus Herb. 14: 1–292.

Verdcourt, B. 1971. *Alysicarpus. In*: E. Milne-Redhead and R. M. Polhill (eds.). Flora of Tropical East Africa. Leguminosae. Part 3: 491–501. London: Crown Agents for Overseas Governments and Administrations.

Ward, D. B. 1972. Checklist of the Legumes of Florida. Gainesville: Herbarium Agriculture Experiment Station.

———. 2004. New combinations in the Florida flora II, Novon 14: 365–71.

———. 2006. Keys to the flora of Florida—13, *Vitis* (Vitaceae). Phytologia 88: 216–23.

———. 2009. The typification of *Crotalaria rotundifolia* and *Crotalaria maritima* (Fabaceae). J. Bot. Res. Inst. Texas 3: 219–25.

Ward, D. B., and D. W. Hall. 2004. Keys to the Flora of Florida—10, *Galactia* (Leguminosae). Phytologia 86: 65–74.

Wemple, D. K. 1970. Revision of the genus *Petalostemon* (Leguminosae). Iowa State J. Sci. 45: 1–202.

West, E., and L. E. Arnold. 1946. The Native Trees of Florida. Gainesville: University of Florida Press.

Wilbur, R. L. 1964. A revision of the dwarf species of *Amorpha* (Leguminosae). J. Elisha Mitchell Sci. Soc. 80: 51–65.

———. 1975. A revision of the North American genus *Amorpha* (Leguminosae-Psoraleae). Rhodora 77: 337–409.

Wilhelm, G. S. 1984. Vascular Flora of the Pensacola region. PhD dissertation. Southern Illinois University, Carbondale.

Wilmot-Dear, C. M., and I. Friis. 2006. The Old World species of *Pouzolzia* (Urticaceae, tribus Boehmerieae). A taxonomic revision. Nord. J. Bot. 24: 5–115.

Windler, D. R. 1966. A revision of the genus *Neptunia* (Leguminosae). Austral. J. Bot. 14: 379–420.

———. 1974. A systematic treatment of the native unifoliolate crotalarias of North America (Leguminosae). Rhodora 76: 151–204.

Wood, C. E. 1949. The American barbistyled species of *Tephrosia* (Leguminosae). Rhodora 51: 193–231; 233–302; 305–64; 369–84.

Woodland, D. W. 1982. Biosystematics of the perennial North American taxa of Urtica. II. Taxonomy. Syst. Bot. 7: 282–90.

Woods, M. 2005. A revision of the North American species of *Apios* (Fabaceae). Castanea 70: 85–100.

Wunderlin, R. P. 1982. Guide to the Vascular Plants of Central Florida. Gainesville: University Presses of Florida/University of South Florida Press.

———. 1997. Moraceae. *In*: Flora of North America Editorial Committee. Flora of North America North of Mexico. 3: 388–99. New York/Oxford: Oxford University Press.

———. 1998. Guide to the Vascular Plants of Florida. Gainesville: University Press of Florida.

Wunderlin, R. P., and B. F. Hansen. 2003. Guide to the Vascular Plants of Florida. 2nd ed. Gainesville: University Press of Florida.

Yang, M.-Q., R. van Velzen, F. T. Baker, A. Sattarian, D.-Z. Li, and T.-S. Yi. 2013. Molecular phylogenetics and character evolution of Cannabaceae. Taxon 62: 473–86.

Zohary, M., and D. Heller. 1984. The genus *Trifolium*. Jerusalem: Israel Academy of Science and Humanities.

Index to Common Names

Index to Scientific Names

Accepted scientific names of plants and plant families are in roman type. Synonyms are in *italics*.

Daubentonia, 160
  drummondii, 160
  longifolia, 163
  punicea, 162
  virgata, 163
Delonix, 77
  regia, 78
    var. genuina, 78
Desmanthus, 78
  brachylobus, 78
  depressus, 80
  diffusus, 80
  illinoensis, 78
  leptophyllus, 79
  luteus, 80, 134
  pernambucanus, 80
  virgatus, 79
    var. depressus, 80
Desmodium, 80
  acuminatum, 100
  arenicola, 85
  bracteosum, 83
    var. cuspidatum, 83
  canescens, 82
  canum, 84
  ciliare, 82
  cuspidatum, 83
  dillenii, 87
  fernaldii, 83
  floridanum, 84
  frutescens, 84
  glabellum, 84
  glabrum, 91
  glutinosum, 100
  grandiflorum, 83
  incanum, 84
    var. supinum, 85
  laevigatum, 85
  lineatum, 85
  marilandicum, 86
  molle, 91
  nudiflorum, 100
  nuttallii, 86
  obtusum, 86
  ochroleucum, 87
  ovalifolium, 31
  paniculatum, 87
    var. angustifolium, 87
    var. dillenii, 87
    var. pubens, 88

    var. typicum, 87
  pauciflorum, 101
  perplexum, 88
  pubens, 88
  purpureum, 90
  rhombifolium, 91
  rigidum, 87
  rotundifolium, 88
  scorpiurus, 89
  sessilifolium, 89
  striatum, 106
  strictum, 89
  supinum, 85
  tenuifolium, 90
  thunbergii, 113
  tortuosum, 90
  triflorum, 90
  viridiflorum, 91
Despeleza
  angustifolia, 110
  capitata, 110
  hirta, 111
Dichrostachys, 91
  cinerea
    subsp. africana, 92
    subsp. cinerea, 92
  glomerata, 92
Dimenops lanceolata, 14
Dioclea multiflora, 97
Diplisca elliptica, 255
Disterepta pilosa, 90
Ditremexa, 154
  ligustrina, 156
  marilandica, 156
  medsgeri, 156
  nashii, 156
  occidentalis, 158
Dolicholus, 147
  cinereus, 147
  erectus, 152
  intermedius, 150
  lewtonii, 148
  michauxii, 149
  minimus, 149
  mollissimus, 152
  parvifolius, 149
  precatorius, 150
  simplicifolius, 150
  swartzii, 151
  tomentosus, 151

    var. undulatus, 148
Dolichos
  bulbosus, 136
  ensiformis, 52
  erosus, 136
  helvolus, 166
  hirsutus, 146
  hosei, 194
  japonicus, 196
  lablab, 107
  lobatus, 146
  luteolus, 195
  maritimus, 52
  minimus, 149
  obtusifolius, 52
  polystachios, 141
  pruriens, 133
  purpureus, 107
  regularis, 96
  repens, 195
  roseus, 52
  sinensis, 195
  unguiculatus, 195
Dorstenia, 271
  contrajerva, 271
  quadrangularis, 271
    var. pinnatifida, 271
Duchesnea, 227
  fragiformis, 227
  indica, 227
Dynamidium reptans, 233

Eaplosia ovata, 43
Ecastaphyllum, 71
  brownei, 72
    var. psilocalyx, 72
  ecastaphyllum, 72
Elaeagnaceae, 249
Elaeagnus, 249
  pungens, 249
    subsp. eupungens, 249
    var. typica, 249
  umbellata, 250
    subsp. euumbellata, 250
    var. typica, 250
Emelista tora, 160
Emerus
  grandiflorus, 161
  herbaceus, 161
  marginatus, 163

suffruticosa, 159
Psoralea
americana, 199
canescens, 138
carnea, 74
corylifolia, 199
corymbosa, 76
gracilis, 135
lupinellus, 135
melilotoides, 135
var. gracilis, 135
melilotus, 135
pedunculata, 135
psoralioides, 135
var. gracilis, 135
var. typica, 135
simplex, 136
virgata, 136
Pterocarpus ecastaphyllum, 72
Pueraria, 146
hirsuta, 146
lobata, 146
montana var. lobata, 146
thunbergiana, 146
Pyracantha, 239
coccinea, 240
fortuneana, 239
koidzumii, 239
Pyrus, 240
arbutifolia, 230
var. erythrocarpa, 230
var. melanocarpa, 231
var. oblongifolia, 230
angustifolia, 229
calleryana, 240
communis, 241
var. sylvestris, 241
coronaria var. angustifolia, 229
sylvestris, 241

Quinaria
hederacea, 4
hirsuta, 4

Racosperma
auriculiforme, 22
retinodes, 22
salignum, 22
Rafnia perfoliata, 43
Ramium, 282

cylindricum, 282
niveum, 283
Rehsonia
floribunda, 196
sinensis, 197
×formosa, 198
Resupinaria grandiflora, 161
Reynosia, 256
septentrionalis, 257
Rhamnaceae, 250
Rhamnidium ferreum, 256
Rhamnus, 257
asiatica, 254
caroliniana, 257
colubrina, 253
domingensis, 255
elliptica, 255
ferrea, 256
ferruginea, 254
iguanaeus, 265
micranthus, 268
minutiflora, 258
scandens, 251
volubilis, 251
zizyphus, 259
Rhynchosia, 146
americana, 152
caribaea, 152
cinerea, 147
cytisoides, 148
difformis, 148
erecta, 152
galactoides, 148
intermedia, 150
lewtonii, 148
menispermoidea, 152
michauxii, 149
minima, 149
var. diminifolia, 149
mollissima, 152
parvifolia, 149
pitcheria, 148
precatoria, 150
reniformis, 150
var. intermedia, 152
reticulata, 152
simplicifolia, 150
var. intermedia, 150
swartzii, 151
tomentosa, 151

var. erecta, 152
var. intermedia, 150
var. mollissima, 152
var. monophylla, 150
var. tomentosa, 151
var. volubilis, 148
volubilis, 148
Rhytidomene, 135
lupinellus, 135
Robinia, 152
acacia, 153
fragilis, 153
grandiflora, 161
hispida, 153
var. typica, 153
mitis, 129
pseudoacacia, 153
forma normalis, 153
forma vulgaris, 153
var. typica, 153
sepium, 99
vesicaria, 162
Rosa, 241
bracteata, 241
carolina, 242
cherokeensis, 242
floridana, 244
laevigata, 242
lancifolia, 243
lucida, 241
lucieae var. wichuraiana, 244
multiflora, 243
palustris, 243
serrulata, 242
setigera, 244
var. glabra, 244
var. tomentosa, 244
wichuraiana, 244
Rosaceae, 213
Rousselia, 291
humilis, 292
Rubus, 245
abundifolius, 247
agilis, 248
allegheniensis, 248
argutus, 247
var. floridus, 247
arrectus, 247
audax, 246
betulifolius, 247

cinerea
  var. *cinerea*, 9
  var. floridana, 6
cordata, 2
cordifolia, 8
  var. *sempervirens*, 8
coriacea, 8
floridana, 9
gigas, 6
glareosa, 9
hederacea, 4
  var. *hirsuta*, 4
heterophylla var. *cordata*, 2
illex, 8
incisa, 3
indivisa, 2
inserta, 4
labrusca, 9
  var. *aestivalis*, 5
latifolia, 9
muscadina, 8
munsoniana, 7
palmata, 7
quinquefolia, 4

rotundifolia, 7
  var. *munsoniana*, 7
  var. *pygmaea*, 7
rufotomentosa, 6
shuttleworthii, 8
sicyoides, 3
  forma *ovata*, 3
  var. *ovata*, 3
simpsonii, 5, 6
smalliana, 5
sola, 6
tiliifolia, 9
trifoliata, 3
vinifera var. *aestivalis*, 5
vitiginea
  forma *ovata*, 3
  var. *sicyoides*, 3
vulpina, 8

Wisteria, 196
  chinensis, 197
  floribunda, 196
  frutescens, 197
    var. *macrostachya*, 197

  macrostachya, 197
  sinensis, 197
  speciosa, 197
  ×formosa, 198

Xamacrista triflora, 57

Ziziphus, 258
  celata, 258
  iguanaea, 265
  jujuba, 259
  mauritiana, 259
  officinarum, 259
  sativa, 259
  vulgaris, 259
  zizyphus, 259
Zornia, 198
  bracteata, 198
  tetraphylla, 198
Zygia
  dulcis, 144
  guadalupensis, 145
  unguis-cati, 145
Zygophyllaceae, 9

Richard P. Wunderlin, professor emeritus of biology at the University of South Florida, is the author of *Guide to the Vascular Plants of Central Florida*, *Guide to the Vascular Plants of Florida*, and coauthor of *Guide to the Vascular Plants of Florida*, second edition, *Guide to the Vascular Plants of Florida*, third edition, and *Atlas of Florida Vascular Plants* website (www.florida.plantatlas.usf.edu).

Bruce F. Hansen, curator emeritus of biology at the University of South Florida Herbarium, is coauthor of *Guide to the Vascular Plants of Florida*, second edition, *Guide to the Vascular Plants of Florida*, third edition, and *Atlas of Florida Vascular Plants* website (www.florida.plantatlas.usf.edu).